UNTANGLING MOLECULAR BIODIVERSITY

Explaining Unity and Diversity Principles of Organization with Molecular Structure and Evolutionary Genomics

UNTANGLING MOLECULAR BIODIVERSITY

Explaining Unity and Diversity Principles of Organization with Molecular Structure and Evolutionary Genomics

Editor

Gustavo Caetano-Anollés

University of Illinois at Urbana-Champaign, USA

World Scientific

NEW JERSEY • LONDON • SINGAPORE • BEIJING • SHANGHAI • HONG KONG • TAIPEI • CHENNAI • TOKYO

Published by

World Scientific Publishing Co. Pte. Ltd.
5 Toh Tuck Link, Singapore 596224
USA office: 27 Warren Street, Suite 401-402, Hackensack, NJ 07601
UK office: 57 Shelton Street, Covent Garden, London WC2H 9HE

Library of Congress Control Number: 2020041551

British Library Cataloguing-in-Publication Data
A catalogue record for this book is available from the British Library.

UNTANGLING MOLECULAR BIODIVERSITY
Explaining Unity and Diversity Principles of Organization with Molecular Structure and Evolutionary Genomics

ISBN 978-981-4656-61-0 (hardcover)
ISBN 978-981-4656-62-7 (ebook for institutions)
ISBN 978-981-4656-63-4 (ebook for individuals)

For any available supplementary material, please visit
https://www.worldscientific.com/worldscibooks/10.1142/9506#t=suppl

Typeset by Stallion Press
Email: enquiries@stallionpress.com

Contents

Preface

Untangling molecular biodiversity

The universe appears complex and diverse. These properties materialize for example in entities, processes, and dynamic behaviors of different kinds. We, as humans, have developed a deep capacity for analysis and manipulation of this complexity and diversity. At first, we learned to exploit the environment for survival. The developed awareness accumulated insights that favored persistence. Our inquisitive nature advanced our *Weltanschauung*, our world view, when we learned how to better control our existence. Our ingenuity and the development of technological power transformed mythologies into science, knowledge and action. This unique capacity developed by the human mind is now massively transforming our planet (perhaps the universe) despite viral pandemics and global change instabilities. François Jacob realized that our big and complex brain had an "*in-built need for order*." Indeed, our species dissects, tinkers and manipulates. We dissect by examining how things work but often dissect through classification. We classify everything, from galaxies, stars and planets to molecules, genes, and organisms. We count and measure, store and preserve. We build libraries, museums, repositories, universities, and a world of data. Remarkably, we also retrodict with memories, learning, experiences, analysis and retrospections. We record history and learn from our past. This helps predict future outcomes and become better engineers. In doing so, we continue to embrace the impact of our persistence.

Untangling Molecular Biodiversity presents a unique global framework to explain molecular and organismal diversity at different temporal scales and complexity levels. When exploring complexity and diversity the concept of "*level*," which has been applied to a variety of doctrines, becomes apparent. Mario Bunge in his "*semantical preliminary*" attempts its dissection for science and ontology (*Review of Metaphysics* 13: 396–406, 1960). He finds that levels can be described with nine different meanings: degree, degree of complexity, degree of analytical depth, emergent whole, poistem, rank, layer, rooted layer, and grade. When "level = degree of any kind," levels are considered grades in a static scale or are considered stages in a process. Such is the case of the abundance of a protein structural variant in a proteome (an extensive property), the level of functionality of a system (a processual property), or the level of abstraction or importance in a classification scheme (a complexity property). When "level = emergent whole," levels become players of a parthood relationship that measures integration and diversity. This view is enhanced by interrelated qualities (poistems), hierarchical views (ranks), stratigraphic arrangements (layers) or tree emergences (rooted layers). Finally, levels can represent linear or non-linear grades of order that are part of an evolutionary series and have emerged in time from lower or higher pre-existing levels. This last "level = grade" instantiation appropriately describes what biologists call "*levels of organization*," which is helpful for building ontological definitions and understanding biological complexity.

The theoretical framework we here present is for the most part grounded in evolutionary genomics and systems biology. This book tackles important questions of origin and structure, such as the origin of life, the emergence of biochemistry, the origin of viruses, the nature of the last universal common ancestor of diversified life, the rationale for having three cellular superkingdoms, the role of information and thermodynamics in evolution, and the centrality of modules in biology. It also explores philosophical problems, such as the search for truth in scientific inquiry, the difficulties of temporal parts and retrodiction, Nietzsche's *Wille zur Macht* driving force of power, and the unity-and-diversity dichotomy of the living world that was advanced by Empedocles. In fact, these paradigms are made explicit through the looking-glass of macromolecules, the component parts that carry the genetic and functional information of life. Some principles that are discussed are general and could be applied to selected aspects of our world and universe.

Untangling Molecular Biodiversity explores 13 general themes: *Unity*, *Memory*, *Linkage*, *Structure*, *Origins*, *Language*, *Growth*, *Persistence*, *Panspermia*, *Complexity*, *Power*, *Time*, and *Foresight*. These themes are tightly linked and form a fully-connected conceptual network. The reader can therefore wander through themes by choosing any chapter as starting point and then selecting other chapters through pathways of interest across the thematic network. One major emphasis of modern biology has been the survey of organismal parts, their mechanisms, and their diversity in the living world. In the post-genomic era, this survey has produced a "big data" deluge, which must now be integrated to find meaningful biological knowledge. The different themes address integrative frameworks for biological organization, evolutionary patterns of molecular biodiversity, links between molecular structure and evolutionary genomics, persistence strategies that borrow from engineering and systems biology, the structural diversity of the molecular world, and explorations of the origin and evolution of cells and viruses. Consequently, this book represents a very unique collection of ideas that can attract the attention of a broad readership. I hope the readers will find each and every theme exciting and thought provoking.

I am grateful to all authors that contributed to this editorial project. Their effort and ideas have helped integrate the many aspects of inquiry that are presented in this exceptional compendium. I thank them for their patience during the long time it took to assemble this volume. I thank my son Derek for the artistic renditions of the book cover and for the many figure illustrations of this book. I also thank Jay Mittenthal for prompting the exploration of unity and diversity that we here present. I much appreciate his friendship during the years and the provocative discussions that we shared. Finally, I wish to thank my wife Gloria for her support and patience and the permanent encouragement of my son Derek and daughter Kelsey. All of them, directly or indirectly, have humbly embarked on an exploration of the amazing biological world. Thank you!

Gustavo Caetano-Anollés
Urbana, Illinois
September 2020

List of Contributors

Ascensión **Ariza-Mateos**. *Laboratory of RNA Archaeology, Instituto de Parasitología y Biomedicina 'López-Neyra', Consejo Superior de Investigaciones Científicas, and Centro de Investigación Biomedicina En Red de enfermedades hepáticas y digestivas (CIBERehd), Armilla, Granada, Spain*

M. Fayez **Aziz**. *Evolutionary Bioinformatics Laboratory, Department of Crop Sciences, University of Illinois, Urbana, Illinois, USA*

Isabel **Cacho**. *Laboratory of RNA Archaeology, Instituto de Parasitología y Biomedicina 'López-Neyra', Consejo Superior de Investigaciones Científicas, and Centro de Investigación Biomedicina En Red de enfermedades hepáticas y digestivas (CIBERehd), Armilla, Granada, Spain*

Derek **Caetano-Anollés**. *Broad Institute of MIT and Harvard, Cambridge, Massachusetts, USA*

Gustavo **Caetano-Anollés**. *Evolutionary Bioinformatics Laboratory, Department of Crop Sciences, University of Illinois, Urbana, Illinois, USA*

Kelsey **Caetano-Anollés**. *Division of Biomedical Informatics, College of Medicine, Seoul National University, Republic of Korea*

Richard **Egel**. *Department of Biology, University of Copenhagen Biocenter, Copenhagen, Denmark*

Jordi **Gómez**. *Laboratory of RNA Archaeology, Instituto de Parasitología y Biomedicina 'López-Neyra', Consejo Superior de Investigaciones Científicas, and Centro de Investigación Biomedicina En Red de enfermedades hepáticas y digestivas (CIBERehd), Armilla, Granada, Spain*

Richard **Janko**. *Department of Classical Studies, University of Michigan, Ann Arbor, Michigan, USA*

Ibrahim **Koç**. *Department of Molecular Biology and Genetics, Gebze Technical University, Gebze, Turkey*

Liudmila S. **Mainzer**. *National Center for Supercomputer Applications (NCSA), University of Illinois, Urbana, Illinois, USA*

Jay E. **Mittenthal**. *Department of Cell and Developmental Biology, University of Illinois, Urbana, Illinois, USA*

Fizza **Mughal**. *Evolutionary Bioinformatics Laboratory, Department of Crop Sciences, University of Illinois, Urbana, Illinois, USA*

Arshan **Nasir**. *Theoretical Biology and Biophysics Group, Los Alamos National Laboratory, Los Alamos, New Mexico, USA*

Santiago **Núñez-Corrales**. *Illinois Informatics and National Center for Supercomputing Applications (NCSA), University of Illinois, Urbana IL, USA*

Fengjie **Sun**. *School of Science and Technology, Georgia Gwinnett College, Lawrenceville, USA*

On unity in organisms and its evolution

J. E. Mittenthal[1]
G. Caetano-Anollés[2]

Summary. We interpret unity in an organism as coordination that promotes economy, flexibility, and robustness, in a combination that aids persistence in a niche. We survey aspects of coordination, at the molecular level and in the physiology and development of cells and multicellular organisms. Combinations of economy, flexibility, and robustness can be used in a top-down strategy for designing networks of modules and regulation that produce appropriate coordination. Selection for these combinations has driven the evolution of modules through competitive optimization.

[1]Department of Cell and Developmental Biology
[2]Evolutionary Bioinformatics Laboratory, Department of Crop Sciences, University of Illinois, Urbana, Illinois, USA

1. Introduction

"Collections: wholes and not wholes;
brought together, pulled apart;
sung in unison, sung in conflict;
from all things one and from one all things."

— Heraclitus of Ephesus (B10)

During the late 20th century biologists aimed at reduction — splitting organisms into ever-smaller parts and determining how the parts work and interact. The task of understanding the integration of parts remains. How do all the parts of an organism cooperate to coordinate behavior? Watching a graceful dancer, we perceive unity, an organic wholeness. The dancer has balance. The positions, orientations, and movements of her body parts are coordinated, at one instant and over time. In this chapter we explore coordination and its evolution. Box 1 provides a glossary of terms.

2. Unity as coordination

We associate unity with coordination. Consider a company that builds houses. (An organism and a company have analogous functional components, connected in analogous ways; *see* Miller, 1995.) The contractor knows that the foundation must be made before the roof, and he needs the materials to make the foundation first. The business manager, a novice, first provides materials for the roof; these must be stored, and construction is delayed. Coordination among processes is essential for effective performance.

2.1. *Linkage, constraints, and degrees of freedom*

Unity depends on an appropriate network of links among parts. With too little linkage the network fragments. Disunion and dysfunction follow — a protein denatures, an organism dies. In the biblical story, construction of the tower of Babel stops. Excessive programmed increase in linkage can increase the cost of encoding, increase inhibitory links that interfere with performance, and decrease evolvability. Random increases in linkage oppose flexible responses. In aging, cross-linking of molecules and stiffening of joints increase rigidity and degrade capacity.

BOX 1: Glossary of terms

Capacity: The variety of activities an organism can perform.
Constraints: Restrictions on the activity of parts.
Coordination: Coordination of parts when their activity has fewer degrees of freedom than the parts would have independently.

Diversification: Appearance of variant parts.
Economy: Use of resources.
Flexibility: Adjustment of activity in response to signals from receptors.
Linkage: Pattern of connections (links) among parts. Links are constraints.
Module: A group of structures or processes linked more extensively to each other than to the rest of the system.
Persistence: Continuance of a feature or a lineage of organisms.
Robustness: Persistence of a feature despite perturbation, without perturbation-specific receptors.
Unification, Unity: Integration among parts.

Parts are coordinated when their activity has fewer degrees of freedom than the parts would have independently. Linkage constrains the activity of parts, reducing their degrees of freedom. Informally, degrees of freedom are the number of independent alternative behaviors a system can have. More formally, "degrees of freedom equal the number of variables involved minus the number of independent equations (constraints) relating them. This is the number of variables whose values may be arbitrarily specified …, thereby determining the values of the other variables" (Dixon, 1972). This definition extends the meaning of degrees of freedom from the conventional meaning for molecules to a meaning for far-from-equilibrium systems such as organisms. The equations governing the dynamics of the system come from physics and chemistry; they constrain variables. The constraints include boundary conditions — for organisms, limits on the availability of resources and interactions with other organisms.

2.2. *Economy, flexibility and robustness*

Some patterns of linkage do not provide coordination that promotes the persistence of a lineage of organisms. What distinguishes patterns that do so? We suggest that connections in an organism promote unity if they contribute to economy, flexibility, and robustness, in a combination specific to the life strategy of the species (Yafremava *et al.*, 2012). Why is this so? An economical organism is frugal in its use of resources — space, time, and chemicals. Integration among its parts directs resources to parts that need them, and reduces resources available to less active parts, thereby reducing the resource budget the organism needs to persist. Thus, unity is associated with economy.

As we use the terms flexibility and robustness, an organism can make flexible responses to signals it perceives with signal-specific receptors, while its generic properties enable robust responses without receptors. Robustness can be defined as maintained capacity for behavior despite perturbation. We include reliability, redundancy, and resilience in robustness, though these terms differ in usage. A reliable system has a high probability of functioning, with or without perturbation. A system with redundancy achieves reliability using equivalent copies of parts. A system is resilient when it can recover from external insults or mitigate internal shortcomings readily.

Robustness and flexibility require unity — integration among parts, to adjust the rates of processes when the supply of resources or the demand for performance changes. The construction and maintenance of a flexible and robust organism require investment of resources; integration can direct this investment. Integration achieves balance, the flexible allocation of resources among an organism's tasks. Organisms manifest balance in a more general sense: Flexibility and robustness require resources, so they compete with each other and with economy. These tradeoffs produce a balance which is different in each of the six kingdoms of organisms, and which differs among species within a kingdom (Yafremava *et al.*, 2012).

2.3. *Other meanings of unity*

Unity has meanings other than coordination, which we do not explore. Like a contractor, an architect who designs houses also seeks unity — an aesthetic unity that will be functional. An architect designs layouts, which a contractor implements. A choreographer is an architect of dance. An organism has aesthetic unity in the body plan, which is not just a collage of parts assembled from a Chinese menu. The design for a house, a dance or an organism may be beautiful, but we are concerned with its functionality.

There is also unity in an organism because, in general, all of its cells (except the microbiome) arise from a zygote and share the same genome. Reuse of molecular networks in diverse contexts produces a kind of unity, akin to reuse of a theme in a musical composition. There is unity among organisms, evident in their descent from a common ancestor and the use of common parts. Basic components of the cell include the use of common sets of components of macromolecules (e.g. amino acids, nucleotides, domains), the use of common metabolites to manage matter and energy constraints in different organisms (e.g. ATP, coenzymes), and common elements of cellular design (e.g. tensegrity, transport and bioenergetics across membranes, assembly designs). These commonalities arise from the nesting processes of evolution. As organisms diversify they retain features that are highly linked and are almost immutable. These features spread through life.

3. Kinds of coordination

We survey kinds of coordination that promote economy, flexibility, and robustness, looking first at the molecular level and then at organisms.

3.1. *Molecular coordination*

In organisms, molecules mediate processes. Coordination operates through links within parts of a molecule or between molecules. There are cooperative interactions within macromolecules, including those that mediate the formation of helices and sheets, folding, and allosteric effects. These interactions are the structural and dynamical prerequisites for coordination. Macromolecules interact with small molecules and with each other by matching at complementary interfaces, through many weak bonds (Root-Bernstein, 2012).

Several aspects of molecular coordination contribute to economy. By analogy with the contractor example, providing molecules in the order in which they are needed contributes to economy of time (speed) and of space (reduced storage). Without such order, molecules needed later must be stored, and earlier processes must be delayed. The number and rate of processes that can run simultaneously would decrease. Providing molecules in the relative concentrations in which they are needed decreases waste, by reducing the error rate in syntheses. Otherwise a molecule abundant in excess is more likely to be used in place of a correct molecule, or side reactions are more likely. Limiting the set of simultaneous processes also promotes economy. Competition among more processes reduces the resources available to each, and so reduces their rates and makes it more difficult to maintain a favorable balance of abundance among prerequisites.

The spatial arrangement of molecules can also promote economy. Within a spatial compartment the molecules contributing to a process can be localized in high concentration, and interfering molecules can be excluded, increasing the rates of reactions and avoiding side reactions. The cooperation of proteins in multimers reduces the time delays and dilution associated with diffusion. In a bacterium, the sequence of genes on the circular chromosome can increase the availability of enzymes for metabolic pathways and accelerate progress through the cell cycle.

Coordination contributes to flexibility. A flexible molecular system can detect and respond to its circumstances using specific receptors. Many molecules must be flexible; their functioning depends on changes of conformation. Often function proceeds through induced fit of interacting molecules — ligand-receptor, enzyme-substrate, antibody-antigen — a kind of flexibility. The genetic code has evolved to increase the flexibility of proteins (Caetano-Anollés *et al.*, 2013; Debes *et al.*, 2013).

Modules are central to flexibility. A module is a group of structures or processes linked more extensively to each other than to the rest of the system (Simon, 1962). Macromolecules that cooperate in performing a task are often activated or inhibited coordinately, as a module. Examples include genes (as used in operons) or enzyme complexes that mediate a multistep pathway. Flexibility occurs within and between modules; diverse groups of modules can be activated coordinately as higher-level modules. Modularity can also reduce costs, increasing economy (Clune *et al.*, 2014). Overlapping functions of modules can increase robustness.

Within RNAs and proteins, regions of secondary structure are modules. Such regions can decrease frustration, increasing the probability that a macromolecule folds from the extended to the functional folded state without becoming trapped in a dysfunctional state (Ancel and Fontana, 2000). In a higher-level module, frustration is less likely because the degrees of freedom in links between macromolecules allow diverse coordination. Examples include holoenzymes (ribosome, spliceosome, RNA polymerase, ATP synthase) and filaments. A region of a cell within which molecules are localized — a vesicle, nucleus, or nucleolus, for example — is a module; processes within it are more extensively linked to each other than they are to processes outside the region.

Coordination also contributes to robustness (Jen, 2003). A macromolecule is more robust to perturbations when it spends more time in a subset of its possible conformations. Then those conformations persist in time (are more stable) and function longer. The molecule has increased its robustness when it has optimized the interplay of stability and function (Caetano-Anollés and Mittenthal, 2010). Under perturbation, the molecular system will tend to return to its initial state of evolutionarily optimized conformations. A more resilient molecular system is one that upon mutation preserves the set of optimized conformations (Fontana, 2002). Such a system also maintains its thermodynamic stability, measured as the ability of the molecular system to maximize its dissipation of energy — that is, maintain minimum free energy conformations. Thus, mutational and thermodynamic stability are coupled and contribute to robustness when molecules are evolutionarily optimized, a process that is termed "plastogenetic congruence." These molecules are said to be "molecularly canalized" (Fontana, 2002). The occurrence of a subset of conformations that favors mutational and thermodynamic stability is a form of coordination.

3.2. *Coordination in organisms*

Coordination is evident in an organism's structure, physiology and development. Early in the 19th century, the anatomist Cuvier understood a structural manifestation of unity: All the parts of an animal have features that contribute to functioning in its niche. A carnivore's digestion is fitted to meat. The jaws are "constructed for devouring prey; the claws, for seizing and tearing it; the teeth, for cutting and dividing its flesh; the entire system of its locomotive organs, for pursuing and catching it; its sense organs for detecting it from afar" (quoted by Kolbert, 2013). This coordination matches structure to function, one of several kinds of matching.

3.2.1. Physiology

An organism resembles a piano that plays itself and responds to its environment. The analogs of keys are modules. Often each module performs a specific function, and different modules perform different functions. The modules can be activated in many combinations to achieve alternative goals (*cf.* Coen, 1999). In these respects, the set of modules resembles the set of normal modes that characterizes an oscillating system. However, overlapping redundancy in the performance of modules is common; modules may share consequences. Also, the piano metaphor is inappropriate for organisms in that no specific set of keys must be used in music, but specific processes and molecules characteristically occur in organisms. A multicellular organism, like a single cell, consists of parts that perform twenty characteristic functions. (The microbiome participates in the performances.) Collectively, the functions mediate fluxes of matter-energy and information that are mutually supportive (Miller, 1995).

The parts must have appropriate position and activity to perform their functions for the whole. Coordination of parts occurs both by direct interaction among them, as occurs within the heart, gut, and brain, and by signaling to and from regulatory systems — in animals, the neural, immune, and endocrine systems. Each part provides services to other parts and uses services from them, in a network. The network is flexible and robust, in ways somewhat understood for cells (Aldana and Cluzel, 2003) but less so for multicellular organisms.

The input-output relation of a part and its regulation match the supply of its resources to the demand for its services. For example, in a biosynthetic pathway, feedback inhibition limits flux to that meeting the demand for the pathway's end product. After deafferentation of neurons and muscle fibers, they become hypersensitive to the missing neurotransmitters. The matching of supply and demand may be manifest as balance in the functional capacity of cooperating parts, so the parts do not have excess and unusable capacity (Weibel *et al.*, 1991).

Physiological regulatory networks (PRNs) coordinate coherent behavior of an organism, within it and in its environment (Michoel *et al.*, 2011; Cohen *et al.*, 2012). PRNs coordinate spatial and temporal patterns of chemical, mechanical and electrical activity. These activities interact; for example, stress alters gene regulation and *vice versa* (Ingber, 2008). Chemical balance is evident in the homeostatic maintenance of concentrations and structures. Homeostasis stabilizes the internal environment of an organism, maintaining the levels of systemically needed prerequisites by regulating their supply to match demand. Homeostasis promotes economy and flexibility. The balance among parts is most obvious in motor coordination, which adjusts the distribution of positions and tensions to allow movements, from locomotion to dancing. Mechanical coordination operates through PRNs and through tensegrity structures, which balance forces between linear elements in tension and in compression. Tensegrity operates at all levels from individual macromolecules to organisms (Swanson, 2013).

PRNs bundle parts acting together into a module while inhibiting antagonistic parts. Modules can correspond to subgoals that recur in various combinations under changing conditions. This is a form of structure-function matching that can be called subgoal matching (Mittenthal *et al.*, 1992; Kashtan and Alon, 2005; Parter *et al.*, 2007). For example, in the encoding of protein synthesis, codons, transfer RNAs, and aminoacyl-tRNA synthetases are used in adjustable combinations to produce proteins with various amino acid sequences. Thus, subgoal matching promotes flexibility. It also promotes economy: Digestion and locomotion compete for limited resources in blood. Regulation distributes these resources according to need, increasing the supply of blood to the gut for digestion or to muscles for locomotion. Subgoal matching may promote robustness, if a subgoal can be achieved in redundant ways. For example, in human exercise redundant control mechanisms regulate the availability of resources in blood to muscles (Paulev-Zubieta, 2013).

In summary, an organism mediates interacting fluxes of matter-energy and information, which move through its surfaces and through the interfaces between its parts. At these interfaces there may be several kinds of matching. Molecules interact at complementary surfaces. An organism's structure and function can match conditions in its environment, as Cuvier observed and as subgoal matching shows. Within the organism there may be matching of supply and demand between interacting parts. To each part, other parts are environment, so the same kinds of matching can operate at any interface. Properties of parts, adjusted by regulation, coordinate matching.

3.2.2. Development

Structures develop through coordination, which helps produce cells with appropriate numbers, locations and capabilities. As development proceeds, prerequisites activate modules that direct cell behaviors — cytodifferentiation, affinity, motility, proliferation and death. For example, in a fruit fly embryo the respiratory organs (tracheae) must be initiated at appropriate anterior-posterior and dorsal-ventral locations in specific segments. Molecular labels at these locations activate a master regulatory gene that acts together with transcription factors to orchestrate tracheal development, in specific patterns of cell division and migration. Analogous regulation directs the development of eyes, heart, appendages, and other organs of fruit flies and other organisms. Subgoal matching occurs; permutations of a small set of growth factors and transcription factors regulate modules in diversifying contexts to produce many body plans, despite the limited information in genomes. Coordination is chemical, but also electrical and mechanical (Piccolo, 2014). A balance of mechanical forces influences the locations and shapes of cells, through tensegrity and through opposed tensions between adhering cells (Mammoto and Ingber, 2010).

How does matching develop? Coordination between matched features could be fixed genetically through natural selection, but this would be a slow and rigid process, during which function might be inadequate. Exploration with feedback during development is faster and more flexible (Kirschner and Gerhart, 2005). As a nervous system develops, growing axons and dendrites explore alternative connections with target cells. Feedback interactions between neurons and targets, and the functional adequacy of their activity, determine patterns of connectivity that survive. Parts that will interact physiologically interact in development to match each other's properties. Some of these interactions depend on contact between the parts: In a vertebrate limb bud, exploratory ingrowth of blood vessels and neurons accommodates changes in bones and muscles. Reciprocal interactions occur between a muscle and the tendons and bones that support it, and between nerves and muscles. The survival of motor neurons depends on their innervating muscle fibers. Overproduction of motor neurons, with death of those that do not innervate muscle, allows the population of motor neurons to accommodate changes in the number of muscle fibers. Thus, matching can develop through local interactions. It may also involve long-range interactions through chemicals: In humans, pituitary and placental tropins stimulate the development of endocrine organs and then regulate the organs' secretion of hormones (e.g. thyroid hormone, gonadotropins). Such reuse of growth factors suggests reuse of modules, which would unify development and physiology.

3.3. *Synthesis*

We have suggested that the coordination in organisms tends to optimize a combination of the criteria economy, flexibility, and robustness. It may be possible to infer the patterns of coordination expected from particular criteria, and to design biological mechanisms that can produce these patterns. Scheduling algorithms are used in manufacturing, business, and computation to choose patterns of coordination that optimize the performance of tasks, given the criteria and the prerequisites for component processes. (For example, in the case of a contractor discussed above, economy is primary, so the cost of alternative schedules for building the foundation and the roof must be evaluated. The foundation is a prerequisite for building the roof. So, providing materials for building the roof before the foundation costs more, in storage and delays, than building the foundation first.) According to the principle of subgoal matching, a module achieves each subgoal. The organization of processes within a module can be designed by specifying and scheduling their subgoals. Thus, a hierarchical top-down design of scheduling from persistence criteria could be achieved.

This design strategy has been used most effectively for bacteria, where economy — speed of passage through the life cycle — is the primary criterion, and there are relatively few component processes and hierarchical levels. Selection for economy has produced expected patterns of molecular coordination and scheduling of processes through the life cycle (e.g. Mittenthal *et al.*, 1993; Zaslaver *et al.*, 2004; Alon, 2007; McAdams and Shapiro, 2009; Jaeger and Calkins, 2012).

In general designers must consider the spatial and temporal patterns of chemical, mechanical and electrical processes. The proximity of macromolecules affects the speed and accuracy of processes, from metabolic pathways to mitosis and cytokinesis. Prerequisites occur within organisms, but also include parameters of the environment that influence internal processes. For example, the diurnal light-dark cycle regulates the activities of organs (Summa and Turek, 2015).

Mainzer *et al.* (this book) propose use of the generic model of Miller (1995) to analyze the coordination of processes in an organism. The scheduling strategy proposed here is compatible with that model but goes beyond it. Scheduling is important for transitions through the life cycle, including development.

4. The evolution of coordination

Organisms evolve by adding or modifying parts in ways that augment their capacity for performance. The services parts perform become increasingly elaborate through their coordination. However, constraints within and outside an organism limit its capacity. For a constant environment, formal analysis can model the effects of constraints on capacity. For example, the constraints on steady-state metabolism of bacteria define a cone in a flux space, in which each point represents a vector of fluxes through the reactions. The constraints allow a cell to have flux vectors within or on the cone (Price *et al*., 2004). Analysis of optimal growth of bacteria growing in constant culture conditions predicts the flux vector that is observed (Ibarra *et al*., 2002). For multicellular organisms also, optimization of a formal model can predict the outcome of evolution with constant constraints and goals (Alexander, 1982). However, in nature selection often does not occur in a constant environment or for a single function. Selection pressures often vary in space and time, and a structure may perform several functions, complicating the relations between constraints and capacity (Dudley and Gans, 1991).

New structures can increase capacity, not only directly, but also through deconstraint — diminishing the effects of constraints (Kirschner and Gerhart, 2005). In a cell or an embryo, a spatial compartment releases constraints on processes within it. A regulatory network can reduce an organism's use of limited resources. For house building, a manual that specifies the order of operations can reduce the need for storage and time, so a construction company can build more houses per unit time.

4.1. *The evolution of modules through diversification and unification*

New modules evolve to deal with selection pressures, including economy, flexibility, and robustness. Selection pressures are constraints. Organisms evolve through changes in linkage, especially within and between modules. During evolution macromolecules diversify through genetic and epigenetic variations. These variations change the processes available to an organism — chemical, mechanical, and electrical processes, their sequences and spatial patterns. The more activities an organism can perform, the greater is the diversity of variants. Often many alternative combinations of variants can cooperate to perform a function (Wagner, 2014). Neutral variants — those that do not change the response to constraints — come and go, though they may change the accessibility of other variants.

Variants that augment the capacity of an organism to deal with constraints compete to be integrated into new structures — modules and regulatory mechanisms. To form these structures, macromolecules interact through many weak bonds. Enough bonds with appropriate specificity can form new links. As linkage increases, a new network canalizes to form a new structure, in which the remaining degrees of freedom can provide functionality (Mittenthal *et al.*, 2012). Thus constraints filter variants to select for a new structural unification, in competitive optimization. The new structure can increase coordination, within the structure and among structures, enhancing capacity and persistence (Figure 1). If subgoal matching occurs, several submodules can form during a process of competitive optimization.

The balance between the retention and loss of variants depends on the stability of the environment. For example, bioinformatic analysis of the evolution of protein structures indicates that the

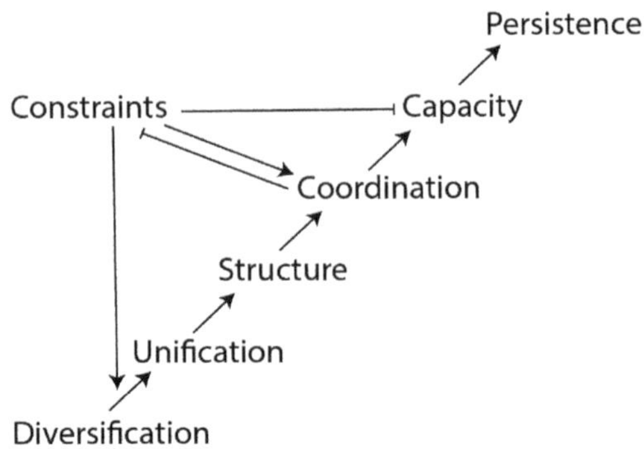

Figure 1. Processes contributing to the evolution of unity through competitive optimization. Point-ended arrow indicates promotion; flat-ended arrow indicates inhibition. See text for discussion.

superkingdoms of organisms arose from a stem lineage of cells. Viruses, then giant viruses, then archaea, and then bacteria diverged from this lineage, which evolved into eukarya (Nasir and Caetano-Anollés, 2015; Nasir *et al.*, 2012; Wang *et al.*, 2007). In the lineage, novel protein architectures accumulated. As cells ventured into challenging niches, they lost folds that were in the stem lineage and retained folds effective in their environment. From this viewpoint, the formation of new modules through competitive optimization may occur most often in environments that change slowly: The more stressful is the environment, the greater the rate at which variants are likely to be lost. Then variants that might contribute to a new module will tend to accumulate more slowly. However, environmental variation selects for flexibility and robustness, which enable a lineage to track correlations in a variable environment, possibly promoting the evolution of modules. Also, a more variable environment tends to fragment or decimate a population; fixation of novelties is more likely in a smaller population. In summary, the influence of environmental variation on the formation of modules probably depends on the relative rates at which the environment and the organisms vary. If the environment varies fast enough, variants in organisms may be destroyed too fast for competitive optimization to generate new modules often. If the environment varies more slowly, selection may canalize modules effectively (see Kashtan *et al.*, 2007, Figure 2).

4.2. *The evolution of networks of modules*

The processes diagrammed in Figure 1 operate at many levels of organization, from macromolecules to codes to embryos (Mittenthal *et al.*, 2012; Harold, 2014). After a module forms it diversifies into many variants, which may be coopted into new modules. A higher-level module forms by association of lower-level modules. A lower-level module may be used in several higher-level modules. Thus, a network of modules with multiple levels evolves, and organisms become increasingly complex (Shapiro, 2011).

In the network of modules, constraint and deconstraint operate in various directions. For example, consider modules in proteins. Protein domains contain secondary structures — alpha helices and beta sheets — which are associations of amino acids. Multi-domain proteins evolved by association of domains. Proteins associate to form multimers (Caetano-Anollés *et al.*, 2009). Lower-level modules constrain a higher-level module, limiting its capacity. The higher-level module also constrains its lower-level modules; the requirement for specific associations among proteins constrains the variation of their complementary surfaces. However, the evolution of a higher-level module can deconstrain its lower-level modules; diverse sequences of amino acids (lower-level modules) can fold into the same architecture of a domain (higher-level module). Lower-level modules can deconstrain their higher-level module: When the amino acid sequence in a domain changes, secondary structures can vary in length without changing the domain's architecture.

The highest-level module of interest here is the organism. How does coordinated evolution of the whole body occur? As discussed here, parts can change through variation and selection, pairwise interaction, and matching. New parts evolve through cooption and aggregation of older parts. Central coordination through growth hormones can adjust the size of the body, and the relative sizes of its parts if they respond to hormones differently. Thus, the appearance of unity results from coordinated changes at many levels — a marvel of parallel processing.

5. Conclusion: Toward a theory of integration in organisms

We have looked for a top-down design process in which a global principle expresses unity. The principle would provide a framework for local algorithms that can generate observed biological organization. The persistence criteria of economy, flexibility, and robustness — weighted according to the life strategy of the species and used with scheduling algorithms — offer such a principle. It could be used to specify candidate networks of component processes, given the prerequisites for the processes. The component processes could then be designed by iterating this procedure. As natural selection has driven the evolution of modules and regulation, computational selection can help to design them.

Cuvier (1798; *see* Rudwick, 1997, p. 36) asserted a principle of the correlation of parts in animals:

> *"Today comparative anatomy has reached such a point of perfection that, after inspecting a single bone, one can often determine the class, and sometimes even the genus of the animal to which it belonged, above all if that bone belonged to the head or the limbs. ... This is because the number, direction, and shape of the bones that compose each part of an animal's body are always in a necessary relation to all the other parts, in such a way that — up to a point — one can infer the whole from any one of them and vice versa."*

Perhaps this principle can be extended. Understanding unity in terms of coordination might aid inference of the networks of modules suitable for an ecological niche. Thus, a more comprehensive understanding of the design and integration of organisms may be possible.

Acknowledgments

We appreciate helpful discussions with Liudmila S. Mainzer and Elbert Branscomb.

References

Aldana M, Cluzel P (2003) A natural class of robust networks. *Proc Natl Acad Sci USA* **100**: 8710–8714.

Alexander RM (1982) *Optima for Animals*. Edward Arnold, London.

Alon U (2007) *An Introduction to Systems Biology*. Chapman & Hall/CRC, Boca Raton.

Ancel LW, Fontana W (2000) Plasticity, evolvability, and modularity in RNA. *J Exp Zool* **288**: 242–283.

Caetano-Anollés G, Wang M, Caetano-Anollés D, Mittenthal JE (2009) The origin, evolution and structure of the protein world. *Biochem J* **417**: 621–637.

Caetano-Anollés G, Wang M, Caetano-Anollés D (2013) Structural phylogenomics retrodicts the origin of the genetic code and uncovers the evolutionary impact of protein flexibility. *PLoS One* **8**: e72225.

Caetano-Anollés G, Mittenthal JE (2010) Exploring the interplay of stability and function in protein evolution. *Bioessays* **32**: 655–658.

Clune J, Mouret J-B, Lipson H (2014) The evolutionary origins of modularity. *Proc Roy Soc B* **280**: 20122863.

Coen E (1999) *The Art of Genes*. Oxford University Press: New York.

Cohen AA, Martin LB, Wingfield JC, McWilliams SR, Dunne JA (2012) Physiological regulatory networks: Ecological roles and evolutionary constraints. *Trends Ecol Evol* **27**: 428–435.

Debès C, Wang M, Caetano-Anollés G, Gräter F (2013) Evolutionary optimization of protein folding. *PLoS Comput Biol* **9**: e1002861.

Dixon DC (1972) Degrees of freedom in dynamic and static systems. *Ind Eng Chem Fundam* **11**: 198–205.

Dudley R, Gans C (1991) A critique of symmorphosis and optimality models in physiology. *Physiol Zool* **64**: 627–637.

Fontana W (2002) Modelling 'evo-devo' with RNA. *Bioessays* **24**: 1164–1177.

Harold FM (2014) In *Search of Cell History*. University of Chicago Press, Chicago.

Ibarra RU, Edwards JS, Palsson BØ (2002) *Escherichia coli* K-12 undergoes adaptive evolution to achieve in silico predicted optimal growth. *Nature* 420: 186–189.

Ingber DE (2008) Tensegrity-based mechanosensing from macro to micro. *Progr Biophys Mol Biol* **97**: 163–179.

Jaeger L, Calkins ER (2012) Downward causation by information control in micro-organisms. *Interface Focus* **2**: 26–41.

Jen E (2003) Stable or robust? What's the difference? *Complexity* **8**: 12–18.

Kashtan N, Alon U (2005) Spontaneous evolution of modularity and network motifs. *Proc Natl Acad Sci USA* **102**: 13773–13778.

Kirschner MW, Gerhart JC (2005) *The Plausibility of Life*. Yale University Press, New Haven.

Kolbert E (2013) *The Lost World*. New Yorker, 12/16/13, pp. 28–38.

Mammoto T, Ingber DE (2010) Mechanical control of tissue and organ development. *Development* **137**:1407–1420.

McAdams HH, Shapiro L (2009) System-level design of bacterial cell cycle control. *FEBS Lett* **583**: 3984–3991.

Michoel T, Joshi A, Nachtergaele B, Van de Peer Y (2011) Enrichment and aggregation of topological motifs are independent organizational principles of integrated interaction networks. *Mol Biosyst* **7**: 2769–2778.

Miller JG (1995) *Living Systems*. University Press of Colorado, Niwot, USA.

Mittenthal JE, Baskin AB, Reinke RE (1992) Patterns of structure and their evolution in the organization of organisms: Modules, matching, and compaction. In *Principles of Organization in Organisms* (eds. Mittenthal J, Baskin AB), pp. 321–332. Addison-Wesley, Reading, MA.

Mittenthal J, Caetano-Anollés D, Caetano-Anollés G (2012) Biphasic patterns of diversification and the emergence of modules. *Front Genetics* **3**: 147.

Mittenthal JE, Clarke, Levinthal M (1993) Designing bacteria. In Stein WD (ed) *Thinking about Biology*. SFI Studies in the Sciences of Complexity, Lect. Note Vol. III, Addison-Wesley, Reading, MA, pp. 65–97.

Nasir A, Kim KM, Caetano-Anollés G (2012) Giant viruses coexisted with the cellular ancestors and represent a distinct supergroup along with superkingdoms Archaea, Bacteria and Eukarya. *BMC Evol Biol* **12**: 156.

Nasir A, Caetano-Anollés G (2015) A phylogenomic data-driven exploration of viral origins and evolution. *Science Adv* **1**: e1500527.

Parter M, Kashtan N, Alon U (2007) Environmental variability and modularity of bacterial metabolic networks. *BMC Evol Biol* **7**: 169.

Paulev-Zubieta G (2015) New human physiology. Chapter 18: Exercise, sports and doping. http://www.zuniv.net/physiology/book/chapter18.html. Accessed 17 January 2020.

Piccolo S (2014) Twists of fate. *Sci Am* **311**: 74–81

Price ND, Reed JL, Palsson BØ (2004) Genome-scale models of microbial cells: Evaluating the consequences of constraints. *Nature Rev* **2**: 886–897.

Root-Bernstein R (2012) A modular hierarchy-based theory of the chemical origins of life based on molecular complementarity. *Accounts Chem Res* **45**: 2169–2177.

Rudwick MJS (1997) Georges Cuvier, *Fossil Bones, and Geological Catastrophes*. University of Chicago Press.

Shapiro JA (2011) *Evolution*. FT Press Science, Upper Saddle River, New Jersey.

Simon HA (1962) The architecture of complexity. *Proc Am Phil Soc* **106**: 467–482.

Summa KC, Turek FW (2015) The clocks within us. *Sci Am* **312**: 5 0–55.

Swanson RL (2013) Biotensegrity: A unifying theory of biological architecture with applications to osteopathic practice, education, and research — a review and analysis. *J Am Osteopathic Assoc* **113**: 34–52.

Wagner A (2014) *Arrival of the Fittest*. Current (Penguin Group), New York.

Wang M, Yafremava LS, Caetano-Anollés D, Mittenthal JE, Caetano-Anollés G (2007) Reductive evolution of architectural repertoires in proteomes and the birth of the tripartite world. *Genome Res* **17**: 1572–1585.

Weibel ER, Taylor CR, Hoppeler H (1991) The concept of symmorphosis: A testable hypothesis of structure-function relationship. *Proc Natl Acad Sci USA* **88**: 10357–10361.

Yafremava LS, Wielgos M, Thomas S, Nasir A, Wang M, Mittenthal JE, Caetano-Anollés G (2013) A general framework of persistence strategies for biological systems helps explain domains of life. *Front Genetics* **4**: 16.

Zaslaver A, Mayo AE, Rosenberg R, Bashkin P, Sberro H, Tsalyuk M, Surette MG, Alon U (2004) Just-in-time transcription program in metabolic pathways. *Nat Genet* **36**: 486–491.

Memory

Retrodiction — Exploring the history of parts and wholes in the biosystems of life

G. Caetano-Anollés[1]

Summary. Unity defines modules in biological structures, which must tightly fit within other modular entities as they unfold in the systems of life. These modules are also observables that can be effectively used to study the evolution of the natural world. Here I explore how tree and network thinking can help explain evolutionary change by building rooted phylogenies of modular parts and wholes. The effort is needed to understand the rise of biological complexity.

[1]*Evolutionary Bioinformatics Laboratory, Department of Crop Sciences, University of Illinois, Urbana, Illinois, USA*

1. Introduction

A quick glance at the natural and man-made world around us reveals complexity. Each and every organism and landscape appear unique and contributes distinct features to the whole. The disposition of veins in a leaf and the textures of a rock are all distinctive and particular. However, and despite their diversity, even leaves and rocks share vein-like features in common if one looks carefully, and the comparative analysis can be unifying. These commonalities or "universals" make the recurrent textures and patterns of Nature, which we sometimes abstract using the tools of mathematics (e.g., branching structures, sigmoid curves, power laws, and networks) and philosophy (e.g. styles, principles and failures of reason, skepticisms). The empirical and rational exploration that focuses on these features is as old as human societies, since it was necessary for survival. For example, the information gathered of plants and animals used for food, shelter and medicine were passed from generation to generation until modern times, simply, because it was adaptive. The recurrent features of the natural world must be however abstracted by the observers. They allow our navigation in it as a species and in the modern times of scientific inquiry, it expresses in language and defines an *ontology* necessary to study beings, entities and existence. The language employed by modern science claims to be objective, yet its impact cannot be fully materialized without human intuition and an effort to integrate knowledge with a goal of explaining overriding order. Therefore, a new approach must be sought that seeks independence from the limitations of language and abandons false belief systems, the shortcoming of apriorism, and reductionist views that plague modern science. Here I will describe the challenges of seeking knowledge and advancing science. My goal is to introduce scientific approaches to *retrodiction*, our ability to go back in time to predict present and future.

2. Scientific progress and the search for truth

One important ongoing controversy is relevant to the enterprise of scientific inquiry and our understanding of unity, diversity and change. It relates to the search for truth and verisimilitude (Niiniluoto, 2015). I will comment on these aspects of scientific inquiry as they impact in particular our understanding of the origin and evolution of life. Before doing so I share the recent opinion of Rowbottom (2015) (in his critique of verisimilitude) and others that scientific progress is both normative and multifaceted. The scientific enterprise, which is a collective social endeavor of humans (in their last few thousand years of history), is constrained by norms of behavior defining what is good or bad science and is pursuing research exploration with a multiplicity of goals (heavily contested at philosophical levels).

The "foundationalist" view of science, shared by rationalists, empiricists and positivists, is a cumulative model of epistemic progress that grows knowledge by accumulating justified truths. Its roots go back to Plato's *episteme*, knowledge as "justified truth belief." This view is opposed by "skepticism," which claims complete certainty can never be achieved by empirical science. The pessimistic views of skepticism gain support directly from the history of science and the rejection of important scientific theories. However, the skeptics view science at best as a mere succession of false theories, a stance vigorously contested by realism and the idea that all statements hold some implicit truth. In contrast to skeptics, Kuhn challenges realism from a different angle. The collective enterprise of scientific communities becomes powerful when searching for "concrete problem-solving solutions." This displaces the cumulative search for truth by a search of problem-solving paradigms (in constant revolution), which act as driver for progress. Similarly, Quine and his ontological relativism consider scientific inquiry as a tool capable of "predicting future experience in the light of past experience." Thus, and within an "instrumentalist" point of view, given empirical evidence there would be always many possible theoretical "instruments" that could explain it, with no realities beyond them. Under this instrumentalist framework, scientific progress becomes an increasingly more predictive endeavor (an overall trend that must materialize in history). A more moderate stance to both the foundationalism and skepticism extremes is "*fallibilism.*" a term proposed by the great American philosopher and father of pragmatism, Charles S. Peirce. Fallibilism acknowledges the limitations of achieving complete certainty but regards scientific exploration as a self-correcting approximation towards the truth. Under this philosophical framework, a succession of false theories can still advance scientific progress, contributing to the quest of increasing understanding. Popper elaborated this view in his notions of "truth-likeness" and "verisimilitude" (Popper, 1963), which have been recently advanced with more adequate technical definitions (see Niiniluoto, 2015).

When dealing with scientific theories, the major controversy between realists and instrumentalists mirrors that of foundationalists and falllibilists. A theory can be defined as a well-substantiated explanation of a selected aspect of the natural world, with validity that can be repeatedly tested and confirmed through observation and experimentation. Realists regard theories as statements having a truth value. Instrumentalists regard theories as instruments of observation that are devoid of truth and are instead measured by other criteria, such as their increasing empirical success. In one extreme, progress is measured by increased capacity to predict, control, and manipulate, which reduces science to technological progress. Most scientists however are fallibilists. They consider theories as hypothetical constructs that are always amenable to revision in the sense of Popper (1963). Some of evolving theories may approach the truth but the task of determining how truthful they are remains always an uncertain or intuitive task. One solution has been to evaluate real and estimated progress with truthlikeness, Tr(H,C^*), and verisimilitude, ver(H/E), where H is a theory represented as a disjunction of its constituents (mutually exclusive and jointly exhaustive) in a language L,C^* is the only true constituent in that language, and E is the evidence at hand (Niiniluoto, 2014, 2015). The unknown degree of truthlikeness Tr(H,C^*) may be estimated by its expected value relative to its constituents C_i and their posterior probabilities given evidence E and for all constituents of the language.

$$ver(H \mid E) = \sum P(C_i \mid E) Tr(H, C_i) \quad (1)$$

This link solves the logical and epistemic problem of determining how close are H to C^*, without knowing it, and the solution is in line with fallibilist belief that all claims of truthlikeness of theories are conjectural as well as claims about their truth. Thus,

(1) *Real* scientific progress occurs when advancing from theory H to H' is progressive iff $\mathrm{Tr}(H,C^*) < \mathrm{Tr}(H',C^*)$
(2) *Estimated* scientific progress occurs when advancing from theory H to H' seems progressive on evidence E iff $\mathrm{ver}(H/E) < \mathrm{ver}(H'/E)$.

I note recent debate by Bird and Rowbottom about the adequacy of semantic or epistemic definitions of truth and truthlikeness (Niiniluoto, 2014; Rowbottom, 2015). I also note that since science is a human endeavor, scientific progress is the result of biology and evolution. Thus, for many evolutionary epistemologists, under both the instrumentalist and realist point of views, science evolves to enhance the collective fitness of human kind. We trust this fitness can be somehow described by the truthlikeness and verisimilitude standards within semantic or epistemic frameworks. However, the evolving multi-goal scientific endeavor may make that idea an idealization.

We can now ascribe to quite a variety of belief frameworks:

(i) Scientific knowledge is simply true.
(ii) Scientific knowledge is true as far as it goes, but always incomplete.
(iii) Scientific knowledge approaches truth as an unattainable goal.
(iv) Scientific knowledge is more and more likely to be correct.

These frameworks must somehow interface with empirical evidence, i.e. the observational statements drawn from experimentation. Popper and his hypothetico-deductive system for overthrowing theories claim this interface involves the important logical tools of *empirical content* and *degree of corroboration* (Popper, 1959). The empirical content of a theory (*empirischer Gehalt*) refers to the breath and scope of evidence the theory purports to explain. It depicts *a priori* the improbability of the theory by showcasing its testability and corroborability qualities, and ultimately its ability to withstand continued refutation. In other words, it measures its *empirical success*. Two important criteria to evaluate the empirical content of a theory are its level of universality (*Allgemeinheit*) and its degree of precision (*Bestimmtheit*). Universality is the scope and range of demands that the theory makes on reality, i.e. the number of different applications the theory can have. Precision refers to the levels of detail in predictions of the theory, i.e. the degree of description of reality. The more universal and precise is the theory the higher is its empirical content, which correlates with its explanatory power. Realists and instrumentalists support *explanatory power* as a criterion of good science derived from the true deductive consequences of the theory and given an observational language. When combined with the predictive power of a theory, the notion transforms into what Hempel calls "*systematic power*" (Hempel, 1965). A good theory with high explanatory power should also "unify" empirical data and theory from many domains of inquiry. These unifying theories can still be powerful conduits within a pluralistic realm of concepts explaining reality. A good theory should also be "economical" in how it explains the evidence. The concept of *simplity* was already prompted by Popper but may be very relevant for maximum parsimony views of retrodiction I will discuss below. Devoid of its aesthetic values, simplicity can provide a relative measure of how the explanatory power of a theory relates to its structural complexity, i.e. the conceptual justifications it requires and/or its information content.

While empirical content describes the explanatory landscape of a theory prior to empirical testing, *degree of corroboration* (Bewährung) provides an estimate of the scientific value of the theory, and with it, a report of its past performance (with all its falsifiers). This central and influential concept of empirical science, which was elaborated by Popper but has roots in the philosophy of "confirmation" (Crupi, 2015), measures the confidence of a theory when subjected to critical testing. Specifically, the measure of corroboration is "the degree to which a hypothesis has been tested, and of the degree it had stood up to the test" (Popper, 1959; p. 415). Popper's corroboration measure C of theory (conjecture) H explains evidence E and background knowledge b within sentences of a first-order logical language L according to the formula

$$C(H,E,b) = \frac{p(E|Hb) - p(E|b)}{p(E|Hb) - p(EH|b) + p(E|b)} \quad (2)$$

where $p(E|Hb)$ denotes the probability (and likelihood) of E given H and b, for pair of sentences in L, $p(E|b)$ is the probability of E *given* b, and $p(EH|b)$ is the probability of E and H unified (their intersection) given b. $p(EH|b) = p(E|Hb)p(H|b)$ from the axioms of probability. A test of a hypothesis must "prove its mettle" by maximizing the numerator of Popper's formula and fulfilling nine adequacy criteria that define a "statistical relevant" condition, including covariance with posterior probability as indicator of statistical relevance (see Sprenger, 2018). I note that Popper's measure of corroboration and its relative, the degree of testability $[1-p(H|b)]$, should be cautiously interpreted within both a logical and statistical framework. In fact, these measures have been shown to be inapplicable if the logical probability of a synthetic universal statement is zero, relative to any evidence it might posses, i.e. when $p(H|b) = 0$ (Rowbottom, 2013). Thus, "stating preference with respect to truth" (Popper, 1959) for universal statements can be problematic, since empirical and logical content of the corroboration equation must be fulfilled. The null hypothesis significance test also benefits from corroboration measures and can illustrate the philosophically challenging problem of non-significant results being unable to refute a conjecture. A mathematically rigorous analysis of confirmation involves specifying (exhaustively) "adequacy conditions" on the concept and then deriving a representation theoretical framework for its measure. Sprenger (2018) for example recently weakened the adequacy conditions of Popper's formula and retained only those adequacies that were indispensable for corroboration. He found that there is no measure of corroboration that fits Popper's description and can amalgamate statistical relevance and the logic-based informativity-testability of a theory. A pluralistic solution however entails defining a partition–relative representational framework of corroboration that results in different corroboration judgments stressing, when necessary, statistical relevance or information logical content. In this regard, it is clear that a focus on Bayesian and frequentist approaches to statistical inference can provide measures of confirmation akin to corroboration. This requires the epistemological position that probability plays a crucial role in rational belief, inference and behavior, which can be problematic. Examples include the *Law of Likelihood* (Hacking, 1965) which assumes conjectures are statistical (Sober 2008) and assigns probabilities to evidence E according to the following statement:

$$\text{Evidence } E \text{ favors } H \text{ over } H' \quad \text{iff } \Pr(E|H) > \Pr(E|H')$$

Similarly, the *Bayesian Confirmation Theory* (Fitelson, 2007) assumes a Bayesian stance in which a measure of incremental confirmation $M(H, E)$ makes explicit the degree to which evidence E increases our confidence in hypothesis H:

$$\text{Evidence } E \text{ confirms } H \text{ iff } M(H, E) > 0 \quad \text{iff } \Pr(H|E) > \Pr(H)$$

$$\text{Evidence } E \text{ undermines } H \text{ iff } M(H, E) < 0 \quad \text{iff } \Pr(H|E) < \Pr(H)$$

and then how evidence E favors one hypothesis H over another H'

$$\text{Evidence } E \text{ favors } H \text{ over } H' \quad \text{iff } M(H, E) > M(H', E)$$

Recent analysis has shown *E* favors *H* when bets on the truth of *H* becomes larger than bets of *H′*, with bets of a same expected utility. The risk inherent to those bets conflicts with the Law of Likelihood but agrees with the Bayesian stance (van Enk, 2015). These approaches do not address problems of the relationship of inductive reasoning and truth (Vickers, 2014), especially when considering evolution (see below).

Corroboration can only be accomplished if *H* is falsifiable, with testable content that is distinct from *b* and makes predictions consistent with *E*, and demanding the minimum number of *ad hoc* hypotheses to fit the evidence with auxiliary assumptions that are maximally explanatory. Both empirical and logical content must be preserved in Equation (2), which rules out keeping only a statistical relevance framework. Positive values of *C* indicate corroboration and negative values indicate falsification. Note that *C* is not a probability (probabilities do not have negative values) but its calculation may encompass frequentist views. While the numerator is of intuitive significance and the denominator only fulfills normalization purposes, avoiding rejection of adequacy conditions and salvaging the problems of conflict previously discussed (Rowbottom, 2013; Sprenger, 2018) may entail focusing only on the testing of a set of best-performing constituents of the disjunction of *H*. This solution may be evident in evolutionary analyses.

Nomothetic and ideographic scientific pursuit

Universals and particulars are manifestations of the kind of science we seek to pursue and Popper's "basic statements" of evidence that need to be falsified. The neo-Kantian philosopher of science Wilhelm Windelband distinguished two general conceptual frameworks of scientific inquiry, the *ideographic* and the *nomothetic* methods. In his Prussian-style "Rectoral Address" at the

Kaiser-Wilhelms-Universität Strassburg on the commemoration day of the university (Windelband, 1894), Windelband made the idea clear:

> *"In their quest for knowledge of reality, the empirical sciences either seek the general in the form of the law of nature or the particular in the form of the historically defined structure. On the one hand, they are concerned with the form, which invariably remains constant. On the other hand, they are concerned with the unique, immanently defined content of the real event. The former disciplines are nomological sciences. The latter disciplines are sciences of process or sciences of the event. The nomological sciences are concerned with what is invariably the case. The sciences of process are concerned with what was once the case. If I may be permitted to introduce some new technical terms, scientific thought is nomothetic in the former case and idiographic in the latter case."*

Thus, the nomothetic (universal, predictive) method is concerned with the search of universal statements that explain present events with high predictive power (e.g. the search of laws) that could then be used to predict (forecast) events in the future. Conversely, the ideographic (historical, retrodictive) method represents the science of process and history. Its goal is explaining how present events have been molded by the past, i.e. forecasting the past. They describe irreducible individual and unrepeatable occurrences. Remarkably, both of these approaches develop explanatory power to describe past, present and future events in natural systems. In other words, both nomothetic and ideographic approaches have the potential to reveal how reality is affected by time, i.e. they both can uncover evolutionary patterns and processes.

Reality can be described as a conjectured state of things, a set of "features," "characters" or "events" describing a particular or a universal. This world of experience is driven by observations. However, reality must be placed on a temporal setting. When reality becomes dependent on time, it can be envisioned as a Markov chain, an ordered set of states $X_1, X_2, \ldots, X_n$, where the probability of evolving from state X_i to state X_j depends on the value of $|i - j|$, the number of time steps. The smaller the difference between i and j the larger the probability that the descendant has state X_j given an ancestor with state X_i. This Markovian property describes how the passage of time erodes information that the present provides about the past, leaving historical traces under different models of evolutionary change (Sober and Steel, 2014, 2015). This framework is ideographic since the present is used to retrodict information about the past. However, if the number of times steps or the time between steps tends to 0, the ability to pinpoint history increases to a point where history becomes irrelevant. Our statement of reality becomes purely nomothetic. For example, let us consider a microbial protein molecule that has been crystallized in the laboratory in multiple occasions during a decade and its structure determined with crystallographic methods. It is highly likely that their folded structures will not differ much and that inferences derived from the crystallized state will give a model of atomic structure with high explanatory power. Given background knowledge (e.g. Förster energy transfer experiments), we could even learn about the dynamics of the molecule's structure and make temporal predictions about molecular change. The model is nomothetic: it describes the present, even though the evidence holds the effect of time. Now, let us consider the existence of a natural process connecting the present to the past that is not historical but is the product of a Laplacian intellect or "demon" of causal determinism, beautifully described in the following quotation (Laplace, 1814):

> *"We may regard the present state of the universe as the effect of its past and the cause of its future. An intellect which at a certain moment would know all forces that set nature in motion, and all positions of all items of which*

> *nature is composed, if this intellect were also vast enough to submit these data to analysis, it would embrace in a single formula the movements of the greatest bodies of the universe and those of the tiniest atom; for such an intellect nothing would be uncertain and the future just like the past would be present before its eyes."*

Applying this demon to the structure of the molecule used as example could describe how the molecule grew from a small peptide into a fully folded protein (and changed) and will continue to grow and unfold in time. We could make inferences about its origin and evolution and about its future. But, would these retrodictions and predictions be historical? They would, *only if* the relation between the past state, the present state and a future state would be deterministic, we would know enough about the evolutionary traces of the present state (our reality), and our demon is biologically realistic. Such optimistic view may be however unwarranted for several reasons that are very general (Sober and Steel, 2014). First, regular Markov chains loose information exponentially when the time between the present and the past increases. Note however that in some cases (e.g. in the presence of drift or balancing selection) information is not completely destroyed (Sober and Steel, 2014, 2015). Second, the *data processing inequality* of causal chains of these kinds guarantees that the information of an effect *E* given its cause *D* is less than or equal to both the information of *E* given a proximate cause *P* and the information of *P* given *D*. Note that this inequality is an internal property of the chain that does not say that recent events *always* provide more information than events that are older. In fact the opposite can be true. Third, the mapping of *D*, *P* and *E* may not be one-to-one, i.e. ancient ancestors may represent ancient relatives and many past events could map to the present. This makes the path from the past to the present an increasingly historical affair and increasingly difficult problem to tackle with statistical solutions. Thus, nomothetic-inspired approaches to history have limited retrodictive power. More powerful scientific methods must be sought.

3. Retrodiction, explanatory power and phylogenetic analysis

So what makes universals from particulars? Or is it particulars from universals? These are the central questions of biology but also of philosophy and the natural sciences. Philosophically, particulars are concrete entities that exist in space and time while universals are commonalities abstracted from the particular things. In other words, particulars appear to be concrete instantiations of universals, though one can assume there are also universals that are not abstract but concrete. Why particulars and universals? Perhaps the only way to explain them is to place particulars and universals within a framework of time and change and find processes responsible for them. This framework, when applied to the problems of life, is often called *evolution* and opinions generated by the framework are often called *evolutionary thinking*. It is noteworthy however that the term evolution is general and also applies to inanimate objects such as stars and galaxies (e.g. Fraix-Burnet *et al.*, 2012), extending the concepts of change it entails to the entire universe.

Unearthing historical traces in the present requires defining both major evolutionary axioms (starting points of reason) and establishing an analytical framework with which to build retrodictive tools. Darwin and Wallace, with background knowledge from many that preceded and followed them (especially Owen and Lenkaster, who elaborated the concept of "homology"), provided the former. Hennig and his cladistic school formalized the latter (see Hennig, 1966). Wiley (1975), in response to Mayr, Bock and Ashlock and their claim that Popper should be adopted for biological classification, made a lucid description of how the hypothetico-deductive method of science interfaces with phylogenetics. There has been very recent clarifying discussions on Popper, explanatory power and their links to frequentist and Bayesian reasoning (see Farris, 2008, 2013, 2014a), all indirectly related to the ongoing contest between parametric (statistical) and non-parametric (parsimony) views of phylogenetic reconstruction (Sober, 2004; Brooks *et al.*, 2007). While some recent views on the subject have been confusing (Vogt 2008) and have been appropriately contested (Farris, 2013, 2014), Wiley's elaborations made 40 years ago remain remarkably current. I will use these elaborations to illustrate the challenges of retrodiction.

Evolutionary axioms. Three nested primary axioms, starting points of reason, support evolution (Wiley, 1975). These axioms of evolutionary theory are inductive statements of the highest level of universality, which have high explanatory power and should be considered fruitful principles of discovery.

> ***Axiom 1 — Continuity:*** *Evolution occurs, with change entailing spatiotemporal continuity.* This axiom complies with Leibniz's principle (law) of continuity *'Natura non facit saltum'*, both in its mathematical and metaphysical inceptions (Leibniz, 1678):
>
> *"When the difference between two instances in a given series or that which is presupposed can be diminished until it becomes smaller than any given quantity whatever, the corresponding difference in what is sought or in their results must of necessity also be diminished or become less than any given quantity whatever. Or to put it more commonly, when two instances or data approach each other continuously, so that one at last passes over into the other, it is necessary for their consequences or results (or the unknown) to do so also. This depends on a more general principle: that, as the data are ordered, so the unknowns are ordered also."*

One formulation of the law is that of spatiotemporal continuity: *"Any change from small to large, or vice versa, passes through something which is, in respect of degrees as well as of parts, in between."* Thus, any change in location or time must be a continuous change. In a letter to Varignon, Leibniz extends this concept to realities (or beings) (Leibniz, 1702):

> *"When the essential determinations of a being approach those of another so that likewise accordingly all the properties of the first must gradually approach those of the last, it is necessary that all the orders of natural beings form only one chain, in which the different classes, like so many links, connect so closely the one to the other, that it is impossible for the senses and the imagination to fix the precise point where any one begins or ends."*

This formulation of the principle of continuity resembles that of evolution, with its axiomatic assumption of succession of events, continuity of existence, and gradual change. Epistemologically, any observation (reality) must comply with the principle of continuity.

Axiom 2 — Singularity: *Only one phylogeny of extant and extinct organisms (and their component parts) exists as a consequence of genealogical descent.* A phylogeny is a hypothesis of genealogical relationships among a group of entities (taxa) in the form of a tree or network with specific connotations of ancestry and an implied time axis (Figure 1A and B). Trees or networks are therefore hypotheses of history. Epistemologically, any observation must be placed within the context of symmetry breaking and irreversibility of time (Brading and Castellani, 2003, 2013). These concepts are central for definitions of unity and diversity. Symmetry, "invariance under a specified group of transformations," in its

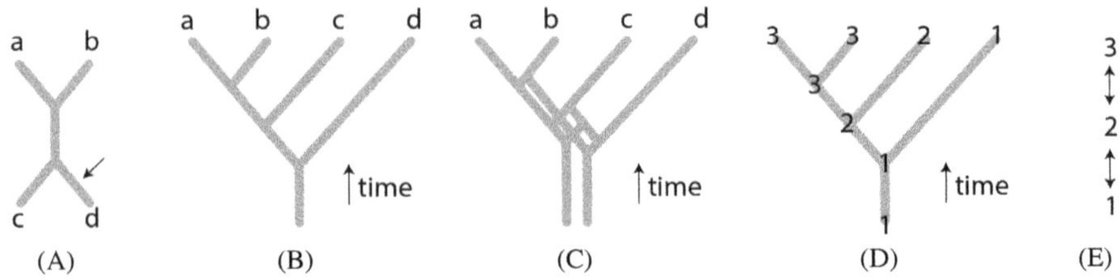

Figure 1. Trees, networks and character change. (A) Unrooted tree (also known as a network) describing the relationship of four taxa (a, b, c and d). The network does not show an arrow of time. If the network is rooted in the branch indicated by an arrow then a rooted tree is produced. (B) Rooted tree resulting from the rooting of the network of panel A. The tree holds time directionality, from its single root to its leaves. Taxa a and b are sisters to each other and derived in relation to c and d. Taxon d is the most ancestral. (C) If the rooted tree holds reticulations, then it becomes a rooted network. The network that is shown has two separate origins. (D) The rooted tree of panel B with character states 1, 2 and 3 labeling terminal, internal and root nodes. Note state 1 is ancestral (plesiomorphic) to state 2 and state 2 is ancestral to state 3. Most importantly, note that once the tree is rooted, the assignment of state 1 to the root implies character information for the branch that roots the tree. (E) Character state transformation series describing a model of evolution of the character depicted in panel D, in which character states are ordered and fully reversible. The rooted tree in panel D polarizes character change in the transformation series, making state 1 ancestral and state 3 derived.

modern philosophical inception, is an abstract group-theory notion bounded by: (i) *equality and interchangeability of parts,* in which equal parts can be exchanged with one another, while preserving the whole, by the use of mathematical operations (e.g. translations, reflections, rotations) or transformations, (i) *invariance of parts,* when the interchangeability of parts or other transformations of the group results in invariance, and (ii) *grouping of parts,* with operations satisfying the algebraic concept of groups. Here, a group **G** is an algebraic structure, a set of elements *e* with an abstract structure determined by the properties of a composition operation (usually called product or "group law" and denoted by ".") that combines pairs of elements into new elements. I note that the group operation is associative, the group contains the identity element, and for each element there is an inverse. These boundaries of symmetry result in the "unity of different and equal elements" of symmetry, and their breaking by the absence of an element of symmetry (asymmetry) results in a phenomenon. This principle, first recognized by Pierre Curie, manifests as a cause-and-effect relationship, a causality principle, in which symmetry breaking is unidirectional:

"[The] conclusions we can draw from symmetry arguments are of two kinds: The first is firm but negative and uncontestably true: There is no effect without cause. The effects are the phenomena that always require a certain dissymmetry to happen. If this dissymmetry does not exist, the phenomenon is impossible. This often prevents us from going astray in search of unattainable phenomena. The considerations on symmetry still allow us to state a second kind of conclusion of positive nature, but that do not provide the same certainty as the results that are negative in nature. They meet the proposal: There is no cause without effect. The effects are the phenomena that can arise in an environment with a certain dissymmetry; they provide precious information for the discovery of new phenomena but the forecast is not as accurate..." (Curie, 1894).

When these aspects of parts and wholes are placed within a temporal framework or are considered within an ontological perspective (Brading and Castellani, 2003), then a mathematical tree or network structure can arise that has an implied time axis and expresses the symmetry breaking principle. Postulating the existence of a tree of life for example, implies the existence of a unique historical sequence of symmetry breaking (cladogenic) events. We note that a network phylogeny with reticulations for example implies the existence of other processes besides genealogical descent that act as noise (homoplasy) and resemble symmetry joining events, including the existence of more that one origin (root), as long as vertical phylogenetic signal unifies history (Figure 1C).

Axiom 3 — Memory: *Characters are transmitted from one generation to the next, modified or unmodified.* In phylogenetics, a character is an observable feature of a biological entity ("primary homology," sensu de Pinna, 1991) that is used to establish history. Characters have alternative mutually exclusive manifestations (character states) and are most powerful when they unfold as "shared and derive" features (synapomorphies) in the branches of evolutionary trees or networks (Figure 1D). Since a character represents a historical event in the evolution of the observable feature as it manifests along the branches of a tree or network hypothesis, defining a character implies establishing a relationship between the character and its states. In other words, defining a character involves mapping correspondences between alternative "homologs" (character states) of the putative "homology" (character). Consequently, a character implies a "transformation series," which provides boundary conditions to the possible historical character state transformations (Grant, 2002) (Figure 1E). Character transformations are therefore the actual character state changes transmitting change as modified or unmodified characters in a phylogeny. Characters also show at least two character states (transformational homologs), one ancestral (a plesiomorphic state) and the other derived (an apomorphic state). Since homologies can only be tested at a same level of universality in which they have been hypothesized to exist as synapomorphies, plesiomorphic or apomorphic character states assignments will depend on the number of taxa they encompass (Figure 1D). Their spread to larger clades implies higher levels of universality and higher number of falsifiers for corroboration. Determining the relative ancestry of alternative character states also defines the polarity of character transformations and roots the trees or networks. Polarity implies the evolutionary specification of the direction of character change. This is a fundamental property of phylogenetic inference. Moreover, polarization not only roots the trees but also makes synapomorphies and their distribution in phylogenies fully explicit. Their goodness of fit to trees and networks and their associated falsifiers serve as corroboration tests of primary homologies.

Brower (2000) has criticized the inclusion of evolutionary axioms (particularly axiom 1) as necessary for cladistics analysis. While some of his arguments may be sound from an instrumentalist point of view, the operations of retrodiction cannot be made in absence of a supporting theoretical framework (Kluge 2001). In particular, the principle of continuity separates evolution from elaborations of creationist agenda that invoke the existence of intelligent design (Caetano-Anollés, 2016). Thus, gradual change must be considered an initial and overarching axiomatic component of retrodiction, which is a condition that may appear far removed from its analytical groundwork. Gradual change is also necessary to explain the rise of biological systems by gradual addition of component parts and interaction of parts of those systems. As I will describe below, the interaction of parts and wholes in biology becomes crucial for the retrodiction enterprise.

The analytical phylogenetic framework: In inductive reasoning (bottom-up logic), the premises (statements) of an argument attempt to supply strong evidence supporting the truth of the argument's conclusion. In other words, induction gathers information necessary for the proposition of useful hypotheses, which are necessary to predict and infer. In contrast, deductive reasoning (top-down logic) links the premises of an argument to its conclusion, hoping that if premises are true the conclusions will be also true. In other words, deduction tests the hypotheses generated by induction and either corroborates or refutes them. Without deductive methods, which at first order-logic are demonstrably complete, induction methods fail to preserve truth (Vickers, 2014). There is no theory that supports induction, which either recursively relies on higher inductive principles, leading to infinite regression, or results in apriorism. For example, if we do not take as true the statement *"all folds of a protein structural domain are homologous,"* we will need to justify the statement by claiming that *"all protein folds with a NAD(P)-binding Rossmann-fold domain are homologous"*.[1] Unwilling to accept this more granular statement as true, we might further qualify this statement as *"all protein folds with a NAD(P)-binding Rossmann-fold that have a alcohol-dehydrogenase-like C-terminal domain are homologous."* Unless we are willing to compromise at some level, which will inevitably lead to apriorism, we will need to continue investigating further details of molecular and atomic structure, adding more and more granular detail *ad infinitum*. To avoid this problem, deductive reasoning provides an objective criterion to determine when to stop induction, and in our example, accept the homology as true. As described earlier, deductive methods inspired by Popper's methodology subject hypotheses to critical testing. If the result of the test is *consistent* with the hypothesis being tested, the hypothesis has been corroborated and is tentatively accepted until further testing; the falsifier now becomes part of the set of falsifiers that corroborates the hypothesis. If that is not the outcome, then the test is either *inconsistent* with the hypothesis (the hypothesis has been falsified and discarded) or the test has been irrelevant. Within this

[1]The statement paraphrases Wiley's (1975) example of morphology: "all feet which are attached to the hind leg are homologous."

framework of scientific progress, the main axioms of retrodiction have already attained high levels of truthlikeness, verisimilitude, empirical content, and degree of corroboration. My concerns however relate to lower levels of inquiry, the observable features we use as phylogenetic characters to generate hypothesis of history that formalize the study of evolution.

Ontologically, characters are class properties that are assumed to be homologies, historical identities (Grant and Kluge 2003). For example, the statement *"all protein folds with a NAD(P)-binding Rossmann-fold domain are homologous"* can take the form of a character. These protein domains can be present in varying numbers in different proteomes (protein collectives in organisms), and these values could represent character states of the character in question. Alternatively, the presence or absence of the protein domain "class" could be coded using a binary system. Characters are the "basic" evidential statements, the premises of the study. These character and character states provides the evidence *E* for evaluating the degree of corroboration of hypotheses of history that is described in Equation (2). They make up columns in data matrices used for tree or network reconstructions (the rows are the taxa labeling their leaves). They are conjectures of perceived similarities (hypotheses of primary homology) that are accepted as empirical facts for the duration of the phylogenetic study. These primary homologies (de Pinna, 1991) can be defined using cladistics, phenetic, evolutionary, classification or other methods. For example, machine learning and data mining methodologies such as hidden Markov models (HMMs) can define protein structural domains at different levels of the structural hierarchy. These statistical Markov models can identify for example protein sequences in a genome harboring the *NAD(P)-binding Rossmann-fold domain* structure. When comparing different sequences that are assumed to be homologous, one could even decide to align the sequences and use amino acid or nucleic acid sites of their sequences as characters. This necessitates for example the use of dynamic homology assessment or the coupling of hidden Markov models and tree reconstruction methods. In deciding what kind of definition of a character one would like to choose, Wiley (1975) indicates that Popper (1959) offers some advice: *"a statement x is said to be 'falsifiable to a higher degree' or 'better testable' than a statement y, ... if and only if the classes of potential falsifiers of x includes the classes of potential falsifiers of y as a proper subset.*" Under Popper's theoretical system the mechanism by which a hypothesis of primary homology is proposed is not important. However, most of these hypotheses are generated from the world of experience of the investigator by induction. In this process, trivial hypotheses are excluded. For example, the criteria of ontogenetic and positional or topographical correspondence are the operational contexts in which most modern homology assessments are undertaken. For example, once the *NAD(P)-binding Rossmann-fold domain* (c.2.1) is defined using a complex set of bioinformatics criteria (and manual curation) in the Structural Classification of Protein (SCOP) database (Murzin *et al.*, 1995), it is an apparently easy task to distinguish the c.2.1 character from another protein domain defined at that same level of structural abstraction (fold superfamily). For example, the c.2.1 fold superfamily holds a 3-layered $\alpha/\beta/\alpha$ core with a parallel β-sheet of 6 β-strands in 321456 order. This structure is quite distinct from the *triosephosphate isomerase (TIM)* (c.1.1), with its parallel β-sheet closed barrel holding 8 β-strands in 12345678 order and a common set of phosphate-binding sites. It is also quite distinct from the other 1,984 superfamilies of SCOPe 2.05. However, most of these kinds of potential falsifiers forming proper subsets of the hypothesis of homology are given without reference to a phylogeny. They are similarities and dissimilarities that may or may not be the result of genealogical descent (axiom 2). Furthermore, establishing how similar is a character from another can be in some cases a difference of opinion. It is therefore important to apply any form of falsifiers available to the investigator at this very early stage of phylogenetic analysis.

Once primary homology hypotheses have been defined, and to avoid circular logic in the hypothetico-deductive framework, the exercise of phylogenetic reconstruction allows testing their validity. This exercise involves discovering historical hypotheses (phylogenies) that take for example the form of relationships between taxa and map character changes and their transformation series to the branches of trees and networks. These hypotheses H are represented by a disjunction of its constituents (phylogenies and transformation series) in Equation (2) (Lienau and DeSalle, 2009). This stems from the simple property that a phylogenetic 2-way and 2-mode matrix (e.g. an alignment of protein sequences), with its collection of primary homology statements, implies the existence of a tree or network structure explaining it, very much like a network can be represented by an adjacency matrix or a phenetic tree can be represented by its distance matrix.

Phylogeny reconstruction requires a formalization of retrodiction in the form of selecting a phylogeny out of all possible phylogenies according to an explicit criterion. There are several possible optimality criteria, including maximizing parsimony or maximizing likelihood. Maximum parsimony has been the workhorse of the analytical framework of cladistics (Kluge, 1984). Its epistemology is rooted in simplicity and the principle of Ockham's Razor (OR), which favors simpler theories over those that have superfluous information (Baker, 2010). OR supports formulations of ontological parsimony, interpreting the evidential support of hypotheses. Its most simple formulation is:

OR1: Entities are not to be multiplied beyond necessity

Current philosophical interpretation of OR impinge on ontological parsimony and the interactions of OR and theories *T*.

> *OR2: Other things being equal, if T_1 is more ontologically parsimonious than T_2 then it is rational to prefer T_1 to T_2*

Theories that commit to smaller ontologies are therefore preferred, or in terms of Quine's ontological commitments, those theories that are less ontologically committed are preferred. In other words, *entities should not be multiplied beyond necessity other things being equal.* OR2 is therefore weaker than OR1, since this interpretation considerably restricts the application of the principle. There are very few cases in biology where OR2 can be applied and justified. One of them is the reconstruction of phylogenies by exploring the space of possible trees or networks with maximum parsimony methods. Note that the space of possible bifurcating trees for example is defined by the number of taxa considered and represents a closed set of hypotheses of history that will be tested by phylogenetic analysis. Optimal trees are those that entail the least number of character state changes (steps). Thus, an algorithm must compute character state changes in the branches of each possible tree, selecting those that are most parsimonious and storing them in memory for further analysis. One widely used method of tree reconstruction is Wagner optimization for ordered characters (Farris, 1970), but other alternatives are available, including Fitch, Camin-Sokal, Dollo, and generalized parsimony optimizations, depending on the *a priori* auxiliary assumptions of the evolutionary model being used (Figure 2). Wagner optimization makes no assumptions about the reversibility of characters and can be applied to continued valued observables through gap-recoding techniques or special algorithmic implementations. The Wagner method requires two passes through each tree, the first computes minimum tree lengths and the second in opposite direction makes optimum character state assignments at

Wagner

TO:	0	1	2	3
FROM: 0	-	1	2	3
1	1	-	1	2
2	2	1	-	1
3	3	2	1	-

Fitch

TO:	0	1	2	3
FROM: 0	-	1	1	1
1	1	-	1	1
2	1	1	-	1
3	1	1	1	-

Camin-Sokal

TO:	0	1	2	3
FROM: 0	-	1	2	3
1	∞	-	1	2
2	∞	∞	-	1
3	∞	∞	∞	-

Dollo

TO:	0	1	2	3
FROM: 0	-	N	2N	3N
1	1	-	N	2N
2	2	1	-	N
3	3	2	1	-

Figure 2. Character state matrices of typical character state transformation models used in phylogenetic analysis. The matrices show transformation costs from one character state to another. In Camin-Sokal optimization, reversals are prohibited by taxing them with an infinite cost. In Dollo optimization, N is such that each gain of a character occurs only once on a phylogeny. The Wagner transformation model describes the character state transformation of Figure 1E.

internal nodes. Finding optimum trees is an NP-hard problem. Exhaustive or branch-and-bound algorithmic implementations allow exact solutions when the number of taxa is less than 20. However, the dimensionality of the space of unrooted and rooted trees increases with number of taxa n according to the $N_u = (2n-5)!/2^{n-3}(n-3)!$ and $N_r = (2n-3)!/2^{n-2}(n-2)!$, respectively. A space of rooted trees with only 50 taxa contains 2.7×10^{76} possible trees, which exceeds Eddington's number of electrons in the visible universe. Therefore, reconstructing larger phylogenetic trees is a computational challenging and time-consuming endeavor that will never arrive at exact solutions. It requires heuristic searches with hill-climbing algorithms to dissect small regions of tree space and find the most-parsimonious trees. One widely used strategy for example is to build a suboptimal most-parsimonious tree, which is then subjected to branch swapping, such as the subtree pruning or regrafting (SPR) or tree bisection and reconnection (TBR) operations. However, other strategies are available (e.g. fast character optimization, parsimony ratchet). None of these algorithms however guarantee exact optimal solutions. We note that modern cladistics analysis favors reconstruction of unrooted trees, since the space of unrooted trees is smaller and there is no commitment to a rooting strategy during tree reconstruction. As I will describe below, this practice only completes part of the retrodiction exercise, which goes full circle only when trees are rooted and the "evolutionary arrow" is made explicit in the phylogenies.

Given a most-parsimonious tree or set of trees, with or without reticulations, the testing of phylogenetic hypotheses according to Hennig (1966) clarifies the relationship between homology and phylogeny. The best test of homology is common ancestry (Wiley, 1975), which is already displayed by the reconstruction of the most-parsimonious unrooted trees or networks. Common ancestry also manifests in the shared and derived patterns of synapomorphies. These patterns can be evaluated with congruence tests, the assessment of how changes (homologs) of individual characters agree or disagree with each other along the branches of the most parsimonious phylogenies. The behavior of these homologs also enables definition of "secondary homologies," homologies that have been "legitimized" by being the subject of corroboration (de Pinna, 1991). Phylogenetic reconstruction will accommodate all available evidence brought by useful characters to phylogenetic analysis, but their goodness of fit to the trees will vary. Each primary homology will transform into one or multiple statements of secondary homology depending on the way character states distribute in the phylogenies. If statements of primary and secondary homologies coincide there will be perfect fit and primary homologies will have been fully corroborated. If primary homologies have to be subdivided into several statements of secondary homologies (by lowering the level of universality of the observed similarities), the mismatch will require *ad hoc* explanations of change, reversals, convergences or parallelisms, which are known as *homoplasies*. Higher homoplasy levels will therefore decrease the degree of corroboration of primary homologies. Further addition of taxa and characters will change the level of universality of the hypotheses of homology and will demand additional optimizations and testing. This dynamics of the explanatory power of the historical framework is strongly grounded in "reciprocal (or mutual) illumination" (Hennig, 1966), a principle that evaluates how each "primary homology" statement of character evolution agrees with the overall favored evolutionary hypothesis obtained from all available data. This successive approximation route for developing scientific theories of evolution, adds additional evidence in the form of more informative phylogenetic characters and taxa to a corpus of ideographic evidence with the goal of supporting the validity of phylogenetic hypotheses. Since agreements reformulate homology hypotheses iteratively to maximize explanatory power through test and corroboration (Farris, 2008), phylogenetic analysis operates under the hypothetico-deductive method and its iterative underpinnings supports the use of degree of corroboration as a tool for exploration of the natural world and for scientific progress.

Farris (2014a) recently debunked some myths about the association between Popper's corroboration C and probabilistic arguments in tree reconstruction. First, probabilities in Equation (2) are not related to prior probabilities of hypotheses. Frequentists consider probabilities as frequencies of empirical outcomes in evidence. For example, they will estimate parameters of a statistical model using some criterion (e.g. maximum likelihood) given data E. In most cases, frequentists will not assign probabilities to hypotheses. In turn, Bayesians seek highly probable hypotheses, formulating how a subjective degree of belief of a hypothesis could change when accounting for evidence. This establishes an inference framework. Popper's probabilities are not Bayesian but are logical and based on empirical content, which increases with degree of falsifiability (Popper 1959):

> *"Science does not aim, primarily, at high probabilities. It aims at a high informative content, well backed by experience [i.e. well corroborated]. But a hypothesis may be very probable simply because it tells us nothing, or very little. A*

> *high degree of probability is therefore not an indication of goodness."*

Thus, Popper's probabilities may remain low while seeking high information content and high corroboration. While corroboration cannot be a probability (*C* can be negative while probabilities cannot), there is a link between corroboration and likelihood. *C* increases monotonically with $p(E|Hb)$ of Equation (2), which denotes the probability of *E* given both *H* and *b*. Since $p(E|Hb)$ is Fisher's likelihood of *H* given *E* and *b*, then the tree with strongest corroboration is the maximum likelihood tree, which expresses as a probability distribution. These facts that were highlighted by Farris (2014a) "fictionalize" recent claims about the unfalsifiability of phylogenies (Vogt 2008). In other words, Popper is safe and sound in phylogenetic analyses.

To summarize, the essence of retrodiction is the search of "discovery operations" by inductive reasoning necessary to study phenomena that vary extensively across time and space and their confirmation with deductive methods. These discovery operations are *"sets of decision rules used to generate and choose among competing empirical claims,"* which select empirical tests capable of decisive falsification of competing explanatory hypotheses (Grant, 2002). Within the systematic framework of Willi Hennig, discovery operations make use of "shared and derived" character states to test evolutionary hypotheses in the form of homologies, transformations between homologs, and phylogenies. This trilogy of data-observations, models of change, and phylogenetic tree or network representations of change defines the fundamental operations of retrodiction. I have stressed the maximum parsimony view in my explication of ideographic thinking, understanding that there is a contest between non-parametric and parametric models of change that affect this trilogy (Sober, 2004), which is deeply entrenched in philosophical views.

4. Rooting undirected networks

Retrodiction is not complete if a tree or network has not been rooted (Caetano-Anollés *et al.*, 2018). Without placing a root, the direction of evolution cannot be made explicit nor the plesiomorphic or apomorphic states of a character in the transformation sequence or in the trees or networks. Brower and de Pinna (2012) in recent argumentation about homology (*see also* Farris, 2014b and other exchanges) made clear the importance of rooting, stressing its recent neglect in phylogenetic analysis:

> *"Rooting is not a neutral procedure ... and may exert an impact on homology assessments previously inferred on an unrooted tree. Early formalizations of rooting, such as by Farris (1970) and Lundberg (1972), were keenly aware of and realistic about the practical implications of rooting an undirected network. The length of a tree is unaffected by the position of the root, but it is certainly not unaffected by the inclusion of a root. Subsequent developments in the field, however, mostly neglected those initial considerations. Gradually, rooting came to be seen as a rather bureaucratic step to crown a phylogenetic analysis. Thus, the root is nowadays often treated as an immaterial entity detached from the rest of phylogenetic analysis, or simply as a marker that indicates where the root would go if it actually existed."*

When rooted phylogenies are desired, one can polarize character state transformations by invoking one of the following sets of *auxiliary assumptions*: (i) adding one or more outgroup taxa assumed to be ancestral, (ii) using character state distributions in taxa to comply with the nested taxic hierarchy that exist in phylogenies, or (iii) imparting a direction of change directly in the character state transformation model. The applicability of these approaches depends on the veracity of the auxiliary assumptions, which are crucial for retrodiction. Two general kinds of methods of rooting exist, depending on the use of taxa outside those of relevance to the study:

Indirect methods. The normal practice in cladistics is rooting optimal trees or networks with outgroups (Bryant, 2001). An outgroup is an external taxon or set of taxa, which is not part of the taxa being studied (the ingroup), that is included in the phylogenetic analysis. Consequently, outgroup addition increases the number of taxa and therefore increases the level of universality of the historical hypothesis. Since character polarity has no effect on the length of trees or networks, outgroup addition roots the undirected trees or networks reconstructed following optimization by placing the outgroup at their base. The method however is *indirect*, since it polarizes character change using the fundamental assumption of a higher-level relationship between the outgroup and the ingroup. This decreases the explanatory power of retrodiction by adding additional and crucial *ad hoc* hypotheses of plesiomorphy or *auxiliary assumptions* of what should be considered unproblematic background knowledge. But are these *ad hoc* or auxiliary hypotheses unproblematic? *Ad hoc* hypotheses are usually added *a posteriori* to account for mismatches of the hypothesis and evidence (homoplasy) and to account for inconsistency (Kluge, 2001). Similarly, additional background assumptions should be minimized, because they weaken the capacity of the empirical evidence to discriminate among competing theories of retrodiction. These additional hypotheses do not help decisive falsification.

> *"Science requires that choice among theories be decided by evidence, and the effect of an ad hoc hypothesis is precisely to dispose of an observation that otherwise would provide evidence against a theory. If such disposals were allowed freely, there could be no effective connection between theory and observation, and the concept of evidence would be meaningless."* (Farris, 1983)

The explanatory content of a hypothesis is function of the simplicity and clarity with which it can be described. The higher its explanatory content the bolder is the hypothesis. The most highly corroborated hypothesis of history with highest explanatory power involves the least number of *ad hoc* hypotheses of homoplasy. These however can increase with number of taxa and level of universality of the analysis. Similarly, the auxiliary assumptions that constitute unproblematic background knowledge must be kept to a minimum, including those related to the rooting of undirected networks. They must also justify the simplest and clearest test of the hypothesis, increasing the degree of falsifiability or testability of hypotheses (Popper, 1959).

Adding an outgroup *a priori* as an auxiliary hypothesis brings with it a minimum of an additional character state vector that is assumed to be ancestral, which can be highly problematic in the presence of uncertainty about the relation of the outgroup to the ingroup. First, the outgroup method's dependence on *a priori* assumptions of relationship could represent a fatal flaw from an epistemologically point of view (*see* Lundberg, 1972). Its inductive reliance on assumptions of higher-level relationships when advancing the exploration of the tree of life can lead to infinite regress or apriorism (similar to the protein structure example for homology). Second, the additional branches of the outgroup carry with them character state information for optimization during tree reconstruction, which could affect ingroup relationships and homoplasy levels (especially if the outgroup is phylogenetically distant from the ingroup). Unless well justified by biology or a realistic model of evolution, outgroups can be problematic or impossible (e.g. they cannot root a tree of life). In fact, outgroup taxa have also been used *a posteriori* as an auxiliary hypothesis that pinpoints the branch that roots the tree, without seeking the benefits of optimization of any kind (a notorious apriorism used in the tree of life field; e.g. Williams *et al*., 2013). Thus auxiliary assumptions, which may or may not be justified, infinite regress and apriorism of rooting can overshadow the entire retrodiction enterprise.

Direct methods. A conceptually distinct and more powerful approach involves *direct* polarization methods that involve only the ingroup (Bryant, 2001). The Lundberg rooting method for example adds a hypothetical ancestor defined by all-ancestral state characters derived from comparative analysis to the branch in an optimal undirected tree or network that is most parsimonious (Lundberg, 1972). This is done without formally adding an extra taxon. Similarly, an all-ancestral state vector polarizing change in the characters state transformation series can be used to pinpoint the optimal root. These methodologies help implement direct polarization methods, which confer polarity through the distribution of the states of homologous characters of the ingroup, generally by focusing on the generality of character states. Kluge and Farris (1969) were first to provide criteria for such generality (in order of reliability):

"(1) *The primitive state of a character for a particular group* [taxa] *is likely to be present in many of the representatives of closely related groups.*
(2) *A primitive state is more likely to be widespread within a group than in any one advanced state.*
(3) *The primitive state is likely to be associated with states of other characters known from other evidence to be primitive.*"

Criteria focus on the commonality or frequency of character states in taxa, which is difficult to justify from an evolutionary point of view when sampling restricted levels of biodiversity. Note how higher levels of universality in the rooting criteria provide background assumptions that are increasingly less problematic for explanatory power. Increasing more and more groups of taxa in the analysis will strengthen the commonality criteria by increasing the ubiquity of the ancestral character state in the nested taxic distribution of phylogenies. Thus, the validity of commonality strengthens with its universality as we seek to explore the entire tree of life.

Weston (1988) recognized the importance of character state distributions and extended Nelson's ontogenetic criterion of character polarization by postulating a "generality criterion," in which the taxic distribution of a character state is a subset of the distribution of another. This hierarchical concept transforms commonality of character states into commonality of character state changes. Paraphrasing Weston's rule: "Given a distribution of two character states in which one, x, is possessed by all of ingroup taxa that possess its homolog character state y, and by at least one other taxa that does not, then y may be postulated to be derived (apomorphous) relative to x." Weston's character polarization ability to distinguish ancestral and derived states is most powerful when characters are multistate and accumulate changes "iteratively" in evolution (e.g. gene paralogs via duplication, protein folds in proteomes). In such cases, polarization equates to additive phylogenetic change in abundance of useful biological features (reviewed in Caetano-Anollés *et al.*, 2014). Since homologies always involve synapomorphies at some point of the taxic hierarchy, nested homologies of these kinds add iteratively to changes and strengthen the polarization scheme. de Pinna (1994) applied Weston's generality criterion in its special ontogenetic formulation in a different way, by proposing that most parsimonious networks be rooted based on the direction of character transformations inferred using the generality criterion in terminal taxa. None of these direct methods have been broadly applied in the molecular biosciences and their application in the analysis of morphological characters has been limited.

In absence of highly dynamic and undetermined processes of change (such as the random and stochastic processes use to model evolution of biomolecular sequences, which change at relatively fast pace in nature), character polarity and direction of change can be derived from an axiom of evolutionary biological process. However, the model of character state transformation must comply with evolutionary axiom 1 mentioned above. We recently polarized character state transformation series that described evolutionary accretion in the form or molecular abundance (e.g. number of protein structural domains holding a specific fold in proteomes; Caetano-Anollés and Caetano-Anollés, 2003) or molecular growth (e.g. number of substructures in nucleic acid molecules; Caetano-Anollés, 2002a). We used an all-ancestral state vector defined by a biological model and the Lundberg rooting method for that purpose (reviewed in Caetano-Anollés *et al.*, 2014). In these cases, the auxiliary assumptions used to polarize multistate ordered characters are well justified based on knowledge from genomic science or the biophysics of biomolecules. Remarkably, the rooting obtained by polarizing the transformation series matched the rooting obtained using Weston's generality criterion, mutually supporting the biological process and phylogenetic methods. It is worth mentioning however that character polarization using models describing evolutionary process must be carefully evaluated and such evolutionary models should be biologically realistic (Kim *et al.*, 2014a; Caetano-Anollés *et al.*, 2019).

5. Characters must serve as independent evolutionary hypotheses of homology

Phylogenetic analyses using optimality methods such as maximum parsimony or maximum likelihood require that phylogenetic characters be independent pieces of evidence (Kluge and Farris, 1969; Felsenstein, 1973). In fact, characters are first selected assuming that they are evolving independently of one another even if they are part of the same evolving organismal system. This is what Brooks and McLennan (1991) call the "Kluge's auxiliary principle":

"Always presume character independence in the absence of evidence to the contrary." When one character interacts with another in evolution, both characters are overweighed in the analysis and the resulting phylogeny fails to represent true history (Chippendale and Wiens, 1994). Possible dependencies could be of many kinds, from structural to functional, from physiological to developmental, from organismal to ecological. They can also stem from logical correlations arising from the definition of characters, including their ontology. These dependencies distort and obscure phylogenetic signal and must be either encoded into the phylogenetic model through parameters or weight corrections or avoided by excluding at least one of the interacting characters from the data matrix.

Determining interactions between characters is a difficult endeavor. Suites of correlated characters have been identified before tree building using character compatibility analyses by constructing an association matrix between characters, deriving eigenvectors and identifying characters with similar patterns of compatibility (O'Keefe and Wagner, 2001). However, dependency and co-dependencies may not be fully extracted from the data with these methods. A number of promising data mining techniques have been applied *a priori* to determine the possible existence of character non-independence (Magglia *et al.*, 2003). For example, a rule-based data mining approach called PT-RICO identified dependencies and co-dependencies responsible for the secondary structure of proteins with high precision (Lee *et al.*, 2010). These kinds of techniques could be used to explore possible non-independences in protein sequence due to higher order structure of the molecules, but could also be extended to the analysis of other molecular and morphological characters. However, even if dependencies and co-dependencies are identified and even if they are proven to be evolutionary, how do we use this information? Molecular and cellular interactions generally result in dependencies that are evolutionarily conserved.

Moreover, plesiomorphic character states are generally more conserved, since it is assumed that they appeared earlier than the apomorphic derived states (conservatism of characters; Hennig, 1966). This assumption supports Weston's generality criterion of rooting. However, deleting these non-independent conserved characters eliminates useful features that are most valuable in phylogenetic analysis of deep and more general relationships. Alternatively, describing the dependencies and co-dependencies for their use in evolutionary models as parameters or as weighting schemes is highly problematic. Calculations of character change probabilities for specified models of evolution rely on the evolutionary realism of the parametric models and a number of crucial assumptions, some of which are invalid. For example, probabilities should be calculated on an individual basis since parameters depend on a unique course of history for the set of taxa beings studied. However, this requires a priori knowledge of the history of the system and a mechanism to calculate the probability of new events, which is computationally impossible (e.g. Felsenstein, 1973). A different strategy that avoids the non-independence problem must be sought.

6. Parts and wholes: A new phylogenetic framework

Nobody will doubt that biological systems are by definition made of parts, regardless of the way parts are defined. Examples include atomic parts of molecular modules such as amino acid residues of polypeptide chains, modular parts of molecules such as NAD(P)-binding Rossmann-fold structural domains of proteins, parts of molecular ensembles of cells such as ribosomal complexes of α-proteobacterial microbes, cellular parts of organisms such as Purkinje cells in the nervous system of whales, organismal parts of populations of marine microbes such as *Vibrio fischeri* communities that are housed in internal light organs of bobtail squids, population parts of ecosystems that for example define mangrove habitats, and so on, from lower to higher levels of hierarchical complexity. Modules are sets of parts that interact more strongly with each other than with other parts of a system. In evolution, diversification and integration of parts unify parts into modules, which then diversify (Mittenthal *et al.*, 2012). This process and the rise of modules may explain evolutionary waves of complexity and organization and the emergence of hierarchical structure (defined broadly) in biology. Since biological systems evolve and carry common ancestry, can parts of these systems by definition evolve and by themselves carry common ancestry? In other words, are the histories of parts embodied in the ensemble of lineages of the tree of life?

Superficially the answer to this question seems trivial. Negating the origin and evolution of parts would equate to negating the evolution of systems, and therefore accepting the stance of "intelligent design," where parts are so interwoven within the fabric of systems that can only be explained by invoking an external entity of divine creation. However, and as I will explain, the answer to the question is far from trivial. Evolutionary biology and its allies, molecular evolution and systematic biology, are well-established fields of scientific inquiry in the life sciences. Their focus however has been overwhelmingly the organism, as testified by the effort devoted to taxonomical classification and building of the tree of life. The focus is therefore the biological system. What about the molecular and cellular parts of organisms? Do these evolve in the same way as systems? Are they by themselves systems? Can these parts be studied with the same phylogenetic, stratigraphic and evolutionary bioinformatic approaches that are used for example in systematic biology or paleontology? Can there be a systematic biochemistry or a molecular paleontology? I argue that the answer is an emphatically yes, though many in the field of molecular evolution may think otherwise. The driver for these questions is my interest in the origin of modern biochemistry and principles of emergence that could help in efforts of synthetic biology. Biochemistry is essentially the science of parts, explaining the patterns and processes of biochemical components that populate organismal systems. Biochemistry however cannot yet be explained in evolutionary terms using standard phylogenetic methods borrowed from sequence analysis. However, the imperfect reductionistic knowledge of genetics and other vocabularies, which seem responsible for the biochemical make up, may give us some clues.

I start by defining the system. According to Collier (2003), a *dynamical system* can be defined as a natural object delimited by a set of interacting component subsystems, which is characterized and individuated from other systems by its cohesion. Cohesion refers to the dynamical stabilities of the components of the system, its parts, when these are constrained. While the stability of parts allow for them to be decomposed into spatially bounded or spatially unlinked components, cohesion enables the integrated mechanical operation of the dynamical system through forces and flows. The dynamic behavior causes systems to be organized entities, which express *structure*, i.e., organizational complexity in a manifold of levels. Previously, we have described components in an organized system as distinct interrelated entities inducing multidimensional ordering relationships of component parts (Caetano-Anollés *et al.*, 2010). These ordering relationships are layers of organization (levels and hierarchies) that express redundancy, a desirable characteristic of systems, which endows both insensitivity to perturbation (robustness) and reuse of existing parts in different contexts (economy). For the purpose of my description, redundancy can be interpreted as the repeated instantiation of parts in quantities greater than necessary for the performance of a system's activity and fulfilling constraints (e.g., functional constraints). In Figure 3A, we describe how realized and non-realized structural states of a system can be coarse-grained into levels, which are generally hierarchical. For

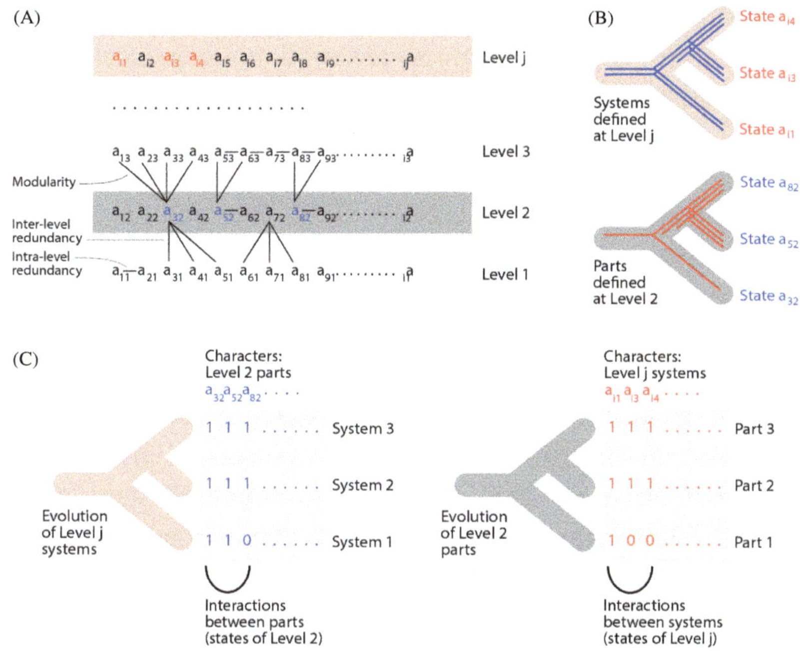

Figure 3. A model of structural evolution of biological systems. (A) The structural states of a meta-system (a_{ij}) are coarse-grained into levels, which are generally hierarchical and express intralevel redundancy (reuse of components with a same state of structure within a level) or interlevel redundancy (mapping relationships between states in different levels) (Caetano-Anollés *et al.*, 2010). Redundancy relationships are depicted with lines connecting states. Interlevel redundancy refers to different lower level states that manifest the same higher-level state (e.g., different protein sequences expressing a same fold structure). Alternatively, modularity, a special kind of interlevel redundancy, refers to different higher-level states including components with the same lower-level state, which now become a "module" (e.g., protein domain combining with others to form multidomain proteins). Note that interlevel relationships need not occur between adjacent levels. (B) The evolution of structural states are traced along a phylogenetic tree describing the diversification of levels of the meta-system defined at some level of granularity. These trees are generally reconstructed by selecting structural states that are useful for the description of change. Structural states at different levels define characters and taxa for analysis. (C) When structural states at high levels are defined as taxa, states at lower level are used to define characters, and vice versa. When taxa share characters that are phylogenetically informative, a phylogenetic tree can be constructed that describes the evolution of a level. These trees describe the evolution of systems and the evolution of parts of the meta-system.

example, the structure of proteins can be described at increasing hierarchical levels of structural abstraction: sequences, motifs, loops, domains, families, superfamilies, folds, architectures, classes. Two accepted gold standards of protein classification, SCOP (Murzin *et al.*, 1995) and CATH (Orengo *et al.*, 1997), use parts of this incomplete scheme to describe the atomic complexity of the molecules. We note that these classifications do not consider unrealized structural states, such as protein folds that are possible but that have never been identified in the natural world of protein structures, though it includes engineered proteins and peptides. In my model, redundant relationships of structural states can be established between and within levels (Figure 3A). One special kind of redundancy, modularity, is established when a structural state at lower level establishes mapping relationships with states at higher levels. For example, a structural state describing a fold domain can be found represented in many proteins and can act as a structural and evolutionary module. This stems from the evolutionary combinatorics of fusions and fissions that is responsible for the diversity of multidomain proteins (Wang and Caetano-Anollés, 2009).

Parts such as structural domains defined at fold family level and systems such as proteomes belong to the meta-system of life. Consequently, structural states of parts and systems can be compared. When these states are observable, they provide an opportunity to study the evolution of parts and wholes (Figure 3B). A higher level of the hierarchy always acts as source of systems for the phylogenetic study. A lower level of the hierarchy acts as source of systems parts. Structural states must comply with the epistemological requirements of phylogenetic characters that we have described above. They must be homologous, heritable, and independent and must show distinct states (e.g., parts at one level must be present or absent in a system defined at another). However, parts, levels and systems of a meta-system are all interrelated to each other through interactions and redundancy, and in all comparisons, there will be violation of the requirement of character independence (Figure 3C). The task is to seek best characters and taxa and choose tree reconstruction strategies that would minimize the problem of character non-independence. I here propose that the best way to achieve this purpose is to choose structural states at high level (systems) as characters and structural states at lower level (parts) as taxa. Parts that are modular become discrete units and are preferred. Similarly, drawing structural states from levels that are distant in the hierarchy are better at minimizing direct mapping interactions. I therefore propose that building trees of parts minimizes the problem of character independence. I use examples of protein sequences and protein structural domains to illustrate the benefits of the strategy:

(i) *Trees of systems:* A protein sequence is composed of amino acid parts, which are generally fit into a data matrix (alignment) and used to construct trees of genes and organisms using the methods of phylogenetic analysis (Brooks *et al.*, 2007). These trees are effectively trees of systems (Figure 3C). However, proteins have complicated 3-dimensional structures that result from interactions between amino acids at the atomic level. These intramolecular interactions, or at least some of them (Sathyapriya *et al.*, 2009), induce protein folding, molecular function and protein stability. They are responsible for protein secondary, supersecondary, domain and tertiary structure, and by definition, their mere existence induces violation of character independence. Penny and Collins (2010) proposed the simple thought-experiment in which the bioinformatician exchanges rows of sequence sites and asks what was lost in the process. Randomization of characters (columns) in the data matrix does not change the phylogenetic tree. However, randomization destroys the structure of the molecule and very likely its function. This confirms that reconstruction of trees from information in sequence sites violates character independence, and in the process, ignores structure and biology. The effects of violation of character independence may be negligible for trees that compare sequences that are closely related. However, trees describing deep historical relationships require that sequences be divergent and this maximizes the chances of even wider divergences in molecular structure that are not being accounted for by the models of sequence evolution. Recent studies have modeled the effect of structure on phylogenetic analysis of sequences and showed it can be substantial (Nasrallah *et al.*, 2011). Strong character dependencies, short internal branches, large overall tree length and simple models of dependencies lead to inaccurate estimation of tree topologies (Nasrallah *et al.*, 2011). Eliminating one of the interacting characters can alleviate some of these problems (Sherbakov and Triboy, 2007). However, there could be more than one-to-one interactions occurring between characters, which need to be sorted out *a priori* through inductive discovery operations. These complexities exacerbate when exploring the deep past. Phylogenetic relationships that describe deep evolutionary events can be driven by a handful of sequence sites in a handful of conserved genes of concatenated sequence alignment (Shen *et al.*, 2017). Incongruent phylogenetic relationships and violations of character independence must be carefully evaluated in these cases to avoid contentious evolutionary descriptions.

(ii) *Trees of parts:* Proteomes are composed of collection of protein domains with well-defined structures and functions. Protein domain counts in proteomes have been used to generate trees of domains (reviewed in Caetano-Anollés *et al.*, 2009). These trees describe the evolution of protein structure at global level. They are effectively trees of parts. While domains interact with each other in multidomain proteins or establish protein-protein interactions with the domains of other proteins, these interactions of parts do not violate character independence. This is because phylogenetic characters are actually proteomes, systems defined by structural states that exist at much higher levels than the protein domains, not far away from the organism level. Remarkably, nothing is lost when character columns in the matrix are randomized in the thought experiment. The order of proteome characters in the matrix does not follow any rationale. Characters are not ordered by lifestyles or trophic levels of the organisms. Interactions between organisms will seldom bias their domain make up, and if so, those characters can be excluded from analysis. Even the establishment of symbiotic or obligate parasitic interactions, such as the nodule-forming symbioses between rhizobia and legumes, may have little impact on character independence.

These examples make clear the evolutionary interrelationship of parts and wholes and prompt: (i) the generation of trees of parts, (ii) devising methods that use information in trees of parts to study the evolution of wholes; (ii) a focus on higher-level structure, and (iii) a decrease in confidence in the power of trees of systems. These concepts have been applied to the study of the history of nucleic acid and protein structure (reviewed in Caetano-Anollés *et al.*, 2014), but can be useful in many phylogenetic applications.

7. Occurrence and abundance

Given a hierarchical biological system (α), its historical memory unfolds by increasing the abundance of both its nested parts and wholes, especially when these parts and wholes are modules. Equation (3) describes the process.

$$a \overset{\alpha 1}{\rightarrow} A \tag{3}$$

a is the number of parts and wholes. A is the number of parts and wholes that have attained higher abundance levels ($a' > a$, with a' reaching a value A in the present). Highly abundant parts and wholes have higher chances of remaining persistent and by doing so enhancing the survival and historical memory of the system under consideration. For that reason, it is generally unlikely that once high abundance levels A are achieved these levels will return to basal levels by loss, unless strong reductive evolutionary forces are at play that are beneficial to the system. This is particularly so when the views of hierarchical systems are global and focus on the higher level rather than the local and lower level. Thus, abundance provides under appropriate conditions a powerful criterion for character state polarization of transformation series.

Abundance can be coded as ordered multistate characters. They represent "serial" homologies, typical of many morphological characters. These characters behave as nested hierarchical homologies when the transformation series are mapped to the phylogenetic trees. Multistate characters of these kinds can be decomposed into sets of binary characters without changing tree length or character state distribution (Farris *et al.*, 1970). Since each binary encoding represents a separate statement of homology, the ordered multistate characters are hypotheses of homology that are compound, decomposable and powerful. Their nested hierarchical nature and their additivity allow matching the generality criterion of rooting to the polarization of the transformation series. Their decomposability links abundance and occurrence. In fact, coarse graining abundance by gap recoding results in effective coarse graining of the decomposability of the nested hierarchical homology statements. Maximum coarse graining turns abundance into occurrence. In turn, when ordered multistate characters result from gap recoding of quantitative valued characters, information in statements of homology can be further increased.

Historical memory can also unfold by recruitment and/or accretion, the ability to use existent parts in new different contextual environments. Equation (4) summarizes how the recruitment and/or accretion process increases the numbers of parts *a* when these are recruited by parts *b*.

$$a + b \xrightarrow{\alpha 2} Ab \tag{4}$$

Since abundance of older parts that are coopted by newer ones follows the trends of Equation (3), recruitment increases abundance of *a* to levels *A*. Note that parts need not be physically associated to fulfill Equations (3) and (4) and enable increases of abundance by recruitment. Analysis of the distribution of protein fold structures in metabolic pathways for example suggests metabolic networks grow piecemeal and form evolutionary patchworks (Teichmann *et al.*, 2001). This patchwork model of metabolic growth through recruitment has been confirmed by phylogenomic reconstruction (Kim *et al.*, 2006; Caetano-Anollés *et al.*, 2007). Networks are defined by collections of enzymes that are not necessarily physically linked in the cell. In contrast to recruitment, accretion occurs when parts are recruited pervasively into a system, one at a time, and remain tightly linked with each other. In many cases the recruited parts become physically associated with the growing system, with b of Equation (4) representing the growing system and a the accreted part. An example of accretion is the evolutionary growth of large macromolecular complexes such the ribosome (Harish and Caetano-Anollés, 2015), which increase in size by addition of amino acids and nucleotides to their molecules and by recruitment of new protein and nucleic acid molecules. Phylogenomic methods revealed that the ribosome developed gradually by addition and growth of rRNA helical segments and ribosomal proteins, which co-evolved to form the complex ribonucleoprotein ribosomal complex (Caetano-Anollés and Caetano-Anollés, 2015). Similarly, phylogenomic dissection revealed that the protein world developed structural innovation piece meal, complying with the axiom of continuity (Nasir *et al.*, 2017).

8. Levels of universality and corollaries

Figure 4 shows few examples illustrating the generation of trees of parts and wholes from occurrence and/or abundance of useful features of biological systems. For instance, molecular parts in molecules can be used to generate trees of molecular parts and trees of molecules (Figure 4A). Similarly, protein structural domains in proteomes can be used to generate trees of domains and trees of proteomes (Figure 4B). At even higher levels of the organismal hierarchy, molecular functions in functionomes can generate trees of functions and trees of functionomes (Figure 4C). Once the

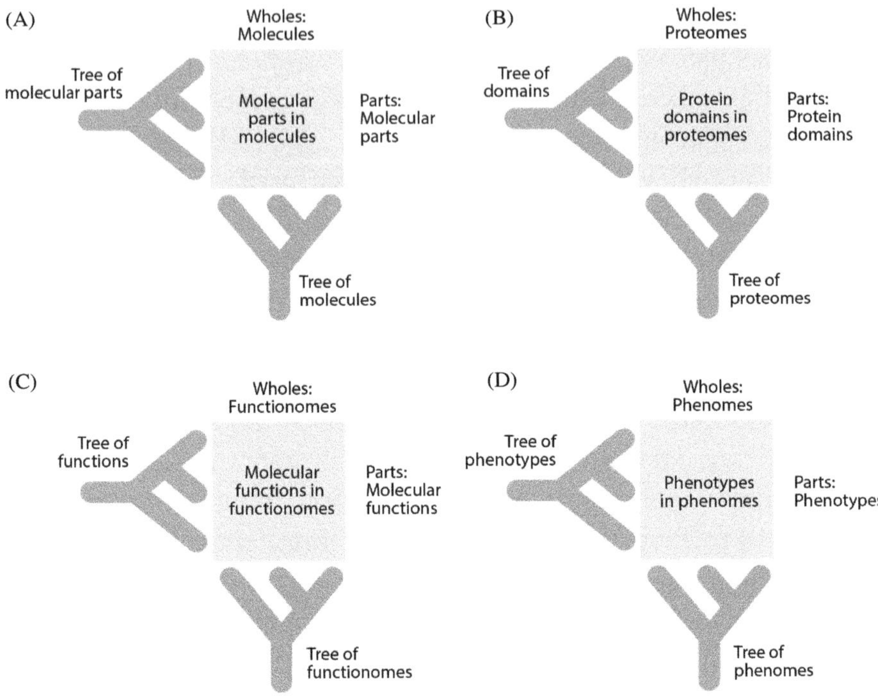

Figure 4. Building trees of parts and wholes from occurrence and abundance of useful features. (A) Molecular parts and molecules (Caetano-Anollés, 2002a,b). (B) Protein domains and proteomes (Caetano-Anollés and Caetano-Anollés, 2003). (C) Molecular functions and functionomes (Kim and Caetano-Anollés, 2010; Kim *et al.*, 2014b). (D) Phenotypes and phenomes (Salazar and Caetano-Anollés, ms. in preparation).

hierarchy materializes in a higher-level phenotype, matrices of phenotypes in phenomes can be used to reconstruct trees of phenotypes and trees of phenomes (Figure 4D). In these examples, the characters that are used to study the history of parts and wholes represent open (technically unlimited) and closed (finite) systems, respectively, while the taxa analyzed represent closed and open systems, respectively. This is because genomic and other "omic" revolutions have been able to provide enough knowledge to describe complete repertoires of molecular and functional parts (suitably defined) with increasing completeness and precision (Doolittle, 2005). Such features that can be exhaustively enumerated make retrodiction a powerful enterprise from a "level of universality" point of view. Figure 5 describes a phylogenomic approach that mines genomic sequence from information in the structure and molecular function of proteins and RNA and generates trees of parts and wholes. The approach has been used with high reliability for almost two decades of steady research.

Popper (1959) made clear that within a theoretical system there exist statements of various levels of universality. While statements of the highest level of universality are the axioms of the theoretical system, lower level statements retain the character of being hypotheses amenable to corroboration, even at the lowest level that may be attainable. For example, homology statements supported by evolutionary axioms 1, 2 and 3 arise via induction from discovery operations, which are in themselves hypotheses of common ancestry that are testable by the hypothetico-deductive framework. When evaluating hypotheses of history, the number of taxa considered is proportional to the level of universality of the theoretical system of history. Thus, every new taxon added to the ingroup or to the outgroup adds character state vectors to terminal and internal nodes that extend both the hypothesis of history and the hypotheses of homology being considered. Any additional character added to an analysis also extends the universality of the system by providing further testing of the historical hypothesis. These facts were already made evident by Wiley (1975), who elaborated on the importance of the level of universality of a phylogenetic hypothesis and the concept of homology, and Kluge (1985), who elaborated on ontogeny and its relation to assessment of homology and phylogenetic reconstruction. While in some cases falsification of a less universal statement can entail falsification of the whole system, by no means this is so in phylogenetic analysis. Many statements of homology can be falsified without falsifying the historical hypothesis that the

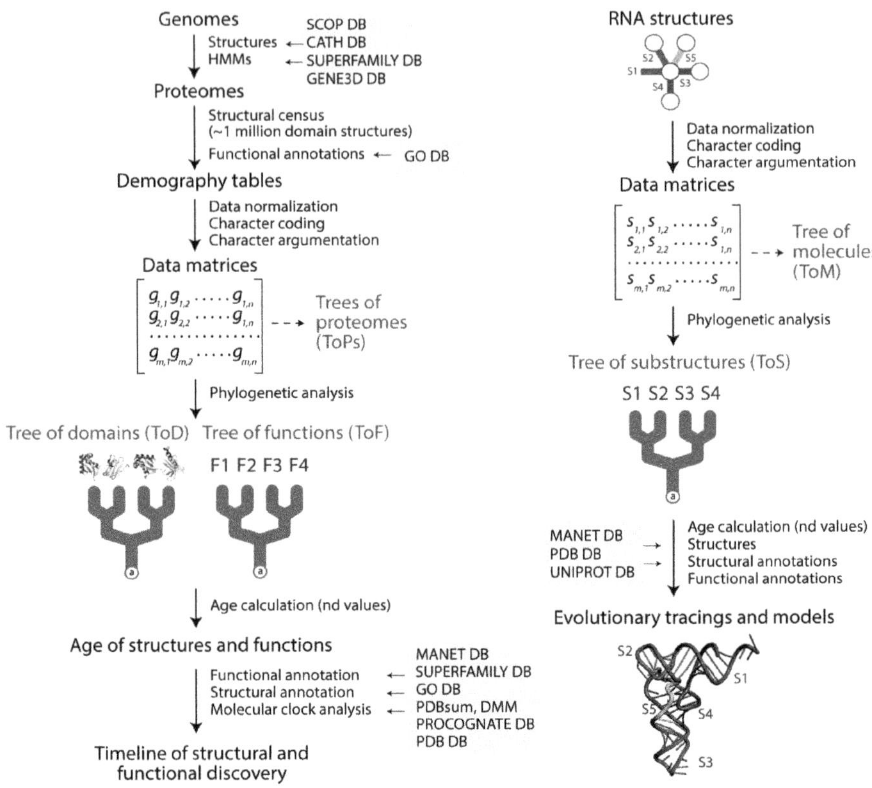

Figure 5. Phylogenomic analyses of protein and RNA structures and functions. The flow diagram *in the left* shows the reconstruction of trees of protein domain structures (ToDs) and molecular functions (ToFs). A census of domain structures and gene ontology (GO) annotations in proteomes of thousands of completely sequenced organisms and viruses is used to compose data matrices, which are then used to build phylogenomic trees describing the evolution of individual protein structures and associated molecular functions, respectively. Elements of the matrix (*g*) represent genomic abundances of domains in proteomes, defined at different level of classification of domain structure, including SCOP folds (Fs), superfamilies (FSFs), and families (FF) or CATH architectures (A), topologies (T), and homologous superfamilies (H). The flow diagram *in the right* describes the phylogenetic reconstruction of trees of RNA substructures (ToSs). The structures of rRNA molecules were first decomposed into substructures, including helical stem tracts and unpaired regions. Structural features of these substructures (e.g., length)

character originally supported. This mainly stems from the principle of character independence I described above. Furthermore, falsifiers that are proper subsets of the historical hypothesis should not intersect with potential falsifiers of other homologies used to further test the hypothesis (Wiley, 1975). In other words, falsification of homology statements may not affect the historical hypothesis that is being tested, as long as such hypothesis is supported by an increasingly supportive corpus of homology statements. This is particularly the case with historical hypotheses of the highest level of universality.

Trees of parts are axiom-like in that that they embody a finite system of study that approaches the highest level of universality attainable. For example, for any set of proteomes sampled from organisms that have been sequenced and used as characters, the taxa defining the tree of structural domains (the historical statement) must be considered finite and of the highest level of universality for that character set. Since characters in this case represent an open system, there would be always opportunity to add additional proteomes to the phylogenetic study, which could potentially contribute new useful homology statements capable of strengthening corroboration of history. The open system however may also reach boundary

Figure 5. (Figure on facing page) were coded as phylogenetic characters and assigned character states according to an evolutionary model that polarizes character transformation towards an increase in conformational order (character argumentation). Coded characters (*s*) are arranged in data matrices, which can be transposed for further cladistic analyses (e.g., to produce trees of molecules; ToMs). Phylogenetic analysis generates rooted phylogenetic trees of RNA substructures (ToS). Embedded in ToDs and ToSs are timelines that assign age to molecular structures. These ages can be "painted" onto 2D or 3D structural models of RNA or proteins, ribonucleoprotein complexes, or protein complexes, generating evolutionary heat maps. Modified from Caetano-Anollés and Seufferheld (2013).

conditions in which additional homologies do not add significantly to the testability of the historical statement under consideration. In those circumstances, the theoretical system truly approaches its status as an evolutionary axiom. For example, if indeed a set of proteomes encases most of the structural diversity of proteins that exists in the natural world, then the tree of structural domains will represent an axiomatic theory of evolution for protein structure. In this regard, Popper (1959) urges scientists to strive for the rigor of axiomatized systems by reducing the scientific enterprise to minimally sufficient and necessary logical terms. This facilitates streamlining of background knowledge when axiomatic theories of evolution are used as supporting evidence for further scientific exploration.

Similar levels of universality arguments can be used with trees of wholes, with the caveat that their supporting homology statements are more prone to falsification because of the problems of character non-independence. Moreover, the hypothesis of history of these trees can be open-ended and its universality always amenable to increases by addition of taxa. In the case of trees of proteomes, new proteomes can be derived from analysis of organisms sampled from the "dark matter" of unfolding biodiversity (Nasir *et al.*, 2015). These new samplings can extend the hypothesis of history (the tree of life) to new levels of knowledge. Alternatively, if dark matter does not hold novel proteomic make up then even the tree of wholes approaches an axiomatic-like statement with which to venture into new discoveries.

To summarize, levels of universality in ideographic hypothesis testing involves the definition of phylogenetic characters and the definition of taxa and the tree in which the character states unfold. Within these constraints: (i) supposed homologies take the form of axiomatic conventions at the level of universality at which they exist as symplesiomorphies, and (ii) the only valid test of homology that exists is to hypothesize that the supposed homology is a synapomorphy.

9. The power of "serial" homologies

The use of serial homologies has been rare in modern phylogenetic analyses, which largely study molecular sequences. In contrast, serial homologies have been common in the study of morphological characters important for paleontology or systematic biology. The unfamiliarity with this kind of evidence, which is powerful, often brings misunderstandings of the interplay of model of character state change, tree optimization, rooting with Weston's generality criterion, and global trends of character change in reconstructed phylogenies. For example, Egel (this volume) argues there cannot be any *"intrinsic coupling between the evolutionary age of a particular domain and its numeric abundance in modern genomes"* when discussing the construction of trees of domains (ToDs) from structural domain counts in proteomes (ToDs are trees of parts). He also perceives there is no *"mechanistic rationale of universal phylogenetic significance behind a tree-building algorithm for all the protein structures..., which a priori presupposes there must be a particular 'first' domain from which even structurally unrelated other domains were 'derived' later on."* He mistakenly states that ToDs are given *"apodictic validity in terms of phylogenetic descendance and relationships.;"*

Kim *et al.* (2014a) responds to concerns of this type by clarifying the phylogenetic model and the algorithmic rationale: *"The model does not force irreversibility of character state change, which violates the triangle inequality and the validity of phylogenetic reconstructions. Instead, it progressively weights change against loss with each new gain of domain abundance using the Wagner optimization algorithm for linearly ordered character states. We note that transformation between ordered states implies a distance relationship of costs (Euclidean) in which costs related to the transformation of two non-neighboring states are larger than one step. Phylogenetic characters of abundance are therefore multistate, linearly ordered (additive) and 'forward' polarized and do not necessarily produce rooted trees, unless rooting is invoked using Lundberg or by considering a hypothetical ancestor as reference. When studying the evolution of proteomes, however, the statement of 'character argumentation' (which relates to characters describing domain structures of the entire world of proteins and proteomes) induces gradual growth of the protein repertoire in compliance with the principle of spatiotemporal continuity that is backbone to Darwinian evolution, from very simple proteomes at the start of life prior to the universal cellular ancestor to the highly complex ensembles that are present in microbial life and the advanced multicellular organisms of today. The hypothetical ancestral state of characters of domain abundance and occurrence is therefore the minimum possible character state (e.g. the absence of the domain structure). We stress that the model is generic, applies to evolution of all proteins (even those that appear before the diversification of organisms), and does not force compliance with any trend for any lineage, allowing genomes to gain and lose abundance at will during exploration of the space of trees and in search of minimal change (if parsimony is used as optimality criterion)."*

Thus, mapping Wagner characters describing domain abundance onto the space of unrooted trees does not "couple" domain age and domain abundance nor gives "apodictic validity" to any phylogenetic reconstruction. It simply finds the best tree by tracing changes of multi-state characters (serial homologies) along the branches of the trees. The exercise shows numerous gains and losses of domain abundance in the branches of trees of life describing for example the evolution of proteomes (Nasir *et al.*, 2014). Figures 6 and 8 of Caetano-Anollés *et al.*, (2018) showcase rooting and character state reconstructions. Remarkably, rooting optimal trees *a posteriori* with the Lundberg method to enforce Weston's generality criterion reveals a direction of hierarchical nesting that complies with retention of serial homologs in trees and higher gain-to-loss ratios inferred using iterative character state reconstruction methods (Nasir *et al.*, 2014). Further approaches of phylogenomic analysis uncover patterns of origin and continuous accretion of fold novelties (Nasir *et al.*, 2017; Caetano-Anollés *et al.*, 2019). In other words, while domain variants accumulate in proteomes by gene duplication and associated diversification processes of subfunctionalization and neofunctionalization (which occur at rates of the same order as mutation rates per nucleotide site), novel fold structures unfold as new domain taxa (and new branches) in ToDs at times scales of million years of evolution. The process was recently modeled (Tal *et al.*, 2016): structural explorations of fold space occur through coarse-grained discoveries that undergo fine-grained elaboration.

For readers that are interested in recent clarifications of the interplay of models, tree optimization, rooting, and global phylogenetic trends see exchanges triggered by the Kim *et al.* (2014a) manuscript (Harish *et al.*, 2016; Nasir *et al.*, 2017) and recent phylogenomic analyses of models of character state reconstruction (Caetano-Anollés *et al.*, 2019).

10. Effects of organismal lifestyles, problematic taxa and character ontology in tree reconstruction

The definition of taxa is particularly problematic for phylogeny reconstruction, including the building and rooting of trees of life. Molecular characters are parts that benefit from machine learning or supervised approaches applied to the study of genomes. In contrast, the taxa of a tree of life embody species and lineages, notions that remain controversial and can alter the meaning of phylogenetic statements. In particular, the concept of the "holobiont" (Zilber-Rosenberg and Rosenberg, 2008; Rosenberg and Zilber-Rosenberg, 2013) has blurred the definition of an organism (Bordenstein and Theis, 2015). An holobiont is a community of organisms consisting of a host and all of its associated symbionts. Examples are numerous and include humans interacting with their associated microbiomes or coral invertebrate animals associating with zooxanthellae, endolithic algae and numerous bacteria. Holobionts have challenged the "biological species concept" (Mayr, 1942) while accommodating "multilevel selection theory" (Sober and Wilson, 1997). Since holobionts are ill-defined collectives that are highly dynamic, their "hologenomes" encode molecular repertoires that are themselves ill-defined and highly dynamic. This complicates definition of taxa and genomic repertoires used for their evolutionary analysis (Bordenstein and Theis, 2015). Most importantly, it devalues the species as a unit of biological organization. By the same token, holobionts and hologenomes prompt revisiting the idea that selection operates at multiple levels in the nested hierarchy of life. Therefore, selection not only applies to the effects of the entire organism but also to the effects of its individual components (genes, domains, network motifs, etc) and to the effects of highler level organization (communities, populations, ecosystems, etc). In particular, the holobiome with its multiple organismal make up acts as cohesive system for definition of novel kinds of taxa and prompts exploration and re-definition of cohesiveness as argument for definition of systems used as taxa. Similarly, the holobiome prompts examination of what is "organism" and what is "environment" when defining systems.

Besides symbiosis, obligate parasitism also poses problems, especially when the genomic make up of parasites is highly reduced (Andersson and Andersson, 1999; Keeling and Slamovitz, 2005). For example, *Cand.* Tremblaya princeps is part of a three-pronged endosymbiotic organismal system (bug in a bug in a bug) (McCutcheon and von Dohlen, 2011). Its genome is minimalistic. It encodes only 55 universal domain superfamilies. It is not considered an independent organism since it depends on its host (*Planococcus citri*) and its endosymbiont (*Cand.* Moranella endobia) to synthesize essential metabolites (López-Madrigal *et al.*, 2011). *Cand.* Tremblaya princeps in association with its γ-proteobacterial symbiont should therefore be considered an organismal consortium or a new living entity. Similarly, *Cand.* Nausia deltocephalinicola is an obligate endosymbiont of leafhoppers, which harbors the smallest known bacterial genome (Bennett and Moran 2013). It encodes only 53 universal domain superfamilies. These hologenomes with extreme proteomic outliers arise from relatively modern genomic losses, exchanges and recruitments likely resulting from complex trade-off relationships that complicate the dissection of their evolutionary origin and their definition as single taxon in a phylogenetic data matrix. The complexity is particularly highlighted by obligate relationships that do not cause genomic reduction, such as that of *Helicosporidium* green algae and animals (Pombert *et al.*, 2014), suggesting not all obligate organismal interactions are driven by strong reductive evolutionary drivers. Phylogenetically speaking and when building trees of species, ill-defined holobionts represent "problematic" taxa that should be excluded from analysis pending further understanding of their genetic make up. Their inclusion generates biased reconstructions (Nasir *et al.*, 2017). Similarly, taxa should be sampled randomly, equally, and densely from each major group of organisms being analyzed for reliable tree reconstruction (Zwickl and Hillis, 2002; Heath *et al.*, 2008). When building trees of life from structural phylogenomic data, for example, the exclusion of organisms that do not engage in free-living relationships avoids many pitfalls (Kim and Caetano-Anollés, 2012). Their inclusion, however, can be justified if the obligate parasitic or symbiotic mode is a hallmark of entire groups of taxa that are being sampled. Such is the case of viruses, which, as a group, harbor life cycles with strict dependence of the host (Nasir *et al.*, 2015, 2017).

The definition of taxa in a tree of life is even more problematic when individual molecular parts of organisms (e.g. 16S rRNA or protein-encoding genes) are used as characters to study their evolution. For example, a tree of life reconstructed from a concatenated sequence alignment of genes is built by defining sequences as taxa and sequence sites as characters. Taxa are generally considered representative of organisms and the tree of life is said to represent the evolution of life. However, the "sequence-equals-organism" assumption is a "leap of faith." since characters describe properties related to the sequence of the molecules and not properties of organisms. The issue makes itself evident when focusing on the models of change used for tree reconstruction. For example, the "advanced" model of sequence evolution (PROTGAMMALG) used by Spang *et al.* (2015) to build trees of life from a concatenated sequence alignment of universal genes simply describes the process of replacement of amino acids in a sequence site. It does not describe processes of molecular diversification of the actual structure that embeds the sequence, which likely involves a multiplicity of structural domain recruitments. It does not describe processes responsible for long-range atomistic patterns of evolutionary constraint in structural domains induced by their functional sites (Jack *et al.*, 2016). More importantly, it does not describe the actual process of organismal diversification that the "sequence-equals-organism" paradigm implies, which involves selection acting at multiple levels of the nested hierarchy of life. Instead, a tree of life generated from these concatenated sequence alignments better describes the diversification of biases and differential dynamics of amino acid use in organisms by the structural domains of a handful of sampled proteins. One argument often used to validate the "sequence-equals-organism" ad hoc is the universality of the molecules being studied. While all proteins in the alignment of Spang *et al.* (2015) are considered universal, (i) not all of their structural domain components are shared by all sampled organisms, (ii) not all universal domains hold the same lower-level structural motifs in the molecules sampled, and (iii) not all aligned sequence sites contribute equally to that universality (Shen *et al.*, 2017; Da Cunha *et al.*, 2017; Nasir *et al.*, unpublished). These considerations question the definition of what is "universal" in biology and phylogenetic reconstruction.

Hennig's definition of taxa was associated with the "semaphoront" concept, the concept of being a "character bearer" (Hennig, 1966). A semaphoront is an organism at a point of time in its development and evolution harboring a complete set of characters describing it, i.e. its "holomorph." The concept is relevant for definition of taxa, since semaphoronts are dynamic entities composed of holomorphs that are themselves dynamic. The ontological complexity of the semaphoront-holomorph relationship can be dissected with computer-aided ontologies for semantic organization, but the challenge of doing so is significant for both phenomics and evolutionary biology (Deans *et al.*, 2015; Göpel and Richter, 2016). This brings us back to the problem of parts and wholes of systems.

11. Levels of selection, evolutionary units and the primacy of cellular processes over replicators

The past century has seen an almost exclusive focus on both the organism as the biological system and the organism as the unit of natural selection. These mechanistic reductionist views that stress effects of environment on molecular biodiversity materialized in model systems for genomic exploration, organism-centered systematic biology, origin of life and tree of life research, and a field of molecular evolution that stressed diversification of organisms more than macromolecules. Similarly, the major focus has been the molecular biology of the gene as the "unit" of evolution and the gene as the molecular genomic part contributing functions that are the subject of natural selection. The reductionist view of the selfish gene (e.g. Dawkins, 1982) as the "inheritable unit" or "replicator" guided the genomic revolution and promoted the neo-Darwinian gene-centered theory of evolution. Both theoretical frameworks have been myopic for several reasons:

(i) They prompted the existence of a "code" responsible for a "genetic program" or "blueprint" of life that dissociated the genome and its genetic information from the cellular vehicle (Noble, 2015). The cellular vehicle appeared disposable when in fact it was central for life and functioned even in the absence of a genome. The frameworks negated the integrative aspect of molecular, cellular and organismal function. They considered the genome was isolated from the lifestyle of the organism and its environment. Genomes were viewed as passive repository (templates) of genetic information while its component parts, the genes, embodied active and privileged causes.

(ii) They neglected epigenetic mechanisms, trans-generational inheritance, viral and horizontal transfers and innovations, and other cellular processes that actively mediated genetic change (Shapiro, 2013). Genetic change is not a passive process prompted by mutation and damage of the DNA repository as the frameworks assumed. Genomes are permanently inscripted by cellular activities, including nucleoprotein complex formation, epigenetic chromatin modification and DNA restructuring and other active manipulations of genomic information that are short term or epigenetic. These activities unfold over distinct time-scales encompassing reproduction, development and long-term evolutionary change.

(iii) They neglected questions about the origin of molecular and cellular biodiversity that supported organismal biodiversity. For example, a focus on how the environment weeds out variation at molecular level does not address how that variation was generated in the first place (Brooks, 2000). Such questions are ideographic in nature and must be addressed through deep and complete retrodiction methods.

(iv) They neglected the hierarchy of biological organization and the integrative aspects of parts and wholes that I have illustrated in Figure 3A. Both organisms and genes are complicated systems that are themselves composed of parts that behave as evolutionary units. These are subjected to natural selection at different levels of biological organization. While the organism acts as cellular unit (or vehicle) of evolution, change also manifests by reticulation processes that are not vertical.

The views of the neo-Darwinian frameworks explain why systematic biology has focused on the organism and why we seek trees of life but neglect global evolutionary views of molecular repertoires, molecular interactions, and biological functions. While there is a need to switch gears and focus on evolution of biological parts as a way to fully access the evolution of biological wholes, there is still the problem of how parts, modules, and systems come into being. In contrast to mechanistic reductionism, holistic views regard the whole as an integrated collective that goes beyond the collection of parts and is responsible for levels of hierarchical organization. However, holistic views are not synthetic in that they cannot explain how levels of organization have emerged. In turn, synthetic biology views are synthetic in that they experimentally explore the interrelationship between the organizational levels (Danchin, 2012). This exploration however misses the evolutionary dynamics required to explain the origin and evolution of molecular and cellular organization. A much more integrated framework is therefore needed. An increasing number of researchers are recognizing the difficulties of the reductionist neo-Darwinian framework and seek novel routes to explain evolution and complexity in life (The Third Way, 2019).

12. Conclusions

Computations in evolutionary biology have focused almost exclusively on the whole by using parts as phylogenetic characters and constructing trees with taxa representing organisms or species. The rise of molecular systematics used the sequences of nucleic acid or protein molecules as parts to build histories of wholes, the molecules themselves. In the process, the molecules were considered abstractions of a higher whole, the organismal system. Thus, the small subunit rRNA molecule of an organism was assumed to carry the history of that organism by invoking its evolutionary centrality. But are there evolutionarily central parts in a fully integrated system? What about the large subunit rRNA of the ribosome? What about the ribosomal proteins of the ribosomal core? Are their histories congruent or divergent? Which is central and which is core? Currently, trees of species derived from the sequence of single molecules or handful of molecules give rise to trees of life in a quest to understand organismal biodiversity. These approaches have overshadowed more integrative attempts that use genomes to reconstruct the history of life. This probably stems from the limitations of computation and the conceptual problems of homology and homoplasy in complex systems.

Using wholes as taxa has the implication of using parts as characters, which by definition cannot be independent pieces of evidence. This is because parts interact with each other to make the whole. In fact, systems biology recognizes this fact when using network theory to understand organismal complexity. Networks describe complex tangles of interactions between parts (e.g. protein–protein interactions, gene co-expression, metabolite pathways). Thus, problems of character non-independence plague canonical phylogenetic methods that attempt to build trees of species and trees of life. In contrast, a focus on parts as evolutionary units of wholes diminishes or avoids violations of character independence in phylogenetic analysis and simplifies the task of retrodiction. In particular, the use of higher-level biological modules as characters has the advantage of facilitating the arduous task of defining modules in biology. Their modularity can now be tested by retrodiction.

BOX 1: Definitions

Phylogenies: Historical account describing the evolution of the biological entities. Phylogenies are multidimensional statements of relationship of the entities that are being studied (phylogenetic taxa) and the biological attributes (phylogenetic characters) used to construct them.

Evolutionary models: Evolutionary transformations of the biological attributes examined in data (phyogenetic characters), which define the relationships of taxa in trees.

Character transformation: A series of character states that transform into each other in evolution.

Character polarization: Assignment of polarity to a character transformation. Polarity implies specifying direction of character change and which states are evolutionarily ancestral and which are derived. Character polarization makes explicit the rooting of phylogenetic trees or networks and the identification of synapomorphies. Since this process is usually frustrated by the interaction of homology and homoplasy, it is the global patterns arising from the interplay of characters and trees that establishes direction of change.

Matrix transposition: Mathematical operation in which a matrix is converted into a new matrix whose rows are the columns of the original.

References

Andersson JO, Andersson SG (1999) Insights into the evolutionary process of genome degradation. *Curr Opin Genet Dev* **9**: 664–671.

Baker A (2010) Simplicity. In: Zalta EN (ed) *The Stanford Encyclopedia of Philosophy*, Fall 2013 edition. URL: http://plato.stanford.edu/entries/simplicity/

Bennett GM, Moran NA (2013) Small, smaller, smallest: the origins and evolution of ancient dual symbioses in a phloem-feeding insect. *Genome Biol Evol* **5**: 1675–1688.

Bordenstein SR, Theis KR (2015) Host biology in light of the microbiome: Ten principles of holobionts and hologenomes. *PLoS Biol* **13**: e1002226.

Brading K, Castellani E (2003) *Symmetries in Physics: Philosophical Reflection.* Cambridge University Press, Cambridge, UK.

Brading K, Castellani E (2013) Symmetry and symmetry breaking. In: Zalta EN (ed) *The Stanford Encyclopedia of Philosophy*, Fall 2013 edition. URL: http://plato.stanford.edu/archives/spr2013/entries/symmetry-breaking/

Brooks DR (2006) The nature of the organism: Life has a life of its own. *Ann NY Acad Sci* **901**: 257–265.

Brooks DR, McLennan DA (1991) Phylogeny, *Ecology and Behavior: A Research Program in Comparative Biology*. University of Chicago Press, Chicago.

Brooks DR, Bilewiych J, Condy C, Evans DC, Folinsbee KE, Fröbisch J, Halas D, Hill S, McLennan DA, Mattern M, Tsuji LA, Ward JL, Whalberg N, Zamparo D, Zanatta D (2007) Quantitative phylogenetic analysis in the 21st Century. *Revista Mexicana Biodiversidad* **78**: 225–252.

Brower AVZ (2000) Evolution is not a necessary assumption of cladistics. *Cladistics* **16**: 143–154.

Brower AVZ, de Pinna MCC (2012) Homology and errors. *Cladistics* **28**: 529–538.

Bryant HN (2001) Character polarity and the rooting of cladograms. In: Wagner GP (ed) *The Character Concept in Evolutionary Biology*. Academic Press, New York, USA, pp. 319–338.

Caetano-Anollés G (2002a) Evolved RNA secondary structure and the rooting of the universal tree of life. *J Mol Evol* **54**: 333–345.

Caetano-Anollés G (2002b) Tracing the evolution of RNA structure in ribosomes. *Nucleic Acids Res* **30**: 2575–2587.

Caetano-Anollés G (2015) Creationism and intelligent design are incompatible with scientific progress: A response to Shanta and Vedanta. *Commun Integr Biol* **9**: e1123356.

Caetano-Anollés G, Caetano-Anollés D (2003) An evolutionarily structured universe of protein architecture. *Genome Res* **13**: 1563–1571.

Caetano-Anollés G, Seufferheld MJ (2013) The coevolutionary roots of biochemistry and cellular organization challenge the RNA world paradigm. *J Mol Microbiol Biotechnol* **23**: 152–177.

Caetano-Anollés G, Kim HS, Mittenthal JE (2007) The origin of modern metabolic networks inferred from phylogenomic analysis of protein architecture. *Proc Natl Acad Sci USA* **104**: 9358–9363.

Caetano-Anollés G, Wang M, Caetano-Anollés D, Mittenthal JE (2009) The origin, evolution and structure of the protein world. *Biochem J* **417**: 621–637.

Caetano-Anolles G, Yafremava L, Mittenthal JE (2010) Modularity and dissipation in evolution of macromolecular structures, functions, and networks. In: Caetano-Anollés G (ed) *Evolutionary Bioinformatics and Systems Biology*. Wiley-Blackwell, Hoboken, NJ, USA, pp. 443–449.

Caetano-Anollés G, Nasir A, Zhou K, Caetano-Anollés D, Mittenthal JE, Sun F-J, Kim KM (2014) Archaea: the first domain of diversified life. *Archaea* **2014**: 590214.

Caetano-Anollés G, Nasir A, Kim KM, Caetano-Anollés D (2018) Rooting phylogenies and the Tree of Life while minimizing *ad hoc* and auxiliary assumptions. *Evol Bioinfo* **14**: 1176934318805101.

Caetano-Anollés D, Caetano-Anollés G (2015) Computing the origin and evolution of the ribosome from its structure — Uncovering processes of macromolecular accretion benefiting synthetic biology. *Comput Struct Biotech J* **13**: 427–447.

Caetano-Anollés D, Nasir A, Kim KM, Caetano-Anollés G (2019) Testing empirical support for evolutionary models that root the Tree of Life. *J Mol Evol* **87**: 131–142.

Chippendale PT, Wiens JJ (1994) Weighting, partitioning, and combining characters in phylogenetic analysis. *Syst Biol* **43**: 278–287.

Collier J (2003) Hierarchical dynamical information systems with a focus on biology. *Entropy* **5**: 100–124.

Crupi V (2015) Confirmation. In: Zalta EN (ed) *The Stanford Encyclopedia of Philosophy*, all 2014 edition. URL: http://plato.stanford.edu/entries/confirmation/

Curie P (1894) Sur le symétrie dans les phénomènes physiques, symétrie d'un champ électrique et d'un champ magnétique. *J Phys Theor Appl* **3**: 393–415.

Da Cunhha V, Gaia M, Gadelle D, Nasir A, Forterre P (2017) Lokiarchaea are close relatives of Euryachaeota, not bridging the gap between prokaryotes and eukaryotes. *PLoS Genetics* **13**: e1006810.

de Pinna MCC (1991) Concepts and tests of homology in the cladistic paradigm. *Cladistics* **7**: 361–394.

de Pinna MCC (1994) Ontogeny, rooting, and polarity. In: Scotland RW, Siebert DJ, Williams DM (eds) *Models in Phylogeny Reconstruction, Systematics Association Special Volume No. 52*. Clarendon Press, Oxford, pp 157–172.

Danchin A (2012) Scaling up synthetic biology: Do not forget the chassis. *FEBS Lett* **586**: 2129–2137.

Dawkins R (1976) *The Selfish Gene*. Oxford University Press, Oxford.

Deans AR, Lewis SE, Huala E, Anzaldo SS, Ashburner M, *et al.* (2015) Finding our way through phenotypes. *PLoS Biol* **13**: e1002033.

Doolittle RF (2005) Evolutionary aspects of whole-genome biology. *Curr Opin Struct Biol* **15**: 248–253.

Farris JS (1970) Methods for computing Wagner trees. *Syst Zool* **19**: 83–92.

Farris JS (1983) The logical basis of phylogenetic analysis. In: Platnick NI, Funk VA (eds) *Advances in Cladistics II*, Columbia Univ. Press, New York, pp 7–36.

Farris JS (2008) Parsimony and explanatory power. *Cladistics* **24**: 1–23.

Farris JS (2013) Popper: not Bayes or Rieppel. *Cladistics* **29**: 230–232.

Farris JS (2014a) Popper with probability. *Cladistics* **30**: 5–7.

Farris JS (2014b) Homology and misdirection. *Cladistics* **30**: 555–561.

Farris JS, Kluge AG, Eckhardt MJ (1970) A numerical approach to phylogenetic systematics. *Syst Zool* **19**: 172–191.

Felsenstein J (1973) Maximum-likelihood and minimal-steps methods for estimating evolutionary trees from data on discrete characters. *Syst Zool* **22**: 240–249.

Fitelson B (2007) Likelihoodsim, Bayesianism, and relational confirmation. *Synthese* **156**: 473–489.

Fraix-Burnet D, Chattopadhyay T, Chattopadhyay AK, Davoust E, Thuillard M (2012) A six-parameter space to describe galaxy diversification. *Astron Astrophys* **545**: A80.

Göpel T, Richter S (2016) The word is not enough: on morphemes, characters and ontological concepts. *Cladistics* **32**: 682–690.

Grant T (2002) Testing methods: the evaluation of discovery operations in evolutionary biology. *Cladistics* **18**: 94–111.

Grant T, Kluge AG (2003) Data exploration in phylogenetic inference: Scientific, heuristic, or neither. *Cladistics* **19**: 379–418.

Hacking I (1965) *Logic of Statistical Inference*. Cambridge University Press.

Harish A, Caetano-Anollés G (2012) Ribosomal history reveals origins of modern protein synthesis. *PLoS One* **7(3)**: e32776.

Harish A, Abroi A, Gough J, Kurland C (2016) Did viruses evolve as a distinct supergroup from common ancestors of cells? *Genome Biol Evol* **8**: 2474–2481.

Heath TA, Hedtke SM, Hillis DM (2008) Taxon sampling and the accuracy of phylogenetic analyses. *J Syst Evol* **46**: 239–257.

Hempel CG (1965) *Aspects of Scientific Explanation*. The Free Press, New York. USA

Hennig W (1966) *Phylogenetic Systematics*. University of Illinois Press, Urbana, USA.

Jack BR, Meyer AG, Echave J, Wilke CO (2016) Functional sites induce long-range evolutionary constraints in enzymes. *PLoS Biol* **14(5)**: e1002452.

Keeling PJ, Slamovitz CH (2005) Causes and effects of nuclear genome reduction. *Curr Op Genet Dev* **15**: 601–608.

Kim HS, Mittenthal J, Caetano-Anollés G (2006) MANET: tracing evolution of protein architecture in metabolic networks. *BMC Bioinf* **7**: 351.

Kim KM, Caetano-Anollés G (2010) Emergence and evolution of modern molecular functions inferred from phylogenomic analysis of ontological data. *Mol Biol Evol* **27**: 1710–1733.

Kim KM, Caetano-Anollés G (2012) The evolutionary history of protein fold families and proteomes confirms that the archaeal ancestor is more ancient than the ancestors of other superkingdoms. *BMC Evol Biol* **12**: 13.

Kim KM, Nasir A, Caetano-Anollés G (2014a) The importance of using realistic evolutionary models for retrodicting proteomes. *Biochimie* **99**: 129–137.

Kim KM, Nasir A, Caetano-Anollés G (2014b) A tree of cellular life inferred from a genomic census of molecular functions. *J Mol Evol* **79**: 240–262.

Kluge AG (1984) The relevance of parsimony to phylogenetic inference. In: Duncan T, Stuessy TF (eds) *Cladistics: Perspectives on the Reconstruction of Evolutionary History*. Columbia Univ. Press, New York, USA, pp 24–38.

Kluge AG (1885) Ontogeny and phylogenetic systematics. *Cladistics* **1**:13–27.

Kluge AG (2001) Parsimony with and without scientific justification. *Cladistics* **17**: 199–210.

Kluge AG, Farris JS (1969) Quantitative phyletics and the evolution of anurans. *Syst Zool* **18**: 1–32.

Laplace PS (1814) *A Philosophical Essay on Probabilities*. Trans. from the French, 6th ed. Truscott F, Emory F (eds)1951. Dover, New York, USA.

Leibniz GW (1678) Letter to Bayle: Extrait d'une Lettre de M. L. sur un Principe général, utile à l'explication des loix de la nature, par la consideration de la sagesse Divine; pour servir de réplique à la réponse du r. P. M. Nouvelles de la République des Lettres. France.

Leibniz GW (1702) Letter to Varignon. In: New Essays Concerning Human Understanding (Langley, 2nd ed., 1896). Macmillan, New York, USA.

Lee L, Leopold JL, Kandoth C, Frank RL (2010) Protein secondary structure prediction using RT-RICO: A rule-based approach. *Open Bioinfo J* **4**: 17–30.

Lienau EK, DeSalle R (2009) Evidence, content and corroboration and the tree of life. *Acta Biotheor* **57**: 187–199.

López-Madrigal S, Lattore A, Porcar M, Moya A, Gil R (2011) Complete genome sequence of "Candidatus Tremblaya princeps" strain PCVAL, an intriguing translational machine below the living-cell status. *J Bacteriol* **193**: 5587–5588.

Lundberg JG (1972) Wagner networks and ancestors. *Syst Zool.* **21**: 398–413.

Maglia IAM, Leopold JL, Ghatti VR (2004) Identifying character non-independence in phylogenetic data using data mining techniques. In: *Proc Second Asia-Pacific Bioinformatics Conference.* Dunedin, New Zealand.

Mayr E (1942) *Systematics and the Origin of Species*. Columbia University Press, New York.

McCutcheon JP, von Dohlen CD (2011) An interdependent metabolic patchwork in the nested symbiosis of mealybugs. *Curr Biol* **21**: 1366–1372.

Mittenthal JE, Caetano-Anollés D, Caetano-Anollés G (2012) Biphasic patterns of diversification and the emergence of modules. *Front Genet* **3**: 147.

Murzin A, Brenner SE, Hubbard T, Clothia C (1995) SCOP: a structural classification of proteins for the investigation of sequences and structures. *J Mol Biol* **247**: 536–540.

Nasir A, Kim KM, Caetano-Anollés G (2014) Global patterns of domain gain and loss in superkingdoms. *PLoS Comput Biol* **10**: e1003452.

Nasir A, Kim KM, Caetano-Anollés G (2015) Lokiarchaeota: eukaryote-like missing links from microbial dark matter? *Trends Microbiol* **23**: 448–450.

Nasir A, Kim KM, Caetano-Anollés G (2017) Phylogenetic tracings of proteome size support the gradual accretion of protein structural domains and the early origin of viruses from primordial cells. *Front Microbiol* **8**: 1178.

Nasrallah CA, Mathews DH, Huelsenbeck JP (2011) Quantifying the impact of dependent evolution among sites in phylogenetic inference. *Syst Biol* **60**: 60–73.

Niiniluoto I (2014) Scientific progress as increasing verisimilitude. *Studies History Phil Sci* **46**: 73–77.

Niiniluoto I (2015) Scientific progress. In: Zalta EN (ed) *The Stanford Encyclopedia of Philosophy*, Fall 2014 edition. URL: http://plato.stanford.edu/entries/scientific-progress/

Noble D (2015) Evolution beyond neo-Darwinism: a conceptual framework. *J Exp Biol* **218**: 7–13.

O'Keefe FR, Wagner PJ (2001): Inferring and testing hypotheses of cladistic character dependence by using character compatibility. *Syst Biol* **50**: 657–675.

Orengo CA, Michie A, Jones S, Jones DT, Swindells M, Thornton JM (1997) CATH — a hierarchic classification of protein domain structures. *Structure* **5**: 1093–1109.

Penny D, Collins LJ (2010) Evolutionary genomics leads the way. In: Caetano-Anollés G (ed) *Evolutionary Genomics and Systems Biology*. Wiley-Blackwell, Hoboken, NJ, USA, pp 1–16.

Pombert J-F, Blouin NA, Lane C, Boucias D, Keeling PJ (2014) A lack of parasitic reduction in the obligate parasitic green alga *Helicosporum*. *PLoS Genet* **10**: e1004355.

Popper K (1959) *The Logic of Scientific Discovery*. Hutchinson, London.

Popper K (1963) *Conjectures and Refutations: The Growth of Scientific Knowledge*. Hutchinson, London.

Rosenberg E, Zilber-Rosenberg I (2013) The hologenome concept: human, animal and plant microbiota. Springer International Publishing, Switzerland.

Rowbottom D (2013) Popper's measure of corroboration and $P(h|b)$. *Brit J Phil Sci* **64**: 739–745.

Rowbottom D (2015) Scientific progress without increasing verisimilitude: In response to Niiniluoto. *Studies History Phil Sci* **51**: 100–104.

Shen X-X, Hittinger CT, Rokas A (2017) Contentious relationships in phylogenomic studies can be driven by a handful of genes. *Nature Ecol Evol* **1**: 0126.

Teichmann SA, Rison SCG, Thornton JM, Riley M, Gough J, Chothia C (2001) Small-molecule metabolism: an enzyme mosaic. *Trends Biotechnol* **19**: 482–486.

Sathyapriya R, Duarte JM, Stehr H, Filippis I, Lappe, M (2009) Defining an essence of structure determining residue contacts in proteins. *PLoS Comput Biol* **5**(12): e1000584.

Shapiro JA (2013) How life changes itself: The Read-Write (RD) genome. *Phys Life Rev* **10**: 287–323.

Shen X-X, Hittinger CT, Rokas A (2017) Contentious relationships in phylogenomic studies can be driven by a handful of genes. *Nature Ecol Evol* **1**: 0126.

Sherbakov Y, Triboy TI (2007) Effct of co-evolving amino acid residues on topology of phylogenetic trees. *Biochemistry* (*Moscow*) **72**: 1363–1367.

Sober E (2004) The contest between parsimony and likelihood. *Syst Biol* **53**: 644–653.

Sober E (2008) *Evidence and Evolution. The Logic Behind the Science.* Cambridge University Press, Cambridge.

Sober E, Steel M (2014) Time and knowability in evolutionary processes. *Phil Sci* **81**: 537–557.

Sober E, Steel M (2015) Similarities as evidence for common ancestry: A likelihood epistemology. *British J Phil Sci* **68**: 617–638.

Sober E, Wilson DS (1997) *Unto others: the evolution of altruism.* Harvard Univ Press, Cambridge, Mass.

Spang A, Saw JH, Jørgensen SL, Zaremba-Niedzwiedzka K, Martijn J, Lind AE, van Eijk R, Schleper C, Guy L, Ettema TJ (2015) Complex archaea that bridge the gap between prokaryotes and eukaryotes. *Nature* **521**: 173–179.

Sprenger J (2018) Two impossibility results for measures of corroboration. *Brit J Phil Sci* **69**: 139–159.

Tal G, Boca SM, Mittenthal J, Caetano-Anollés G (2016) A dynamic model for the evolution of protein structure. *J Mol Evol* **82**: 230–243.

The Third Way (2019) Evolution in the era of genomics and epigenomics. URL: https://www.thethirdwayofevolution.com

Van Enk SJ (2015) Betting, risk, and the Law of Likelihood. *Ergo* **2**: 105–121.

Vickers J (2014) The problem of induction. In: Zalta EN (ed) The Stanford Encyclopedia of Philosophy, Fall 2014 edition. URL: http://plato.stanford.edu/entries/induction-problem/

Vogt L (2008) Popper and systematics. *Syst Biol* **52**: 259–271.

Wang M, Caetano-Anollés G (2009) The evolutionary mechanics of domain organization in proteomes and the rise of modularity in the protein world. *Structure* **17**: 66–78.

Weston H (1988) Indirect and direct methods in systematics. In: Humphries CJ (ed) *Ontogeny and Systematics*. Columbia University Press, New York, USA, pp 27–56.

Wiley EO (1975) Karl R. Popper, systematics, and classification: a reply to Walter Bock and other evolutionary taxonomists. *Syst Zool* **24**: 233–243.

Williams TA, Foster PG, Cox CJ, Embley TM (2013) An archaeal origin of eukaryotes supports only two primary domains of life. *Nature* **504:** 231–236.

Windelband W (1894) *Geschichte und Naturwissenschaft*. Heitz, Strassburg.

Zilber-Rosenberg I, Rosenberg E (2008) Role of microorganisms in the evolution of animals and plants: the hologenome theory of evolution. *FEMS Microbiol Rev* **32**: 723–735.

Zwickl DJ, Hillis DM (2002) Increased taxon sampling greatly reduces phylogenetic error. *Syst Biol* **51**: 588–598.

A "double tale" of module creation in evolving networks

G. Caetano-Anollés[1]
F. Mughal[1]
M. F. Aziz[1]
I. Koç[2]
K. Caetano-Anollés[3]
D. Caetano-Anollés[4]
J. E. Mittenthal[5]

Summary. In biology, structure is driven by accretion and change. Accretion brings together disparate parts to form bigger wholes, while change provides opportunities for growth and innovation. Here we review patterns and processes that are responsible for a "double tale" of evolutionary accretion and change at various levels of complexity. We focus on proteins and nucleic acids. In our "linkage" model, parts are at first weakly linked and associate variously. As they diversify, they compete with each other and are selected for performance. The emerging interactions constrain their structure and associations. This causes parts to self-organize into modules with tight linkage. In a second phase, variants of the modules evolve and become new parts for a new generative cycle of higher-level organization.

[1]*Department of Crop Sciences, University of Illinois, Urbana, Illinois, USA*
[2]*Department of Molecular Biology and Genetics, Gebze Technical University, Gebze, Turkey*
[3]*Division of Biomedical Informatics, College of Medicine, Seoul National University, Republic of Korea*
[4]*Broad Institute of MIT and Harvard, Cambridge, Massachusetts, USA*
[5]*Department of Cell and Developmental Biology, University of Illinois, Urbana, Illinois, USA*

As a way forward, we formalized the concept of linkage with a network in which nodes represent parts and links describe node interactions, predicting that the "double tale" induces both hierarchy and modularity in evolving networks. Remarkably, evolutionary genomics and network biology support the linkage hypothesis and validate an evolutionary "principle of granularity" that drives the gain and loss of modules.

1. Introduction

> *"A double tale I'll tell. At one time one thing grew to be just one from many, at another many grew from one to be apart. Double the birth of mortal things, and double their demise."*
>
> — Empedocles, On Nature, P. Strasb. Gr. Inv. 1665–6, lines 233–235 (Janko, 2004)

Biological systems show patterns of organization that are far from homogeneous. Their make up is complex and structured. For example, metabolism is a set of enzyme-catalyzed reactions that transfer and modify metabolites to carry mass and energy through pathways of a network structure (Caetano-Anollés *et al.*, 2009). These pathways sometimes form cycles that link to other reactions and pathways. The resulting flow of matter and energy occurs along the many routes (paths) of the metabolic system creating dynamic behaviors that are highly heterogeneous (Papin *et al.*, 2003). Similarly, databases such as KEGG divide metabolism into distinct subnetworks according to metabolic functionalities (Kanehisa *et al.*, 2017). For example, the 17 subnetworks of carbohydrate metabolism and the

7 subnetworks of energy metabolism gather around network cycles, the citrate (TCA) cycle subnetwork (map0020) and the reductive carboxylate cycle subnetwork (map00720), respectively. Their make up and dynamics are highly inhomogeneous. These kinds of metabolic subnetworks behave as functional modules when patterns of inter-module and intra-module connectivity are used to build modular or cartographic representations of metabolism (Ravasz *et al.*, 2002; Guimerà and Amaral, 2005). Remarkably, metabolites that participate in only a few reactions but that connect to different metabolic subnetworks are more evolutionarily conserved than hubs with links connecting to a single module (e.g. pyruvate, acetyl-CoA). This finding suggests evolution actively tailors the structure of the metabolic network, which is highly hierarchical.

At a global level, metabolism appears to be structured as an hourglass (bow-tie) in such a way that many inputs are processed into many outputs by constraining metabolic flow through a network "waist" (Ma and Zheng, 2003; Tanaka *et al.*, 2005; Zhao *et al.*, 2006). This structure of complex systems is quite general. In fact, the same bow-tie structure exists in socio-technical systems such as the world wide web (WWW) (Broder *et al.*, 2000). When the hypertext transfer protocol (HTTP) is used to exchange information between computers, the information flow grows and diversifies generating a "bow-tie" structure with external "tubes" and "tendrils" that acts as information shunts. Here we explore the underlying processes that define the complex structure of biological systems and are responsible for evolutionary innovations.

2. Understanding biological systems with network abstractions

Biological systems are made of parts — components that combine with other components to make up a whole (Simon, 1962). When parts interact with other parts of the system their interactions are constrained by space, time or function, all of which are influenced by the external environment. Interactions are usually modeled with *graphs*, mathematical constructs that connect points known as vertices with lines (Barabasi and Oltvai, 2004). Vertices describe parts of a system and lines describe pairwise interactions between them. For example, a graph describing the combination of structural domains in multidomain proteins will connect vertices describing structural domains with lines describing the presence of domains in proteins. When connections of vertices are undirected, lines fail to point in any direction; each connection involves an unordered pair of (end) vertices. These lines are called *edges*. When connections are directed, lines point in one direction; each connection involves an ordered pair of vertices (an initial vertex and a terminal vertex). These lines are called *arcs*. Graphs become *networks* whenever value functions (properties or weights) are mapped onto the vertices and lines of the graphs. We will call the vertices of the networks *nodes* and the lines that connect the vertices the *links* of the networks. Figure 1 summarizes the anatomy of a network and its mathematics.

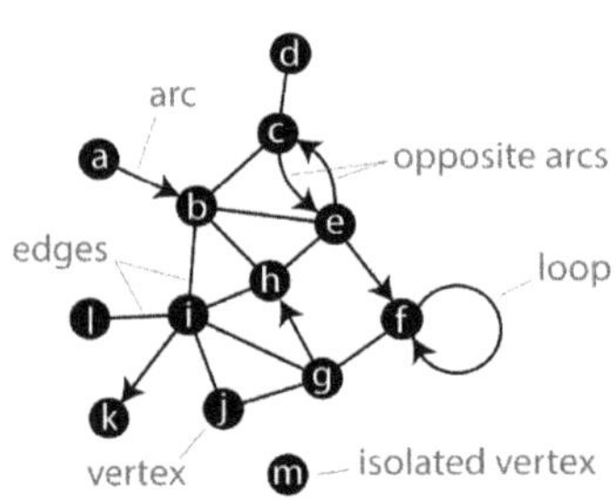

Network: A network $N = (V,L,P,W)$ consists of:
(1) A ***graph*** $G = (V,L)$, where V is a set of vertices (*nodes*), and L is a set of lines (*links*) in which each element of L is a pair $\{u,v\}$ of elements of V, with $u \neq v$ and $L \subseteq \binom{V}{2}$
$L = E \cup A$, where E is the set of *edges* and A is the set of *arcs*
$n = \text{card}(V)$, $m = \text{card}(L)$
(2) A set P of ***vertex value functions*** – properties: $p: V \rightarrow A$
(3) A set W of ***line value functions*** – weights: $w: L \rightarrow B$

Figure 1. The anatomy of a network. A network N is a combination of four sets, a set V of vertices (nodes), a set L of lines (links), and sets of vertex and line *value functions* that are mapped onto the V and L sets, respectively. The cardinality of set V, *card*(V), is simply the number of elements n in V. The cardinality of set L, *card*(L), is simply the number of elements m in L. Lines are 2-element subsets of V (i.e. each line is associated with a pair of vertices) representing *edges* or *arcs* if lines are undirected or directed, respectively. Loops are lines with identical endpoints. The network is a "mixed network" because it contains both arcs and edges.

Some network properties help visualize and study network structure and make up (Wasserman and Faust, 1994; Newman, 2003). First, networks can be represented by a matrix called the *adjacency matrix*, A, an $n \times n$ symmetric matrix in which n represents the number of nodes. A has elements:

$$A_{ij} = \begin{cases} 1 & \text{if there is an link between nodes } i \text{ and } j \\ 0 & \text{otherwise} \end{cases} \tag{1}$$

Symmetry of undirected networks implies $A_{ij} = A_{ji}$; if there is a link between i and j then there must be a link between j and i. In contrast, a directed network can be represented with an asymmetric adjacency matrix, where $A_{ij} = 1$ indicates the existence of a link pointing from j to i that is generally independent of the existence of a link from i to j. In metabolic networks, for example, many reactions are irreversible, i.e. the products of the enzymatic reactions cannot revert back into reactants. This unique directionality can be modeled with arcs and asymmetric adjacency matrices.

Second, network structure can be studied with measures of network *centrality*, the identification of the most important nodes of a network. For example, measures of *degree centrality* (or simply *degree*) describe patterns of node connectivity. Two nodes are *adjacent* in the network when they share a link. The *degree* of node i, $deg(i)$, is simply the set of nodes that are adjacent to it:

$$deg(i) = \sum_{j=1}^{n} A_{ij} \tag{2}$$

Nodes that have higher degrees are more central. In the case of directed networks, the degree can be decomposed into an indegree (*indeg*) and an outdegree (*outdeg*) measure of centrality, i.e. links that point into a node or out of a node, respectively. For example, node e of the mixed network of Figure 1 that contains both directed and undirected links has $deg(e) = 5$, $indeg(e) = 3$, and $outdeg(e) = 4$.

The *average degree* $\bar{d}$ of a network is a measure of how many links m there are in the set of links L [i.e. *card*(L)] compared to the number of nodes n in the set of nodes V [i.e. *card*(V)]. Because each line is incident to two vertices in an undirected graph, the average degree is:

$$\bar{d} = 2\frac{m}{n} = D[1-n] \tag{3}$$

which is proportional to the network's *density* D that measures the number of links of a network with n nodes given the maximum number of links that is possible. Similarly, an *eigenvector centrality* considers that not all links are equal in a network; some connections will be more central than others. Since the *neighbors* of a node belong to the set of nodes adjacent to it, the centrality of node i can be made proportional to the average of centralities of the neighbors of that node, making the vector of centralities the eigenvector of the adjacency matrix A_{ij}. Many variants of eigenvector centrality are possible (e.g. Google's *PageRank* centrality for ranking web pages). Other popular measures of centrality include *closeness centrality* and *betweenness centrality*. These measures take advantage of *walks* along the network diagram, w, which represent ordered lists of nodes with lengths $|w|$ corresponding to the number of links the walks contain. There are three kinds of walks: trails, paths and cycles. A *trail* is a walk with all links being different. A *path* is a walk with all nodes different. A *cycle* is a path in which the initial and terminal node of the walk are the same. Closeness and betweenness centrality use one special kind of path, the geodesic path. A *geodesic path* is the shortest path between two nodes, the path that requires the least number of links to traverse. The length of a geodesic path between generic nodes i and j, $d(i,j)$, is the number of links in that path. For example, the length $d(j,e)$ in the network of Figure 1 is $d(j,e) = |(j,i,h,e)| = 3$; note that several paths with a same length are possible. *Closeness centrality* of node i is the mean length of geodesic paths (geodesic distances) from node i to every other node. More central nodes will show lower distances. Mathematically,

$$closeness(i) = \frac{1}{\Sigma_{i \neq j} d(i, j)} \tag{4}$$

In contrast, *betweenness centrality* of node i is the fraction of geodesic paths between other nodes that contains that node. In other words, betweenness centrality quantifies the number of times a node acts as a "bridge" along shortest paths between any two other nodes in the network. It provides a measure of how relevant is node i over the flow of communication between nodes. Mathematically,

$$betweenness(i) = \sum_{u \neq i \neq v} \frac{P_{uv}(i)}{P_{uv}} \tag{5}$$

where P_{uv} is the total number of geodesic paths from node u to v and $P_{uv}(i)$ is the number of those paths that go through node i. Betweenness can be normalized by the number of pairs of vertices that exclude i.

Finally, some networks show a "*small world*" property when geodesic paths along networks are studied relative to the size of the entire network (Newman, 2001). This property follows the *"six degrees of separation"* idea in which a social world of *"friends of a friend"* chains are formed when all information is less than six steps away. Small world networks materialize when the mean geodesic distances between node pairs is small compared to the size of the network.

3. Network modularity

In general, biological systems are structured, i.e. the interactions, behavior and goals of subsets of parts may be different that the rest.

One kind of structure that is common in biological systems is the *module*. Modules are sets of integrated (coordinated) parts that cooperate to perform a task and interact more extensively with each other than with other parts and modules of the system (Hartwell *et al*., 1999). Modules are generally recognized by the property of *modularity*, the degree to which a system's components can be separated and rearranged (Simon, 1962). Modularity is important, as it relates to Herbert Simon's concept of the "*near-decomposability*" of a system that interacts with its environment: *"Each of the parts of a nearly-decomposable system has strong internal links among its sub-parts, but the several top-level parts are bound together with each other only by comparatively weak linkages"* (Simon, 1997). Thus, near-decomposability allows *"long-term behavior to be studied on an aggregative basis without concern for internal details of the parts, and allows the short-term behavior of each part to be studied independently of the behavior of the other parts"* (Simon, 1997). This concept has been seminal not only for biology and philosophy of science but also for computation and economics (Callebaut, 2007). It is now centrally important for network science. When systems are modeled with networks, network modularity can be detected and measured with a variety of methods. A primary index of modularity of the network is the *clustering coefficient* — a metric that describes the average probability that two neighbors of a node are also connected (Wasserman and Faust, 1994; Watts and Strogatz, 1998; Ravasz *et al*., 2002; Barrat *et al*., 2004). The *local clustering coefficient* C_i measures for example the degree to which nodes cluster together around node i, which can be calculated with the formula:

$$C_i = \frac{no.\ closed\ triplets\ connected\ to\ node\ i}{no.\ triplets\ connected\ to\ node\ i} = \frac{2L_i}{\deg(i)[\deg(i)-1]} \qquad (6)$$

where triplets are groups of three nodes connected by either two (open triplet) or three (closed triplet) links, L_i is the number of links established with the neighbors of node *i* and *deg*(*i*) is its degree. The index measures the tendency of nodes of a network to cluster together forming "triangles" (closed triplets) of links that comply with the adage, *"the friend of my friend is also my friend."* The clustering coefficient over all of the nodes of the network is defined as:

$$C = \frac{no.\ closed\ triplets}{no.\ triplets} = \frac{1}{n}\sum_{i=1}^{n} C_i \qquad (7)$$

where *n* is the number of nodes of the networks. The clustering coefficient ranges from 0 (no clustering) to 1 (maximal clustering with disjoint cliques). Networks with the largest possible values of *C* exhibit both a modular structure and the smallest possible average distance among nodes. In addition, higher values indicate the presence of the "small world" effect when compared to a version of the network where all links have been shuffled. Besides the clustering coefficient there are other statistics that measure the cohesiveness of network structure. For example, sets of nodes that are densely connected with each other are said to form a "community." Communities can be detected with hierarchical clustering methods such as the *Fast-Greedy-Community* (*FGC*) detection algorithm (Clauset *et al.*, 2004), the path-pruning *Newman-Girvan* (*NG*) algorithm (Newman and Girvan, 2004), or the maximization of modularity functions (e.g. the Louvain method; Blondel *et al.*, 2008). The identification of community structure has become an important endeavor for network biology.

4. A structural morphospace of networks

Biological networks are structured differently than man-made networks but share some common properties that demand discussion. One way to dissect network structure is the use of a *morphospace*, a phenotypic space defined by a limited number of variable properties that account for the most salient features of a system (Niklas *et al.*,

1994; Shoval *et al.*, 2012; Schuetz *et al.*, 2012). Theories of optimality (such as natural selection) maximize a fitness function for a system, which generally results in an optimal point or area in the morphospace. However, a system depends on the performance of multiple tasks, which complicates definition of the fitness function and requires trade-off solutions of complex optimization.

Biological and engineered systems exhibit trade-offs between three very important properties: flexibility, economy and robustness (Yafremava *et al.*, 2013). *Economy* manages the system's matter-energy budget to maximize performance while streamlining cost. *Flexibility* takes advantage of structural and functional information-processing mechanisms, which respond to internal and external changes imposed on the system. *Robustness* channels information to maintain structure and function while resisting damage and change.

A general persistence triangle of *economy-flexibility-robustness* (Figure 2) has been supported by ecological and molecular data and has been applied to persistence strategies of cellular organisms (Yafremava *et al.*, 2013) and viruses (Nasir *et al.*, 2017) and the organization of structural domains in proteins (Shahzad *et al.*, 2015). Here, we extend the triangle to describe trade-off solutions for network structure. In these triangles of network persistence, optimal trade-off solutions for a network can locate in Pareto fronts according to crucial factors that affect the systems portrayed by the network. Pareto fronts help find best trade-offs for all tasks, i.e. optimization solutions that cannot improve all sets of fitness functions of the morphospace.

Solé and Valverde (2004) proposed a morphospace that explores the "limits of the possible" in the structure of networks. Their morphospace measures the modularity, heterogeneity, and randomness of the networks. We find that the persistence triangle of *economy-flexibility-robustness* can be adapted to explain the three network

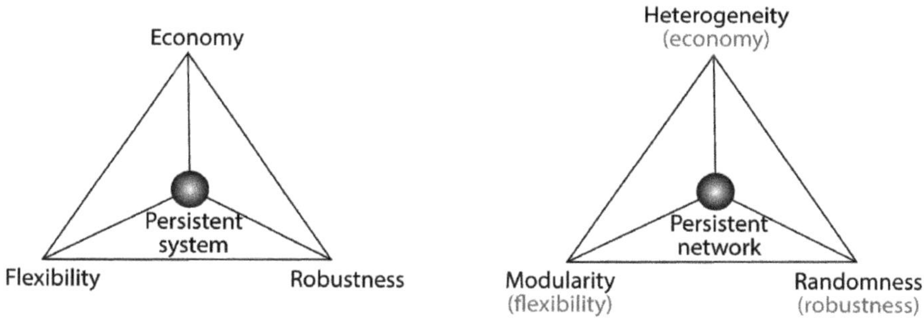

Figure 2. A morphospace of economy, flexibility and robustness in the structure of networks. An *economy-flexibility-robustness* triangle of biological persistence explains trade-off solutions for the engineering of a biological system (left). This generic persistence triangle can explain the persistence of network structure by defining modularity, heterogeneity and randomness as proxies for flexibility, economy and robustness, respectively (right). The spheres at the center of the triangles represent optimal trade-off solutions for a system or network. Solutions with maximal fitness locate in Pareto fronts within a 4-dimensional fitness landscape (a 2-polytope) according to factors that affect the systems or networks.

variables (Figure 2). Modularity embodies network *"flexibility"* in the structuring of communities with dense and sparse connections. Heterogeneity describes the scale-freeness of network connectivity, which measures the *"economy"* of traversing paths along network structure (Barabasi and Oltvai, 2004). Randomness entails uniform connectivity of nodes throughout the network, a property of *"robustness"* that confers a network fault tolerance to stochastic error (Albert *et al.*, 2000).

(i) *Modularity*: Modularity improves the flexibility and evolvability of a system (its ability to adapt and innovate) and can be measured in networks with a maximum modularity score, such as those obtained with clustering methods (*FGC* or *NG*). Modularity compartmentalizes the structure and function of the system by generating small functional substructures that can be reused many times in different contexts. In absence of modularity, the structure and function of the system are monolithic and must be built or modified from scratch for improvement.

Simon's famous parable of the "Hora" and "Tempus" watchmakers (Simon, 1962) illustrates the evolutionary implications of modularity (Figure 3). Both Hora and Tempus were highly skilled watchmakers and applied the same levels of effort to build equally complex watches. However, Hora prospered because his watches were built in such a way that they could be dissected into individually functioning components, i.e. into watch modules. When Hora's watchmaking was for some reason interrupted (e.g. because of a distraction or a faulty component part), he simply reassembled the

p = probability of no interruption

Watchmaker	Tempus	Hora
Assembly design	Integrated	Modular
Watch parts	1,000 parts	1,000 parts
Assembly/1,000 parts	1	1
Sub-assemblies/100 parts	0	10
Sub-assemblies/10 parts	0	100
Assemblies/watch	1	111
Probability of assembly without interruption	$(1-p)^{1000}$	$(1-p)^{10}$
If $p = 0.01$ (one chance in 100 of an interruption)		
Relative assembly attempts per watch	20,000	1
Relative assembly time per watch	4,000	1

Figure 3. Simon's parable of watchmaking. The parable shows that building a watch from 1,000 parts by hierarchically assembling groups of 10 elements each takes 4,000 less time than assembling the watch with a monolithic integrated design. This calculation assumes there is one chance in 100 of an interruption (or failure) during assembly. Note that Tempus builds each watch in one assembly step, while Hora require 111 modular assembly steps. However, Tempus must invest 20 times more work than Hora to compensate for interruptions (failures). The parable teaches the benefits of modularity as a design principle.

individual watch component and not the entire watch. This increased his watchmaking efficiency and the functionality of his watch collection. In contrast, Tempus assembled every watch as a fully integrated piece. Each time the assembly was interrupted he had to start from scratch. His watchmaking efficiency and the functionalities of his watch collection decreased with time.

When the Hora and Tempus parable is applied to networks, networks that are structured will partition the flow of information (or other properties such as matter/energy or time) among the varying betweenness centralities of their nodes. Since betweenness measures how an individual node controls information flow through the network, partitioning the flow into clusters with high betweenness divides the risks of centralizing information control in the network. We note that the environment perturbs the information flow of a system. In turn, the system responds by modifying its make up and functionality to make the system more flexible. Thus, varying environments (goals) alter a system's modularity (generally promoting it; Kashtan and Alon, 2005; Kashtan *et al.*, 2007; Lorenz *et al.*, 2011), improving the speed, stability and quality of information transfer through networks (Solé and Valverde, 2004).

(ii) *Heterogeneity*: Heterogeneity describes the small world "scale-freeness" property of network connectivity (Solé and Valverde, 2004). It is a measure of the economy of information flow through a system. Heterogeneity can be identified with the degree distribution p_k, the probability that a node has k links:

$$p_k = k^{-\gamma}\phi\left(\frac{k}{\xi}\right), \tag{8}$$

where $\phi(k/\xi)$ represents a cut-off at scale ξ. (Amaral *et al.*, 2000). Scale-free networks are inhomogeneous and have p_k distributions that follow a power law, with γ values ranging 2–3 and heavy tailed distributions lacking a significant cut-off, such as those of metabolic networks (Jeong *et al.*, 2000), the WWW (Broder *et al.*, 2000), or the internet (Faloutsos *et al.*, 1999). In these networks, few nodes have many links. For example, ATP, Google, and AT&T (tier 1) are highly connected nodes for metabolism, the WWW, and the internet, respectively. For other networks, p_k can be better fit with a power law that has an exponential cut-off k_c (e.g. co-authorship networks with cut-off $\phi(k/\xi) = \exp(-k/k_c)$; Newman *et al.*, 2001) or with exponential or gaussian distributions (Strogatz, 2001). Network models offer mechanisms for the generation of network structure. Such generative models for example can explain the heavy-tailed degree distribution of scale-free networks.

Simon (1955), de Solla Price (1960) and Barabási and Alberts (1999) described a *"cumulative advantage"* or *"preferential attachment"* model, in which nodes are gradually added to a network one at a time and linked "preferentially" to nodes with high connectivity (high degree). When new nodes are attached to an existing node with probability proportional to the number of links connected to that node then a *"rich-gets-richer"* scenario occurs in which few nodes harbor many links and the vast majority of nodes are poorly connected (Figure 4). The tail of the distribution tends towards $p_k \sim k^{-3}$ with large k, which strongly matches known γ values of real scale-free networks. More sophisticated generative models exist in which the connectivity of nodes follow different attachment rules, is altruistic, is constrained by spatial growth, or decreases with time (Albert and Barabási, 2000; Dorogovtsev and Mendes, 2000; Krapivsky *et al.*, 2000; Bianconi and Barabási, 2000; Kaiser and Hilgetag, 2004; Small *et al.*, 2015; Pachon *et al.*, 2018). Some of these models predict exponential or truncated power-law distributions. Experimentally, the Kolmogorov-Smirnov (KS) statistical test of power law fit can be used to scrutinize the scale-free behavior of simulated and real networks (Newman, 2005; Clauset *et al.*, 2009). Lower KS fit values (≤ 0.1) and higher p-values of the KS test (> 0.05) support a power-law fitting of network degree distribution.

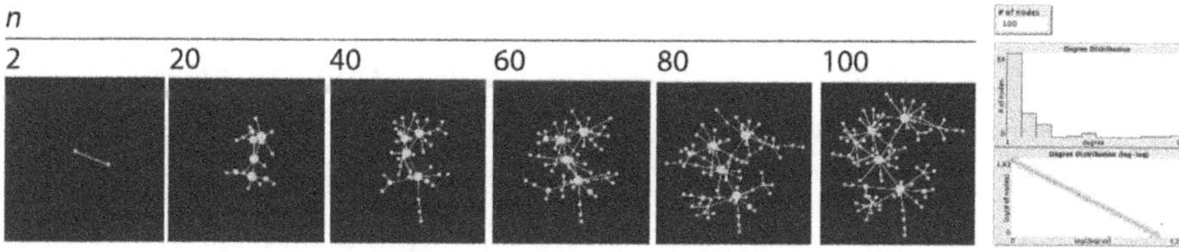

Figure 4. Studying the growth of a scale-free network with a "preferential attachment" model. The growth of a scale-free network was modeled with the "preferential attachment" implementation of NetLogo 6.0.4 (Wilensky, 2005), using default parameters. Nodes are attached one-by-one and networks sampled every 20 rounds of node attachment. Network size is measured in number of nodes n. Node size was made proportional to its degree (number of links) to visualize the power-law degree distribution. Plots in the right describe the power-law degree distribution for $n = 100$ as an exponentially decreasing histogram or as a log-log plot showing a decreasing straight line with slope γ.

Alternatively, the exponent (α) of the fitted power law distribution $P(X = x^{-\alpha})$ is usually $\alpha > 1$ when the fitting is significant. In addition, the log-likelihoods and coefficient of determination (R^2) of the fitted power law parameters can be monitored to ensure data quality. In fact, heterogeneity can be effectively measured with the maximum likelihood scaling exponent. α Alternatively, parameters of entropy of the degree distribution H describe long tailed distributions, which can be considered a measurement of both uncertainty and heterogeneity (Solé and Valverde, 2004).

(iii) *Randomness*: Randomness describes the stochastic effect of randomly adding nodes and links to a network. The process can gradually transform regular networks (e.g. chains, grids, lattices, trees) lacking randomness and heterogeneity into network structures that are random. For example, lattices form a regular tilling with fixed number of links per node and Cayley trees have nodes with exactly the same degree. These regular networks, which are fully non-random and homogeneous, can be gradually randomized and made small-world by adding "short-cut" connections between remote network regions (Watts and Strogatz, 1998). On the other end of the spectrum, the Erdös-Rényi (ER) "random graph" (also known as the Poisson random graph) describes an idealized network in which sets of two out of n nodes are connected with probability p or $(1-p)$ (Solomonoff and Rapoport, 1951; Erdös and Rényi, 1960). Given a large number of nodes $n \gg 1$, the topology of the random network changes as a function of the number of links m that are added at random, which is proportional to p. When m is small, most nodes of the network exist in isolation or form small clusters of connected nodes, which are known as *components*. As m grows, components grow at first by linking to isolated nodes and then by associating with other components. However, when $m = n/2$ and $p = 1/n$ a phase transition occurs in which all components are brought together into a single 'giant component'. Further increases of m and p simply grow the giant component with scaling proportional to n. While many real networks follow this model of growth, ER networks do not portray the properties of real networks. They are highly democratic: most nodes will end having the same number of links and connectivity follows a Poisson bell-shape distribution. Note that random networks are 'exponential' because the degree distribution p_k decreases exponentially for large k degree values according to a binomial distribution, which becomes a Poisson distribution for large n:

$$p_k = \binom{n-1}{k} p^k (1-p)^{kn-1-k} \cong \frac{z^k e^{-z}}{k!} \tag{9}$$

where $z = (n-1)\,p$ is the mean degree of the network. These random networks can be treated as null hypotheses in the exploration of the effect of randomness on the morphospace of networks. However, their bell-shaped degree distributions cannot accommodate real network distributions. A different class of random networks must be employed that is more flexible and can model real networks. Molloy and Reed (1995, 1998) proposed a *generalized random network* model in which any degree distribution p_k (i.e. any fraction of nodes in the network with degree k) can be specified. Given a p_k, a set of n values with degrees k_i are chosen for nodes $i = 1, 2, \ldots, n$ as a *degree sequence* of non-negative integers drawn from the distribution. Then, pairs of this set of n nodes are randomly linked to produce an ensemble of networks, the *"configuration model."* The model generates a giant component with size that is known and matches behavior of real networks (Molloy and Reed, 1995, 1998). For $n \gg 1$, the giant component exists if and only if:

$$\langle k^2 \rangle - 2\langle k \rangle > 0 \tag{10}$$

where $\langle k \rangle = z = \Sigma_i k_i / n$ is the mean degree and $\langle k^2 \rangle = \Sigma_i k_i^2 / n$ is the mean square degree. This random network modeling approach can be applied to scale-free networks (Aiello *et al.*, 2001). The giant component exists for $\gamma < 3.47$ values and all nodes are connected for $\gamma < 1$. A generating function formalism creates an equally mathematically rigorous random network model that can also be applied to directed and bipartite networks (Newman *et al.*, 2001).

Randomness can push uniform connectivity of nodes throughout any network structure and can define how components break down or build up by addition or removal of the nodes of a network. This democratic property provides a measure of network resilience (e.g. a resistance to failure due to error or deliberate attack), which confers robustness to the network and the system (Newman, 2003).

Network robustness can be modeled by node removal (Albert *et al.*, 2000). For example, node removal by random failure or deliberate attack of the most highly connected nodes makes explicit the vulnerability of a network structure (Figure 5). Increasing the fraction of nodes that are removed (f) cause the network to fragment into increasing numbers of isolated node clusters (components) that become smaller with increased fragmentation. In random networks, the average size of the largest component S decreases with

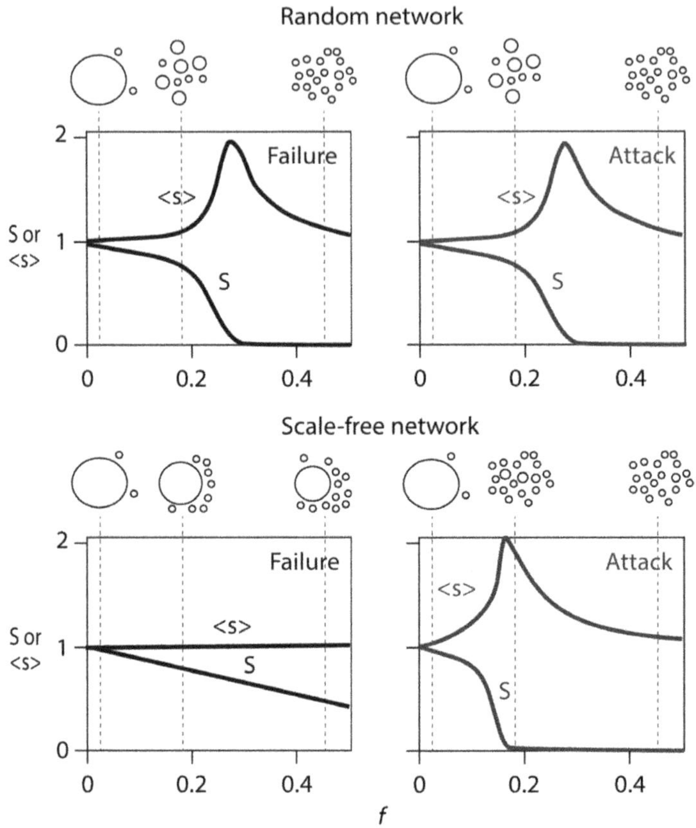

Figure 5. Failure and attack of network structure. Removal of nodes by failure due to random error or by deliberate attack of the most highly connected nodes causes the fragmentation of networks into components (isolated node clusters) of varying size. The relative size of the largest component S and the average size of components <s> are functions of the fraction of removed nodes *f* and are useful proxies of network fragmentation. Analysis of random and scale-free networks shows that random networks are more resilient to deliberate attack, i.e. that it takes a larger fraction of removed nodes for networks to fall apart. In contrast, scale-free networks are higher resilient to random failure. They only fall apart when most nodes are removed ($f = 0.8$; not shown). Data from Albert *et al.* (2000).

sigmoidal behavior reaching a threshold analogous to a "percolation critical point," while the average size of components <s> increases to a maximum and then decreases. S vanishes to 0 and <s> reaches its maximum when $f \sim 0.28$. This causes the network to fall apart. This behavior occurs regardless of nodes being removed by random failure or deliberate attack. A scale-free network subjected to deliberate attack responds similarly to a random network. However, the network falls apart when a smaller fraction of nodes is attacked. It is therefore less fault tolerant to attack than a random network. In contrast, scale-free networks display a surprisingly degree of tolerance to random failure. No threshold of fragmentation is observed and the size of the largest component slowly decreases with increases of node removal.

Robustness can also be modeled (more effectively) with percolation, the random occupation of links on a network with probability *p* (Cohen *et al.*, 2000; Callaway *et al.*, 2000; Holme *et al.*, 2002). In fact, percolation can be reformulated as a message-passing or information-passing process to study thresholds and percolating and non-percolating network clusters as averages of all possible realizations of the randomness of percolation (Karrer *et al.*, 2014). One way to measure randomness is with the logarithm of Bartel's test statistic, which can help define the morphospace of network structure.

Real networks are heterogeneous, modular, and strike a balance between order and randomness (Watts, 1999). The morphospace of Solé and Valverde (2004) makes this clear by measuring the three important network parameters we discussed, modularity, heterogeneity, and randomness (Figure 6). The approach explores heterogeneity of the degree distribution with entropy measures and correlations of degree distributions with conditional probabilities. All real networks had scale-free configurations and their mappings showed fundamental topological constraints imposed on network structure that made most of the morphospace inaccessible, which is

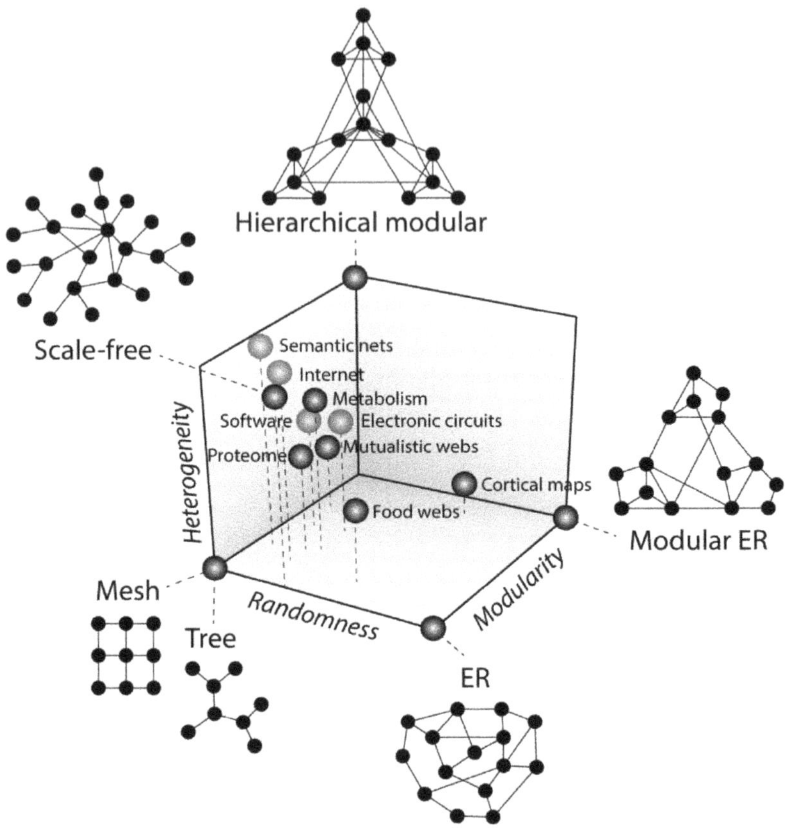

Figure 6. Morphospace of network structure showing toy examples of typical graphs describing archetypes of the phenotypic landscape. In the morphospace, Erdös-Rényi (ER) random graphs transform into regular graphs by decreasing randomness, or into modular ER graphs by increasing modularity. Hierarchical modular structure requires both increasing modularity and heterogeneity and decreasing randomness. Modeled scale-free networks map onto a heterogeneity-randomness projection plane with high heterogeneity. Different man-made and natural networks are mapped onto the morphospace with light and dark colored spheres, respectively. As expected, man-made networks are less random (more regular) and more heterogeneous (more scale-free). Morphospace and data modified from Solé and Valverde (2004).

an expected property of morphospaces in general. However, man-made and natural networks map to distinct regions of the morphospace. Man-made networks (e.g. semantic nets, internet) appear more ordered than natural networks (e.g. metabolism, the proteome, food webs). To benchmark properties of the morphospace, idealized networks such as ER random graphs, modular ER networks, and regular networks (e.g. meshes, trees) are also mapped onto the morphospace. ER random graphs transform into regular graphs by decreasing randomness. They transform into modular ER graphs by increasing modularity (Figure 6). Hierarchical modular structure requires both increasing modularity and heterogeneity and decreasing randomness. Real networks appear to establish trade-offs in the structural morphospace of networks necessary to balance the economy, flexibility and robustness of an evolving system.

5. Hierarchical modularity

Modular networks harbor "communities," groups of nodes that are more densely connected with themselves than with the rest of the network (Newman and Girvan, 2004). Modularity usually counterbalances patterns of growth by preferential attachment of new nodes to highly connected nodes, which generally results in the power-law behavior of scale-free networks (Jeong *et al.*, 2000). However, both modularity and scale-freeness often coexist in networks when modules coalesce hierarchically (Ravasz *et al.*, 2002).

To understand hierarchical modularity of networks we must first explain what is a hierarchy and how it evolves. Simon (1962) defined a hierarchical system as a *"system that is composed of interrelated subsystems, each of the latter being, in turn, hierarchic in nature until we reach some lowest level of elementary subsystem,"* and more narrowly, as a *"complex system in which each of the subsystems is subordinated by an authority relation to the system it belongs to."* This nesting relationship has been elaborated further by

Salthe (1985, 2012), who divided them into *scalar* (compositional) and *subsumption* (specification) hierarchies.

Scalar hierarchies nest differently sized dynamical entities into each other by mereologically defining parts-and-wholes as [[part] whole] statements and invoking 'is-a-part-of' relationships between them (e.g. atoms, molecules, cells, organisms, populations). A relevant example is the successive nesting of amino acid residues, secondary structures, loop structures, and domains in protein molecules. In these hierarchies, parts are treated as modules. Nesting relationships are compositional in their logic; wholes made out of parts are successively nested in expanding manner both outward (towards higher levels of the hierarchy) or inward (towards lower levels) from a focus level of organization without an integrative or historical rationale. In sharp contrast, subsumption hierarchies nest entities by defining taxonomies that describe "general-to-specific" or "coarse-to-fine" properties of a system as [general[specific]] statements and invoking a "is-a-kind-of" ordering principle between them (e.g. physics, chemistry, biology, sociality). A typical example is the placement of man among great apes according to the NCBI taxonomy database, from the general to the specific: Hominidae, Homininae, Homo, *Homo sapiens*. A relevant example is the classification of the family of aminoacylation domains of aaRS enzymes according to SCOP classification (Murzin *et al.*, 1995): "Class I aminoacyl-tRNA synthetases (RS), catalytic domain" fold family ⊂ "Nucleotidylyl transferase" fold superfamily ⊂ "Adenine nucleotide alpha hydrolase-like" fold ⊂ "Alpha and beta (a/b)" class ⊂ SCOP root, where ⊂ indicates "is a subset of." Such hierarchies integrate all aspect of the world and their definition demands establishing history, i.e. establishing "intermediate forms" or "prior forms" of a refining, developing or evolving system. Figure 7 illustrates the two complementary logical forms of hierarchical structure, one focusing on the nesting and the other on the evolving network. These two forms capture more broadly Simon's view of hierarchies being

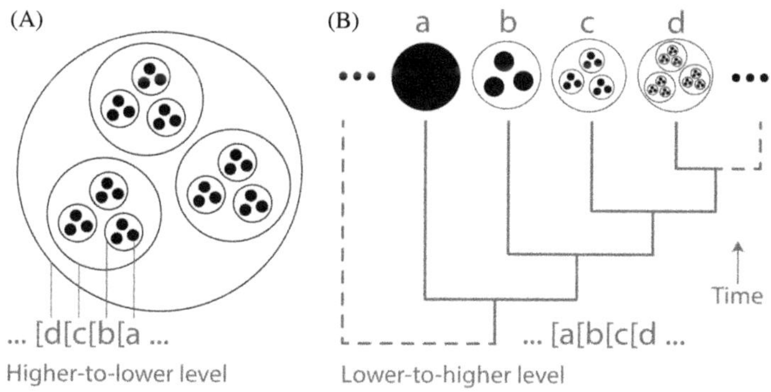

Figure 7. The two faces of hierarchy. (A) A *scalar hierarchy* stresses the mereological aspect of a system, with parts embedded in wholes at each level of organization (a, b, c and d, from lower to higher levels). Note that under this view the system expands from a focal view, inwards and outwards. **(B)** A *specification hierarchy* involves a taxonomy that explains the refinement, development or evolution of levels of organization as time unfolds. Note how under this view the system becomes more granular and expands from the highest level, which is always in focus.

described by *"states"* (the world as sensed), which defines observables, and hierarchies described by *"processes"* (the world as acted upon), which prompt the system to act *"purposefully upon its environment"* (Simon, 1962). The two logical forms have been recently integrated into a *model hierarchy* that is also described by two partial hyperscalar hierarchies, one focusing on the nested entities, the other focusing on the context of that nesting (Cottam *et al.*, 2016). These two views are unified into an information-based metascale description. Here we contend that the *"scalar"* or *"states"* view stresses a unification of the system while the *"subsumption"* or *"process"* view stresses the diversification of the system. This will become explicit below.

In the context of networks, hierarchy involves the fractal-like reuse or embedding of simpler network modules into modules of higher complexity. Thus, lower level network modules are subordinated by an *"authority relation"* to higher network modules. This relation can be imposed externally by classifying modules into levels of a hierarchy according to knowledge outside the realm of the interrelationship of nodes and links (e.g. some ontological description) or internally by using features of structure of the network that dictate how the hierarchy is organized (e.g. a clustering strategy). If there is an *"organizing principle"* for the structure of systems, the latter internal descriptor will tend to organize the network according to the former external descriptor. To establish such tendency, hierarchical modularity must be measured.

One strong indicator of hierarchical modularity is the power law scaling of the clustering coefficient with node degree (Ravasz, 2009). While in many scale-free models (e.g. the Barabási model) the clustering coefficient $C(k)$ is independent of the degree of a specific node k, a hierarchical modular network structure shows $C(k) \sim k^{-\beta}$, where $\beta \sim 1$ in deterministic scale-free networks (Dorogotsev *et al.*, 2002). It has been demonstrated that the clustering coefficient scales negatively with node degree (Barabási and Oltvai, 2003) for metabolic networks (Ravasz *et al.*, 2002) and protein–protein and genetic regulatory interaction networks (Ravasz, 2009), and does so with $\beta \sim 1$. We note however that such scaling can also occur in absence of a hierarchy when applying the Klemm and Eguíluz (2002) model of network growth. This model *"de-activates"* with probability $p \sim 1/k$ the addition of links to a set of less connected nodes. This causes the generation of a central core in the growing network and peripheral clusters that are not nested in larger structures. To avoid this kind of false positive, any power law scaling of the clustering coefficient should be confirmed by a visual examination of the hierarchy of network modules. Hierarchical modularity must show the integration of smaller modules into larger ones, in fractal-like manner (Figure 8). In order to identify the hierarchically embedded modules, a node-to-node similarity distance can be used to define a topological overlap measure, which can then be coded in

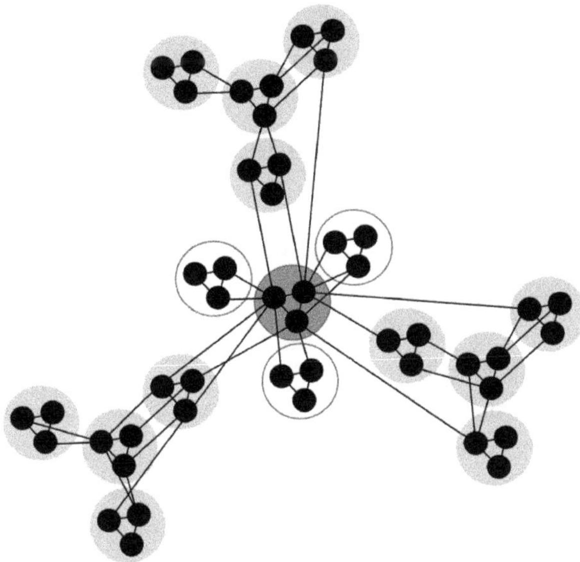

Figure 8. A model of a minimalistic hierarchical network. In the model, a small core of three nodes tightly connected to each other by a triangle (highlighted in grey) grows by triplicating the triangle modules in two successive steps (highlighted in light peach and blue) and connecting them at random. These steps define hierarchical levels of network organization.

the adjacency matrix for analysis with a clustering algorithm that groups nodes onto a hierarchical tree (e.g. the UPGMA method; Ravasz *et al.*, 2002). Alternatively, links can be removed along shortest paths to reveal the nesting of functional modules (Girvan and Newman, 2002; Holme *et al.*, 2002).

The fact that most real-world networks show modularity and hierarchy suggests that these properties provide numerous benefits to a system over network designs that are monolithically integrated. In fact, network modules, and as we will later discuss, hierarchy, can increase evolvability (the ability to adapt and innovate; Kashtan and Alon 2005; Kashtan *et al.*, 2007; Lorenz *et al.*, 2011) and robustness (the maintenance of function in light of internal or external perturbation; Kitano, 2004), while decreasing the costs of establishing network connections (Mengitsu *et al.*, 2016).

6. A morphospace of hierarchy describes flows along wiring diagrams

Directly or indirectly, Salthe's *"subsumption"* or Simon's *"process"* interpretations of hierarchy embody a temporal context. This context generally materializes as either flows of matter/energy, information, and/or signs of language, or if "time event" information (dating of node entities) is available, directly as flow of time. Flow hierarchies channel the temporal flux along the wiring diagrams of a directed or semi-directed network (Luo and Magee, 2010). They do so through arcs, which define the directionality of travel (Figure 9A), or directly by labeling nodes or links with time (Figure 9B). When traveling ("walking") along the wiring diagram, walks become "paths" in cases where every node is different. In particular, paths that begin and end in a same node form "cycles." return structures that interfere with flows. A flow hierarchy will often harbor the three ideal properties of a hierarchical system: (i) order, (ii) reversibility, and (iii) pyramidal structure, which will be maximized in hierarchical systems such as directed acyclic graphs (DAGs). Order describes a tendency of nodes to be arranged unambiguously without being compromised by cycles. Reversibility is a property that defines *"only one commander for any commanded."* Pyramidal structure is a feedforward tree-like pattern in which the *"commander commands more than one node"* and lower-level nodes are subjected to chains of commands. Because hierarchies often show varying levels of network organization, these three properties of flow hierarchies can be evaluated with a number of metrics, including the percentage of links not included in cycles (Luo and Magee, 2010), measures of flow heterogeneity such as global and local 'reaching centralities' (Luo and Magee, 2010; Mones *et al.*,

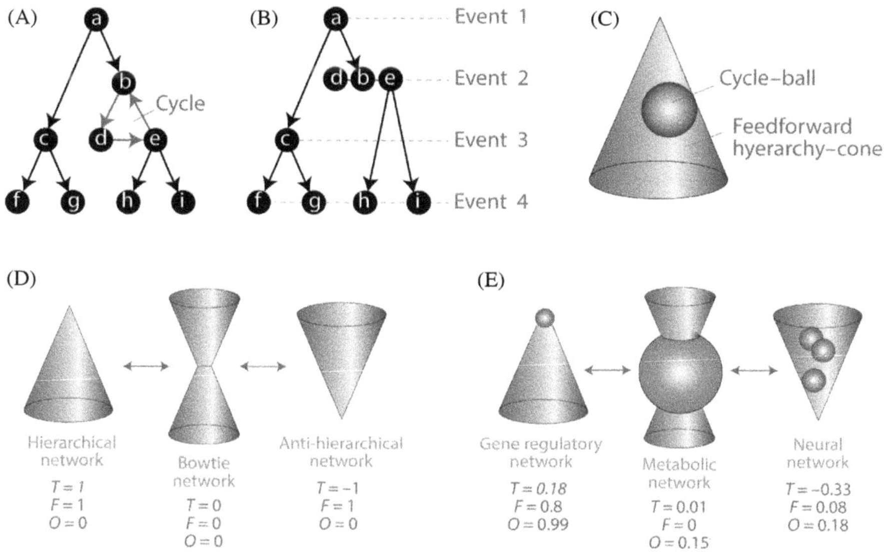

Figure 9. The hierarchical organization of flow networks. (**A**) A feedforward tree-like network structure with a cycle. Note how arcs flow in one direction except for those in cycles. (**B**) The direction of temporal flow now complies with *time events*. (**C**) An icon of cones and balls describes the feedforward and cycle structure of the network. (**D**) The parameters of *treeness* (*T*), *feedfordwardness* (*F*) and *orderability* (*O*) describe the hierarchical, bow-tie or anti-hierarchical feed-forward structure of the flow hierarchy. (**E**) Three examples of biological flow networks are summarized with icons and hierarchy statistics: *Bacillus subtilis* gene regulatory network, the human metabolic network, and the *Caenorhabditis elegans* neural network. Data from Corominas-Murtra *et al.* (2013).

2012), or the use of morphospaces (Corominas-Murtra *et al.*, 2013) or random walks on the network (Czégel and Palla. 2015).

Corominas-Murtra *et al.* (2013) presented a formal approach to dis-sect network hierarchies with three coordinates, *orderability* (O), *feedforwardness* (F) and *treeness* (T). The three metrics define a morphospace of pyramidal structure, reversibility and order of net-work hierarchical relationships (Figure 10). O measures order by calculating the fraction of nodes that do not form cycles ($0 \leq O \leq 1$,

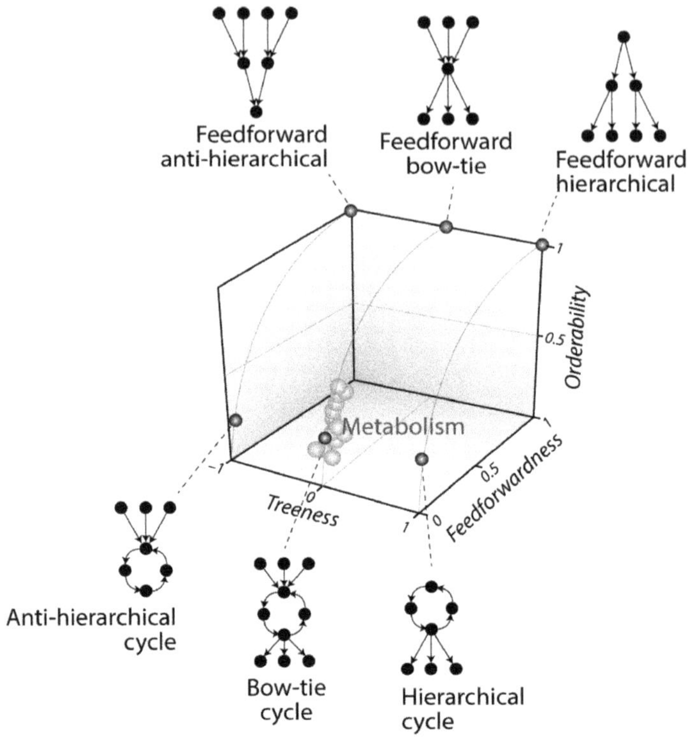

Figure 10. Morphospace of network hierarchies. Typical morphologies are shown for specific locations of the morphospace. Coordinates for metabolic networks are identified with spheres colored in light gray. Treeness defines the unification or diversification of hierarchical signal in the network, while orderability defines the centrality of cycles in network structure. Modified from Corominas-Murtra *et al.* (2013).

with 0 describing maximum disorder induced by network cycles). F measures the impact of network regions (modules) that cannot be ordered (i.e. cycles) on feedforward hierarchical structure, maximally penalizing modules of cycles closer to the top of the hierarchy ($0 \leq F \leq 1$, with 0 describing minimum order). Together, O and F define the number and location of cyclic regions in the network; increasing F values push modules from the top to the bottom of the hierarchy. Finally, T accounts for both reversibility and pyramidal structure of the network by measuring how unambiguous is its chain of command. Sliding the *O–F* plane along the *T* axis gradually transforms "anti-hierarchical" pyramidal structure ($T = -1$) to "hierarchical" structure ($T = 1$) by transitioning through "bow-tie" symmetric structures ($T \sim 0$), with negative and positive values describing anti-hierarchical and hierarchical feedforward structure, respectively. This transition of hierarchical structure is illustrated with icons of cones and balls in Figures 9C–9E. In bow-ties, the diversity of downstream paths is cancelled by the uncertainty of reversing those paths. Many real networks show bow-tie hierarchical structures, including metabolic and linguistic networks (Figure 9E). An analysis of the metabolic networks from 19 organisms showed similar bow-tie structures mapping to a very confined region of the morphospace of hierarchy (Figure 10). This region contains also linguistic and social networks, and more distantly, neural networks.

We note that the first report of the "bow-tie" structural arrangement was described for the World Wide Web (WWW) using the hypertext transfer protocol (HTTP) that exchanges information between computers (Broder *et al.*, 2000). In these networks, static web pages have links defining arcs pointing to other static pages. The bow-tie structure of the WWW has a strongly connected cycle component (a "giant" component; G_s) stabilized by a balanced feedforward structure of inputs (G_{in}) and outputs (G_{out}). However, its structure has additional complications, including external "tendrils" and "tubes" (*T*) that connect inputs and outputs bypassing the giant component, and isolated ("disconnected") components (*D*). Out of 203.5 million web pages, roughly an eighth of this number made up the strong component and additional eighths made up the "in" and "out" structures (Figure 11). A similar bow-tie structure was later reported for metabolism (Ma and Zeng 2003). For example, the 821 metabolites of the metabolic networks of *Escherichia coli* define a flow of mass and energy in which a small number of input metabolites (93) engage in walks through a giant component of 273 metabolite

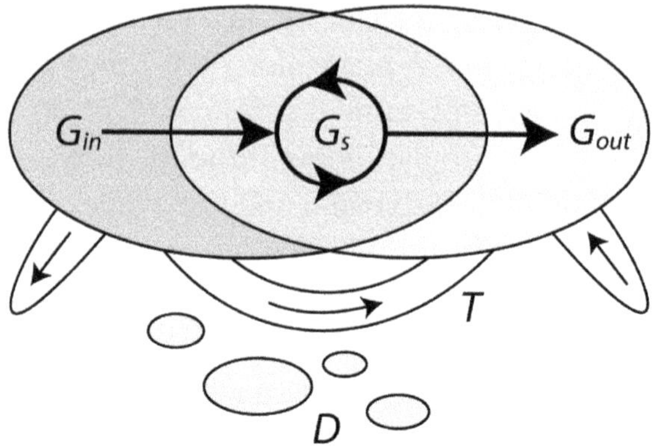

	G_{in}	G_s	G_{out}	T	D
WWW ($\times 10^6$)	43.3	56.5	43.2	43.8	16.8
Metabolism	93	274	161	—	283

Figure 11. The bow-tie hierarchical structure of directed networks. These networks have a giant strongly connected component (G_s), giant "in" component (G_{in}), giant "out" component (G_{out}), tendrils and tubes (T) and disconnected components (D). The number of nodes present in these subgraphs are listed for the WWW (Broder *et al.*, 2000) and the metabolic networks of *Escherichia coli* (Ma and Zeng, 2003).

precursors that reuse and recycle molecules to produce 161 products (Figure 11). Yet, 283 metabolites remained isolated from the main network. More recent studies of the WWW revealed a hierarchical multi-layer organization of tendrils, tubes and isolated components, which is also present in the neural network of *Caenorhabditis elegans* (Timár *et al.*, 2017). The bow-tie biphasic structure, which is balanced by input and outputs and/or the degeneracy (many-to-one) and pluripotentiality (one-to-many) of the pyramidal structure, is therefore complex and mysterious.

7. Evolutionary drivers of network structure in biological systems

What underlying processes explain the structuring of networks? Can they be explained by trade-offs between hierarchy and modularity?

Considerable discussion exists related to the origins of modularity (Wagner *et al.*, 2007), hierarchical modularity (Clune *et al.*, 2013; Mengitsu *et al.*, 2016) and bow-ties in flow hierarchies (Sabrin and Dovrolis 2017), both from systems and network biology perspectives. It is highly likely that multiple evolutionary drivers actively contribute to the emergence of network structure. By definition, drivers should further or at least maintain the persistence of the biological system as the system evolves. If the system is visualized as a network, they should further the persistence of nodes and links, heterogeneities that exist in network connectivity (e.g. motifs, communities), and network dynamics, all of which perfect the workings of the system and its component parts.

One general evolutionary strategy is to advance fitness through natural selection or competitive optimization (defined later). These forces enhance the adaptation of the system to changing environments. Selection or optimization can be driven by direct or indirect causes, sometimes making use of mutational biases. In direct models, network structure directly contributes to higher fitness, for example, by being the direct target of individual-level selection through epistatic or pleiotropic "constraints," including those important for morphology or development (Leroi, 2000). Network motifs, such as feedforward loops that are common in signaling networks, could offer an epigenetic advantage by for example fostering higher information flow through motif clustering (Gorochowski *et al.*, 2018). However, Cordero and Hogeweg (2006) showed that a hierarchical feedforward structure with overrepresentation of feedforward loops emerges in simulation of evolving transcriptional regulatory networks of yeast with models that do not select for functional motifs. In contrast, indirect models consider network structure as an adaptation to changing external environments (Kashtan and Alon, 2005; Kashtan *et al.*, 2007). Under these conditions, the system that is being optimized responds to different goals. This often enhances network modularity. Indirect models of this kind are driven by evolvability, the adaptation of a system to the changing environments, or specialization, the specific gain of functions (Espinosa-Soto and Wagner, 2010), both of which are enhanced as networks become structured.

An alternative evolutionary strategy takes advantage of non-adaptive biases in mutational mechanisms as these are affected by the networks themselves. Here, network structure is driven by a natural selection process that approaches "neutrality." For example, modularity can emerge through patterns of duplication and differentiation in growing networks (Solé *et al.*, 2002; Solé and Fernandez, 2003). The random duplication of nodes and their diversification results in both a small world structure and a scale-free distribution. However, as a network grows and nodes gain or lose links, a phase transition occurs in which network modularity arises spontaneously "for free," i.e. without incurring in explicit costs. This model is biologically plausible under a number of evolutionary models. It reproduces properties of real protein–protein interaction networks modeled from genomic data. Hierarchy appears also similarly driven. The hierarchical bow-tie structure can be explained by a preference to "reuse" modules of similar complexity (Sabrin and Dovrolis, 2017). A focus on source-to-target paths traversing each node identified a core of nodes that covered almost all paths and revealed there was a bias of new modules reusing existent modules. This fostered the emergence of a central bottleneck of modules responsible for the bow-tie structure, which could limit the flow of information and/or its time duration in evolving networks (as long as tendril connectivity remains constrained; Broder *et al.*, 2000). Thus, biases in the reuse of nodes, motifs and communities could represent an important "neutral" force driving the evolution of network structure.

Another popular related mechanism supported by computational simulation merges indirect selection for evolvability and indirect selection pressures associated with adding and maintaining links (Clune *et al.*, 2013; Mengitsu *et al.*, 2016). Convincingly, minimizing connection costs increases both network modularity and evolvability over levels induced by selection pressures that maximize performance (Clune *et al.*, 2013). This same approach was later extended to the emergence of network hierarchy (Mengitsu *et al.*, 2016). Thus, reducing connection costs promotes hierarchy and modularity in network structure. Note that connection costs include costs incurred in the manufacture of a link, its maintenance, or its ability to transmit information, all of which are important for maintaining a matter/energy budget. Naturally, costs involve nodes that participate in establishing the links, i.e. the interacting parts (actors) of the system. For example, links that involve connecting molecules or molecular parts require the development of interacting surfaces with costs constrained for example by molecular surface area (e.g. in protein-protein interaction networks or networks of protein domain organization) or chemical functionality (e.g. molecular functions in metabolic networks). Costs incurred in furthering the stability and robustness of interacting molecules must also be minimized (e.g. allosteric interactions, pocket surfaces of active sites). Note that adding or subtracting connections can hinder the flow of matter-energy in metabolic networks or the delivery of information in wiring diagrams such as those of signaling networks and neural connections in brains.

The impact of evolutionary agents on network structure can be explored using the rationale of the persistence triangle of *economy-flexibility-robustness* (Figures 2 and 12A). A biological system maximizes fitness by complex optimization of a multiplicity of tasks, many of which have been dissected with computational simulations. As previously discussed, the optimization exercises involve the popular evolutionary drivers of cost, evolvability/specialization and reuse, which nicely correspond to goals needed for a Pareto front of network persistence. Streamlining cost impacts the *economy* vertex, which manages the system's matter-energy budget for nodes, links, motifs, communities and even network dynamic behaviors, including the collective internal state of the evolving network.

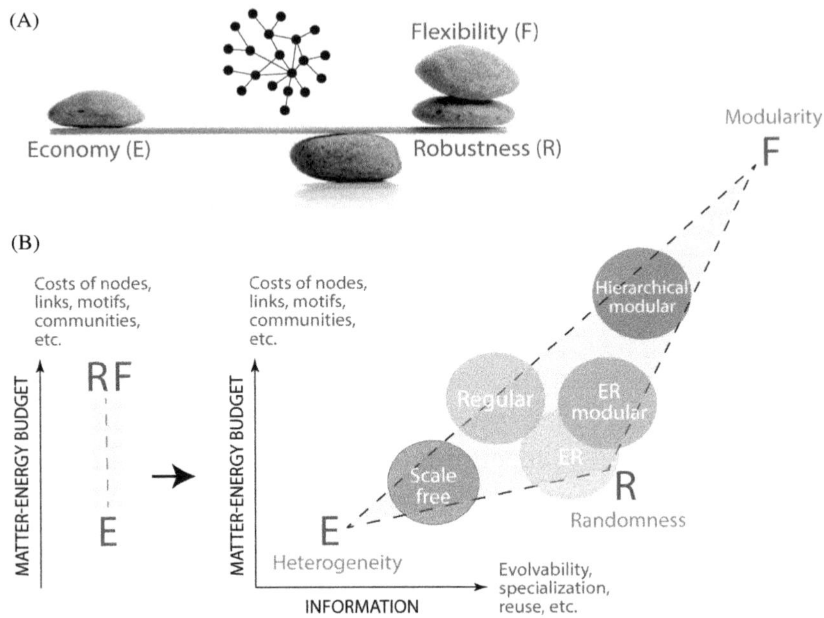

Figure 12. The persistence triangle and how it impacts network structure. (**A**) The diagram shows the existence of a balance between economy (E) and both flexibility (F) and robustness (R) when a network system seeks persistence. (**B**) The networks show different propensities towards E, F and R and find optimal trade-off solutions given a matter-energy budget reflecting the costs of nodes, links, motifs, communities, etc. and information needed to unfold evolvability, specialization, reuse, novelty, etc. Possible patterns of trade-off solutions for reference networks are mapped onto a polytope (shaded in yellow) that when fully-fledged becomes a triangle of persistence.

Tailoring evolvability/specialization impacts the *flexibility* vertex, which takes advantage of information-processing mechanisms needed for the system to adapt to changing environments. Here, the networks rewire to efficiently respond to a multiplicity of environmental changes to gain functional specializations by differentiation, exercises that are most likely frustrated. Finally, reuse (recycling) of nodes, links, motifs, communities and even network dynamic behaviors impact the *robustness* vertex, which channels information to maintain structure and function while responding to damage (attack, failure, suboptimality) and change. Figure 12B unfolds this mapping of evolutionary drivers onto vertices of the persistence triangle. First, a molecular matter-energy budget defines a central transect of the triangle, sorting trade-offs imposed by the matter-energy budget of the networks. Second, the full triangle unfolds once a new transect of information is projected perpendicular to the matter-energy budget. This new transect dissects the contributions of flexibility and robustness. Figure 12B illustrates a Pareto polytope involving the three goals, which embody the possible drivers of cost, evolvability/specialization and reuse and the network archetypes of heterogeneity, modularity and randomness. Note that the polytope is likely tailed and may involve a multidimensional landscape that is driven by more than the three goals of the persistence triangle we here propose. Remarkably, most real networks have been shown to display a balance between integration of information and control over multiple goals under a bow-tie network pattern (Corominas-Murtra *et al.*, 2013). This pattern manifests in the morphospace within a cloud of random graphs. Indeed, the bow-tie pattern of metabolic networks has a large central cycle component that reflects the metabolic advantage of reusing and recycling metabolites (Ma and Zeng, 2003; Corominas-Murtra *et al.*, 2013), which enhances the flexibility and robustness of the metabolic network given a matter-energy budget.

8. Temporal biphasic patterns of diversification are widespread in evolution

The observation that the form of vertebrate embryos converges to a common embryonic design before diversifying and the discovery of evolutionarily conserved clusters of gene expression of *Hox* genes helped define a developmental "hourglass" model (Duboule, 1994; Raff, 1996). This model attempts to explain the basic body plan shared by vertebrates, which is highly conserved and originated at least half a billion years ago (Gya), ~0.55 Gya for vertebrates. The developmental "hourglass" revisited, with a twist, the *"ontogeny recapitulates phylogeny"* concept of Haeckel, which advanced von Baer's links between embryogenesis and taxonomy that were put forth in the 1800s. The morphologically-conserved "phylotypic stage" of embryonic development, the temporal waist of the *evo-devo* hourglass, has been interpreted with a multiplicity of hypotheses and still remains an open question (Irie and Kuratani, 2014). Interpretations include hypotheses of modularity of developmental signals, *hox*-gene collinearity, developmental burden, and developmental constraint, none of which we will here discuss. The hourglass is a biphasic pattern that assumes the age of genes encoding phenotypes of the mid-embryonic waist are older than those occurring earlier or later in embryonic development. Recent molecular and genomic analyses have shown that indeed this may be the case (Kalinka *et al.*, 2010; Domazet-Lošo and Tautz, 2010; Quint *et al.*, 2012). Comparison of genome-wide expression patterns across embryonic development in six species of fruit flies (*Drosophila*) showed that gene expression variations were minimal during the phylotypic stage, suggesting strong selective constraints operating at this stage (Kalinka *et al.*, 2010). Similarly, Domazet-Lošo and Tautz (2010) connected morphologies to the evolutionary age of genes inferred using conservation level classification (phylostratigraphy) in zebrafish (*Danio rerio*), nematodes (*Caenorhabditis elegans*) and fruit flies. In all cases, the repertoire of expressed genes was evolutionarily older during the phylotypic stage. The differentially expressed genes responsible for a phylotypic stage of animal development likely describe modules of gene expression networks necessary to establish the basic rules of the body plan that is common between diversifying animal organisms. The fact that these modules are older than modules necessary to establish older and younger embryonic phases demands an explanatory framework of the temporal hourglass pattern. Tracing the evolution of biological innovations in biology with the tools of network biology and evolutionary genomics can be of help.

The biphasic "hourglass" or "bow-tie" pattern is widely present in biology and man-made systems (Mittenthal *et al.*, 2012; Sabrin and Dovrolis, 2017). For example, these patterns have been observed in metabolism (Ma and Zheng, 2003; Tanaka *et al.*, 2005; Zhao *et al.*, 2006; Caetano-Anollés *et al.*, 2011), gene expression (Kalinka *et al.*, 2010; Domazet-Lošo and Tautz, 2010; Quint *et al.*, 2012), signaling networks (Supper *et al.*, 2009), immunology (Beutler, 2004; Oda and Kitano, 2006), and in vision and cognition (Riesenhuber and Poggio, 1999; Quiroga *et al.*, 2005). We previously discussed biphasic patterns of diversification that exist in evolution of macromolecules (protein and RNA structures), networks (metabolism, genetic code), and development and epigenetics (developmental hourglass, transcription patterns in embryogenesis) (Mittenthal *et al.*, 2012). Biological innovations developing at a particular level of biological organization appear gradually and accumulate with time. Examples of molecular innovations include the protein fold structure of protein structural domains defined at fold superfamily (FSF) level of SCOP (the gold standard of protein classification; Murzin *et al.*, 1995), Gene Ontology (GO) definitions of biological functions (Gene Ontology Consortium 2017), the global structure of functional RNA molecules catalogued by the Rfam database (Kalvari *et al.*, 2017), proteins (r-proteins) associated with the ribosome, the central protein biosynthetic machinery of the cell (Lecompte *et al.*, 2002), and the enzymes of metabolism (Mughal and Caetano-Anollés, 2019). However, if a higher hierarchical level apportions innovations into different groups (e.g. groups of diversified cellular organisms defined at superkingdom level of NCBI taxonomical classification; Federhen, 2012)), those that accumulate before the appearance of the new level of organization will be common to all groups while those appearing later will spread (increasingly sparsely) in the growing series of increasingly complicated Venn diagrams defined by the

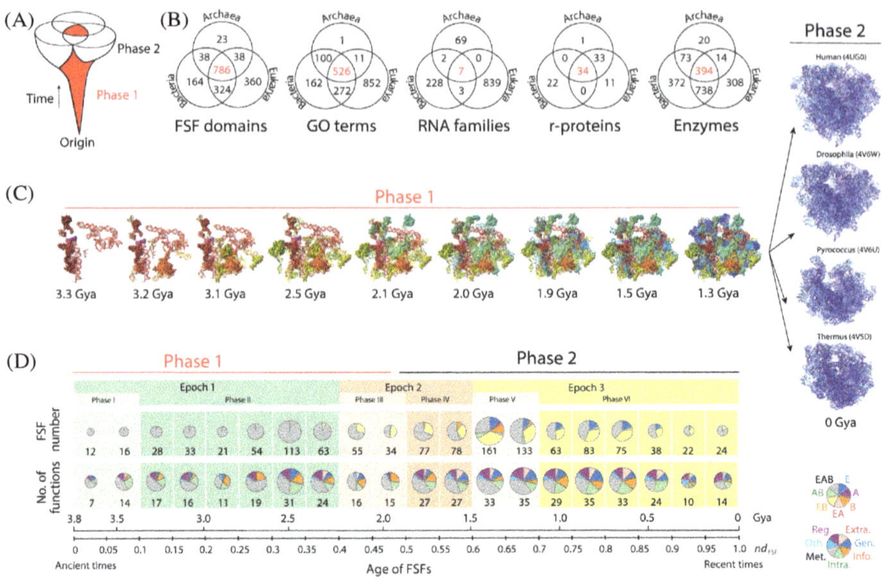

Figure 13. Temporal biphasic patterns revealed by comparative and evolutionary genomic analyses. (**A**) Biphasic patterns can be explained with a biphasic model of unification and evolutionary change. In phase 1, parts of biological sub-systems unify into a universal conserved core illustrated with a red-shaded growing funnel. In phase 2, the growing core diversifies by adding parts specific to the evolving organismal systems. (**B**) Ancient universal cores and derived peripheries support the biphasic model. The Venn diagrams describe censuses of protein structural domains defined at fold superfamily (FSF) level of structural classification of the SCOP taxonomy, Gene Ontology (GO) terms of molecular functions, RNA families defined by the RFAM database, homologies of ribosomal proteins (r-proteins), and metabolic enzymes. The universal repertoires shared by cellular superkingdoms Archaea (A), Bacteria (B) and Eukarya (E) are colored in red. Those colored in black are specific to individual or groups of superkingdoms. (**C**) The history of the ribosome is biphasic. An evolutionary timeline of co-evolving ribosomal RNA (rRNA) and proteins (r-proteins) inferred directly from phylogenomic data shows two evolutionary phases. During phase 1, helical structures of rRNA and r-proteins accumulated to form an increasingly larger universal conserved ribosomal core. During phase 2, the universal ribosomal core diversified alongside with evolving organismal lineages. This phase started ~1.3 billion years ago (Gya) or earlier. Data from Caetano-Anollés (2002), Harish and Caetano-Anollés (2012) and Caetano-Anollés Caetano-Anollés (2015).

new level of organization. This induces a biphasic process (Figure 13A). During a first phase, accumulating innovations are increasingly unified into a module as they unfold in a growing funnel. In a second phase, the rise of a new level of organization provides opportunities for further diversification of those innovations. Comparative genomics makes this unification-diversification biphasic process clear using the examples of molecular innovations we described above (Figure 13B). Venn diagrams of the distribution of protein structural domains, GO molecular functions, functional RNA, r-protein and enzyme repertoires in the three cellular superkingdoms, Archaea, Bacteria and Eukarya, show that a substantial core of the repertoire of innovations is universally present in all superkingdoms (numbers colored in red), counteracting a significant repertoire of innovations present in individual superkingdoms or pairs of them (Figure 13B). As expected, innovations that are known to be ancient are generally universal. For example, universal RNA families include central

Figure 13. (Figure on Facing Page) A molecular clock of folds links structural and geological timescales (Wang *et al.*, 2011a). **(D)** The biphasic patterns of unification and diversification results in an hourglass of proteome evolution revealed by the evolutionary appearance of SCOP FSF domains, their distribution in the organismal world, and the rise of associated biological functions. FSF numbers and biological functions are plotted as a function of their age (nd_{FSF}), with ages binned into 0.05 increments within a relative 0–1 scale. The different evolutionary epochs and phases of the protein world are shaded in the timeline and follow definitions by Wang *et al.* (2007). Pie charts above bins describe either the percent distribution of FSFs that are unique (A, B and E) or shared between two (AB, EB and EA) or all (EAB) superkingdoms, or the distribution of the 50 subcategories that are grouped into seven broad categories of biological functions in SUPERFAMILY, general (Gen), metabolism (Met), information (Inf.), regulation (Reg.), intracellular processes (Intra.), extracellular processes (Extra.), and other (Oth.). The relative area of each pie is proportional to the number of FSFs or functional subcategories in each bin. Data was derived from dataset A184 (Caetano-Anollés *et al.*, 2011).

molecules such as transfer RNA (tRNA), ribosomal RNA (rRNA), and RNase PRNA molecules, all of which are believed to have originated very early in evolution of life.

The general comparative genomic analysis approach illustrated in the Venn diagrams of Figure 13B, however, does not reconstruct history. Any evolutionary conclusions derived from it are simply indirect inferences made from extant data. In contrast, powerful methods of phylogenetic reconstruction analysis can be used to assign relative ages of first appearance to evolutionary innovations that are present in protein and RNA macromolecules (Caetano-Anollés, 2002; Caetano-Anollés and Caetano-Anollés, 2003). A molecular clock can then link chronologies of innovations to the geological record (Wang *et al.*, 2011a). These evolutionarily conserved innovations, which must carry deep phylogenetic history, include structural, topological, and thermodynamic-driven features of RNA molecules and protein domains that are encoded in the sequence of thousands of genomes (recently reviewed in Caetano-Anollés and Caetano-Anollés, 2015). The phylogenomic trees can be used to trace the gradual evolutionary addition (i.e. the accretion) of component parts of central macromolecular complexes of the cell or entire repertoires representing for example the proteome or the full set of metabolic enzymes. Every new branch reflects the addition of a new part to the whole. Figure 13C shows the existence of an evolutionary biphasic pattern in evolution of the ribosome, with a Phase 1 responsible of generating a universal core and Phase 2 diversifying that core in organismal lineages.

The ribosome is a central molecular machine that is universally present in cells and is responsible for the biosynthesis of proteins. It contains a small subunit (SSU) with typically 30–40 r-proteins and one rRNA molecule with ~50 universal helical stems that fold independently into 3 major domains. It also contains a large subunit (LSU) with typically 30–45 r-proteins and 2–3 rRNA with ~100 universal helices that fold into 6 domains (5S rRNA is the seventh). The history of the ribosome revealed the gradual accretion of central structures such as the ribosomal head and ratchet, the central protuberance, and stalks, all of which are involved in ribosomal dynamics (Caetano-Anollés, 2002; Harish and Caetano-Anollés, 2012; Caetano-Anollés and Caetano-Anollés, 2015). Early structures incorporated into the complex were also involved in energetics, decoding, helicase activity and translocation. In contrast, the peptidyl transferase center (PTC) that is responsible for protein biosynthesis appeared late, 2.8–3.1 Gya, together with a fully functional processivity core and a set of inter-subunit bridges and r-proteins that stabilized the interaction of the two subunits (Harish and Caetano-Anollés, 2012; Caetano-Anollés and Caetano-Anollés, 2015). This initial proto-ribosome served as center for coordinated ribonucleoprotein accretion to form a highly processive universal ribosome core during the Great Oxygenation Event (GOE) of our planet ~2.4 Gya. The chronology showed that the universal ribosomal core continued to grow until it was fully formed ~1.3 Gy. After that, a new phase developed in which the structure and organization of the complex diversified in different organismal lineages (Figure 13C). For example, the eukaryotic ribosome added eukaryotic-specific structural appendages that surrounded the universal core, remodeling inter-subunit interfaces and affecting rotational subunit movements (Ben-Shem *et al.*, 2010; Kahtter *et al.*, 2015). Thus, a ribosomal core diversified in organismal lineages by addition of shells, prokaryotic and eukaryotic (Melnikov *et al.*, 2012). The late impact of organismal diversification on the eukaryotic ribosomal core that started ~1.3 Gya, ~1.5 billion years after the rise of aerobic metabolism and the last universal common ancestor (LUCA) of cellular life (Wang *et al.*, 2011a), can be explained by the rise of early decimeter-scale multicellular organisms, 1.56 Gya (Zhu *et al.*, 2016). These large organisms overcame the limitations of microbial life at low Reynolds numbers and the absence of inertia (Purcell, 1977), which probably relieved known reductive evolutionary pressures on molecules imposed by microbial size. In this regard, we note that the length of proteins and structural domains of prokaryotic proteomes evolves reductively while that of eukaryotes does not (Wang *et al.*, 2011b).

The biphasic "hourglass" behavior observed in the phylogenomic studies of the ribosomes becomes even more clear in the phylogenomic study of the proteome and its associated functionome (Figure 13D). Pie charts describing the number of FSF domains and biological functions that appear in each of 20 age bins along the entire timeline of protein domain evolution describe a clear hourglass that peaks ~2.5 Gya and ~1.4 Gya and shows a waist ~2 Gya. Remarkably, the vast majority of FSFs appearing before the waist are universal, making up a Phase 1 period in which the proteome generates a universal core of structures. This initial phase of proteome evolution resembles the initial phase of ribosomal evolution, in which individual RNA substructures and protein domains accumulate to form a universal ribosomal core. In contrast, most FSFs appearing after the waist during Phase 2 are specific to one or more cellular superkingdoms. Along the timeline, domains were first lost in primordial archaeal lineages (EB FSFs) and then in eukaryal (AB) and bacterial (EA) lineages. In turn, the rather late gain of Bacteria-specific domains (B), and then Eukarya-specific (E) and Archaea-specific (A) domains, signal the emergence of diversified superkingdoms. Further details can be found in Caetano-Anollés *et al.* (2011). Thus, parts of a system first unify and then diversify as they grow.

A phylogenomic analysis of metabolic enzymes also showed a temporal biphasic pattern of growth (Figure 14). Metabolic networks are organized hierarchically, with enzymes grouping into subnetworks, and subnetworks into mesonetworks. Subnetworks are sets of biochemical reactions and molecular interactions that make up the metabolic system, grouping enzymes responsible for functionally related

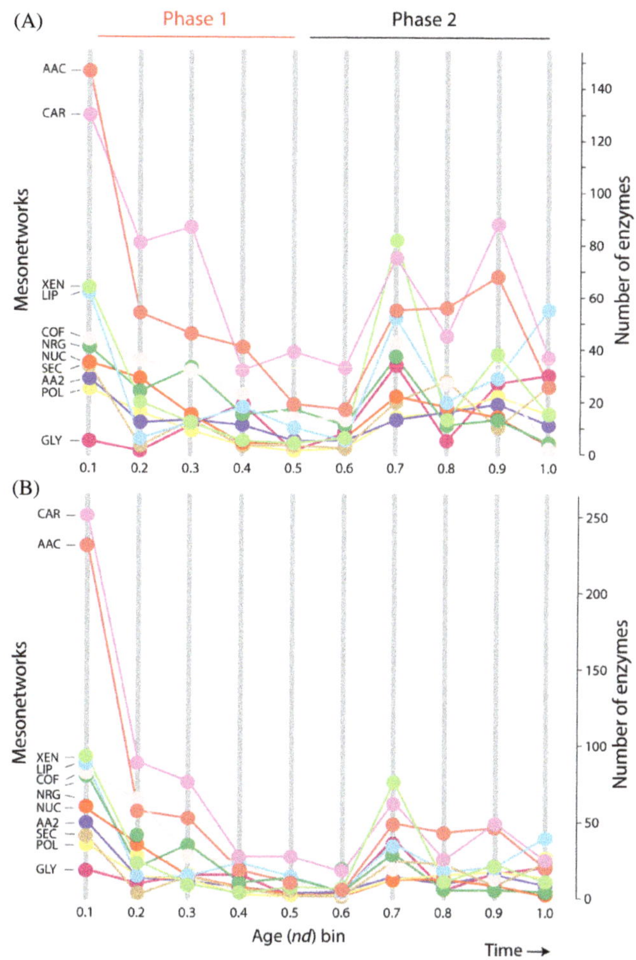

Figure 14. Run charts of enzymes in mesonetworks of metabolism. The line charts describe the evolutionary appearance of enzymes in the eleven mesonetworks of metabolism at different time intervals, i.e. at bins of ages (*nd*) stratified in 0.1 *nd* intervals, from *nd* = 0 (origin of proteins) to *nd* = 1 (the present). The age of enzymes was defined as the age of the second oldest structural domain (**A**) or as the age of the oldest domain (**B**) that is present in the protein, when domains are studied at fold family level of SCOP classification (Mughal and Caetano-Anollés, 2019). A biphasic pattern of enzyme growth is clearly evident.

metabolic pathways. Mesonetworks are upper level categorizations of subnetworks holding similar functional capacities. The metabolic *Molecular Ancestry Networks* (MANET) database traces the age of metabolic enzymes in the subnetworks and mesonetworks of metabolism with two algorithms (MANET, 2006; Kim *et al.*, 2006; Mughal and Caetano-Anollés, 2019). One considers that the age of a multidomain enzyme is that of the second oldest domain while the other considers it is the age of the oldest domain. Regardless of algorithmic implementation and assumptions, a clear biphasic pattern of growth is visible. Thus, the world of enzymes very much behaves as the world of protein domains.

Recall that the interrelated subsystems of a biological system are organized hierarchically into levels of organization (Simon, 1962; Salthe, 1985, 2012). If hourglasses are pervasive, then it is expected that the nesting relationship of levels of organization should also exhibit temporal hourglass patterns. The SCOP classification (Murzin *et al.*, 1995) groups protein structural domains into families (FFs), superfamilies (FSFs), folds (Fs) and classes (C) in a hierarchy of decreasing granularity (Figure 15, inset). Domains with pairwise amino acid sequence identities of more than 30% (exceptionally with lower sequence identities but conserved functions and structures) are unified into FFs, which by definition are phylogenetically closely related at the sequence level. Those FFs that share similar structural and functional features suggesting a common origin are further unified into FSFs. FSFs sharing secondary structures that are similarly arranged and connected in 3-dimensional space are further unified into folds. Finally, those that share similar overall designs are further grouped into protein classes (*a* through *l*; only 7 are true classes). Indeed, Figure 15 describes an evolutionary mapping of SCOP domains defined at two levels of structural abstraction in two-dimensional plots, which clearly unfolds a temporal hourglass (data from Caetano-Anollés *et al.*, 2011). The plots compare rooted trees of Fs and FSFs and the ages of leaves derived from them, with horizontal

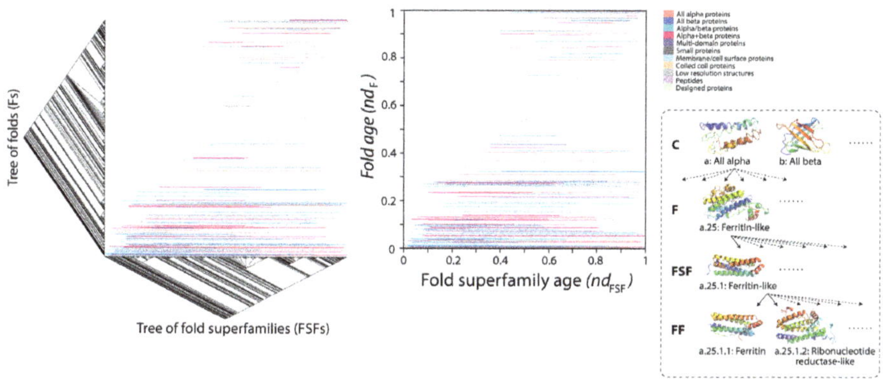

Figure 15. Temporal hourglasses describing the hierarchical complexity of the protein world. SCOP is a taxonomical classification of proteins that unifies fold families (FF) into fold superfamilies (FSF), FSFs into folds (F), and Fs into classes (C). The inset box illustrates the SCOP hierarchy with an example structure of the "all alpha" class, the ferritin-like fold a.25 (core: four helices; bundle, closed, left-handled twist; one crossover connection), ferritin-like superfamily a.25.1 (contains bimetal-ion center in the middle of the bundle), and ferritin family c.25.1.1 (10 protein domains; e.g. apoferritin, bacterioferritin). The figure shows two-dimensional plots that map two contiguous levels of structural organization, FSFs and Fs, in evolutionary timelines defined by the order of taxa in ladderized phylogenomic trees (left) or by the age of individual domain structures (right). Dots represent individual domain structures and horizontal lines linking dots define the evolutionary spread of FSFs within Fs. They are colored according to protein class. These lines unfold a "z-diagram" that resembles an hourglass, which is similar in both plots, that indicates Fs are composed of FSF of different ages that are not clustered in the phylogenomic trees.

lines colored according to protein class and defining the evolutionary spread of FSFs within Fs. They reveal that Fs are heterogeneous entities often composed of FSFs of different age that are undergoing divergent evolution and harbor multiple functions. Remarkably, the plots reveal an hourglass in which Fs of intermediate age (nd_F = 0.55–0.8) are overwhelmingly composed of single FSFs. These domain architectures probably forfeit hierarchical diversification to gain in specificity, as they are usually associated with single functional subcategories. While the evolutionary relatedness of FSFs within an F has been proven for only a handful of cases, the hourglass pattern of phylogenetic diversification shows hierarchical organization is constrained half-way in evolution of the protein world, at a time when organismal lineages were diversifying. Similar hourglasses were observed in plots that compare the ages of FSFs and FFs (data not shown), suggesting that evolutionary constraints express at different levels of granularity.

9. A linkage-based biphasic model of module creation

Mittenthal *et al.* (2012) elaborated a biphasic (bow-tie) theory of module emergence that explains temporal biphasic patterns of unification and diversification in biological systems. The theory centers on *linkage*, the extent of interaction between parts. It predicts the emergence of nested hierarchies of modules, in which diversifying parts of a system converge under optimization or selection into tightly linked groups, which subsequently diversify. In a first phase of unification, parts are weakly linked and associate variously. As parts change by processes such as mutation, reassortment and/or recruitment (BOX 1), they become more numerous and diverse. The chance of establishing interactions also increases when the number of parts grows. In particular, links that help perform tasks that are useful to the system persist because they increase fitness or functional capabilities. However, emerging interactions are costly and limited by the space of possible interactions; parts therefore compete with each other to meet optimization or selection criteria and their associated interactions constrain the structure and organization of the system. This forces the system to undergo *competitive optimization*, the selection or optimization of both parts that cooperate more strongly and interactions that are more stabilizing for the system. The competitive process decreases diversity and increases the chances of tighter linkages. The modules that emerge from the growing interactions become dynamically resistant to fluctuations and change. In a second phase, the modules diversify. At this stage, linkage is tighter within modules than between modules. Modules are therefore free to diversify by mutation, reassortment and/or recruitment, often becoming new building blocks for a new generative cycle of higher-level organization.

BOX 1: Genetic processes driving change

Mutation: Altering of nucleotide sequence by substitutions, insertions and deletions.

Reassortment: Mixing of genetic material into new combinations including mutational rearrangements through chromosomal crossovers, translocations and inversions, deletions of gene segments causing for example domain rearrangements, de novo gene creation, and a host of horizontal gene transfer mechanisms that place genetic material in different genetic contexts.

Recruitment: Accretion of genetic material by co-option or subsumption into genetic modules, such as placing a duplicated gene under a different regulatory context.

The unification of increasingly diversified parts by integration and the subsequent diversification of unified parts was illustrated by Mittenthal *et al.* (2012) with a system of Lego® blocks (Figure 16A). A Lego® block can be regarded as an automaton, a self-operating control mechanism (a machine) that responds to predetermined instructions. It can be a digital logic machine, a more advanced finite-state machine, a stack-driven pushdown automaton, or even a Turing or Oracle machine. Blocks are also intuitive toys, parts of a whole. Predetermined instructions are given in the Lego® model through stud-hole interactions of neighboring blocks, interactions of blocks with walls, frequencies of change of the shape of blocks and frequency of replacement of blocks. Note that blocks interact only with neighbors and attempt to maximally occupy a space defined by their functionality. This is

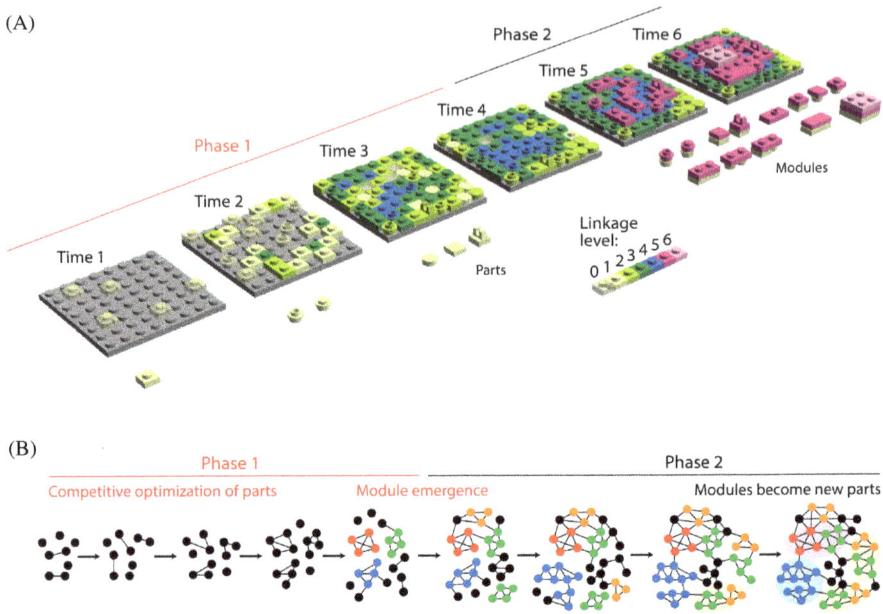

Figure 16. Biphasic theory of module generation. (**A**) The simplified linkage model of evolution of a Lego® block system used by Mittenthal *et al.* (2012) to illustrate the emergence of modules. (**B**) A generic biphasic model of module creation illustrates the emergence of network structure in evolution. Nodes and links of the network are parts of a growing system of entities and interactions. The larger number of links, the more cohesive and stable is the structure of a subnetwork. The rise of hierarchical modularity during phase 1 results in small highly connected subnetworks. These subnetworks become modules, which in phase 2 coalesce by combination into higher modules of network structure (highlighted with shades of yellow and blue).

usually accomplished by maximizing information through interactions. In this toy model, blocks are parts that have the potential to build hierarchical levels of organization of the Lego® system, i.e. multidimensional spaces in which parts diffuse as they change.

When constructing Lego® block structure, construction occurs in layers materializing in two dimensions (a gray-colored base plate and 3 additional layers in Figure 16). Each layer reflects a level of hierarchical organization that is similar to levels of structural abstraction in proteins: amino acid sequences, secondary structures, tertiary structure of domains, and quaternary structures. As the Lego® system evolves, linkage levels (numbers of interactions per part) increase (from 0 to 6, which are color coded) as well as the number and repertoire of parts that accumulate in each layer. While the increasing interaction of neighboring blocks opens opportunities for accumulation of novelties through innovation and advancement to upper levels of organization, increases in occupancy of levels fosters competition during block replacement. This early competitive optimization phase often reduces the diversity of parts and the rate of innovation. Despite the shortcoming, early interactions enable new extents of connectivity among blocks within and between hierarchical levels and overall increase in the information of the system. The process generates modules, groups of blocks that become independent entities with the property of becoming new parts for a next level of organization.

Figure 16A shows a clear hourglass pattern of "block" novelties appearing for the first time along the timeline. The number of new Lego® parts reaches a peak at Time 3, then decreases to zero at Time 4, and finally explodes in a combination of modules once the third hierarchical level is established at Time 5. This combinatorial explosion facilitates the establishment of a fourth hierarchical level, which is colored light pink in the figure. Note how early discovered block novelties are the most abundant and linked, how lower hierarchical levels tend to be more densely occupied (i.e. more granular), and how certain novelties (blocks with protrusions or without studs or holes) can curtail formation of modules and interactions with top hierarchies. These behaviors of the toy model are all plausible mechanisms operating in biology.

The basic premise of the linkage model is that a new hierarchical level can only be added if linkage has increased to levels that originate modules (networks of block parts enriched in blue or dark pink), and that once the level is added, both linkage and competitive optimization actively act upon it to form additional inter-linked hourglasses. These hourglass patterns stem from percolation in the network of interacting parts due to: (i) increases of linkage, (ii) limits of space accessibility for parts (competitive optimization), and (iii) the rise of modules and new levels of organization, which provides further opportunities for system's diversification.

10. Phylogenomic analyses of biological networks test the linkage hypothesis

Mittenthal *et al.* (2012) suggested a way forward: *"To test a linkage hypothesis it is necessary to formalize the concept of linkage. In a network where nodes represent parts, the extent and pattern of connectivity among nodes provides indices of linkage."* Indeed, the generation of modules arising from the biphasic model can be described with networks in which nodes are parts of the system and links define the extent of interaction between those parts, i.e. linkage (Figure 16B). The unification and diversification forces driving the biphasic theory of module emergence take advantage of the stability that accretion of parts and their interactions provide to a growing biological system. For example, adding inter- and intramolecular interactions increases the cohesiveness of molecular machinery in a highly dynamic and crowded molecular environment (as revealed with single-molecule approaches; Monachino *et al.*, 2017). Similarly, the addition of nodes and links enhance the structure of a network by advancing the many possible evolutionary agents we previously described, many of which increase network stability. We already discussed how removal of nodes affects the integrity of random and scale-free networks (Figure 5). A similar effect on network integrity results from removal of links. The larger the number of links, the more cohesive and stable is the structure of an evolving network.

The biphasic model explicitly unfolds in the structure of networks and can be elaborated within a network biology perspective. In the first phase, the accretion of nodes and links is highly dynamic (Figure 16B). However, while nodes and links are added to the network only those parts and interactions that benefit the structure and stability of the system are dynamically stabilized by competitive optimization. Given enough time, the network becomes structured with the rise of highly connected and stable subnetworks. These "canalized" subnetworks become modules. In the second phase, the emergent modules of the network diversify and coalesce by combination into higher level modules of network structure. The process is cyclic and gives rise to the fractal-like pattern of connectivity that

Table 1. Phylogenomic-based studies supporting the evolutionary rise of hierarchical modularity in biological networks.

Level	Process	Network	Timescale*	Ref. **
Molecular behavior	Conformational change of protein loops	Networks describing protein dynamics	ns	1
Molecular behavior	Metabolite rewiring	Metabolomic networks	s to hr	2
Cellular behavior	Changes in gene expression and metabolic pathway activities	Tripartite metabolic networks	min to hr	3
Molecular evolution	Evolution and recruitment of enzyme activities	Metabolic networks	Gy	4
Molecular evolution	Emergence of loop and domain make up	Elementary functionome network	Gy	5
Molecular evolution	Evolution of protein organization	Protein domain organization network	Gy	6

**Timescale:* Time allowed for a process or sequence of events to unfold.

***References:* 1, Mughal and Caetano-Anollés, 2017; Mughal *et al.*, ms. in preparation.; 2, Aziz *et al.*, 2012; 3, Koç *et al.*, 2018; 4, Mughal and Caetano-Anollés, 2019; 5, Aziz *et al.*, 2016; 6, Aziz and Caetano-Anollés, 2013; Aziz and Caetano-Anollés, 2020.

is typical of hierarchical modularity (Ravasz, 2009). However, we note that unlike Ravasz's generating model, the modules at different levels does not need to be necessarily identical nor the whole assemblage strictly fractal.

To test the linkage theory and study how networks become structured in evolution we evaluated the hierarchy and modularity of networks at different timescales (Table 1). We revealed temporal biphasic patterns (hourglasses) describing restrictions to variation in the behavior, development or evolution of a system:

(1) *Behavior:* Behavioral changes can be defined very broadly as internally coordinated responses (actions or inactions) of a system triggered by internal or external causes. These changes often occur once or repeatedly in the life cycle of an organism, can be adaptive, and are restricted by evolutionary constraints, other systemic physiological responses and levels of organization. For example, metabolic networks quickly rewire in the presence of stress, changing the repertoire of metabolites that is available to the cell (Aziz *et al.*, 2012). Metabolomic data retrieved using gas chromatography coupled to mass spectrometry prior and after perturbation revealed that while the dynamic rewiring of bacterial metabolomic networks was largely random, the networks of metabolites remained tightly hardwired to the scale-free networks of metabolic reactions and exhibited a biphasic behavior. Similarly, cold stress triggered coordinated responses of gene expression, metabolites and metabolic pathways in the model plant system *Arabidopsis* (Koç *et al.*, 2018). Tripartite networks of transcripts, metabolites and pathways that describe how gene expression controls plant metabolism revealed that the plant first mobilized energy from glycolysis and ethanol degradation to enhance the activity of the tricarboxylic acid (TCA) cycle via acetyl-coA. This then triggered a host of other metabolic processes (e.g. energy,

signal, carbon and redox metabolisms), including a significant change in the Rubisco shunt, which likely balances carbohydrate and oil production to make efficient use of the carbon pool under cold stress. While these tripartite networks lacked scale-free connectivity their rewiring resembled that of metabolomic networks and gained modularity. For example, a subnetwork of carbohydrate metabolites revealed the formation of tightly knit metabolic clusters with increases in cold acclimation (Figure 17).

(2) *Development:* Developmental changes describe for example how cells and organisms grow, differentiate and become structured as they develop. Developmental hourglasses bundle diverse developmental trajectories, which are constrained by both physiological and evolutionary restrictions imposed by the developing biological system. Developmental studies of model animal systems such as fruit flies, nematodes and zebrafish allow to track changes in experimental time. For example, the temporal waist of the developmental hourglass of embryonic development we previously discussed is controlled by the expression of genes that are older and more evolutionarily conserved in all model systems (Kalinka *et al.*, 2010; Domazet-Lošo and Tautz, 2010; Quint *et al.*, 2012).

(3) *Evolution:* Evolutionary changes are protracted and of lasting impact at all levels of organization. They are constrained by canalization (entrenchment), the invariance of phenotype in light of genotypic and environmental change. Without a phylogenetic retrodictive framework, it is impractical to observe network structure evolving in a biological system. With exceptions, such as cases of viral evolution, change occurs at timescales that are exceedingly longer than that of the experimenter. Consequently, much of the effort invested in the study of network evolution has been devoted to simulations (e.g. Clune *et al.*, 2013; Sabrin and

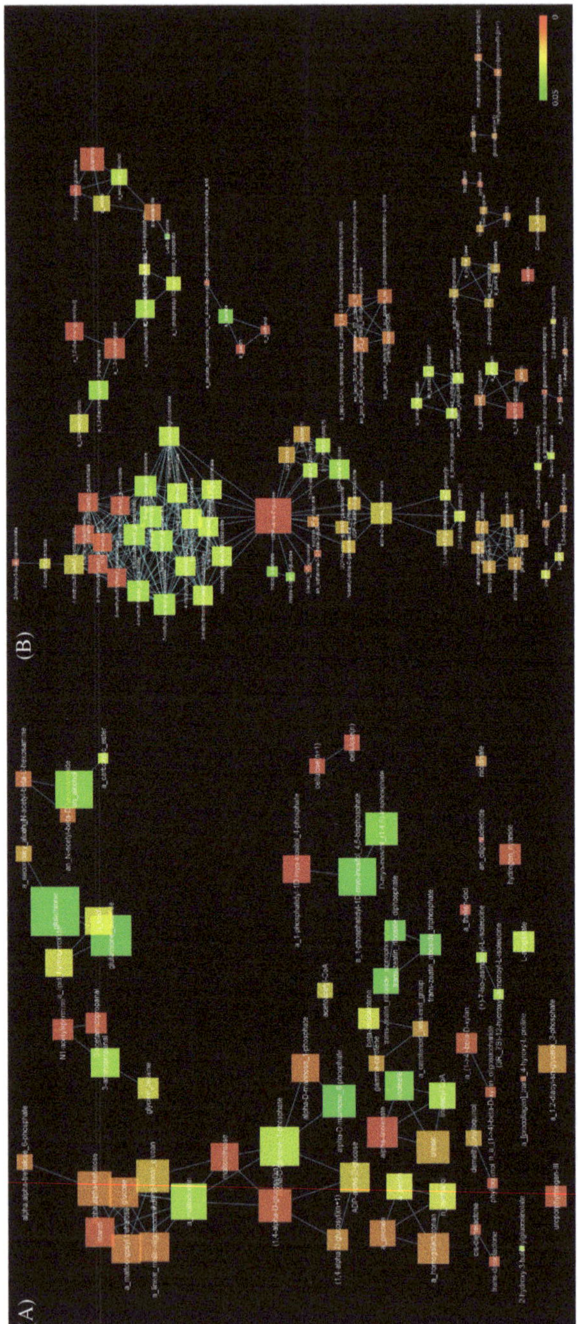

Figure 17. Subgraph of tripartite networks showing significantly changed metabolites (P ≤ 0.05) in 3 h (A) and 24 h (B) cold acclimated *Arabidopsis* plants. Nodes represent metabolites. Their sizes were scaled to number of neighbors. Red color describes most significantly associated metabolites in the tripartite network. Note how increased cold acclimation results in the rise of modular structure in the subnetworks. Data from Koç *et al.* (2018).

Dovrolis, 2017; Caetano-Anollés *et al.*, 2019). However, the approach of simulating network evolution rarely unfolds hierarchy or modularity on its own and the emergence of network structure in a system still remains contentious (Flack *et al.*, 2013). One alternative is to use the tools of phylogenomic analysis to study the deep evolution of networks important for molecular biology. Using the phylogenomic trees of structural domains we previously described, we find for example temporal biphasic patterns describing evolutionary hourglasses of the diversification of metabolism (Mughal and Caetano-Anollés, 2019), elementary functionomes (Aziz *et al.*, 2016), and domain organization in proteins (Aziz *et al.* 2013, 2016; Aziz and Caetano-Anollés, 2020). In all cases, networks become more hierarchical and modular with evolutionary time. Networks that are highly dynamic in nature also exhibit patterns of evolution in their structural make up. For example, the speed with which protein molecules fold as well as their flexibility constitute beneficial traits that are evolutionarily conserved (Debès *et al.*, 2013; Marsh and Teichmann, 2014). Similarly, protein complexes assemble through ordered pathways with strong conservation tendencies (Ahnert *et al.*, 2015). We constructed networks that captured the conformational dynamic behavior of proteins at nanosecond (ns) timescales with a cross-correlation matrix of motions of residues derived from molecular dynamics (MD) simulations (Caetano-Anollés *et al.*, 2019; Mughal *et al.*, ms. in preparation). Remarkably, a deep historical trend was evident in the ns-level physical movement of the loop atoms of aminoacyl-tRNA synthetase (aaRS) enzymes when the age of functional loop-containing structural domains was mapped onto a morphospace describing the heterogeneity, randomness and modularity of the dynamic networks. As younger networks transition into older networks, modularity and randomness increased, while network heterogeneity (scale-freeness) generated a noisy "bow-tie" pattern.

Thus, a phylogenomic analyses of the 3-dimensional (3D) structure of millions of protein domains encoded in the genomes of thousands of cellular organisms and viruses provide powerful evidence in support of the rise of both hierarchy and modularity in the structure of biological networks. In doing so, the coupling of evolutionary and network biology fails to falsify the linkage hypothesis that explains temporal biphasic patterns unfolding at behavioral, developmental or evolutionary levels. Since the rise of hierarchical modularity occurs at different timescales, from networks with behaviors that are highly dynamic and stochastic (ns-to-hrs) to networks that are biologically entrenched by deep evolution (Gy), we here contend that the theory of linkage has high explanatory power.

11. A "principle of granularity" in the hierarchical structure of networks

One central assumption embraced by network biology is Simon's near-decomposability of systems. Parts are considered elements of a whole that at any time can be represented with a set of nodes. This set must be appropriately defined. For example, when using Hidden Markov Models (HMMs) of structural recognition to classify the 3D structure of protein domains that are encoded in genomic sequence we assume that these powerful machine learning models do not fail to appropriately catalog the protein world into a finite set of fold structures. However, regions of protein space can be continuous and under certain circumstances it may be difficult to appropriately dissect structural variants (Harrison *et al.*, 2002). Also, nodes are assumed to represent more or less independent entities that are only constrained by an explicit set of links. This set must exhaustively portray possible interactions occurring at any time in the system. The problem however is that "interdependence" forms the basis of living systems as they diversify. This "*unitas multiplex*" manifests at the level of the individual or the system that subsumes that individual but materializes at all levels of organization, both integrating and separating parts in the process. It would be possible to quantify the information in the system beyond the information that is present in its parts, so as to measure how much the state of each part affects the state of other parts (Tononi, 2008). This "qualia" space of integrated information *"above and beyond the information generated by its parts"* could measure a system's non-decomposability, which from a neurobiology perspective has been associated with consciousness. One interesting aspect of the theory of *"consciousness as integrated information"* is that integration at neurobiological level acts very much as entrenched information at evolutionary level. In other words, the hierarchical modular make up generated at different timescales and levels of organization defines an integration space that goes beyond the simple structure of the network. Timescales of integration can range from the ns-level dynamics of molecules to the Gy-entrenchment of metabolic pathways. Levels of organization range from stress-induced metabolic rewiring to for example experience-induced rewiring of neuronal networks.

One way to assess integration space is to bind networks with external information such as physical or functional constraints and ask how hierarchical modularity unfolds under those conditions. One interesting line of exploration that highlights integration space is the study of Rentian scaling of networks (Bassett *et al.*, 2010; Ho and Nvlakha, 2018). In the 1960s, IBM scientist E.F. Rent discovered a peculiar scaling relationship between the number of logic gates (internal components acting as network nodes) in a logical block of a computer circuit (a piece of circuit resembling a network module) and the number of circuit connections between circuit blocks (Landman and Russo, 1971). This empirical relationship followed a power law with an exponent that generally ranged $0.5 < p < 0.8$, the Rent's exponent. Circuits with larger logical capacity have higher exponents. Rent's law can be formalized in terms of networks. Given a network with k modules, one can plot the number of connected nodes that each module holds (m_i) and the number of external links that connect modules (e_i), for each module, in a log-log scale and find a linear relationship for all *i*:

$$e = cm^p \tag{11}$$

in which *c* represents an integration constant proportional to the average number of links between modules. The calculation is straightforward as long as modules do not overlap. If they do, the calculation can still proceed given some algorithmic decomposition of modules and external links (e.g. Ho and Nvlakha, 2018). Regression analysis verifies compliance with Rent's law via R^2 coefficients of determination and *p*-values. When the *p* exponent approaches 1, the links between modules become random and exhibit cliquish complexity. When it approaches 0, modules become sparsely linked by simple designs. Larger *p* exponents imply larger topological connection complexity of links between modules. Figure 18 shows Rentian scaling relationships for a very large scale integrated (VLSI) circuit and a number of biological networks, including neural networks. Neurons in the brain form a complex pattern of connections. These neural networks make up the most complex biological machinery that is known. They show high levels of hierarchical modularity and Rentian scaling (Figure 18B) when topology and physical partitioning of grey and white matter volumes are applied to the human brain (Bassett *et al.*, 2010; Meunier *et al.*, 2010). Remarkably, the scaling compares to that of computer circuits (Figure 18B) showing that connections in brains and computers have similar allometric scaling properties. These properties appear general, highly conserved across species, and dependent on physical constraints imposed by the physical environment. For example, analyses of distribution networks such as mycelial fungi and rodent brain vasculature networks (Figures 18E and 18F) reveal structures that respond to correlates of material costs, efficiency and robustness, which reflect differences in function (Papadopulos

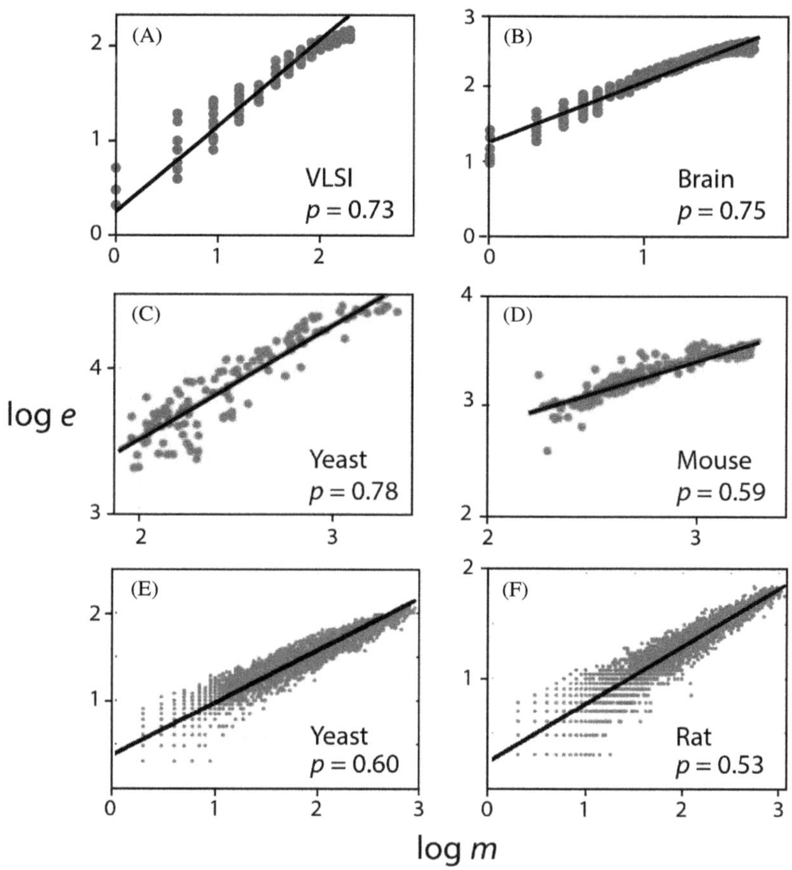

Figure 18. Rentian scaling of networks. Physical scaling relationships are described in plots of the log of the number of links connecting modules (e) as a function of the log of the number of nodes within modules (m). Rentian scaling is described for a very large scale integrated (VLSI) circuit (**A**), the cortical anatomical network of the brain analyzed with MRI in 259 human volunteers (**B**), protein interaction networks of yeast and mouse (**C and D**), and distribution networks of yeast mycelia and rat vasculature (**E and F**). Rent's exponents (p) ranged 0.53 to 0.78. Coefficient of determinations (R^2) ranged from 0.8 to 0.92. Data from Bassett *et al.* (2010), How and Navlakha (2018), and Papadopoulos *et al.* (2018).

et al., 2018). In all cases, a physical constraint induces modules, which then unfold fractal-like embeddings in networks that result from both hierarchy and the community structure of modules. However, Rent's power law describes a scaling of inter-module relationships that predicts it is a consequence of statistically homogeneous wiring of modules (Christie and Stroobandt, 2000).

Another way to assess integration in complex networks is to classify nodes into universal roles according to patterns of intra- and inter-module connections, with network modules identified through simulated annealing (Guimerà and Amaral, 2005). This method produces a "cartographic" depiction that assumes nodes with the same role should have similar topological properties. The role of a node is determined with two measures, a within-module degree z_i and a "participation" coefficient P_i of the node. z_i measures how well connected is a node to other nodes of the module. In turn, P_i measures how well distributed are connecting links among modules. When the coefficient is close to 1, links are uniformly distributed among all the modules. When it is close to 0 then all its links are within its own module. Using these two parameters, nodes can be divided into 7 different roles within a z–P_i parameter space. At low z_i measures ($z_i < 2.5$), these non-hub nodes are considered "ultra-peripheral" ($P_i \leq 0.05$), "peripheral" ($0.5 < P_i \leq 0.62$), "connectors" ($0.62 < P_i \leq 0.80$), and "kinless" ($P_i > 0.80$) according to increasing P_i values, ranging from nodes with most links within their module to nodes with links homogeneously distributed among modules. At high z_i measures ($z_i \geq 2.5$), hub nodes are considered "provincial" ($P_i \leq 0.30$), "connector" ($0.3 < P_i \leq 0.75$) and "kinless" ($P_i > 0.75$) with increasing P_i. The approach was applied to metabolic networks that had nodes describing metabolites and links describing enzymatic reactions (Guimerà and Amaral, 2005). Most modules contained mostly metabolites from one mesonetwork of KEGG. However, a minority of modules (two in *Escherichia coli*) could not be trivially associated with a single mesonetwork. Instead, their metabolites corresponded to a collective associated with carbohydrate and amino acid metabolisms. Thus, modules are associated with metabolic functions, though sometimes loosely. Most metabolites (76–86%) were non-hub ultra-peripheral nodes with no inter-module connections, followed in number by non-hub peripheral nodes and provincial hubs. Remarkably, only a minority of non-hub connector nodes that take part in a small but fundamental set of reactions was found to be evolutionarily conserved, i.e. present in a majority of metabolic networks from different species. This suggests a central evolutionary role for nodes that have many links to other modules.

A third way to assess integration is to define modules according to function and then study the complexity of connections between them. In other words, bind networks with functional constraints and study module inter-connectivity. This approach makes sense, especially because both Rentian scaling and cartographic approaches reveal links between modular structure and function. For metabolism, the wiring diagrams of biochemical reactions and molecular interactions that make up the metabolic system have been carefully annotated by the KEGG database with information stemming from more than 100 years of biochemical research (Kanehisa *et al.*, 2017). These diagrams group enzymes responsible for functionally related metabolic pathways into reference network maps (subnetworks), which are further grouped according to similar functional capacities (mesonetworks). The subnetworks and mesonetworks are lower and upper level categorizations of the hierarchical system, respectively. This system does not describe metabolites. Instead, it portrays the metabolic network as a collection of enzymatic catalysts that are responsible for energy interconversion and the biochemical transformation of small molecules. Figure 19 describes mesonetworks and subnetwork categorizations in KEGG and illustrates the metabolic hierarchy with an example, the amino

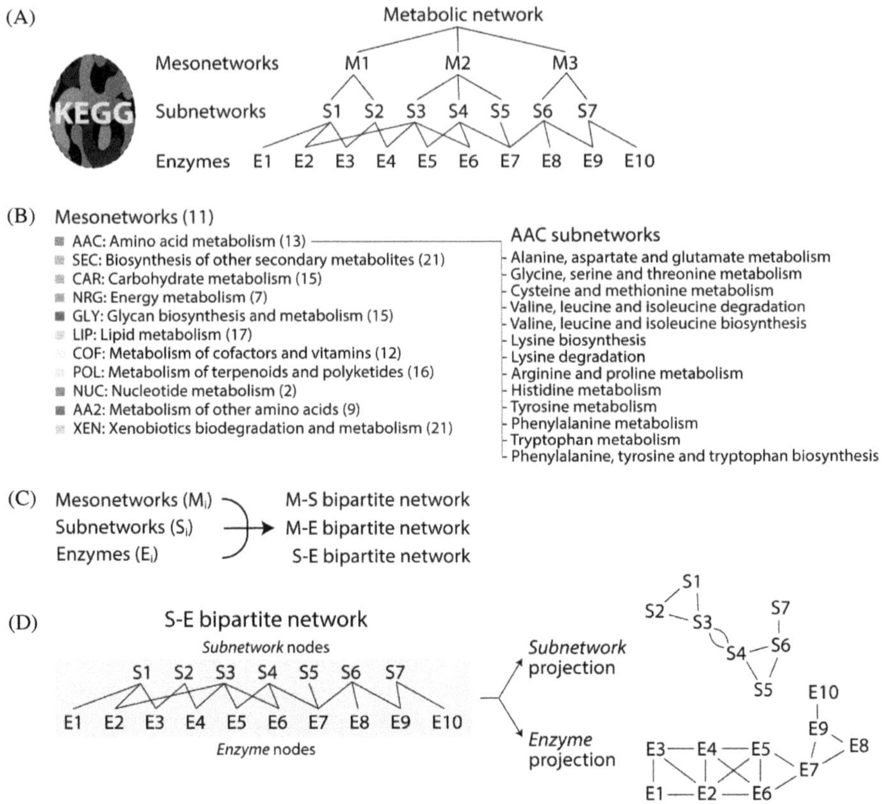

Figure 19. Decomposing the hierarchical structure of metabolic networks. (**A**) The Kyoto Encyclopedia of Genes and Genomes (KEGG) provides a global, functional and hierarchical view of metabolism. The enzymatic activities (E) of metabolism are dissected into a hierarchical system of subnetworks (S) and mesonetworks (M), which act as modules of metabolic activity. (**B**) The KEGG database unifies 148 subnetworks with associated enzymes into 11 mesonetworks. The 13 subnetworks of the AAC mesonetwork are shown. (**C**) The relationship between any two of these levels of metabolic organization can be studied with bipartite networks. The three levels of KEGG allows from the construction of three types of bipartite networks. (**D**) Each bipartite network can be decomposed into two one-mode projections. For example, the subnetwork-enzyme bipartite network can be dissected into a projection describing how enzymes link subnetworks to each other and another describing how subnetworks link enzymes to each other.

acid metabolism (AAC) mesonetwork and its 13 subnetwork components.

To study the evolutionary structuring of a system and untangle module integration between levels of metabolic functional organization, the hierarchy of functional modules can be dissected with undirected bipartite networks that link mesonetworks and subnetworks to enzymes (Mughal and Caetano-Anollés, 2019) (Figure 19C). Bipartite graph theoretic representations can be decomposed into dual one-mode projections using mathematical properties of finite graphs (Chung *et al.*, 1983). One-mode network projections of these bipartite networks are ideally suited to study how individual levels of network organization contribute to overall network structure. Three possible bipartite networks arise from the top-down hierarchy of mesonetworks, subnetworks and enzymes (Figure 19C). As expected for a hierarchical system, networks describing the relationship of neighboring hierarchical levels are linked more densely at lower levels of the hierarchy (e.g. enzyme-subnetwork) than at higher levels (e.g. subnetwork-mesonetwork). Non-neighbor relationships also result in density increase. However, the value of bipartite networks is that their projections describe how neighboring levels interact with each other from two vantage points, a top-down view and a bottom-up view. The top-down view entails a process of commanding lower authorities. In contrast, the bottom-up view entails a process of subordination to higher authorities. For example, the subnetwork-enzyme bipartite network can be dissected into a projection describing how enzymes link subnetworks to each other (a top-down subnetwork projection) and another describing how subnetworks link enzymes to each other (a bottom-up enzyme projection)(Figure 19D). These two opposing processes are informative of the fractal-like integration space of modules that determines the creation of levels of organization. For example, Figure 20 shows a modularity matrix of a bipartite network projection of how enzymes

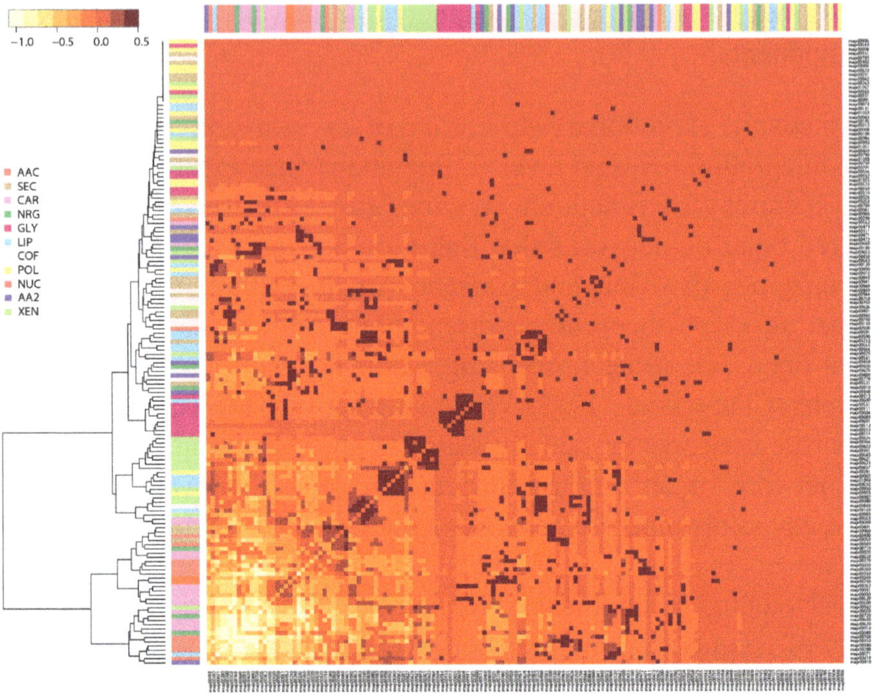

Figure 20. A heatmap describing the modularity matrix and dendrogram of hierarchical clustering of the subnetwork one-mode projection of the subnetwork-enzyme bipartite network of metabolism. In order to identify the fractal-like arrangement of hierarchically embedded modules, a modularity adjacency matrix was created by computing node-to-node dissimilarities (squared Euclidean) distances clustered with the Ward's algorithm. The heatmap was generated using the *Fast-Greedy-Community* (*FGC*) detection algorithm and scaled to range –1 to 1 (see scale). The fractal-like arrangement of modules onto modules suggest disassortative mixing across modules, i.e. nodes tend to connect to dissimilar nodes. Leaves of the dendrogram are metabolic subnetworks color-coded according to the mesonetwork they belong. Data from Mughal and Caetano-Anollés (2019).

link subnetworks to each other. This matrix unfolds fractal-like patterns by clustering relationships with dendrograms and visualizing linkage with heatmaps. The matrix revealed modules embedded into modules, scaling relationships, and patterns of 'disassortative mixing' across modules in which nodes tend to connect to other dissimilar nodes. While most subnetwork modules clustered in a heterogeneous manner, only the nucleotide metabolism mesonetwork (NUC) and a few subnetworks from the carbohydrate (CAR), glycan biosynthesis (GLY) and xenobiotics (XEN) mesonetworks clustered into groups. Thus, the function of mesonetworks is somehow unlinked to subnetwork module make up. This fuzziness appears important.

As we previously discussed, phylogenomic analysis of metabolic networks (Caetano-Anollés and Caetano-Anollés, 2013), elementary functionome networks (Aziz *et al.*, 2016) and networks of domain organization in proteins (Aziz *et al.*, 2013, 2016; Aziz and Caetano-Anollés, 2020) showed that hierarchy and modularity increased in time. The same tendency was uncovered when tracing the history of metabolic networks using a new release of the MANET database (Mughal and Caetano-Anollés, 2019). The evolutionary history of the protein structural domains of enzymes was traced onto subnetwork-enzyme or mesonetwork-enzyme bipartite networks. Structural domains identified in thousands of genomes with HMMs defined at the FF level of structural abstraction in SCOP. Figure 21A illustrates the gradual evolutionary growth of enzymes and metabolic pathways with snapshots of the subnetwork projection taken 3.3 Gya, 2.4 Gya, 1.5 Gya and the present. Evolving networks show how subnetwork nodes became increasingly connected to each other through shared enzymes via enzymatic recruitment. Similarly, the enzyme projection (not shown) revealed growth of enzyme communities through sharing of subnetworks.

Algorithmic analyses of network structure revealed that hierarchical modularity of both network projections gradually increased as networks became more structured and recruitment more pervasive with evolutionary time. More importantly, the power law scaling of the clustering coefficient with node degree, which is a hallmark of the hierarchical structure of modules (Ravasz, 2009), increased in

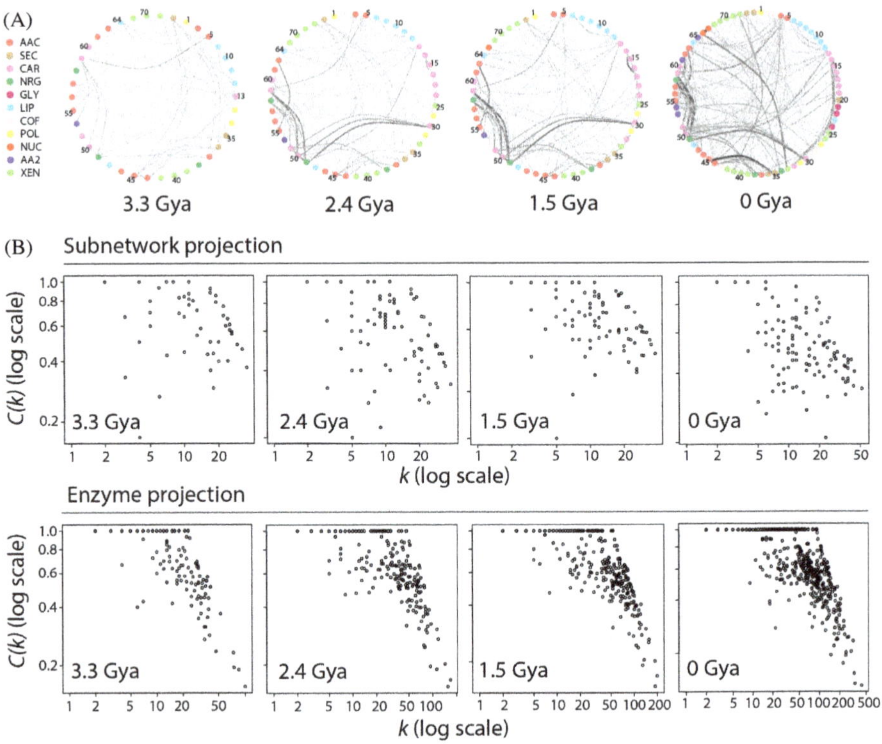

Figure 21. Metabolic network evolution. (**A**) Evolution of a reduced subnetwork one-mode projection of the subnetwork-enzyme bipartite network of metabolism, describing how subnetworks and enzymes grow in evolutionary time, billions of years ago (Gya). The projection shows major nodes (subnetworks) connecting to each other through links (shared enzymes). Greyscale values of links indicate the number of enzymes shared among the subnetworks. (**B**) Log-log plots of the clustering coefficient *C(k)* as a function of the number of links *k* for evolving subnetwork and enzyme projections of the subnetwork-enzyme bipartite network. The scaling is the hallmark of hierarchical modularity. It increases in evolution and is stronger at lower levels of metabolic organization. Abbreviations: AAC, Amino acid metabolism; SEC, Biosynthesis of other secondary metabolites; CAR, Carbohydrate metabolism; NRG, Energy metabolism; GLY, Glycan biosynthesis and metabolism; LIP, Lipid metabolism; COF, Metabolism of cofactors and vitamins; POL, metabolism of terpenoids and polyketides; NUC, Nucleotide metabolism; AA2, Metabolism of other amino acids; XEN, Xenobiotics biodegradation and metabolism. Data from Mughal and Caetano-Anollés (2019).

strength for both projections (Figure 21B). Remarkably, evolutionary constraints on network structure were stronger at the "enzyme" lower level and weaker at the "subnetwork" higher level of metabolic organization. The lower level network structure was maintained tightly knit by a set of enzymes with significant centralities in the metabolic network, including enzymes with transferase and oxidoreductase activities (mainly EC 2.3.1.9 and EC 1.2.1.3) leading to production of both acetoacetyl-coA and NADH hub metabolites that are crucial to the lipid and purine biosynthetic metabolisms, respectively. Note that together with pyruvate, acetyl-CoA was identified by the cartographic representation of metabolism (Guimerà and Amaral, 2005) as a central connector hub between the citrate TCA cycle and the metabolisms of carbohydrates, amino acids and lipids, pathways that embed an ancient prebiotic "energy amphiphile" core (Morowitz, 1999).

In contrast, tests of network randomness showed that the upper level structure of the subnetwork projection was bound together by linkages that were weak and more chaotic. This looseness in the establishment of pathways of chemical reactions manifested in significant heterogeneities of enzyme sharing across superkingdoms and viruses, diverse mapping of enzymatic functionalities, and cooption of catalytic site diversity along the evolutionary timeline (details in Mughal and Caetano-Anollés, 2019). All of these processes resulted in an evolutionary patchwork that is clearly evident when browsing metabolic pathways in MANET. The evolutionary entrenchment of catalytic functionalities at the enzyme lower level of metabolic organization that loosens at the subnetwork higher level supports a *"principle of granularity,"* an increase of the cohesiveness of lower-level parts of an evolving system. It also supports Simon's prediction that parts are nearly-decomposable and show stronger internal links the lower the level in the "*parts-within-parts*" structure of systems (Simon, 1997). This tendency causes parts to act almost independently from each other. Intuitively, this is plainly evident in the organization of companies or in the structure of proteins. For example, when considering conformational diversity, there is little fuzziness in the definition of how amino acid modules are integrated into primary structure. However, as one travels to upper levels of organization (secondary structure, super-secondary motifs, folded tertiary structure, etc.) conformational fuzziness increases considerably, as conformational spaces of each level add to the global conformational landscape of the protein molecules. This is true without considering a possible mutational landscape that would contribute to conformational diversity at longer time scales of change. The principle of granularity that we here propose therefore explains near-decomposability as the result of a trend of the system's lower levels to become increasingly more granularly entrenched with time.

12. Conclusions

We explore the underlying processes that define the complex structure of biological systems and give rise to evolutionary innovations. Biological systems are made of parts, components that combine with other components to make up a whole (Simon, 1962). Networks are convenient representations of how parts interact and integrate with each other. A number of network properties help understand the structure and dynamics of networks. These include measures of network centrality (degree, closeness centrality, betweenness centrality), community structure (clustering, modularity), and hierarchy. A structural morphospace of heterogeneity, modularity and randomness conveniently describes the structure of networks. Biological networks occupy an area of this phenotypic space with significant heterogeneity and modularity but lacking substantial levels of randomness. Conversely, a morphospace of hierarchy describes flows along wiring diagrams. They reveal that the bow-tie is the optimal flow network structure.

This bow-tie is also hinted by temporal biphasic patterns of diversification that are widespread in behavior, development and evolution. Since these patterns demand explanation, we here highlight a biphasic linkage theory of unification and change that explains the rise of modules in biology (Mittenthal *et al.*, 2012). In a first phase, parts are weakly linked and associate variously. As they diversify, they compete with each other and are selected for performance. The emerging interactions constrain their structure and associations. This causes parts to self-organize into modules with tight linkage. In a second phase, variants of the modules evolve and become new parts for a new generative cycle of higher-level organization.

The linkage hypothesis can be tested by studying the rise of hierarchy and community structure in networks. Indeed, phylogenomic analyses supports the linkage hypothesis, since hierarchical modularity was found to increase in evolution. The rise of hierarchical modularity can be seen as a "nucleation" process, i.e. as an emergence of cohesion in selected parts of the make up of a molecular system. Cohesion refers here to the dynamical stabilities of component parts of the system when their dynamics are constrained by the system as a whole. These dynamical stabilities ultimately determine if the system can be easily decomposed into parts. Remarkably, phylogenomic analysis also revealed increasing granularity of network structure at lower levels of network organization. Agents of evolutionary change must now explain how networks become evolutionarily entrenched. Computational simulations have shown that several evolutionary agents further the persistence and structure of a system, including adaptive drivers such as natural selection or competitive optimization, and non-adaptive (neutral) drivers such as reuse, connection costs, or information transmittal. One challenge that will be worth pursuing is testing how these evolutionary agents can be modeled and integrated with phylogenomic data in order to advance our evolutionary understanding of complex biological systems.

References

Ahnert SE, Marsh JA, Hernandez H, Robinson CV, Teichmann SA (2015) Principles of assembly reveal a periodic table of protein complexes. *Science* **350**: aaa2245.

Aiello W, Chung F, Lu L (2001) A random graph model for power law graphs. *Experiment Math* **10**: 53–66.

Albert R, Barabási A-L (2000) Topology of evolving networks: Local events and universality. *Phys Rev Lett* **85**: 5234–5237.

Albert R, Jeong H, Barabási A-L (2000) Error and attack tolerance of complex networks. *Nature* **406**: 378–382.

Amaral LAN, Scala A, Barthélemy M, Stanley HE (2000) Classes of behavior of small- world networks. *Proc Natl Acad Sci USA* **97**: 11149–11152.

Aziz MF, Caetano-Anollés G (2013) Emergence of power law behavior in evolution of protein domain networks. Paper presented at: Annual Meeting of the Society for Molecular Biology and Evolution; July 7–12, 2013; Chicago, IL.

Aziz MF, Caetano-Anollés G (2020) Emergence of preferential attachment and hierarchical modularity in evolution of networks of protein domain organization. Ms. in preparation

Aziz MF, Caetano-Anollés K, Caetano-Anollés G (2016) The early history and emergence of molecular functions and modular scale-free network behavior. *Sci Rep* **6**: 25058.

Aziz MF, Chan P, Osorio JH, Minhas BF, Parekatt V, Caetano-Anollés G (2012) Stress induces biphasic-rewiring and modularization patterns in metabolomics networks of *Escherichia coli*. IEEE *Intl Conf Bioinf Biomed* 2012: 593–597.

Barabási A-L, Albert R (1999) Emergence of scaling in random networks. *Science* 286: 509–512.

Barabási A-L, Oltvai ZN (2004) Network biology: Understanding the cell's functional organization. *Nat Rev* **5**: 101–113.

Barrat A, Barthelemy M, Pastor-Satorras R, Vespignani A (2004) The architecture of complex weighted networks. *Proc Natl Acad Sci USA* **101**: 3747–3752.

Bassett DS, Greenfield DL, Meyer-Lindenberg A, Weinberger DR, Moore SW, Bullmore ET (2010) Efficient physical embedding of topologically complex information processing networks in brains and computer circuits. *PLoS Comput Biol* **6(4)**: e1000748.

Ben-Shem A, Jenner L, Yusupova G, Yusupov M (2010) Crystal structure of the eukaryotic ribosome. *Science* **330**: 1203–1209.

Beutler B (2004) Inferences, questions and possibilities in toll-like receptor signalling. *Nature* **430(6996)**: 257–263.

Bianconi G, Barabási A-L (2001) Competition and multiscaling in evolving-networks, *Europhys Lett* **54**: 436.

Blondel VD, Guillaume J-L, Lambiotte R, Lefebvre E (2008) Fast unfolding of communities in large networks. *J Stat Mech* **2008**: P10008.

Broder A, Kumar R, Maghoul F, Raghavan P, Rajagopalan S, Stata R, Tomkins A, Wiener J (2000) Graph structure in the Web. *Comp Networks* **33**: 309–320.

Caetano-Anollés D, Kim KM, Caetano-Anollés G (2011) Proteome evolution and the metabolic origins of translation and cellular life. *J Mol Evol* **72**: 14–32.

Caetano-Anollés G (2002) Tracing the evolution of RNA structure in ribosomes. *Nucleic Acids Res* **30**: 2575–2587.

Caetano-Anollés G, Aziz FM, Mughal F, Gräter F, Koç I, Caetano-Anollés K, Caetano-Anollés D (2019) Emergence of hierarchical modularity in evolving networks uncovered by phylogenomic analysis. *Evol Bioinform* **15**: 1–18.

Caetano-Anollés G, Caetano-Anollés D (2003) An evolutionary structured universe of protein architecture. *Genome Res* **13(7)**: 1563–1571.

Caetano-Anollés G, Caetano-Anollés D (2015) Computing the origin and evolution of the ribosome from its structure — Uncovering processes of macromolecular accretion benefiting synthetic biology. *Comput Struct Biotech J* **13**: 427–447.

Caetano-Anollés G, Yafremava LS, Gee H, Caetano-Anollés D, Mittenthal JE (2009) The origin and evolution of modern metabolism. *Int J Biochem Cell Biol* **41**: 285–97.

Caetano-Anollés K, Caetano-Anollés G (2013) Structural phylogenomics reveals gradual evolutionary replacement of abiotic chemistries by protein enzymes in purine metabolism. *PLoS One* **8(3)**: e59300.

Callaway DS, Newman MEJ, Strogatz SH, Watts DJ (2000) Network robustness and fragility: Percolation on random graphs. *Phys Rev Lett* **85**: 5468–5471.

Callebaut W (2007) Herbert Simon's silent revolution. *Biol Theory* **2**: 76–86.

Christie P, Stroobandt D (2000) The interpretation and application of Rent's rule. *IEEE Trans VLSI Syst* **8(6)**: 639–648.

Chung FRK, Erdös P, Spencer J (1983) On the decomposition of graphs into complete bipartite subgraphs. In: Erdös P, Alpar L, Halasz G, Saeközy A (eds). In *Studies in Pure Mathematics: To the Memory of Paul Turán*. Bikhäuser Verlag, Budapest, pp 95–101.

Clauset A, Newman MEJ, Moore C (2004) Finding community structure in very large networks. *Phys Rev E* **70**: 66111.

Clauset A, Shalizi CR, Newman MEJ (2009) Power-law distributions in empirical data. *SIAM Rev* **51(4)**: 661–703.

Clune J, Mouret J-B, Lipson H (2013) The evolutionary origins of modularity. *Proc R Soc B* 280: 20122863.

Cohen R, Erez K, ben-Avraham D, Havlin S (2000) Resilience of the internet to random breakdowns. *Phys Rev Lett* **85**: 4626–4628 (2000).

Cordero OX, Hogeweg P (2006) Feed-forward loop circuits as a side effect of genome evolution. *Mol Biol Evol* **23**: 1931–1936

Corominas-Murtra B, Goñi J, Solé RV, Rodríguez-Caso C (2013) On the origins of hierarchy in complex networks. *Proc Natl Acad Sci USA* **10**: 13316–13321.

Cottam R, Ranson W, Vounckx R (2016) Hierarchy and the nature of information. Information **7**: 1.

Czégel D, Palla G (2015) Random walk hierarchy measure: What is more hierarchical, a chain, a tree or a star? *Sci. Rep* **5**: 17994.

de Solla Price DJ (1976) A general theory of bibliometric and other cumulative advantage processes. *J Amer Soc Inform Sci* **27**: 292–306.

Debès C, Wang M, Caetano-Anollés G, Gräter F (2013) Evolutionary optimization of protein folding. *PLoS Comput Biol* **9**: e1002861.

Domazet-Lošo T, Tautz D (2010) A phylogenetically based transcriptome age index mirrors ontogenetic divergence patterns. *Nature* **468(7325)**: 815–818.

Dorogovtsev SN, Goltsev A, Mendes JFF (2002) Pseudofractal scale-free web. *Phys Rev E* **65**: 066122.

Dorogovtsev SN, Mendes JFF (2000) Evolution of networks with aging of sites. *Phys Rev E* **62**: 1842–1845.

Duboule D (1994). Temporal colinearity and the phylogenetic progression: A basis for the stability of a vertebrate Bauplan and the evolution of morphologies through heterochrony. *Development Suppl* 135–142.

Erdös P, Rényi A (1960) On the evolution of random graphs. *Publ Math Inst Hung Acad Sci* **5**: 17–61.

Espinosa-Soto C, Wagner A (2010) Specialization can drive the evolution of modularity. *PLoS Comput Biol* **6**: e1000719.

Faloutsos M, Faloutsos P, Faloutsos C (1999) On power-law relationships of the internet topology. *Comp Comm Rev* **29**: 251–262.

Federhen S (2012) The NCBI taxonomy database. *Nucleic Acids Res* **40**: D136–D143.

Flack JC, Erwin D, Elliot T, Krakauer DC (2013) Timescales, symmetry, and uncertainty reduction in the origins of hierarchy in biological systems. *Evol Cooperation Complexity* **22**: 45–74.

Girvan M, Newman MEJ (2002) Community structure in social and biological networks. *Proc Natl Acad Sci USA* **99**: 7821–7826.

Gorochowski TE, Grierson CS, di Bernardo M (2018) Organization of feedforward loop motifs reveals architectural principles in natural and engineered networks. *Sci Adv* **4**: eaap9751

Guimerà R, Amaral LAN (2005) Functional cartography of complex metabolic networks. *Nature* **433**: 895–900.

Harish A, Caetano-Anollés G (2012) Ribosomal history reveals origins of modern protein synthesis. *PLoS One* **7**: e32776.

Harrison A, Pearl F, Mott R, Thornton J, Orengo C (2002) Quantifying the similarities within fold space. *J Mol Biol* **323**: 909–926.

Hartwell LH, Hopfield JJ, Leibler S, Murray AW (1999) From molecular to modular cell biology. *Nature* **401**: c47–c52.

Ho JJ, Navlakha S (2018) Evidence of rentian scaling of functional modules in diverse biological networks. *Neural Comput* **30(8)**: 2210–2244.

Holme P, Kim BJ, Yoon CN, Han SK (2002) Attack vulnerability of complex networks. *Phys Rev E* **65**: 056109.

Irie N, Kuratani S (2014) The developmental hourglass model: A predictor of the basic body plan? *Development* **141**: 4649–4655.

Janko R (2004) Empedocles, On nature I 233–364: A new reconstruction of p. Strasb. Gr. Inv. 1665–6. *Zeitschrift für Papyrologie und Epigraphik* **150**: 1–26

Jeong H, Tombor B, Albert R, Oltvai ZN, Barabási A-L (2000) The large-scale organization of metabolic networks. *Nature* **407**: 651–654.

Kahtter K, Mysanikov AG, Natchiar SK, Klaholz BP (2015) Structure of the human 80S ribosome. *Nature* **520**: 640–645.

Kaiser M, Hilgetag CC (2004) Spatial growth of real-world networks. *Phys Rev E* **69**: 036103.

Kalinka AT, Varga KM, Gerrard DT, Preibisch S, Corcoran DL, Jarrells J, Ohler U, Bergman CM, Tomancak P (2010) Gene expression divergence recapitulates the developmental hourglass model. *Nature* **468(7325)**: 811–814.

Kalvari I, Argasinska J, Quinones-Olvera N, Nawrocki EP, Rivas E, Eddy SR, Bateman A, Finn RD, Petrov AI (2018) Rfam 13.0: Shifting to a genome-centric resource for non-coding RNA families. *Nucleic Acids Res* **46(D1)**: D335–D342.

Kanehisa M, Furumichi M, Tanabe M, Sato Y, Morishima K (2017) KEGG: New perspectives on genomes, pathways, diseases and drugs. *Nucleic Acids Res* **45**: D353–D361.

Karrer B, Newman MEJ, Zdeborová L (2014) Percolation on sparse networks. *Phys Rev Lett* **113**: 208702.

Kashtan N, Alon U (2005) Spontaneous evolution of modularity and network motifs. *Proc Natl Acad Sci USA* **102**: 13773–13778

Kashtan N, Noor E, Alon U (2007) Varying environments can speed up evolution. *Proc Natl Acad Sci USA* **104**: 13711–13716.

Kim HS, Mittenthal JE, Caetano-Anollés G (2006) MANET: Tracing evolution of protein architecture in metabolic networks. *BMC Bioinformatics* **7**: 351.

Kitano H (2004) Biological robustness. *Nature Rev Genet* **5**: 826–837.

Klemm K, Eguíluz VM (2002) Growing scale-free networks with small-world behavior. *Phys Rev E* **65**: 057102.

Koç I, Yuksel I, Caetano-Anollés G (2018) Metabolite-centric reporter pathway and tripartite network analysis of Arabidopsis under cold stress. *Front Bioeng Biotechnol* **6**: 121.

Krapivsky PL, Redner S, Leyvraz F (2000) Connectivity of growing random networks. *Phys Rev Lett* **85**: 4629–4632. Albert R. Jeong H, Barabási. A-L (2000) error and attack tolerance of complex networks. *Nature* **406**: 378–382.

Landman BS, Russo RL (1971) On a pin versus block relationship for partitions of logic graphs. IEEE T Comput C-**20(12)**: 1469–1479.

Lecompte O, Ripp R, Thierry J-C, Moras D, Poch O (2002) Comparative analysis of ribosomal proteis in complete genomes: An example of reductive evolution at the domain scale. *Nucleic Acids Res* **30**: 5382–5390.

Leroi AM (2000) The scale independence of evolution. *Evol Dev* **2**: 67–77.

Lorenz DM, Jeng A, Deem MW (2011) The emergence of modularity in biological systems. *Phys Life Rev* **8**: 129–160.

Luo X, Magee CL (2010) Detecting evolving patterns of self-organizing networks by flow hierarchy measurements. *Complexity* **16**: 53–61.

Ma H-W, Zeng A-P (2003) The connectivity structure, giant strong component and centrality of metabolic networks. *Bioinformatics* 19(11): 1423–1430.

MANET (2006) Molecular Ancestry Networks, University of Illinois. https://manet.illinois.edu

Marsh JA, Teichmann SA (2014) Protein flexibility facilitates quaternary structure assembly and evolution. *PLoS Biol* **12**: e1001870.

Melnikov S, Ben-Shem A, Garreau de Loubresses N, Jenner L, Yusupova G, Yusupov M (2012) One core, two shells: Bacterial and eukaryotic ribosomes. *Nature Struct Mol Biol* **19(6)**: 560–567.

Mengistu H, Huizinga J, Mouret J-B, Clune J (2016) The evolutionary origins of hierarchy. *PLoS Comput Biol* **12**: e1004829.

Meunier D, Lambiotte R, Bullmore ET (2010) Modular and hierarchically modular organization of brain networks. *Front Neurocience* **4**: 1–11.

Mittenthal JE, Caetano-Anollés D, Caetano-Anollés G (2012) Biphasic patterns of diversification and the emergence of modules. *Front Genet* **3**: 147.

Molloy M, Reed B (1995) A critical point for random graphs with a given degree sequence. *Random Struct Algor* **6**: 161–179.

Molloy M, Reed B (1998) The size of the giant component of a random graph with a given degree sequence. *Combin Probab Comput* 7: 295–305.

Monachino E, Spenkelink LM, van Oijen AM (2017) Watching cellular machinery in action, one molecule at a time. *J Cell Biol* **216**: 41–51.

Mones E, Vicsek L, Vicsek T (2012) Hierarchy measure for complex networks. *PLoS One* 7: 1–10.

Morowitz HJ (1999) A theory of biochemical organization, metabolic pathways, and evolution. *Complexity* **4**: 39–53.

Mughal F, Caetano-Anollés G (2019) MANET 3.**0**: Hierarchy and modularity in evolving metabolic networks. *PLoS One* 14(10): e0224201.

Mughal F, Gräter F, Caetano-Anollés G (2017) How function shapes dynamics in protein evolution. In: Szuch S, Watkins C (eds). Blue Waters Annual Report. Champaign, IL: *National Center for Supercomputer Applications*, **2017**: 198–199.

Murzin A, Brenner SE, Hubbard T, Chothia C (1995) SCOP: A structural classification of proteins for the investigation of sequences and structures. *J Mol Biol* **247**: 536–540.

Nasir A, Kim KM, Caetano-Anollés G (2017) Long-term evolution of viruses: A Janus-faced balance. *Bioessays* 39(8): 1700026.

Newman MEJ (2001) The structure of scientific collaboration networks. *Proc Natl Acad Sci USA* **98**: 404–409.

Newman MEJ (2003) The structure and function of complex networks. *SIAM Rev* **45**: 167–256.

Newman MEJ (2005) Power laws, Pareto distributions and Zipf's law. *Contemporary Phys* 46(5): 323–351.

Newman MEJ, Girvan M (2004) Finding and evaluating community structure in networks. *Phys Rev E* **69**: 26113.

Newman MEJ, Strogatz SH, Watts DJ (2001) Random graphs with arbitrary degree distributions and their applications. *Phys Rev E* **64**: 026118.

Niklas KJ, Wright S, Simpson GG (1994) Morphological evolution through complex domains of fitness. *Proc Natl Acad Sci USA* **91**: 6772–6779.

Oda K, Kitano H (2006) A comprehensive map of the toll-like receptor signaling network. *Mol Syst Biol* **2**: 2006.0015.

Pachon A, Sacerdote L, Yang S (2018) Scale-free behavior of networks with the copresence of preferential and uniform attachment rules. *Physica D* **37**: 1–12

Papadopoulos L, Blinder P, Ronellenfitch H, Klimm F, Katifori E, Kleinfeld D, Bassett DS (2016) Comparing two classes of biological distribution systems using network analysis. *PLoS Comput Biol* 14(9): e1006428.

Papin JA, Price ND, Wiback SJ, Fell DA, Palsson BO (2003) Metabolic pathways in the post-genome era. *Trends Biochem Sci* **28**: 250–258.

Purcell EM (1977) Life at low Reynolds number. *Am J Phys* **45**: 3–11.

Quint M, Drost H-G, Gabel A, Ullrich KK, Bönn M, Grosse I (2012) A transcriptomic hourglass in plant embryogenesis. *Nature* **490(7418)**: 98–101.

Quiroga RQ, Reddy L, Kreiman G, Koch C, Fried I (2005) Invariant visual representation by single neurons in the human brain. *Nature* **435(7045)**: 1102–1107.

Raff R (1996) *The Shape of Life*. University of Chicago Press, Chicago, IL

Ravasz E (2009) Detecting hierarchical modularity in biological networks. *Methods Mol Biol* **541**: 145–160.

Ravasz E, Somera AL, Mongru DA, Oltvai ZN, Barabási A-L (2002) Hierarchical organization of modularity in metabolic networks. *Science* **297**: 1551–1555.

Riesenhuber M, Poggio T (1999) Hierarchical models of object recognition in cortex. *Nature Neurosci* **2(11)**: 1019–1025.

Sabrin KM, Dovrolis C (2017) The hourglass effect in hierarchical dependency networks. *Network Sci* **5**: 490–528.

Salthe SN (1985) *Evolving Hierarchical Systems: Their Structure and Representation*. Columbia University Press, New York.

Salthe SN (2012) Hierarchical structures. *Axiomathes* **22**: 355–383.

Schuetz R, Zamboni N, Zampieri M, Heinemann M, Sauer U (2012) Multidimensional optimality of microbial metabolism. *Science* **336**: 601–604.

Shahzad K, Mittenthal JE, Caetano-Anollés G (2015) The organization of domains in proteins obeys Menzerath-Altmann's law of language, *BMC Syst Biol* **9**: 44.

Shoval O, Sheftel H, Shinar G, Hart Y, Ramote O, Mayo A, Dekel E, Kavanagh K, Along U (2012) Evolutionary trade-offs, pareto optimality, and the geometry of phenotype space. *Science* **336**: 1157–1160.

Simon HA (1955) On a class of skew distribution functions. *Biometrika* **42**: 425–440.

Simon HA (1962) The architecture of complexity. *Proc Am Phil Soc* **106**: 467–482.

Simon HA (1997) *Models of Bounded Rationality: Empirically Grounded Economic Reason*, Vol 3. MIT Press, Cambridge, MA.

Small M, Li Y, Stemler T, Judd K (2015) Growing optimal scale-free networks via likelihood. *Phys Rev E* **91**: 042801.

Solé RV, Fernández P (2003) Modularity "for free" in genome architecture. arXiv: q-bio/0312032[q-bio.GN].

Solé RV, Pastor-Satorras R, Smith E, Kepler TB (2002) A model of large-scale proteome evolution. *Adv Compl Syst* **5**: 43–54.

Solé RV, Valverde S (2004) Information theory of complex networks: On evolution and architectural constraints. *Lect Notes Phys* **650**: 189–207.

Solomonoff R, Rapoport A (1951) Connectivity of random nets. *Bull Math Biophys* **13**: 107–117.

Strogatz SH (2001) Exploring complex networks. *Nature* **410**: 268–176.

Supper J, Spangenberg L, Planatscher H, Dräger A, Schröder A, Zell A (2009) Bowtiebuilder: Modeling signal transduction pathways. *BMC Systems Biology* **3(1)**: 1.

Tanaka R, Csete M, Doyle J (2005). Highly optimised global organisation of metabolic networks. IEE *Proc Syst Biol* **152(4)**: 179–184.

Timár G, Goltsev AV, Dorogovtsev SN, Mendes JFF (2017) Mapping the structure of directed networks: Beyond the Bow-tie diagram. *Phys Rev let* **118**: 078301.

Tononi G (2008) Consciousness as integrated information: A provisional manifesto. *Biol Bull* **215**: 216–242.

Wagner GP, Pavlicev M, Cheverud JM (2007) The road to modularity. *Nature Rev Genet* **8**: 921–931.

Wang M, Jiang Y-Y, Kim KM, Qu G, Ji HF, Mittenthal JE, Zhang HY, Caetano-Anollés G (2011a) A universal molecular clock of protein folds and its power in tracing the early history of aerobic metabolism and planet oxygenation. *Mol Biol Evol* **28**: 567–82.

Wang M, Kurland CG, Caetano-Anollés G (2011b) Reductive evolution of proteomes and protein structures. *Proc Natl Acad Sci USA* **108**: 11954–11958.

Wang M, Yafremava LS, Caetano-Anollés D, Mittenthal JE, Caetano-Anollés G (2007) Reductive evolution of architectural repertoires in proteomes and the birth of the tripartite world. *Genome Res* **17**: 1572–158.

Wasserman S, Faust K (1994) *Social Network Analysis: Methods and Applications*. Cambridge University Press, New York

Watts DJ (1999) *Small Worlds: The Dynamics of Networks between Order and Randomness*. Princeton University Press, Princeton, NJ.

Watts DJ, Strogatz SH (1998) Collective dynamics of 'small-world' networks. *Nature* **393**: 440–442.

Wilensky U (2005) NetLogo preferential attachment model. Center for connected Learning and Computer-Based modeling, Northwestern University, Evanston, IL, USA. URL: http: //ccl.northwestern.edu/netlogo/models/PreferentialAttachment.

Yafremava LS, Wielgos M, Thomas S, Nasir A, Wang M, Mittenthal JE, Caetano-Anollés G (2013) A general framework of persistence strategies for biological systems helps explain domains of life. *Front Genet* **4**: 16.

Zhao J, Yu H, Luo J-H, Cao Z-W, Li Y-X (2006) Hierarchical modularity of nested bow-ties in metabolic networks. *BMC Bioinfo* **7(1)**: 386.

Zhu S, Zhu M, Knoll AH, Yin Z, Zhao F, Sun S, Qu Y, Shi M, Liu H (2016) Decimeter-scale multicellular eukaryotes from the 1.56-billion-year-old Gaoyuzhuang Formation in North China. *Nature Comm* **7**: 11500.

Structure

RNA structure and diversified life

F.-J. Sun[1]
G. Caetano-Anollés[2]

Summary. We discuss the reconstruction of global phylogenies that incorporate RNA structural evidence and how they potentially reveal the evolutionary patterns associated with the emergence of modern biochemistry and early life. Since evolutionary direction and the rooting of the tree of life are fundamental and inevitable issues in evolutionary biology, we also discuss how the use of RNA structural evidence can be established and improved in order to gain insights into deep evolution of ancient molecules. We show that studies consistently and congruently identify the same or similar basal groups on the tree of life derived by using various molecular systems. We summarize the results and discuss the rooting strategy in studying the tree of life.

[1]*School of Science and Technology, Georgia Gwinnett College, Lawrenceville, Georgia, USA*
[2]*Department of Crop Sciences, University of Illinois, Urbana, Illinois, USA*

1. Introduction

Phylogenetic methods have been proved to be effective ways to study the diversification of organisms. A phylogenetic tree displays leaves representing "ingroup" taxa, while "outgroups" are distantly related outliers representing external hypotheses of relationships that need to be amply justified and supported by external assumptions (Maddison *et al.*, 1984). Because selection of outgroups can alter optimization of characters in the reconstruction of ingroup relationships (Dalevi *et al.*, 2001; O'Brien *et al.*, 2002; Rota-Stabelli and Telford, 2008), it is always important to correctly select an appropriate outgroup. This is especially true when the tree of life is reconstructed at the superkingdom level (i.e., Archaea, Bacteria, and Eukarya) because it is difficult, if not impossible, to identify an outgroup and subsequently determine the evolutionary direction on the unrooted global phylogenetic tree (Baldauf *et al.*, 1996). Thus, the root of the tree of life is a long-standing controversy in the study of the origins and evolution of early life on Earth (Forterre and Philippe, 1999). Various rooting scenarios have been explored mainly based on the primary structures of DNA sequences (Leffers *et al.*, 1987; Olsen *et al.*, 1994; Brown and Doolittle, 1995; Baldauf *et al.*, 1996; Xue *et al.*, 2003, 2005; Di Giulio, 2007; Wong *et al.*, 2007) or secondary structures of either ribonucleotide or protein molecules (Caetano-Anollés and Caetano-Anollés, 2003; Wang and Caetano-Anollés, 2006; Wang *et al.*, 2007; Sun and Caetano-Anollés, 2008a–e; Kim *et al.*, 2014). Furthermore, the origin of eukaryotes and their evolutionary relationship with the Archaea is still a fundamental biological question seeking an answer (Gribaldo *et al.*, 2010). To date, the rooting of the tree of life and the origins of diversified life remain controversial, mostly due to our inability to root trees and distinguish between base and crown group taxa that are both extinct and extant, but also because of limited phylogenetic signal in sequences. With the provocative interpretation of the tree of life (O'Malley and Koonin, 2011) and the sustained robustness of the original tree of life reconstructed from rRNA (Pace *et al.*, 2012) under constant severe scrutiny and criticism (Doolittle, 1999), the establishment of the root of the tree of life possesses fundamental evolutionary and conceptual significance.

Molecular structure evidence has been used in phylogenetic analyses either by using RNA secondary structures to improve the alignment of sequences (Higgs, 2000; Higgs *et al.*, 2003) or by directly embedding structural characters in phylogenetic analysis (Billoud *et al.*, 2000; Collins *et al.*, 2000; Caetano-Anollés 2001, 2002a, 2002b, 2005; Pollock 2003; Swain and Taylor 2003; Grajales *et al.*, 2007; Sun *et al.*, 2007; Sun and Caetano-Anollés, 2008a–e, 2009, 2010). These applications of RNA structures are made possible and benefit from three fundamental biological tendencies, which we consider here axiomatic: (i) RNA structure is far more conserved than sequence and has advantage when used to study ancient events in the history of life; (ii) there is a universal tendency toward molecular order that is supported by thermodynamic, statistical, and phylogenetic arguments; and (iii) successfully implemented biological designs tend to be reused in nature, generally through recruitment by takeovers or cooptions. These tendencies are initially exploited in an award-winning phylogenetic approach that uses RNA structural information to reconstruct evolutionary history of macromolecules such as rRNA (Caetano-Anollés, 2002a, 2002b). This method has been extensively applied to tRNA and other RNA molecules (Collins *et al.*, 2000; Caetano-Anollés, 2002a, 2002b, 2005; Swain and Taylor, 2003; Grajales *et al.*, 2007; Sun *et al.*, 2007; Sun and Caetano-Anollés, 2008a–e, 2009, 2010) to study long-standing evolutionary questions, and protein domains to uncover patterns in the diversification of cellular life (Caetano-Anollés and Caetano-Anollés, 2003; Wang and Caetano-Anollés, 2006, 2009; Wang *et al.*, 2007; Wang *et al.*, 2011; Kim and Caetano-Anollés, 2012; Caetano-Anollés and Nasir, 2012; Nasir *et al.*, 2014), protein biosynthesis (Caetano-Anollés *et al.*, 2012;

Harish and Caetano-Anollés, 2012), viruses (Nasir *et al.*, 2012; 2015), and molecular networks (Caetano-Anollés *et al.*, 2007; Aziz *et al.*, 2016).

2. A novel phylogenetic approach based on macromolecular structure

RNA molecules are mostly composed of stem regions (helices) that stabilize the molecule by establishing hydrogen binding pairing interactions and unpaired regions (hairpin and internal loops, bulges, multiloops, and free ends) that generally destabilize the molecules (Figure 1). In the process of reconstructing phylogenetic trees of organisms (Caetano-Anollés, 2002a), RNA structural characters are coded as multiple discrete states, linearly ordered, and polarized by fixing the direction of character state change using a transformation sequence that distinguishes ancestral states as those

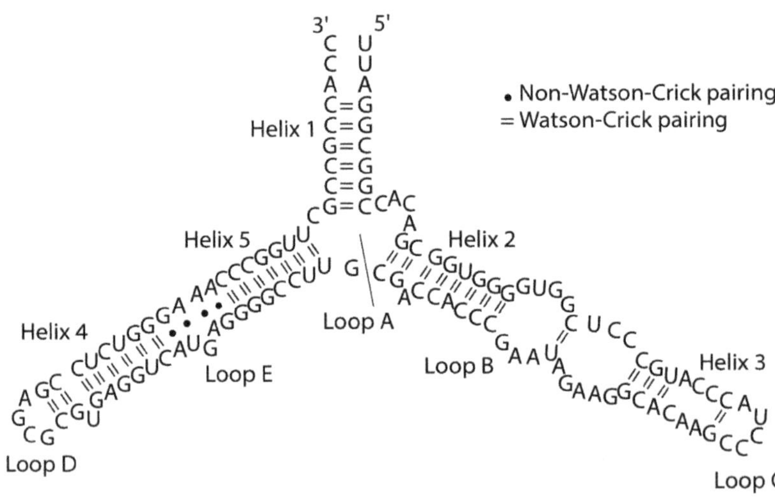

Figure 1. The secondary structure of 5S rRNA from *Holoarcula marismortui* (Ban *et al.*, 2000) showing different stem (helices) and unpaired (loops) substructural components. Unpaired segments include hairpin loops (loops C and D), internal loops (e.g., loop B), multiloops (loop A), and bulges (loop E).

more thermodynamically stable (e.g., larger state values for stems and smaller state values for hairpin, bulges, and unpaired regions), a tendency that complies with Weston's generality criterion of rooting. The model of character state transformation is based on the hypothesis that evolved RNA molecules are optimized to produce highly stable folded conformations with increased conformational order. This optimization process increases favorable and decreases nonfavorable inter- and intra-molecular interactions and restricts alternative outcomes of the RNA folding process. The model assumes a priori that evolution of individual substructures is independent of each other. Results in this seminal cladistic method show that considerable phylogenetic signal is present in RNA secondary structure and that this information is capable of grouping organisms in highly divergent lineages (Caetano-Anollés, 2002a). The phylogenies reconstructed from evolved structure match those derived from nucleic acid sequences at all taxonomical levels and group organisms in concordance with traditional classification, especially in superkingdoms Archaea and Eukarya. Results challenge the "prokaryotic dogma" stating that the last universal common ancestor, the "cenancestor" proposed by Fitch and Upper (1987), was bacterial-like in genomic and cellular organization, and support an eukaryotic-like ancestry (Reanney, 1974; Darnell, 1978; Doolittle, 1978) and the hypothesis that many akaryotic features originated by simplification through gene loss and nonorthologous displacement (Forterre and Philippe, 1999).

One of the unique features of this structure-based cladistic method is its ability to produce rooted topologies capable of establishing direction of evolutionary change at the molecular level. The phylogenies reconstructed from evolved structure are intrinsically rooted. This can be very useful in applications where suitable outgroups or paralogous sequences are not available. The true potential of the method has since been explored with several important RNA molecules, including tRNA, short interspersed element (SINE) RNA,

5S rRNA, ribonuclease (RNase) P RNA, and rRNA based on comprehensive data matrices derived from a large number of sequences. Many of these RNA molecules are evolutionarily, structurally, and/or functionally closely related to each other, and their molecular functions are either directly or indirectly linked to the process of protein synthesis. We review the evolutionary models of evolutionary growth for these molecules proposed based on phylogenetic trees of substructures.

2.1. *Structural characters and derived phylogenetic trees*

A flow diagram is given in Figure 2 to illustrate the essential justification and steps involved in an evolutionary study of RNA structure. In the example, tRNA structures are collected, coded, and used to generate data matrices with fundamental elements describing geometrical or statistical features of structure. Phylogenetic trees are then reconstructed to depict the evolution of both molecules and substructures. The trees are finally used to build evolutionary heat maps, in which ancestries are traced directly onto two-dimensional (planar representations) or three-dimensional models of structure. Both the character coding (character attributes such as stems or loops used to describe RNA structures coded using an alphanumerical format) and character polarization (the evolutionary tendency toward molecular order determined by ancestral and derived character states) are discussed in details elsewhere (Caetano-Anollés, 2000, 2001, 2002a, 2002b, 2005; Sun *et al.* 2007; Caetano-Anollés *et al.*, 2008; Knudsen and Caetano-Anollés, 2008; Sun and Caetano-Anollés, 2008a–e, 2009, 2010). More importantly, these techniques are also supported by a considerable body of theoretical and experimental evidence (reviewed in Sun *et al.*, 2010).

2.2. *Phylogenetic analysis of RNA structural characters*

All data matrices for both geometrical and statistical characters are analyzed using equally weighted maximum parsimony (MP) as the

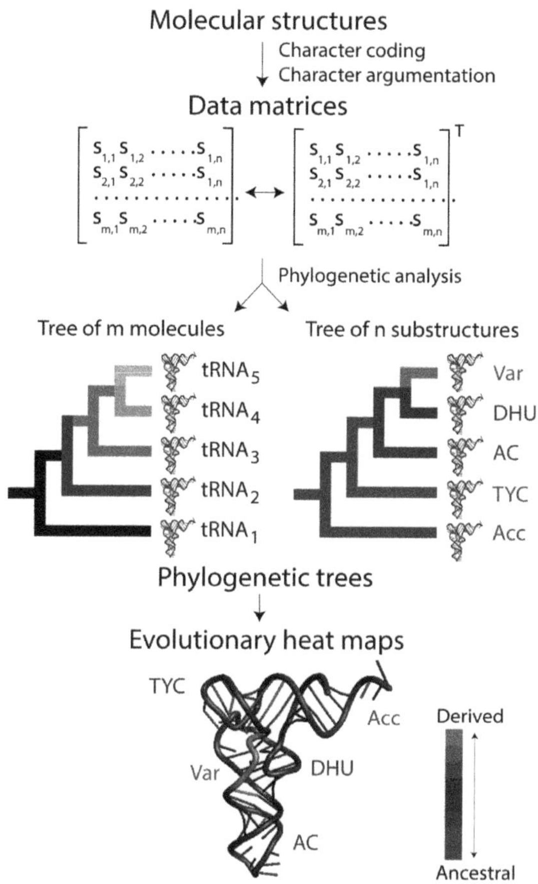

Figure 2. Phylogenetic reconstruction of trees of molecules and substructures. The structure of an RNA molecule (illustrated with tRNAs) can be decomposed into substructures, such as coaxial stem tracts and unpaired regions, which can be studied using features (characters) that describe their geometry (e.g., length of stems or unpaired regions) or their branching, stability, and uniqueness (e.g., statistical parameters describing a morphospace). These "shape" and "statistical" characters are coded and assigned "character states" according to an evolutionary model that polarizes character transformation toward an increase in molecular order (character argumentation). Coded characters (s) are arranged in data matrices, which can be transposed and subjected to cladistic analyses to generate phylogenies of either molecules or substructures. Rooted trees are used to generate evolutionary heat maps of secondary structure that color 3D structural models with molecular ancestries.

optimality criterion (Swofford, 2003). Note that a more realistic weighting scheme should consider, for example, the evolutionary rates of change in structural features. However, this requires the measurement of evolutionary parameters along individual branches of the tree and the development of an appropriate quantitative model. In the absence of this information, it is most parsimonious and preferable to give equal weight to the relative contribution of each character. The preference of solutions that require the least amount of change in MP is particularly appropriate and can outperform maximum likelihood (ML) approaches in certain circumstances (Steel and Penny, 2000). Indeed, MP is precisely ML when character changes occur with equal probability but rates vary freely between characters in each branch. This model is useful when there is limited knowledge about underlying mechanisms linking characters with each other (Steel and Penny, 2000). Furthermore, the use of large multistep character state spaces decreases the likelihood of revisiting a same character state on the underlying tree, making MP statistically consistent.

Structural attributes that describe the geometry or the molecular mechanics of RNA molecules were used to generate phylogenetic trees of molecules that portray the evolution of molecular lineages (Figure 2). The general approach has been used for example to reconstruct a tree of life from rRNA (Caetano-Anollés, 2002a; Harish and Caetano-Anollés, 2012), trace the evolution of RNA structures in ribosomes (Caetano-Anollés, 2002b), establish deep phylogenetic relationships in Poaceae (Caetano-Anollés, 2005), and study the structural evolution of retrotransposable elements (Sun *et al.*, 2007), 5S rRNA (Sun and Caetano-Anollés, 2009), and RNase P RNA (Sun and Caetano-Anollés, 2010), in addition to the studies of tRNA structural evolution (Sun and Caetano-Anollés, 2008a), the diversification of cellular life and viruses (Sun and Caetano-Anollés, 2008b), the evolution of amino acid charging and the genetic code (Sun and Caetano-Anollés, 2008c), and the evolutionary significance of the long variable arm in tRNA (Sun and Caetano-Anollés, 2008d).

The method also produces phylogenetic trees of molecular substructures, the histories of the structural evolution of RNA molecules (Figure 2). These trees are atypical. They do not describe the evolution of organismal taxa or molecules embedded in them. Instead, they describe the evolution of molecular repertoires of RNA components that are present in the organismal world. Very much as with the trees of molecules (Caetano-Anollés, 2002a), these trees of molecular substructures are derived directly from RNA structure and are intrinsically rooted. Consequently, the relationships of ancestral or derived substructures embedded in the phylogenies can be used to establish which of them are the most ancient (defining a structural origin) and how individual substructures were evolved or added (sequentially or in groups) into an evolving RNA molecule. Therefore, these trees of molecular substructures define a molecular chronology.

2.3. *Major properties of phylogenetic trees derived from RNA structure*

Phylogenetic trees of molecules and of substructures have several desirable properties that are salient and enable direct inferences about not only the phylogenetic relationships in biodiversity but also the evolutionary origin and history of RNA molecules.

(a) Phylogenetic trees are intrinsically rooted, establishing evolution's arrow. Phylogenetic reconstruction produces trees that are rooted according to models of character transformation that delimit how individual characters transform from one character state to another along the branches of the trees (Caetano-Anollés, 2001, 2002a). Consequently, the derived trees of molecules and substructures are rooted without the need and associated uncertainties of local external hypotheses of relationship, e.g., outgroup or paralogous sequences.

(b) Terminal leaves in trees of substructures represent a finite repertoire of RNA substructural components. Many kinds of RNA substructures can be recognized as taxa, such as structural domains (e.g., arms in tRNA), stem loops, stems, hairpin loops, internal loops, bulges, unpaired segments in multiloops or external segments, base pairs (e.g., weak G:U pairs), and modified bases, which are commonly identified in tRNAs. To ensure homology, each kind of RNA substructure is used separately in tree reconstruction. In contrast to the traditional phylogenetic and phylogenomic approaches that generally deal with taxa that cannot be considered finite, RNA substructures represent necessarily a finite set, which is defined by the repertoire present in sampled RNAs. For example, when organismal phylogenies are reconstructed using either morphological or molecular characters, taxa represent species drawn from millions of species that inhabit earth (Bull *et al.*, 1992).

(c) Internal nodes in trees of substructures determine relative chronologies of structural diversification. An internal node delimits the emergence of substructures in the context of other substructures of the molecular repertoire. Consequently, branches in trees describe network-like genealogies in which the evolution of substructures provides the foundation for evolutionary changes. Furthermore, branch lengths measure character state change at the structural level.

(d) Structural characters fulfill the requirement of character independence needed in phylogenetic analysis. Characters used to build trees of substructures represent molecular lineages associated with RNA evolving within organismal lineages, defining, for example, a species or tRNA molecules coding for either individual amino acids or with specific codon specificities. Therefore, molecular lineages and associated characters should be relatively independent of each other, satisfying the requirements of character independence in phylogenetic analyses. It is well known that character independence is difficult to achieve in traditional phylogenetic analyses. In contrast, characters used to build trees of substructures represent features in molecular lineages that are, for the most part, evolving independently of each other. Therefore, we argue that in this regard, structural characters are more advantageous than other types of characters that are generally used in phylogenetic reconstructions.

(e) The criterion of primary homology is determined by the feature of structure being studied and its associated evolutionary model, and how this feature relates to the substructural taxa analyzed. Features can describe the geometry (e.g., shape characters) or the branching, stability, and plasticity (e.g., statistical characters) of homologous substructural components. Homologous substructures represent those that are of the same kind (e.g., domains, stems, nucleotide base pairs) and respond to the same evolutionary model defining the character transformation sequences. For example, we reconstruct trees of coaxial stems corresponding to arms in tRNA, separate from trees of hairpin loops or trees of tRNA arms using statistical characters (Sun and Caetano-Anollés, 2008a). For a specific type of RNA molecule, we generally examine the entire group of RNA sequences representing the structures of the complete set of known sequences acquired at the RNA level. This will guarantee an extensive sampling of molecular variants in our studies.

(f) Substructural taxa representing molecular features larger than a nucleotide are necessarily the subject of information compression, which is generally dependent on how substructures are defined. The negative consequences of information compression would be loss of information or differential weighting of taxa. However, this impact decreases if the lost information constitutes noise and if models are derived from substructures of one kind. The conversion between amino acid and nucleotide sequences represents a typical example of information compression that has a natural rationale delimited by the genetic code. Indeed, the compression of RNA sequences into structures also has a natural rationale, the mapping of genotype (sequence) into phenotype (structure) enabled by the unique chemistry and folding of the RNA biopolymers (Ancel and Fontana, 2000; Higgs, 2000; Fontana, 2002). Because the distribution of sequences that fold into a same structure within neutral networks in RNA is approximately random, the mapping has "space covering" properties. In other words, all structures can materialize within relatively few mutational changes in sequence space, a property that has been confirmed experimentally using RNA functional switches (Schultes and Bartel, 2000). Based on these considerations, information compression in RNA is natural and should not bias significantly the exercise of phylogenetic reconstruction. Instead, we argue that it will uncover deeper phylogenetic signals that are embedded at higher levels of biological organization and are difficult to retrieve in an analysis of primary sequence.

(g) Trees of substructures provide by definition models of RNA structural evolution. When using shape characters, models describe evolutionary processes in which RNA molecules evolve from an originating substructure by addition of nucleotides and base pairs to substructural components. This process manifests as a sequence of events ordered in time. Consequently, the topologies of trees of substructures can be used to build models of RNA evolution. Generally, the phylogenetic relationship of stems determines the evolution of the overall shape of the RNA molecule that is typical of the molecular repertoire studied, and should be considered first. This means that trees of stems will set a framework in the model for the chronological addition of major substructures to the evolving

RNA. Once this framework is in place, trees of unpaired structural components define patterns of diversification that do not result in molecular multifurcation (branching), and therefore, provide evolutionary patterns of decoration of the evolving molecule.

2.4. *Potential limitations of the methodology*

In contrast to traditional phylogenetic trees reconstructed from nucleic acid sequences, trees derived from RNA structure establish ancestral–descendant relationships between the individual structural components of a nucleic acid molecule. Instead of nucleic acid bases, quantitative and geometrical attributes of RNA secondary structure are the characters analyzed. Superficially, when such kind of quantitative descriptions are chosen, the exercise of distinguishing among character states is to some extent subjective. Moreover, this method has also to deal with the trade-off between the accuracy of the characters that are defined and the amount of phylogenetic signal embedded in them. In other words, the amount of phylogenetic signal may be influenced by the way structural characters are selected and coded. However, studies have shown little if no discrepancy between results obtained using alternative quantitative character definitions (Caetano-Anollés, 2002b; Sun and Caetano-Anollés, 2008a–e, 2009, 2010). Furthermore, phylogenetic analysis of structure may also suffer from the same problems that affect reconstructions of phylogenies from sequences, such as mutational saturation, variation of evolutionary rates across sites, and covarion structure.

Character selection and scoring are also influenced by the structural complexity of the RNA molecules that are studied because the number of character states and the sparse evolutionary models for quantitative characters using structural characters in phylogenetic analysis are limited by currently available phylogenetic software, e.g., MacCalde (Maddison and Maddison, 2003), PAUP* (Swofford, 2003), and MrBayes (Huelsenbeck and Ronquist, 2001; Ronquist and Huelsenbeck, 2003). One way to deal with this problem is to combine character states so that the total number of character states is decreased and can be appropriately processed with currently available phylogenetic software by applying various weighting schemes. However, this practice sacrifices the amount of phylogenetic signal embedded in the combined character states.

Finally, we assure the accuracy of RNA secondary structures during character selection and coding by using robust structural models. The structural characters are coded based on available secondary structures, which are supported by structural information derived from nuclear magnetic resonance (NMR) spectroscopy, crystallography, genetics, biochemistry, and comparative sequence analysis. In some cases, RNA secondary structure can be defined by sequence and structural alignments and further complemented by the use of RNA folding software (Lorenz *et al.*, 2011).

3. Applications

Since its inception, our phylogenetic method has been applied to the study of many noncoding RNA molecules. Here we summarize results derived from our investigations of the structure and evolution of five types of RNA molecules, tRNA, 5S rRNA, RNase P RNA, SINE RNA, and rRNA.

3.1. *tRNA*

Generally, tRNA molecules are short and are composed of runs of ~73–95 nucleotides. The molecules generally adopt a typical cloverleaf-shaped secondary structure and fold into a L-like conformation in 3-dimensional (3D) space (Figure 2). This cloverleaf structure is made up of two structural and functional domains. The acceptor (Acc) and the pseudouridine (TCC) arms comprise the "top half" domain, while the "bottom half" domain is composed of the dihydrouridine (DHU) and the anticodon (AC) arms. A variable (extra) arm spans the TCC and AC arms. This molecular adaptor delimits the genetic code and the encoding of amino acids in proteins by establishing numerous interactions with many other macromolecules, many of which are associated with the ribosome. For each specific tRNA, a specific aminoacyl-tRNA synthetase attaches a specific amino acid to the Acc arm at one end of the L-shaped folded molecule, while a three-nucleotide sequence on the AC arm at the other end recognizes the complementary codon sequence in messenger RNA (mRNA) when embedded in the ribosomal ensemble. Detailed information regarding the number and organization of transfer RNA (tRNA) genes at the genome level using bioinformatics and comparative genomics approaches is becoming readily available with the increase of DNA sequencing of whole genomes (e.g., Tang *et al.*, 2009). New non-canonical functions of tRNAs and physiological roles for post-transcriptional modifications on tRNA residues have been demonstrated (Yona *et al.*, 2013; Torres *et al.*, 2014; Kirchner *et al.*, 2015; Albers and Czech, 2016; Huang and Hopper, 2016). For example, the studies of tRNAs and their modifications have played important roles in our understanding of biology, from the proposal of an ancient RNA world to their present role in the development and maintenance of complex organisms and even highly sophisticated features such as human cognition (Balke *et al.*, 2016). On the one hand, tRNA molecules contain conserved secondary structures. On the other hand, tRNA gene families show diverse evolutionary patterns identified in various lineages (e.g., Dowton *et al.*, 2003). In light of the identification of small (~42–66 nucleotides), armless tRNAs (Klimov and Oconnor, 2009; Watanabe and Yokobori, 2011; Jühling *et al.*, 2012; Wende *et al.*, 2014), it is therefore important to revisit the definition of functional tRNAs and their evolution.

The origins and early evolution of tRNA have been constantly explored from various perspectives, including nucleotide ligation, chiral selectivity of amino acids, genetic code evolution, and the organization of the ribosomal peptidyl transferase center (PTC). The proto-tRNA molecules may have evolved not only as adaptors but also as contributors to the composition of the ribosome (reviewed in Tamura 2015). Many hypotheses have been proposed to decipher the origin and evolution of tRNA molecules. For example, it is suggested that tRNA molecules originated by the assembly of two hairpin-like structures of RNA and the origin of tRNAs is polyphyletic (Di Giulio, 2012, 2013). Based on the reconstruction of ancestral sequences of many types of tRNAs, it is proposed that the evolutionary diversification of tRNA molecules is driven by the mutations in the second base of the anticodon sequence of tRNA but not correlated with the characteristics of the specified amino acid (Wang and Lavrov, 2011; Farias, 2013). This further indicates the co-evolutionary pattern between tRNAs and aminoacyl-tRNA syntheses to give rise to the system correlation between anticodons and amino acids (Farias *et al.*, 2014a). By reconstructing the ancestral sequences from all types of tRNA, Farias *et al.* (2014b) demonstrated that tRNA gene gave origin to ribosomes, particularly the PTC of the 23S rRNA, which could have originated by the fusion of ancestral sequences of proto-tRNAs.

Because of its central role in biology, it is believed that tRNA carries deep phylogenetic signal and its evolution reflects the evolution of the entire protein biosynthetic apparatus (Dick and Schamel, 1995). Indeed, this fundamental molecule has been the focus of many evolutionary investigations that explore the origins and evolution of life, such as the genetic code (Wong, 1975; Eigen and Winkler-Oswatitsch, 1981a, 1981b; Fitch and Upper, 1987; Di Giulio, 2000; Widmann *et al.*, 2005) and the ancient RNA world (Zuckerkandl *et al.*, 1971; Brooks *et al.*, 2002; Trifonov, 2004; Jordan *et al.*, 2005). The tRNA "core hypothesis" (Farias *et al.*,

2016) emphasizes the central role of tRNAs molecules in the origin and evolution of fundamental biological processes. The hypothesis focuses specifically on the transitions from the RNA world (Gilbert, 1986; Lazcano, 2012; Reyes-Prieto *et al.*, 2012; Neveu *et al.*, 2013) to the ribonucleoprotein world (Farias *et al.*, 2007; Guimarães, 2011), and from this world to modern biological systems. These transitions were driven by three kinds of tRNAs transformations, to wit, tRNAs leading to both mRNA and rRNA.

We retrieved the entire set of 571 tRNA sequences with cloverleaf secondary structures from Part 2 (Compilation of tRNA Sequences) of the September 2004 edition of the Bayreuth tRNA Database (Sprinzl and Vassilenko, 2005; Jühling, *et al.*, 2009) and scored a total of 42 structural characters to infer the evolutionary histories of organisms and tRNA substructures (Sun and Caetano-Anollés, 2008a). Phylogenetic results generated from a total of 56 data matrices sorted according to taxonomic distribution, amino acid specificity, and biological sources show that the monophyly of tRNAs belonging to the three superkingdoms or expressing different amino acid specificities was not revealed in most of the parsimonious trees that were reconstructed. This is probably due to loss of phylogenetic signal or because tRNA diversification predated organismal diversification. The substructure trees derived from both geometrical and statistical characters are congruent revealing the same evolutionary patterns — the Acc arm is the most basal substructure, generally followed in order by the TCC, AC, and DHU and variable arms. With these two congruent lines of phylogenetic evidence, we proposed an evolutionary model that explains the origin and evolution of the major functional and structural components of tRNA (Sun and Caetano-Anollés, 2008a; Figure 2). This model considers that the modern cloverleaf tRNA structure resulted from gradual addition of structural components to the growing molecule, either by insertion of single or multiple nucleotides or by partial or total duplications. In this model, short RNA hairpins with stems homologous to the ACC arm of present-day tRNAs were extended with regions homologous to the TCC and AC arms. The DHU arm was then incorporated into the resulting three-stemmed structure to form a proto-cloverleaf structure. The variable region was the last structural addition to the molecular repertoire of evolving tRNA substructures. The results and the tRNA model support the ancestral nature of the Acc and the TCC arms, the cornerstone of the genomic tag hypothesis that postulates tRNAs were ancient telomeres in the RNA world (Weiner and Maizels, 1987; Maizels and Weiner, 1994).

On the basis of tRNA structural phylogenies, we were able to provide novel insights into the origin and diversification of cellular life and viruses and into the evolution of the genetic code and amino acid charging (Sun and Caetano-Anollés, 2008b, 2008c). The generation of organismal timelines that were based on evolutionary distances among alternative or competing hypotheses of origin and diversification revealed Archaea was the most ancestral superkingdom, followed by viruses, and superkingdoms Eukarya and Bacteria in that order (Sun and Caetano-Anollés, 2008b). These results support conclusions from recent phylogenomic studies of protein architecture (Wang *et al.*, 2007; Wang and Caetano-Anollés, 2009; reviewed in Caetano-Anollés *et al.*, 2014). Strikingly, constraint analyses showed that the origin of viruses was not only ancient but was also linked to Archaea. Our findings have important biological and evolutionary implications. These results support the notion that the archaeal lineage was very ancient, resulted in the first organismal divide, and predated diversification of tRNA function and specificity. Results are also consistent with the concept that viruses contributed to the development of the DNA replication machinery during the early diversification of the living world (Sun and Caetano-Anollés, 2008b).

Timelines of amino acid charging and codon discovery show that charging of selenocysteine (Sec), tyrosine (Tyr), serine (Ser), and leucine (Leu) were ancient, while specificities related to asparagine (Asn), methionine (Met), and arginine (Arg) were derived (Sun and

Caetano-Anollés, 2008c). The timelines also uncovered an early role of the second and then first codon bases, identified codons for alanine (Ala) and proline (Pro) as the most ancient, and revealed important evolutionary takeovers linked to the loss of the long variable arm in tRNA. The lack of correlation between ancestries of amino acid charging and encoding indicated that the separate discoveries of these functions reflected independent histories of recruitment. These histories were probably curbed by cooptions and important takeovers during early diversification of the living world (Sun and Caetano-Anollés, 2008c).

It has been proposed that the class II tRNA structure of $tRNA^{Ser}$ and the arrangement of serine codons in the genetic code provide clues to the early evolution of tRNA and the genetic code (Bernhardt, 2016). For the first time, our method allowed to address the evolutionary significance of the long variable arm in tRNA from a phylogenetic perspective (Sun and Caetano-Anollés, 2008d). We showed that class II tRNA molecules containing a long variable arm, including $tRNA^{Sec}$, $tRNA^{Ser}$, $tRNA^{Tyr}$, and $tRNA^{Leu}$, were ancestral compared to those lacking this structural feature. This suggests that Sec, Ser, Tyr, and Leu were among the first amino acids to be charged by their cognate tRNAs and that they may represent the first group of amino acids with functional specificities linked to modern biochemistry. Results also suggest that the stop codon UGA, which also codes for Sec, may be the oldest codon to have a modern functional role in the history of the genetic code. Finally, the charging of amino acids by cognate tRNAs appear to have occurred once the canonical cloverleaf structure was fully realized in evolution but prior to the diversification of amino acid specificities and superkingdoms of life (Sun and Caetano-Anollés, 2008a, 2008b).

3.2. *5S rRNA*

5S ribosomal RNA (5S rRNA) is a universal integral component of ribosomes of all living organisms. Its corresponding gene is commonly identified in archaeal, bacterial, and nuclear genome sequences. Generally considered an integral component of the large subunit of the ribosome, the 5S rRNA molecule plays fundamental roles during protein synthesis. It probably acts as a signal transducer between the PTC and domain II of the large subunit responsible for translocation (Bogdanov *et al.*, 1995; Dokudovskaya *et al.*, 1996) and between regions of 23S rRNA responsible for ribosomal function (Kouvela *et al.*, 2007). 5S rRNA is also a determinant of stability for the large subunit (Holmberg and Nygard, 2000). It can also act as an evolutionary resource for other genetic components. For example, SINE3 molecules, a class of SINE RNAs, are indeed evolutionarily derived from 5S rRNA (Kapitonov and Jurka, 2003). Despite extensive studies undertaken during the past decades, detailed understanding of 5S rRNA function is still lacking (Bogdanov *et al.*, 1995; Barciszewska *et al.*, 2000, 2001; Szymanski *et al.*, 2003) and the evolutionary histories of the molecular components of this universally conserved molecule are still unclear. It is known that the ribosome without 5S rRNA is functionally inactive. However, the question about the specific role of this RNA in functioning of the translation apparatus is still open. Recent study showed that the 5S rRNA is located in the central protuberance and participates in formation and structure of this projection, which also include the 23S rRNA (or alternatively 28S in eukaryotes) and several proteins (Gongadze, 2012). It is further suggested that 5S rRNA acts as a mediator between the functional centers of the ribosome by forming intersubunit bridges and tRNA-binding sites (Gongadze, 2012), indicating that 5S rRNA acts as a physical transducer of information, facilitating communication between the different functional centers and coordinating of the multiple events catalyzed by the ribosome.

Regardless of its origin, 5S rRNA can always be folded into a common secondary structure (Figure 1). This structure contains five helices (I–V), two hairpin loops (C and D), two internal loops (B and E), and a multiloop (hinge) region (A) connecting helices I, II, and V. This three-branched general structure has been confirmed by a number of structural studies and comparative sequence analyses. Recently, the largest diversity of secondary structures in mitochondrial 5S rRNA is revealed in stramenopiles and the most unconventional case in brown algae, where the helix I is replaced by a closed hairpin resulting in an open three-way junction (Valach, 2014). We explored the evolutionary history of 5S rRNA with structural characters. Our goal was to identify the ancestral functional and structural components or substructures. We retrieved the entire set of 1371 5S rRNA sequences from the September 2005 edition of the 5S Ribosomal RNA Database (Szymanski *et al.*, 2002). It is well known that although programming algorithms for secondary structure prediction have improved, predicted 5S rRNA structures are not always satisfactory (Azad *et al.*, 1998; Ding and Lawrence, 1999; Mathews *et al.*, 1999). We therefore decided to select only those structures that match a set of predefined requirements. We first predicted 5S rRNA secondary structure with RNAFOLD (Hofacker, 2003). Approximately one half of the available sequences (666), representing a comprehensive sampling among the three superkingdoms of life (89 Archaea, 168 Bacteria, and 409 Eukarya), were then selected for further study as 5S rRNA sequences folded into minimum free energy structures that were compatible with phylogeny and known 3D models of structure.

Phylogenetic analysis of 46 structural characters in 5S rRNA showed that, individually, the three superkingdoms of life were not monophyletic (Sun and Caetano-Anollés, 2009). As with tRNA, we reasoned that this is probably due to loss of phylogenetic signal in the secondary structure of 5S rRNA. Alternatively, 5S rRNA diversification could have predated true organismal diversification and structures were the result of primitive processes during evolutionary stages prior to superkingdom differentiation. By applying the total evidence approach, we conducted phylogenetic analysis of combined structure and sequence data. Results showed that both Bacteria and Eukarya were each monophyletic, while Archaea was paraphyletic and occupied a basal position in the universal tree. Interestingly, this archaeal rooting of the tree of life again supports previous studies based on tRNA (Xue *et al.*, 2003, 2005; Di Giulio, 2007; Wong *et al.*, 2007; Sun and Caetano-Anollés, 2008b) and proteomes (Wang *et al.*, 2007; Wang and Caetano-Anollés, 2009).

Based on phylogenetic trees of substructures reconstructed from geometrical characters describing the complete data set of 666 5S rRNA sequences, an evolutionary model of 5S rRNA structure is proposed. According to this model, short RNA hairpins with stems homologous to helix I (the most ancient substructure) of present-day 5S rRNAs were extended with regions homologous to the branch containing helices II and III, then followed by a third branch composed of helices IV and V, to form a proto-three-branched structure. Specifically, the evolutionary history of 5S rRNA proceeded via two routes in Archaea and Bacteria/Eukarya, respectively. Both routes started with helix I, followed by helix III. At this point, the two routes departed. For Archaea, helix III is followed by helix V; helices IV and II are the last components added to the three-branched structure. For Bacteria/Eukarya, the order is helical arms I, III, II, and V and IV; in other words, helix I is followed by the branch harboring helices III and II, then the branch containing helices V and IV. Interestingly, the two-route scenario that distinguishes molecules between Archaea and Bacteria/Eukarya agrees with the two-route scenario of the evolutionary model of tRNA structure (Sun and Caetano-Anollés, 2008a). Both molecules suggest the early divergence of Archaea from Bacteria/Eukarya. More importantly, the existence of these two evolutionary routes is also compatible with whole-genome analysis of protein complements and domain combinations that suggest an early split of the archaeal lineage from a protein architecture-rich communal world by reductive genomic tendencies that were protracted and ultimately led to superkingdom Archaea and then to Bacteria and Eukarya (Wang and Caetano-Anollés, 2006, 2009; Wang *et al.*, 2007).

A terminal stem-loop structure (a hairpin) has also been proposed to be the starting evolutionary component of tRNA molecules (Woese, 1969; Hopfield, 1978; Eigen and Winkler-Oswatitsch, 1981a, 1981b; Bloch *et al.*, 1985; Dick and Schamel, 1995; Tanaka and Kikuchi, 2001; Widmann *et al.*, 2005; Sun and Caetano-Anollés, 2008a). Similarly, a terminal hairpin structure may also be the starting structure in the evolution of RNase P RNA (Sun and Caetano-Anollés, 2010), SINE RNA (Sun *et al.*, 2007), and rRNA (Harish and Caetano-Anollés, 2012). The evolutionary importance of hairpin structures has been emphasized by the genomic tag hypothesis (Weiner and Maizels, 1987; Maizels and Weiner, 1994). Based on these results, we hypothesize that tRNA, SINE RNA, RNase P RNA, 5S rRNA, and rRNA may have shared the same common ancestral structure during early evolutionary history of the RNA world, a primordial stem-loop hairpin structure. We speculate that further studies in other types of RNA molecules may provide further evidence to support this hypothesis.

3.3. *RNase P RNA*

The RNase P enzyme, one of the two natural ribozymes (the other is the ribosome) that are present in all living organisms, is generally composed of one RNA subunit and one or more protein subunits in Bacteria and Archaea/Eukarya, respectively (for exceptions see Manam and Van Tuyle, 1987; Wang *et al.*, 1988; Rossmanith and Karwan, 1998; Thomas *et al.*, 2000b; Salavati *et al.*, 2001; Holzmann *et al.*, 2008; Gobert *et al.*, 2010; Gutmann *et al.*, 2012; Taschner *et al.*, 2012). The RNA-based catalytic activity has been preserved during evolution in all three superkingdoms (Pulukkunat and Gopalan, 2008). The ribonucleoprotein functions as a phosphodiesterase and carries out the endonucleolytic cleavage of the precursor sequence from the 5′-end of a primary tRNA transcript to form the mature 5′-end, a process that catalyzes the hydrolysis of the phosphate backbone of pre-tRNA at the 5′-end (reviewed in Karwan *et al.*, 1995–1996; Pace and Brown, 1995; Frank and Pace, 1998; Hartmann and Hartmann, 2003; Evans *et al.*, 2006; Kazantsev and Pace, 2006; Walker and Engelke, 2006; Ellis and Brown, 2010). The catalytic function can be conducted by the RNA subunit independent of protein subunits, indicating that it resides in the RNA subunit (Guerrier-Takada *et al.*, 1983; Pannucci *et al.*, 1999; Thomas *et al.*, 2000a; Kikovska *et al.*, 2007). Complete functional exchangeability of the highly dissimilar enzymes is demonstrated between *Saccharomyces cerevisiae* and *Arabidopsis thaliana*, suggesting that the RNase P family, with its diverse structure and uniform function, is derived from convergent evolution (Weber *et al.*, 2014).

RNase P RNAs (RPRs) in bacteria have been widely studied and can be divided into two independently folding domains involved in both substrate cleavage, the catalytic (C) domain, and substrate binding, the specificity (S) domain (Pan, 1995; Loria and Pan, 1996; Figure 3). The phylogeny of the structural domains of RNase P RNA was investigated in *Leptospira* (Ravishankar *et al.*, 2014). Structurally, the S domain is composed of helical structure P7 and everything else distal to P7, while the C domain is composed of the rest of the molecule. Furthermore, five universally distinct conserved regions (CR I–V) that are distal to each other in the primary sequence are proposed in RPRs (Chen and Pace, 1997). These five regions consist of the universally conserved core structure. It is further suggested that the S domain comprises CR II and III, and that the C domain comprises CR I, IV, and V (Torres-Larios *et al.*, 2006). Despite the overwhelmingly helical components in the tertiary fold, it is interesting that both domains are nonhelical: CR II

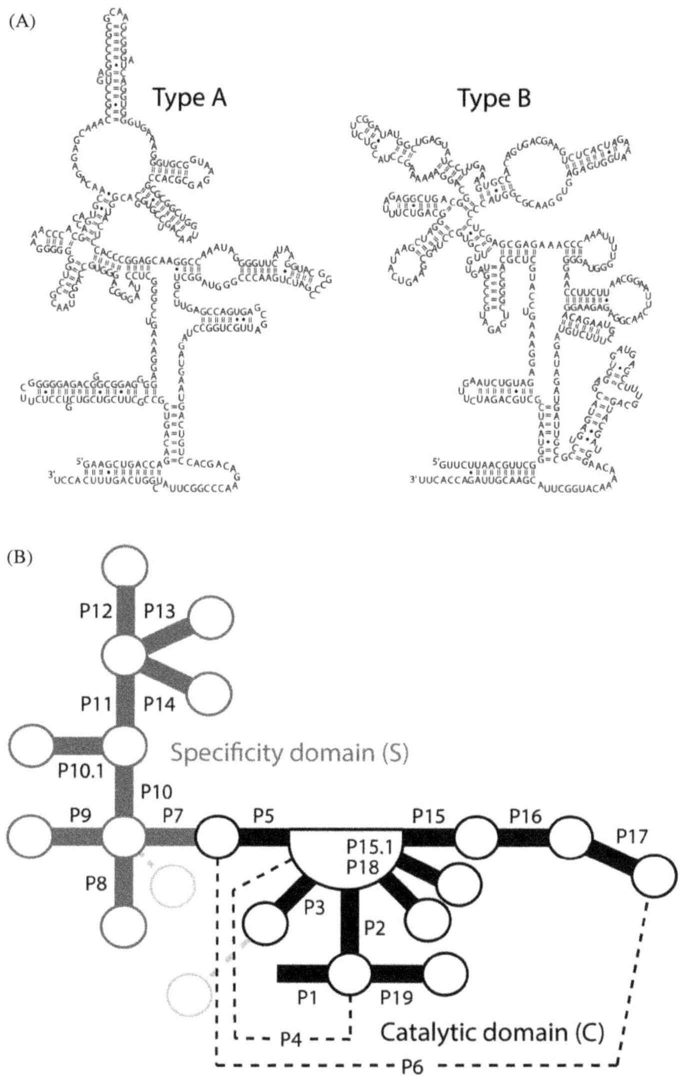

Figure 3. Secondary structure models of RNase P RNA. (A) Bacterial type A and type B molecules from *Escherichia coli* (Harris *et al.*, 2001) and *B. subtilis* (Haas and Brown, 1998), respectively. (**B**) Consensus structure showing the catalytic and specificity domains highlighted with different shades. Extension stems present in RNAse MRP in eukaryotes are shown as dashed lines in light gray. Loops are indicated with circles, stems with solid lines, and pseudo-loops (P4 and P6) with dashed lines connecting interacting segments of the molecule.

and III form two interleaving T-loop motifs, whereas CR I, IV, and V are part of loops and turns. In terms of functions, the C domain contains the entire active site and binds the Acc stem/5′-leader and the ACCA sequence at the 3′-end (by Watson-Crick base pairing mechanism) of the pre-tRNA substrate. It also cleaves the leader sequence of the pre-tRNA in the presence of the bacterial RNase P protein (RPP) cofactor (Guerrier-Takada and Altman, 1992; Pan, 1995; Pan *et al.*, 1995; Loria and Pan, 1996; Christian *et al.*, 2002; Harris and Christian, 2003). In contrast, the S domain binds the TCC stem-loop region of the pre-tRNA substrate and confers the substrate specificity. As found in both RNase P and RNase MRP, yeast telomerase contains a P3-like domain, which binds the Pop6/7 heterodimer and the Pop1 protein and is essential for telomerase activity on telomeres in vivo (Lemieux *et al.*, 2016).

Given the distinct functions of these two structural domains during enzymatic catalysis, the S and C domains may have different evolutionary histories. In fact, Altman and Kirsebom (1999) proposed that an earlier form of the RPR in the RNA world lacked the S domain simply because this domain was not needed for the binding of the substrate. This hypothesis indicates that between the two functional and structural domains of RPR, the C domain is more ancient than the S domain, and the S domain is a rather derived structure. In fact, substrate specificity can be altered without changing the basic cleavage reaction by modifying the S domain of *Bacillus subtilis* (Mobley and Pan, 1999). This strongly suggests that the C domain is ancestral. However, current evolutionary studies do not provide convincing evidence in support of this hypothesis. Pace and coworkers have defined the "minimal" RPR, the smallest molecule needed to carry out the hydrolysis reaction associated with RNase P (Siegel *et al.*, 1996). However, this minimal RPR contains molecular components from both S and C domains. To date, the evolutionary histories of the molecular components of each of these two folding domains remain unclear.

To explore the origin and evolution of these two structural and functional domains in RPR, we reconstructed phylogenetic trees derived from 129 structural characters in 133 RPRs using our cladistic method (Sun and Caetano-Anollés, 2010). These RPR sequences with secondary structures were retrieved from Release No. 12 of the 2005 edition of the RNase P Database (Brown, 1999). Partial RNA sequences were excluded and the remaining sequences folded into secondary structures that were compatible with the RPR phylogeny and the known 3D models of RPR structure (Kazantsev *et al.*, 2005).

Phylogenetic analysis of structure revealed the monophyly of two superkingdoms of life, Archaea and Eukarya. In contrast, molecules of bacterial lineages clustered into two distantly related groups. One group was basal and contained molecules from 15 Gram positive bacteria (low-GC subdivision having type B RPRs; Figure 3) plus one green nonsulfur bacterium (*Thermomicrobium roseum* with type C RPR). The other group, a sister clade of Archaea, was a largely unresolved clade composed of molecules from other bacteria with type A RPRs. Phylogenetic analysis of combined structure and sequence data revealed that both Eukarya and Archaea were monophyletic, while Bacteria remained paraphyletic. Our results support superkingdom monophyly and are in agreement with studies by Collins *et al.*, (2000), who demonstrated that phylogenetically informative characters are indeed embedded in the secondary structure of RPR molecules and produce trees that recover the three superkingdoms. One of the remarkable benefits of using structural characters in these evolutionary studies is that phylogenetic relationships are inferred from distantly related RPR and RNase MRP RNAs with scarcely alignable sequences because structures are more conserved than sequences. Recent studies have indicated that RPR molecules are also suitable for phylogenetic analysis of closely related bacterial taxa and have the potential as a tool for species discrimination (Tapp *et al.*, 2003; Birkenheuer *et al.*, 2002). One distinctive feature of RPR that makes it a reliable molecular marker is that there is only one copy of its gene per genome. This makes its gene less likely to be compromised by interspecific lateral gene transfer.

Phylogenetic trees of substructures allowed proposing an evolutionary model of RPRs structure (Figure 4). Phylogenetic analyses of stem substructures derived from three partitioned data sets according to superkingdoms revealed similar topologies as the topology derived from the complete data set. In contrast, analyses of bulges, hairpin loops, and other unpaired segments produced uninformative patterns. In this model, RPRs start from a hairpin structure equivalent to P12 (a helical substructure in S domain) in present-day RPR molecules. This is followed by four helical regions in the C domain (in the order P1, P3, P4, and P2) and four stems in the S domain (P10/P11, P9, P8, and P7). Five additional stems in the S domain are then added in defined order. Later on, one (P18) and two (P13 and P14) helices in the C and S domains are added followed by a last group of elements composed of eight helices in the C domain and one helix (P10.1) in the S domain.

With the exception of P12, the C domain appears generally more ancient than the S domain on all of the phylogenetic trees of stem substructures. For example, four helical regions in the C domain, P1–P4, which are also reconstructed as the structural components in the conserved minimal RPR (Siegel *et al.*, 1996), are ancient structures in comparison to many other helical structures in both C and S domains. It has been suggested that the S domain facilitates substrate binding in the ribozymic reaction, but is dispensable for catalysis, and the RPPs can compensate to aid in substrate binding in its absence (Krasilnikov *et al.*, 2003, 2004). This indicates the ancestral nature of the C domain, as proposed by Altman and Kirsebom (1999). Our results add an additional line of evidence and are largely in support of this hypothesis.

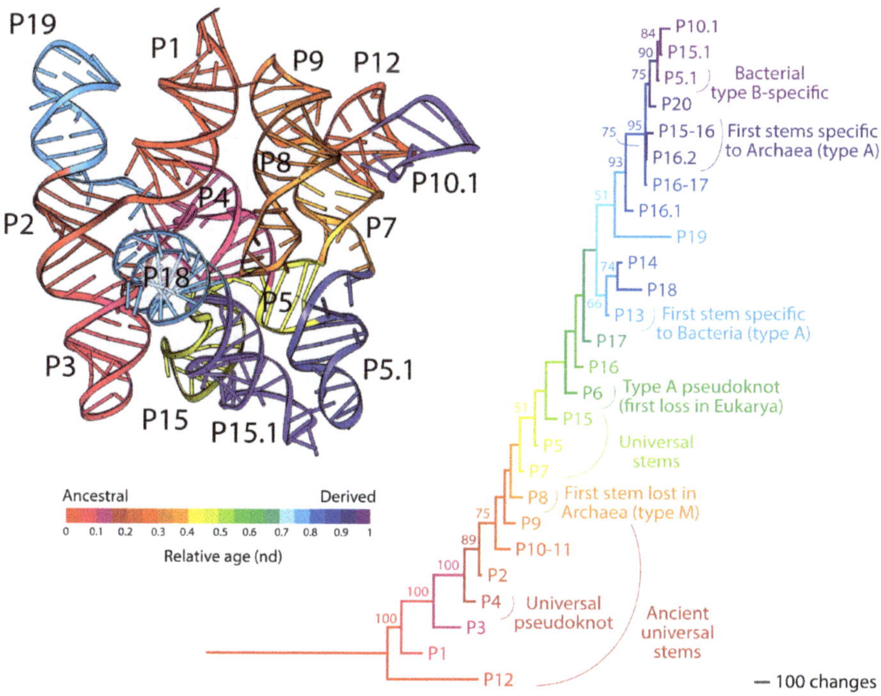

Figure 4. The evolutionary history of the RNase P RNA molecule. A tree of stem substructures was reconstructed from characters describing molecular geometry. The branches of the tree were colored according to the relative age (nd) of substructures on a scale that goes from 0 (origin of the molecule) to 1 (the present). The same ancestry color scale was used to paint an atomic 3D model visualized using the ribbons format. Note that base of the oldest P12 stem structure interacts with tRNA and that the oldest RNase P protein (RPP) cofactor interacts with the very ancient P2 and P3 stems, and laterally with the ancient P4 pseudoknot, all part of the catalytic domain (C) of the molecule. The tree of stem substructures also describes how stem structures lost by Archaea and then Eukarya are more basal than structures specific to Bacteria and Archaea, all of which are more derived than the universal substructures at the base of the rooted tree. The figure was modified from Sun and Caetano-Anollés (2010).

RPRs catalytically cleave the 5′-end of the pre-tRNA sequence to generate the mature tRNA. In doing so, RPRs interact with the ancient top domain of the pre-tRNA composed of the TCC and Acc arms. Previous studies indicate that the top domain of tRNA evolutionarily predated the bottom domain composed of the DHU and AC arms (Maizels and Weiner, 1994; Sun and Caetano-Anollés, 2008a). Based on our study, the phylogenetic trees of substructures show that helices P1–P4 are the ancestral elements. These are the only four conserved stems of the C domain in the universal consensus minimum structure of RPR (Evans *et al.*, 2006). These results suggest a coevolutionary scenario between the top domain of tRNA (the substrate) and the C domain of the RPR (the catalyst). Given the supporting functions (instead of catalytic functions) carried out by the S domain, the evolutionary significance of P12 is worthy of further study. Furthermore, it should also be noted that in comparison to the currently available RPRs (Brown, 1999), especially with the generation of sequences by metagenomic studies (Zhu *et al.*, 2007), the number of RPR sequences sampled in this study is limited. Consequently, conclusions drawn should be taken with caution. Future studies with comprehensive sampling would also provide insights into the evolutionary patterns of various types of RPR molecules, the evolutionary relation between the RPRs and their associated proteins, and the phylogenetic utility of RPR molecules.

3.4. *SINE RNA*

In eukaryotic genomes, SINEs are widely distributed mobile retroelements (Kramerov and Vassetzky, 2005; Fawcett *et al.*, 2006) but as especially effectively propagated in mammalian species (reviewed in Kramerov and Vassetzky, 2011). Approximately 80–500 bp long SINEs normally have genomic copy numbers ranging from a few hundred to more than a million (Jurka, 1995; Okada and Ohshima, 1995; Kramerov and Vassetzky, 2005). They may play important regulatory roles closely related to transcription and translation of higher eukaryotes (Moran and Gilbert, 2002; Dewannieux *et al.*, 2003; Pélissier *et al.*, 2004; Ohshima and Okada, 2005). Novel biological functions of SINEs have been described (Ponicsan *et al.*, 2010). New conserved domains in SINEs and their diversity are demonstrated during animal evolution (Luchetti and Mantovani, 2013). A recent study using 454-pyrosequencing revealed that the genomic landscape of SINEs is lineage specific and evolutionary pathways could rapidly change in any organismal lineage (Gallus *et al.*, 2015). They substantially expand the genomic space and introduce structural variation to the genome. SINEs have the potential to mutate genes, to alter gene expression, and to generate new parts of genes. A balanced distribution and controlled activity of such properties is crucial to maintaining the organism's dynamic and thriving evolution (Schmitz, 2012). These transposable elements can be the causes of cancers because they can produce changes in the genetic make up, such as deleterious mutations, gene disruption, and chromosome rearrangements (Chénais, 2013). The usage of SINEs in phylogenetic studies has also been demonstrated (Shedlock *et al.*, 2000; Ray *et al.*, 2006; Nikaido *et al.*, 2007; Konkel *et al.*, 2010; Ray *et al.*, 2015).

SINEs are believed to be derived from tRNAs or more exceptionally from 7SL or 5S RNAs (Jurka, 1995; Kapitonov and Jurka, 2003). The common composite structure of SINEs usually consists of a 5′-tRNA-related segment followed by a tRNA-unrelated segment of variable size (Okada and Ohshima, 1995; but for an exception in Piskurek *et al.*, 2003). In all cases, SINEs harbor in their tRNA-related segment the internal promoter recognized by the RNA polymerase III (Pol III) machinery (Arnaud *et al.*, 2001). Therefore, SINE-specific transcription completely depends on the Pol III transcription system. Several vertebrate and invertebrate SINE families were grouped into three super-families, CORE-SINE (Gilbert and Labuda, 2000), V-SINE (Ogiwara *et al.*, 2002), and

AmnSINE (Nishihara *et al.*, 2006), based on the conserved nature of the central region. Given the strong conservation of the central domain of these SINEs over a long evolutionary period, it is suggested that this region might have been subject to some form of positive selection (Gilbert and Labuda, 2000; Ogiwara *et al.*, 2002; Nishihara *et al.*, 2006).

To date, little is known about SINE RNA structure and its evolutionary significance. This is especially true for tRNA-related SINEs. Only two tRNA-related SINE structures have been experimentally studied (Rozhdestvensky *et al.*, 2001; Kawagoe-Takaki *et al.*, 2006). Recently, Deragon and Zhang (2006) probed the secondary structure of two tRNA-related plant SINE RNAs from *Brassica oleracea* and *Arabidopsis thaliana* and found that these molecules did not conserve the folding patterns of ancestral tRNAs. However, they still presented similar RNA structures despite being largely unrelated at the primary sequence level. Furthermore, by using a bioinformatic approach, the RNA secondary structure of different SINEs corresponding to known SINE families in Brassica and Arabidopsis were also predicted (Sun *et al.*, 2007) to test if similar RNA folding patterns could be observed for tRNA-related SINE families outside the Brassicaceae. More than 50 tRNA-derived SINE families have now been described in mammals, fishes, and plants (Kramerov and Vassetzky, 2005). Consensus sequences from many of these SINE families are not robust (they were generated from the alignment of a small number of copies) but were nevertheless used to predict the corresponding RNA structures. Surprisingly, the diversity of structures observed for most of these SINE families is similar to that observed for SINEs of Brassicaceae and shows a clear conservation of patterns. These RNAs either fold as two extended stem loops or adopt a structure comprising three stem loops, which is the most common structure for tRNA-related SINEs.

Using our cladistic method, we compared the different SINE RNA structures to test more rigorously the relatedness of the different

SINE RNA structures. We scored 37 substructural characters from 20 representative eukaryotic SINE RNAs. Phylogenetic analysis of these structural characters showed that SINE RNAs started as simple structures with a single tRNA-related segment and became more complex in some lineages or stayed simple in others. It is noteworthy that SINE length associates with increases in structural complexity, suggesting SINE sequences increase in length during evolution.

Based on the phylogenetic trees of substructures from 20 representative eukaryotic SINE RNAs, a general model of structural evolution for tRNA-related SINEs was proposed (Sun *et al*., 2007). Phylogenies describing the evolution of substructures established clearly that the SINE RNA molecule has an origin in a GC-rich tRNA-related segment. A timeline for the evolution of SINE molecules can be inferred directly from the phylogenetic trees of substructures. This timeline represents a global model of evolution of SINE RNA structure that suggests modern SINEs resulted from gradual addition of structural components to a tRNA-related founder. Overall, the proposed model shows that the SINEs could have evolved and amplified from a single-stem loop to two and three, as inferred from the phylogenetic trees of substructures. To accomplish this process, SINEs either amplify the tRNA-related segment or add short or long tRNA-unrelated segments to their molecules.

3.5. *rRNA*

In addition to small molecules such as tRNAs and SINE RNAs, our method has proven useful to study the evolution of structure of complex molecules such as rRNA, providing insights into the evolution of the ribosome and phylogenetic support to theories of ribosome evolution. It is hypothesized that the ribosome is the link between the RNA world and cellular (last universal common ancestor) approaches to the study of cellular evolution based on the evidence that the entire set of tRNAs for all 20 amino acids are encoded in both the 16S and 23S rRNAs of *Escherichia coli* (Root-Bernstein and Root Bernstein, 2015). Recent advances in solving high-resolution structures of the ribosomal complex (Yusupov *et al.*, 2001; Schuwirth *et al.*, 2005; Selmer *et al.*, 2006) resulted in a scaffold for evolutionary tracings (Caetano-Anollés, 2002b). Initial studies showed patterns of evolution in intersubunit bridge contacts and tRNA binding sites that were remarkable (Caetano-Anollés, 2002b). These patterns were consistent with the proposed coupling of tRNA translocation and subunit movement (Yusupov *et al.*, 2001). Results also supported the concerted evolution of tRNA binding sites in the two rRNA subunits and the ancestral nature of both PTC site and the functional relay of the penultimate stem of the small subunit (SSU).

Further advances in our methods allowed reconstruction of trees of substructures of rRNA and evolutionary heat maps of the entire ribosomal ensemble (Harish and Caetano-Anollés, 2012). The tree of substructures provides a chronology that aids our understanding of which parts of the ribosome are ancestral and which are derived. Over 20,000 rRNA sequences from the European ribosomal RNA database (Wuyts *et al.*, 2004) annotated with secondary structures derived from comparative sequence analysis are scored to generate comprehensive data matrices of RNA substructures. Analysis of SSU rRNA confirmed that the penultimate stem (S49 or helix 44 in the bacterial model), the critical SSU component of the subunit interface (Yusupov *et al.*, 2001) and the proposed ribosomal functional relay (Cate *et al.*, 1999; Frank and Agrawal, 2000), was the most ancestral substructure of the ensemble. Figure 5 shows the evolution of the SSU ribosome. Analysis of the LSU rRNA suggested a late and more complicated structural origin. Interestingly, ancient substructures were located at the center of the ribosomal complex surrounding the intersubunit interface and were clearly linked to ratchet mechanics. In other words, the ancestral substructures of the SSU and LSU form a functional core. Our observations are compatible with an origin of

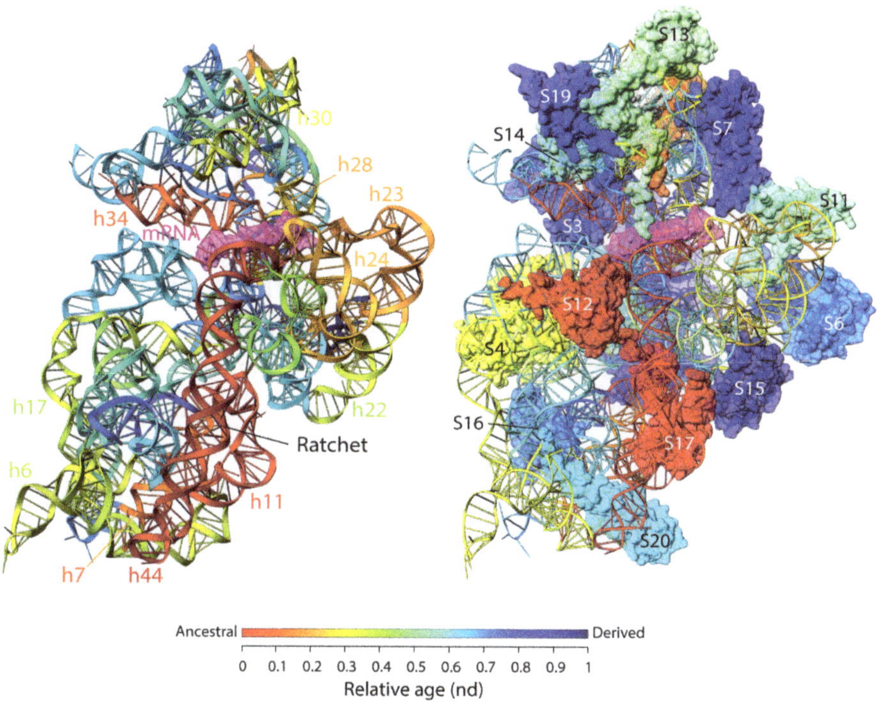

Figure 5. An evolutionary 3D heat map of the small subunit (SSU) of the ribosome showing the relative age of rRNA substructures and ribosomal proteins. Phylogenetic trees of rRNA stem substructures and proteins are reconstructed. Ancestries from the trees are mapped onto the 3D structure to reveal an evolutionary model describing the relative ages of rRNA substructures (left) and ribosomal proteins (right). The penultimate stem in red can be easily identified across most of the surface of the molecule.

the ribosome in structures that were not linked with modern protein synthesis. It is intuitive that an ensemble as complex as the ribosome did not evolve at once. Our trees of substructures and timelines support the progressive evolution of this complex biosynthetic machinery from a much simpler proto-ribosomal structure. However, the origins of the molecule did not lie in the functional core as one might expect. Instead, ancient and derived substructures occupied the same regions of the molecular ensemble. A complicated history of recruitment of substructures seems to dominate evolution of these complex molecules.

Using our method we were also able to test theories of RNA structure evolution (Ancel and Fontana, 2000; Fontana, 2002). We traced the complete repertoire of ribosomal structural characters, lineage-by-lineage, in the universal phylogenetic tree of rRNA molecules (Caetano-Anollés, 2002b; Caetano-Anollés, 2015; Caetano-Anollés and Caetano-Anollés, 2015). This tracing showed patterns of evolutionary change in rRNA and allowed the reconstruction of hypothetical ancestral molecules. The exercise revealed a tendency toward molecular simplification, especially in highly variable regions of the molecules. This tendency was maximal in rRNA from *Encephalitozoon cuniculi*, an amitochondriate microsporidian endoparasite with a highly reduced genome and protein complement. The exercise also showed reduction of ribosomal structural change with time that occurred concomitantly in both ribosomal subunits, which is compatible with plastogenetic congruence and structural canalization (Ancel and Fontana, 2000; Fontana, 2002).

4. Conclusions

In addition to the traditionally known RNA molecules (e.g., rRNA, tRNA, and mRNA), a large number of novel nonprotein coding RNA (ncRNA) systems has been documented in various model organisms in recent years (Berezikov *et al.*, 2006; Ruby *et al.*, 2006; Axtell *et al.*, 2007; Brennecke *et al.*, 2007; Houwing *et al.*, 2007; Kasschau *et al.*, 2007; Lu *et al.*, 2007; Zhao *et al.*, 2007). Transcriptomic and bioinformatic studies suggest the existence of thousands of ncRNAs and many types of novel RNAs in various genomes (Brosius, 2005; Cheng *et al.*, 2005; Birney *et al.*, 2007; Washietl *et al.*, 2007; van Bakel, 2010; Morris, 2012; Palazzo and Lee, 2015). The recent increase in rapid and inexpensive RNA sequencing has led to the discovery of a myriad of novel ncRNA species, shedding light on the variations in small RNA expression amongst different organisms, developmental stages and disease states (Cech and Steitz, 2014). These studies have uncovered a truly remarkable and complex modern RNA world. Functional analysis of these ncRNAs is posing a great challenge and is attracting tremendous attention from scientists in various fields. One excellent example that demonstrates the significance of these ncRNAs is the 2006 Nobel Prize in Medicine and Physiology awarded to Andrew Fire and Craig Mello for providing new insights into ncRNA-based gene regulation. The cladistic method that we here describe has been applied to the evolutionary study of a number of ncRNAs, including rRNA, 5S rRNA, tRNA, RNase P RNA, and SINE RNA. The method embeds structure directly in phylogenetic analysis and generates intrinsically rooted phylogenetic trees without the need of outgroups. We have exemplified the potential of this novel phylogenetic approach by focusing on several fundamental molecules that are involved in or are linked to protein synthesis. These studies provide novel insights into important evolutionary questions surrounding the emergence of cellular life and viruses and the origins and evolution of genetic code and the protein biosynthetic machinery. Because trees of life generated from these ncRNA molecules establish evolution's arrow, it becomes possible to establish the location of the root on the tree of life. We here show that a common topology emerges congruently from the analysis of these ncRNA molecules, which indicates the Archaea is the most ancient lineage on Earth. This result is important because the root of the tree of life has been debated over decades, with controversy largely stemming from the various rooting approaches that have been used and the alternative evolutionary scenarios that had been derived. The rooting strategy that we apply to the study of ncRNA has also been extended to the study of protein domain structure at fold and fold superfamily levels, revealing again the ancestry of the archaeal lineage and uncovering the origin, evolution, and structure of the protein world (Caetano-Anollés *et al.*, 2009). We anticipate future studies will extend analysis to all kinds of ncRNAs, providing further support to the validity of the method and clarifying evolutionary questions surrounding origins of modern biochemistry and diversified life. Phylogenetic analyses will also impact the study of function and structure of ncRNA molecules, placing them within an evolutionary context. Together with evidence derived from molecular, genetic, and biochemical studies, evolutionary insights will surely enhance our understanding of biological functions and how these are linked to mechanisms embodied in molecular repertoires. Given the increasing demand for phylogenetic applications in various scientific fields (Baldauf, 2003), we optimistically foresee the increased phylogenetic utility of establishing evolution's arrow for the study of molecular evolution.

References

Albers S, Czech A (2016) Exploiting tRNAs to boost virulence. *Life* **6**: 4.

Altman S, Kirsebom L (1999) Ribonuclease P. In: *The RNA World*, 2nd ed., Gesteland R, Cech TR, Atkins JF (eds) Cold Spring Harbor Lab Press, Plainview, NY, USA, pp 351–380.

Ancel LW, Fontana W (2000) Plasticity, evolvability, and modularity in RNA. *J Exp Zool (Mol Dev Evol)* **288**: 242–283.

Arnaud P, Yukawa Y, Lavie L, Pélissier T, Sugiura M, Deragon JM (2001) Analysis of the SINE S1 Pol III promoter from Brassica; Impact of methylation and influence of external sequences. *Plant J* **26**: 295–305.

Axtell MJ, Snyder JA, Bartel DP (2007) Common functions for diverse small RNAs of land plants. *Plant Cell* **19**: 1750–1769.

Azad AA, Failla P, Hanna PJ (1998) Inhibition of ribosomal subunit association and protein synthesis by oligonucleotides corresponding to defined regions of 18S rRNA and 5S rRNA. *Biochem Biophys Res Commun* **248**: 51–56.

Aziz MF, Caetano-Anollés K, Caetano-Anollés G (2016) The early history and emergence of molecular functions and modular scale-free network behavior. *Sci Rep* **6**: 25058.

Baldauf SL (2003) Phylogeny for the faint of heart: A tutorial. *Trends Genet* **19**: 345–351.

Baldauf SL, Palmer JD, Doolittle WF (1996) The root of the universal tree and the origin of eukaryotes based on elongation factor phylogeny. *Proc Natl Acad Sci USA* **93**: 7749–7754.

Balke D, Kuss A, Müller S (2016) Landmarks in the evolution of (t)-RNAs from the origin of life up to their present role in human cognition. *Life* **6**: 1.

Ban N, Nissen P, Hansen J, Moore PB, Steitz TA (2000) The complete atomic structure of the large ribosomal subunit at 2.4 A resolution. *Science* **289**: 905–920.

Barciszewska MZ, Szymanski M, Erdmann VA, Barciszewski J (2000) 5S ribosomal RNA. *Biomacromolecules* **1**: 297–302.

Barciszewska MZ, Szymanski M, Erdmann VA, Barciszewski J (2001) Structure and functions of 5S rRNA. *Acta Biochim Pol* **48**: 191–198.

Berezikov E, Thuemmler F, Van Laake LW, Kondova I, Bontrop R, Cuppen E, Plasterk RH (2006) Diversity of microRNAs in human and chimpanzee brain. *Nat Genet* **38**: 1375–1377.

Bernhardt HS (2016) Clues to tRNA evolution from the distribution of Class II tRNAs and serine codons in the genetic code. *Life* **6**: 10.

Billoud B, Guerrucci MA, Masselot M, Deutsch JS (2000) Cirripede phylogeny using a novel approach: Molecular morphometrics. *Mol Biol Evol* **17**: 1435–1445.

Birkenheuer AJ, Breitschwerdt EB, Alleman AR, Pitulle C (2002) Differentiation of Haemobartonella canis and Mycoplasma haemofelis on the basis of comparative analysis of gene sequences. *Am J Vet Res* **63**: 1385–1388.

Birney *et al.* ENCODE Project Consortium (2007) Identification and analysis of functional elements in 1% of the human genome by the ENCODE pilot project. *Nature* **447**: 799–816.

Bloch DP, Mcarthur B, Mirrop S (1985) tRNA–rRNA sequence homologies: Evidence for an ancient modular format shared by tRNAs and rRNAs. *Biosystems* **17**: 209–225.

Bogdanov AA, Dontsova OA, Dokudovskaya SS, Lavrik IN (1995) Structure and function of 5S rRNA in the ribosome. *Biochem Cell Biol* **73**: 869–876.

Brennecke J, Aravin AA, Stark A, Dus M, Kellis M, Sachidanandam R, Hannon GJ (2007) Discrete small RNA-generating loci as master regulators of transposon activity in Drosophila. *Cell* **128**: 1089–1103.

Brooks DJ, Fresco JR, Lesk AM, Singh M (2002) Evolution of amino acid frequencies in proteins over deep time: Inferred order of introduction of amino acids into the genetic code. *Mol Biol Evol* **19**: 1645–1655.

Brosius J (2005) Waste not, want not–transcript excess in multicellular eukaryotes. *Trends Genet* **21**: 287–288.

Brown JW (1999) The ribonuclease P database. *Nucleic Acids Res* **27**: 314.

Brown JR, Doolittle WF (1995) Root of the universal tree of life based on ancient aminoacyl-tRNA synthetase gene duplications. *Proc Natl Acad Sci USA* **92**: 2441–2445.

Bull AT, Goodfellow M, Slater JH (1992) Biodiversity as a source of innovation in biotechnology. *Annu Rev Microbiol* **46**: 219–252.

Caetano-Anollés G (2000) Evolution of RNA secondary structure and genome size in the grasses. *Plant Mol Biol Rep* **18**: S21–S23.

Caetano-Anollés G (2001) Novel strategies to study the role of mutation and nucleic acid structure in evolution. *Plant Cell Tiss Org Cult* **67**: 115–132.

Caetano-Anollés G (2002a) Evolved RNA secondary structure and the rooting of the universal tree of life. *J Mol Evol* **54**: 333–345.

Caetano-Anollés G (2002b) Tracing the evolution of RNA structure in ribosomes. *Nucleic Acids Res* **30**: 2575–2587.

Caetano-Anollés G (2005) Grass evolution inferred from chromosomal rearrangements and geometrical and statistical features in RNA structure. *J Mol Evol* **60**: 635–652.

Caetano-Anollés G (2015) Ancestral insertions and expansions of rRNA do not support an origin of the ribosome in its peptidyl transferase center. *J Mol Evol* **80**: 162–165.

Caetano-Anollés G, Caetano-Anollés D (2003) An evolutionarily structured universe of protein architecture. *Genome Res* **13**: 1563–1571.

Caetano-Anollés G, Caetano-Anollés D (2015) Computing the origin and evolution of the ribosome from its structure — uncovering processes of macromolecular accretion benefiting synthetic biology. *Comput Struct Biotechnol J* **13**: 427–447.

Caetano-Anollés G, Kim KM, Caetano-Anollés D (2012) The phylogenomic roots of modern biochemistry: Origins of proteins, cofactors and protein biosynthesis. *J Mol Evol* **74**: 1–34.

Caetano-Anollés G, Kim HS, Mittenthal JE (2007) The origin of modern metabolic networks inferred from phylogenomic analysis of protein architecture. *Proc Natl Acad Sci USA* **104**: 9358–9363.

Caetano-Anollés G, Nasir A (2012) Benefits of using molecular structure and abundance in phylogenomic analysis. *Front Genet* **3**: 172.

Caetano-Anollés G, Nasir A, Zhou K, Caetano-Anollés D, Mittenthal JE, Sun F-J, Kim KM (2014) Archaea: The first domain of diversified fife. *Archaea* **2014**: 590214.

Caetano-Anollés G, Sun F-J, Wang M, Yafremava LS, Harish A, Kim HS, Knudsen V, Caetano-Anollés D, Mittenthal JE (2008) Origins and evolution of modern biochemistry: Insights from genomes and molecular structure. *Front Biosci* **13**: 5212–5240.

Caetano-Anollés G, Wang M, Caetano-Anollés D, Mittenthal JE (2009) The origin, evolution and structure of the protein world. *Biochem J* **417**: 621–637.

Cate JH, Yusupov MM, Yusupova GZ, Earnest TN, Noller HF (1999) X-ray crystal structures of 70S ribosome functional complexes. *Science* **285**: 2095–2104.

Cech TR, Steitz JA (2014) The noncoding RNA revolution-trashing old rules to forge new ones. *Cell* **157**: 77–94.

Chen J-L, Pace NR (1997) Identification of the universally conserved core of ribonuclease P RNA. *RNA* **3**: 557–560.

Cheng J, Kapranov P, Drenkow J, Dike S, Brubaker S, Patel S, Long J, Stern D, Tammana H, Helt G, Sementchenko V, Piccolboni A, Bekiranov S, Bailey DK, Ganesh M, Ghosh S, Bell I, Gerhard DS, Gingeras TR (2005) Transcriptional maps of 10 human chromosomes at 5-nucleotide resolution. *Science* **308**: 1149–54.

Chénais B (2013) Transposable elements and human cancer: A causal relationship? *Biochim Biophys Acta* **1835**: 28–35.

Christian EL, Zahler NH, Kaye NM, Harris ME (2002) Analysis of substrate recognition by the ribonucleoprotein endonuclease RNase P. *Methods* **28**: 307–322.

Collins LJ, Moulton V, Penny D (2000) Use of RNA secondary structure for studying the evolution of RNase P and RNase MRP. *J Mol Evol* **51**: 194–204.

Dalevi D, Hugenholtz P, Blackall LL (2001) A multiple-outgroup approach to resolving division-level phylogenetic relationships using 16S rDNA data. *Int J Syst Evol Microbiol* **51**: 385–391.

Darnell JE Jr (1978) On the origin of prokaryotes. *Science* **202**: 1257–1260.

Deragon JM, Zhang X (2006) Short interspersed elements (SINEs) in plants: Origin, classification and use as phylogenetic markers. *Syst Biol* **55**: 949–956.

Dewannieux M, Esnault C, Heidmann T (2003) LINE-mediated retrotransposition of marked Alu sequences. *Nat Genet* **35**: 41–48.

Di Giulio M (2000) The RNA world, the genetic code and the tRNA molecule. *Trends Genet* **16**: 17–18.

Di Giulio M (2007) The tree of life might be rooted in the branch leading to Nanoarchaeota. *Gene* **401**: 108–113.

Di Giulio M (2012) The origin of the tRNA molecule: Independent data favor a specific model of its evolution. *Biochimie* **94**: 1464–1466.

Di Giulio M (2013) A polyphyletic model for the origin of tRNAs has more support than a monophyletic model. *J Theor Biol* **318**: 124–128.

Dick TP, Schamel WWA (1995) Molecular evolution of transfer RNA from two precursor hairpins: Implications for the origin of protein synthesis. *J Mol Evol* **41**: 1–9.

Ding Y, Lawrence CE (1999) A Bayesian statistical algorithm for secondary structure prediction. *Comput Chem* **23**: 387–400.

Dokudovskaya S, Dontsova O, Shpanchenko O, Bogdanov A, Brimacombe R (1996) Loop IV of 5S ribosomal RNA has contacts both to domain II and to domain V of the 23S RNA. *RNA* **2**: 146–152.

Doolittle WF (1978) Gene in pieces: Were they ever together. *Nature* **272**: 581–582.

Doolittle WF (1999) Phylogenetic classification and the universal tree. *Science* **284**: 2124–2129.

Dowton M, Castro LR, Campbell SL, Bargon SD, Austin AD (2003) Frequent mitochondrial gene rearrangements at the Hymenopteran nad3–nad5 junction. *J Mol Evol* **56**: 517–526.

Eigen M, Winkler-Oswatitsch R (1981a) Transfer-RNA, an early gene? *Naturwissenschaften* **68**: 282–292.

Eigen M, Winkler-Oswatitsch R (1981b) Transfer-RNA: The early adaptor. *Naturwissenschaften* **68**: 217–228.

Ellis JC, Brown JW (2010) The evolution of RNase P and its RNA. In: *Ribonuclease P*, Liu F, Altman S (eds), *Protein Reviews* **10**. *Springer Science*, pp 17–40.

Evans D, Marquez SM, Pace NR (2006) RNase P: Interface of the RNA and protein worlds. *Trends Biochem Sci* **31**: 333–341.

Farias ST (2013) Suggested phylogeny of tRNAs based on the construction of ancestral sequences. *J Theor Biol* **335**: 245–248.

Farias ST, Moreira CH, Guimarães RC (2007) Structure of the genetic code suggested by the hydropathy correlation between anticodons and amino acid residues. *Orig Life Evol Biosph* **37**: 83–103.

Farias ST, Rêgo TG, José MV (2014a) Evolution of transfer RNA and the origin of the translation system. *Front Genet* **5**: 303.

Farias ST, Rêgo TG, José MV (2014b) Origin and evolution of the Peptidyl Transferase Center from proto-tRNAs. *FEBS Open Bio* **4**: 175–178.

Farias ST, Rêgo TG, José MV (2016) tRNA core hypothesis for the transition from the RNA world to the ribonucleoprotein world. *Life* **6**: 15.

Fawcett JA, Kawahara T, Watanabe H, Yasui Y (2006) A SINE family widely distributed in the plant kingdom and its evolutionary history. *Plant Mol Biol* **61**: 505–514.

Fitch WM, Upper K (1987) The phylogeny of tRNA sequences provides evidence for ambiguity reduction in the origin of the genetic code. *Cold Spring Harb Symp Quant Biol* **52**: 759–767.

Fontana W (2002) Modelling "evo–devo" with RNA. *BioEssays* **24**: 1164–1177.

Forterre P, Philippe H (1999) Where is the root of the universal tree of life. *BioEssays* **21**: 871–879.

Frank J, Agrawal RK (2000) A ratchet-like inter-subunit reorganization of the ribosome during translocation. *Nature* **406**: 318–322.

Frank DN, Pace NR (1998) Ribonuclease P: Unity and diversity in a tRNA processing ribozyme. *Annu Rev Biochem* **67**: 153–180.

Gallus S, Kumar V, Bertelsen MF, Janke A, Nilsson MA (2015) A genome survey sequencing of the Java mouse deer (*Tragulus javanicus*) adds new aspects to the evolution of lineage specific retrotransposons in Ruminantia (Cetartiodactyla). *Gene* **571**: 271–278

Gilbert W (1986) Origin of life: the RNA world. *Nature* **319**: 618.

Gilbert N, Labuda D (2000) Evolutionary inventions and continuity of CORE-SINEs in mammals. *J Mol Biol* **298**: 365–377.

Gobert A, Gutmann B, Taschner A, Göbringer M, Holzmann J, *et al.* (2010) A single Arabidopsis organellar protein has RNase P activity. *Nat Struct Mol Biol* **17**: 740–744.

Gongadze GM (2012) 5S rRNA and ribosome. *Biochemistry (Moscow)* **76**: 1450–1464.

Grajales A, Aguilar C, Sánchez JA (2007) Phylogenetic reconstruction using secondary structures of Internal Transcribed Spacer 2 (ITS2, rDNA): Finding the molecular and morphological gap in Caribbean gorgonian corals. *BMC Evol Biol* 7: 90.

Gribaldo S, Poole AM, Daubin V, Forterre P, Brochier-Armanet C (2010) The origin of eukaryotes and their relationship with the Archaea: Are we at a phylogenomic impasse? *Nature Rev Microbiol* **8**: 743–752.

Guerrier-Takada C, Altman S (1992) Reconstitution of enzymatic activity from fragments of M1 RNA. *Proc Natl Acad Sci USA* **89**: 1266–1270.

Guerrier-Takada C, Gardiner K, Marsh T, Pace N, Altman S (1983) The RNA moiety of ribonuclease P is the catalytic subunit of the enzyme. *Cell* **35**: 849–857.

Guimarães RC (2011) Metabolic basis for the self-referential genetic code. *Orig Life Evol Biosph* **41**: 357–371.

Gutmann B, Gobert A, Giegé P (2012) PRORP proteins support RNase P activity in both organelles and the nucleus in Arabidopsis. *Genes Dev* **26**: 1022–1027.

Haas ES, Brown JW (1998) Evolutionary variation in bacterial RNase P RNAs. *Nucleic Acids Res* **26**: 4093–4099.

Harish A, Caetano-Anollés G (2012). Ribosomal history reveals origins of modern protein synthesis. *PLoS ONE* **7**: e32776.

Harris ME, Christian EL (2003) Recent insights into the structure and function of the ribonucleoprotein enzyme ribonuclease P. *Curr Opin Struct Biol* **13**: 325–333.

Harris JK, Haas ES, Williams D, Frank DN, Brown JW (2001) New insight into RNase P RNA structure from comparative analysis of archaeal RNA. *RNA* **7**: 220–232.

Hartmann E, Hartmann RK (2003) The enigma of ribonuclease P evolution. *Trends Genet* **19**: 561–569.

Higgs PG (2000) RNA secondary structure: Physical and computational aspects. *Quart Rev Biophys* **33**: 199–253.

Higgs PG, Jameson D, Jow H, Rattray M (2003) The evolution of tRNA-Leu genes in animal mitochondrial genomes. *J Mol Evol* **57**: 435–445.

Hofacker IL (2003) Vienna RNA secondary structure server. *Nucleic Acids Res* **31**: 3429–3431.

Holmberg L, Nygard O (2000) Release of ribosome-bound 5S rRNA upon cleavage of the phosphodiester bond between nucleotides A54 and A55 in 5S rRNA. *Biol Chem* **381**: 1041–1046.

Holzmann J, Frank P, Löffler E, Bennett KL, Gerner C, Rossmanith W (2008) RNase P without RNA: Identification and functional reconstitution of the human mitochondrial tRNA processing enzyme. *Cell* **135**: 462–474.

Hopfield JJ (1978) Origin of the genetic code: a testable hypothesis based on tRNA structure, sequence, and kinetic proofreading. *Proc Natl Acad Sci USA* **75**: 4334–4338.

Houwing S, Kamminga LM, Berezikov E, Cronembold D, Girard A, Van Den Elst H, Filippov DV, Blaser H, Raz E, Moens CB, Plasterk RH, Hannon GJ, Draper BW, Ketting RF (2007) A role for Piwi and piRNAs in germ cell maintenance and transposon silencing in zebrafish. *Cell* **129**: 69–82.

Huang H, Hopper AK (2016) Multiple layers of stress-induced regulation in tRNA biology. *Life* **6**: 16.

Huelsenbeck JP, Ronquist F (2001) MRBAYES: Bayesian inference of phylogeny. *Bioinformatics* **17**: 754–755.

Jordan IK, Kondrashov FA, Adzhubei IA, Wolf YI, Koonin EV, Kondrashov AS, Sunyaev S (2005) A universal trend of amino acid gain and loss in protein evolution. *Nature* **433**: 633–638.

Jurka J (1995) Origin and evolution of Alu repetitive elements. In: Maraia R (ed) The impact of short interspersed elements (SINEs) on the host genome. R.G. Landes Co, Austin, USA, pp 25–41.

Jühling F, Mörl M, Hartmann RK, Sprinzl M, Stadler PF, Pütz J (2009) tRNAdb **2009**: Compilation of tRNA sequences and tRNA genes. *Nucleic Acids Res* **37**: D159–D162.

Jühling F, Pütz J, Florentz C, Stadler PF (2012) Armless mitochondrial tRNAs in enoplea (nematoda). *RNA Biol* **9**: 1161–1166.

Kapitonov VV, Jurka J (2003) A novel class of SINE elements derived from 5S rRNA. *Mol Biol Evol* **20**: 694–702.

Karwan R, Pluk H, Van Venrooij WJ (editors) (1995–1996) RNase MRP/RNase P Systems. *Mol Biol Rep* **22** (2–3): 67.

Kasschau KD, Fahlgren N, Chapman EJ, Sullivan CM, Cumbie JS, Givan SA, Carrington JC (2007) Genome-wide profiling and analysis of Arabidopsis siRNAs. *PLoS Biol* **5**: e57.

Kawagoe-Takaki H, Nameki N, Kajikawa M, Okada N (2006) Probing the secondary structure of salmon SmaI SINE RNA. *Gene* **365**: 67–73.

Kazantsev AV, Krivenko AA, Harrington DJ, Holbrook SR, Adams PD, Pace NR (2005) Crystal structure of a bacterial ribonuclease P RNA. *Proc Natl Acad Sci USA* **102**: 13392–13397.

Kazantsev AV, Pace NR (2006) Bacterial RNase P: A new view of an ancient enzyme. *Nat Rev Microbiol* **4**: 729–740.

Kikovska E, Svärd SG, Kirsebom LA (2007) Eukaryotic RNase P RNA mediates cleavage in the absence of protein. *Proc Natl Acad Sci USA* **104**: 2062–2067.

Kim KM, Caetano-Anollés G (2012) The evolutionary history of protein fold families and proteomes confirms that the archaeal ancestor is more ancient than the ancestors of other superkingdoms. *BMC Evol Biol* **12**: 13.

Kim KM, Nasir A, Hwang K, Caetano-Anollés G (2014) A tree of cellular life inferred from a genomic census of molecular functions. *J Mol Evol* **79**: 240–262.

Kirchner S, Ignatova Z (2015) Emerging roles of tRNA in adaptive translation, signalling dynamics and disease. *Nat Rev Genet* **16**: 98–112.

Klimov PB, Oconnor BM (2009) Improved tRNA prediction in the American house dust mite reveals widespread occurrence of extremely short minimal tRNAs in acariform mites, *BMC Genomics* **10**: 598.

Knudsen V, Caetano-Anollés G (2008) NOBAI: A web server for character coding of geometrical and statistical features in RNA structure. *Nucleic Acids Res* **36**: W85–W90.

Konkel MK, Walker JA, Batzer MA (2010) LINEs and SINEs of primate evolution. *Evol Anthropol* **19**: 236–249.

Kouvela E, Gerbanas GV, Xaplanteri MA, Petropoulos AD, Dinos GP, Kalpaxis DL (2007) Changes in the conformation of 5S rRNA cause alternations in principal functions of the ribosomal nanomachine. *Nucleic Acids Res* **35**: 5108–5119.

Kramerov DA, Vassetzky NS (2005) Short retroposons in eukaryotic genomes. *Int Rev Cytol* **247**: 165–221.

Kramerov DA, Vassetzky NS (2011) Origin and evolution of SINEs in eukaryotic genomes. *Heredity* **107**: 487–495.

Krasilnikov AS, Xiao Y, Pan T, Mondragón A (2004) Basis for structural diversity in homologous RNAs. *Science* **306**: 104–107.

Krasilnikov AS, Yang X, Pan T, Mondragón A (2003) Crystal structure of the specificity domain of ribonuclease P. *Nature* **421**: 760–764.

Lazcano A (2012) The biochemical roots of the RNA world: From zymonucleic acid to ribozymes. *Hist Philos Life Sci* **34**: 407–423.

Leffers H, Kjems J, Ostergaard L, Larson N, Garrett RA (1987) Evolutionary relationships amongst archaebacteria: A comparative study of 23S ribosomal RNAs of a sulphur-dependent extreme thermophile, an extreme halophile and a thermophilic methanogen. *J Mol Biol* **195**: 43–61.

Lemieux B, Laterreur N, Perederina A, Noël J, Dubois M, Krasilnikov AS, Wellinger RJ (2016) Active yeast telomerase shares subunits with ribonucleoproteins RNase P and RNase MRP. *Cell* **165**: 1171–1181.

Lorenz R, Bernhart SH, zu Siederdissen CH, Tafer H, Flamm C, Stadler PF, Hofacker IL (2011) ViennaRNA Package 2.0. *Algorithm Mol Biol* **6**: 26.

Loria A, Pan T (1996) Domain structure of the ribozyme from eubacterial ribonuclease P. *RNA* **2**: 551–563.

Lu C, Meyers BC, Green PJ (2007) Construction of small RNA cDNA libraries for deep sequencing. *Methods* **43**: 110–117.

Luchetti A, Mantovani B (2013) Conserved domains and SINE diversity during animal evolution. *Genomics* **102**: 296–300.

Maddison WP, Donoghue MJ, Maddison DR (1984) Outgroup analysis and parsimony. *Syst Zool* **33**: 83–103.

Maddison DR, Maddison WP (2003) MacClade Version 4: Analysis of Phylogeny and Character Evolution. Sinauer Associates, Sunderland, MA.

Maizels N, Weiner AM (1994) Phylogeny from function: Evidence from the molecular fossil record that tRNA originated in replication, not translation. *Proc Natl Acad Sci USA* **91**: 6729–6734.

Manam S, Van Tuyle GC (1987) Separation and characterization of 5′- and 3′-tRNA processing nucleases from rat liver mitochondria. *J Biol Chem* **262**: 10272–10279.

Mathews DH, Sabina J, Zuker M, Turner DH (1999) Expanded sequence dependence of thermodynamic parameters improves prediction of RNA secondary structure. *J Mol Biol* **288**: 911–940.

Mobley EM, Pan T (1999) Design and isolation of ribozyme substrate pairs using RNase P-based ribozymes containing altered substrate binding sites. *Nucleic Acids Res* **27**: 4298–4304.

Moran J, Gilbert N (2002) Mammalian LINE-1 retro-transposons and related elements. In: *Mobile DNA II*, Craig NL, Cragie R, Gellert M, Lambowitz AM (eds), ASM Press, Washington DC, USA, pp 836–869.

Morris KV (2012) *Non-coding RNAs and Epigenetic Regulation of Gene Expression: Drivers of Natural Selection.* Caister Academic Press.

Nasir A, Kim KM, Caetano-Anollés G (2012) Giant viruses coexisted with the cellular ancestors and represent a distinct supergroup along with superkingdoms Archaea, Bacteria and Eukarya. *BMC Evol Biol* **12**: 156.

Nasir A, Kim KM, Caetano-Anollés G (2014) A phylogenomic census of molecular functions identifies modern thermophilic Archaea as the most ancient form of cellular life. *Archaea* **2014**: 706468.

Nasir A, Sun F-J, Kim KM, Gustavo Caetano-Anollés G (2015) Untangling the origin of viruses and their impact on cellular evolution. *Ann NY Acad Sci* **1341**: 61–74.

Neveu M, Kim HJ, Benner SA (2013) The "strong" RNA world hypothesis: Fifty years old. *Astrobiology* **13**: 391–403.

Nikaido M, Piskurek O, Okada N (2007) Toothed whale monophyly reassessed by SINE insertion analysis: The absence of lineage sorting effects suggests a small population of a common ancestral species. *Mol Phylogen Evol* **43**: 216–24.

Nishihara H, Smit AFA, Okada N (2006) Functional noncoding sequences derived from SINEs in the mammalian genome. *Genome Res* **16**: 864–874.

O'Brien MJ, Lyman RL, Saab Y, Saab E, Darwent J, Glover DS (2002) Two issues in archaeological phylogenetics: Taxon construction and outgroup selection. *J Theor Biol* **215**: 133–150.

O'Malley MA, Koonin EV (2011) How stands the Tree of Life a century and a half after The Origin? *Biol Direct* **6**: 32.

Ogiwara I, Miya M, Ohshima K, Okada N (2002) V-SINEs: A new superfamily of vertebrate SINEs that are widespread in vertebrate genomes and retain a strongly conserved segment within each repetitive unit. *Genome Res* **12**: 316–324.

Ohshima K, Okada N (2005) SINEs and LINEs: Symbionts of eukaryotic genomes with a common tail. Cytogenet. *Genome Res* **110**: 475–490.

Okada N, Ohshima K (1995) Evolution of tRNA-derived SINEs. In: *The Impact of Short Interspersed Elements (SINEs) on the Host Genome*, Maraia R (ed), R.G. Landes Co, Austin, USA, pp 61–80.

Olsen GJ, Woese CR, Overbeek R (1994) The winds of (evolutionary) change: Breathing new life into microbiology. *J Bacteriol* **176**: 1–6.

Pace NR, Brown JW (1995) Evolutionary perspective on the structure and function of ribonuclease P, a ribozyme. *J Bacteriol* **177**: 1919–1928.

Pace NR, Sapp J, Goldenfeld N (2012) Phylogeny and beyond: Scientific, historical, and conceptual significance of the first tree of life. *Proc Natl Acad Sci USA* **109**: 1011–1018.

Palazzo AF, Lee ES (2015) Non-coding RNA: What is functional and what is junk? *Front Genet* **6**: 2.

Pan T (1995) Higher order folding and domain analysis of the ribozyme from Bacillus subtilis ribonuclease P. *Biochemistry* **34**: 902–909.

Pan T, Loria A, Zhong K (1995) Probing of tertiary interactions in RNA: 2′-hydroxyl-base contacts between the RNase P RNA and pre-tRNA. *Proc Natl Acad Sci USA* **92**: 12510–12514.

Pannucci JA, Haas ES, Hall TA, Harris JK, Brown JW (1999) RNase P RNAs from some Archaea are catalytically active. *Proc Natl Acad Sci USA* **96**: 7803–7808.

Pélissier T, Bousquet-Antonelli C, Lavie L, Deragon J-M (2004) Synthesis and processing of tRNA-related SINE transcripts in Arabidopsis thaliana. *Nucleic Acids Res* **32**: 3957–3966.

Piskurek O, Nikaido M, Boeadi, Baba M, Okada N (2003) Unique mammalian tRNA-derived repetitive elements in dermopterans: The t-SINE family and its retro-transposition through multiple sources. *Mol Biol Evol* **20**: 1659–1668.

Pollock D (2003) The Zuckerkandl Prize: structure and evolution. *J Mol Evol* **56**: 375–376.

Ponicsan SL, Kugel JF, Goodrich JA (2010) Genomic gems: SINE RNAs regulate mRNA production. *Curr Opin Genet Dev* **20**: 149–155.

Pulukkunat DK, Gopalan V (2008) Studies on Methanocaldococcus jannaschii RNase P reveal insights into the roles of RNA and protein cofactors in RNase P catalysis. *Nucleic Acids Res* **36**: 4172–4180.

Ravishankar V, Ahmed A, Sivagnanam U, Muthuraman K, Karthikaichamy A, Wilson HA, Devendran A, Hartskeerl RA, Raj SML (2014) Evolution of the RNase P RNA structural domain in Leptospira spp. *Res Microbiol* **165**: 813–825.

Ray DA, Pagan HJT, Platt RN II, Kroll AR, Schaack S, Stevens RD (2015) Differential SINE evolution in vesper and non-vesper bats. *Mobile DNA* **6**: 10.

Ray DA, Xing J, Salem A-H, Batzer MA (2006) SINEs of a nearly perfect character. *Syst Biol* **55**: 928–935.

Reanney DC (1974) On the origin of prokaryotes. *Theor Biol* **48**: 243–251.

Reyes-Prieto F, Hernández-Morales R, Jácome R, Becerra A, Lazcano A (2012) Coenzymes, viruses and the RNA world. *Biochimie* **94**: 1467–1473.

Ronquist F, Huelsenbeck JP (2003) MRBAYES 3: Bayesian phylogenetic inference under mixed models. *Bioinformatics* **19**: 1572–1574.

Root-Bernstein M, Root-Bernstein R (2015) The ribosome as a missing link in the evolution of life. *J Theor Biol* **367**: 130–158.

Rossmanith W, Karwan RM (1998) Characterization of human mitochondrial RNase P: Novel aspects in tRNA processing. *Biochem Biophys Res Commun* **247**: 234–241.

Rota-Stabelli O, Telford MJ (2008) A multi criterion approach for the selection of optimal outgroups in phylogeny: Recovering some support for Mandibulata over Myriochelata using mitogenomics. *Mol Phylogenet Evol* **48**: 103–111.

Rozhdestvensky TS, Kopylov AM, Hütenhofer A (2001) Neuronal BC1 RNA structure: Evolutionary conversion of a tRNA (Ala) domain into an extended stem-loop structure. *RNA* **7**: 722–730.

Ruby JG, Jan C, Player C, Axtell MJ, Lee W, Nusbaum C, Ge H, Bartel DP (2006) Large-scale sequencing reveals 21U-RNAs and additional microRNAs and endogenous siRNAs in *C. elegans*. *Cell* **127**: 1193–1207.

Salavati R, Panigrahi AK, Stuart KD (2001) Mitochondrial ribonuclease activity of Trypanosoma brucei. *Mol Biochem Parasit* **115**: 109–117.

Schmitz J (2012) SINEs as driving forces in genome evolution. *Genome Dyn* **7**: 92–107.

Schultes EA, Bartel DP (2000) One sequence, two ribozymes: Implications for the emergence of new ribozyme folds. *Science* **289**: 448–452.

Schuwirth BS, Borovinskaya MA, Hau CW, Zhang W, Vila-Sanjurjo A, Holton JM, Cate JHD (2005) Structures of the bacterial ribosome at 3.5 A resolution. *Science* **310**: 827–834.

Selmer M, Dunham CM, Murphy FV IV, Weixlbaumer A, Petry S, Kelley AC, Weir JR, Ramakrishnan V (2006) Structure of the 70S ribosome complexed with mRNA and tRNA. *Science* **313**: 1935–1942.

Shedlock AM, Milinkovitch MC, Okada N (2000) SINE evolution, missing data, and the origin of whales. *Syst Biol* **49**: 808–817.

Siegel RW, Banta AB, Haas ES, Brown JW, Pace NR (1996) Mycoplasma fermentans simplifies our view of the catalytic core of ribonuclease P RNA. *RNA* **2**: 452–462.

Sprinzl M, Vassilenko KS (2005) Compilation of tRNA sequences and sequences of tRNA genes. *Nucleic Acids Res* **33**: D139–D140.

Steel M, Penny D (2000) Parsimony, likelihood, and the role of models in molecular phylogenetics. *Mol Biol Evol* **17**: 839–850.

Sun F-J, Caetano-Anollés G (2008a) The origin and evolution of tRNA inferred from phylogenetic analysis of structure. *J Mol Evol* **66**: 21–35.

Sun F-J, Caetano-Anollés G (2008b) Evolutionary patterns in the sequence and structure of transfer RNA: Early origins of Archaea and viruses. *PLoS Comput Biol* **4**: e1000018.

Sun F-J, Caetano-Anollés G (2008c) Evolutionary patterns in the sequence and structure of transfer RNA: A window into early translation and the genetic code. *PLoS ONE* **3**: e2799.

Sun F-J, Caetano-Anollés G (2008d) The evolutionary significance of the long variable arm in transfer RNA. *Complexity* **14**: 26–39.

Sun F-J, Caetano-Anollés G (2008e) Transfer RNA and the origins of diversified life. *Sci Prog* **91**: 265–284.

Sun F-J, Caetano-Anollés G (2009) The evolutionary history of the structure of 5S ribosomal RNA. *J Mol Evol* **69**: 430–443.

Sun F-J, Caetano-Anollés G (2010) The ancient history of the structure of ribonuclease P and the early origins of Archaea. *BMC Bioinformatics* **11**: 153.

Sun F-J, Fleurdépine S, Bousquet-Antonelli C, Caetano-Anollés G, Deragon J-M (2007) Common evolutionary trends for SINE RNA structures. *Trends Genet* **23**: 26–33.

Sun F-J, Harish A, Caetano-Anollés G (2010) Phylogenetic utility of RNA structure: evolution's arrow and emergence of early biochemistry and diversified life. In: *Evolutionary Genomics and Systems Biology*, Caetano-Anollés G (ed), Wiley, Hoboken, NJ, USA, pp. 329–360.

Swain TD, Taylor DJ (2003) Structural rRNA characters support monophyly of raptorial limbs and paraphyly of limb specialization in water fleas. *Proc R Soc Lond Ser B* **270**: 887–896.

Swofford DL (2003) PAUP*. Phylogenetic Analysis Using Parsimony* (and Other Methods), Version 4. Sinauer Associates, Sunderland, MA.

Szymanski M, Barciszewska MZ, Erdmann VA, Barciszewski J (2002) 5S ribosomal RNA database. *Nucleic Acids Res* **30**: 176–178.

Szymanski M, Barciszewska MZ, Erdmann VA, Barciszewski J (2003) 5S rRNA: Structure and interactions. *Biochem J* **371**: 641–651.

Tamura K (2015) Origins and early evolution of the tRNA molecule. *Life* **5**: 1687–1699.

Tanaka T, Kikuchi Y (2001) Origin of the cloverleaf shape of transfer RNA — the double-hairpin model: Implication for the role of tRNA intron and the long extra loop. *Viva Origino* **29**: 134–142.

Tang DTP, Glazov EV, McWilliam SM, Barris WC, Dalrymple BP (2009) Analysis of the complement and molecular evolution of tRNA genes in cow. *BMC Genomics* **10**: 188.

Tapp J, Thollesson M, Herrmann B (2003) Phylogenetic relationships and genotyping of the genus Streptococcus by sequence determination of the RNase P RNA gene, rnpB. *Int J Syst Evol Microbiol* **53**: 1861–1871.

Taschner A, Weber C, Buzet A, Hartmann RK, Hartig A, *et al.* (2012) Nuclear RNase P of Trypanosoma brucei: A single protein in place of the multicomponent RNA-protein complex. *Cell Rep* **2**: 19–25.

Thomas BC, Chamberlain J, Engelke DR, Gegenheimer P (2000a) Evidence for an RNA-based catalytic mechanism in eukaryotic nuclear ribonuclease P. *RNA* **6**: 554–562.

Thomas BC, Li X, Gegenheimer P (2000b) Chloroplast ribonuclease P does not utilize the ribozyme-type pre-tRNA cleavage mechanism. *RNA* **6**: 545–553.

Torres AG, Batlle E, Ribas de Pouplana L (2014) Role of tRNA modifications in human diseases. *Trends Mol Med* **20**: 306–314.

Torres-Larios A, Swinger KK, Pan T, Mondragón A (2006) Structure of RNase P, a universal ribozyme. *Curr Opin Struct Biol* **16**: 327–335.

Trifonov EN (2004) The triplet code from first principles. *J Biomol Struct Dyn* **22**: 1–11.

Valach M, Burger G, Gray MW, Lang BF (2014) Widespread occurrence of organelle genome-encoded 5S rRNAs including permuted molecules. *Nucleic Acids Res* **42**: 13764–13777.

van Bakel H, Nislow C, Blencowe BJ, Hughes TR (2010) Most "dark matter" Transcripts are associated with known genes. *PLoS Biol* **8**: e1000371.

Walker SC, Engelke DR (2006) Ribonuclease P: The evolution of an ancient RNA enzyme. *Crit Rev Biochem Mol* **41**: 77–102.

Wang M, Caetano-Anollés G (2006) Global phylogeny determined by the combination of protein domains in proteomes. *Mol Biol Evol* **23**: 2444–2454.

Wang M, Caetano-Anollés G (2009) The evolutionary mechanics of domain organization in proteomes and the rise of modularity in the protein world. *Structure* **17**: 66–78.

Wang MJ, Davis NW, Gegenheimer P (1988) Novel mechanisms for maturation of chloroplast transfer RNA precursors. *EMBO J* 7: 1567–1574.

Wang M, Kurland CG, Caetano-Anollés G (2011) Reductive evolution of proteomes and protein structures. *Proc Natl Acad Sci USA* **108**: 11954–11958.

Wang X, Lavrov DV (2011) Gene recruitment–a common mechanism in the evolution of transfer RNA gene families. *Gene* **475**: 22–29.

Wang M, Yafremava LS, Caetano-Anollés D, Mittenthal JE, Caetano-Anollés G (2007) Reductive evolution of architectural repertoires in proteomes and the birth of the tripartite world. *Genome Res* **17**: 1572–1585.

Washietl S, Pedersen JS, Korbel JO, Stocsits C, Gruber AR, Hackermüller J, Hertel J, Lindemeyer M, Reiche K, Tanzer A, Ucla C, Wyss C, Antonarakis SE, Denoeud F, Lagarde J, Drenkow J, Kapranov P, Gingeras TR, Guigó R, Snyder M, Gerstein MB, Reymond A, Hofacker IL, Stadler PF (2007) Structured RNAs in the ENCODE selected regions of the human genome. *Genome Res* **17**: 852–64.

Watanabe K, Yokobori S (2011) tRNA modification and genetic code variations in animal mitochondria. *J Nucleic Acids* **2011**: 623095.

Weber C, Hartig A, Hartmann RK, Rossmanith W (2014) Playing RNase P evolution: swapping the RNA catalyst for a protein reveals functional uniformity of highly divergent enzyme forms. *PLoS Genetics* **10**: e1004506.

Weiner AM, Maizels N (1987) tRNA-like structures tag the 3′ ends of genomic RNA molecules for replication: Implications for the origin of protein synthesis. *Proc Natl Acad Sci USA* **84**: 7383–7387.

Wende S, Platzer EG, Jühling F, Pütz J, Florentz C, Stadler PF, Mörl M (2014) Biological evidence for the world's smallest tRNAs. *Biochimie* **100**: 151–158.

Widmann J, Di Giulio M, Yarus M, Knight R (2005) tRNA creation by hairpin duplication. *J Mol Evol* **61**: 524–535.

Woese CR (1969) The biological significance of the genetic code. *Prog Mol Subcell Biol* **1**: 5–46.

Wong JT (1975) A coevolution theory of the genetic code. *Proc Natl Acad Sci USA* **72**: 1909–1912.

Wong JT, Chen J, Mat W-K, Ng S-K, Xue H (2007) Polyphasic evidence delineating the root of life and roots of biological domains. *Gene* **403**: 39–52.

Wuyts J, Perriere G, Van de Peer Y (2004) The European ribosomal RNA database. *Nucleic Acids Res* **32**: D101–D103.

Xue H, Ng S-K, Tong K-L, Wong JT-F (2005) Congruence of evidence for a Methanopyrus-proximal root of life based on transfer RNA and aminoacyl-tRNA synthetase genes. *Gene* **360**: 120–130.

Xue H, Tong K-L, Marck C, Grosjean H, Wong JT-F (2003) Transfer RNA paralogs: Evidence for genetic code–amino acid biosynthesis coevolution and an archaeal root of life. *Gene* **310**: 59–66.

Yona AH, Bloom-Ackermann Z, Frumkin I, Hanson-Smith V, Charpak-Amikam Y, Feng Q, Boeke JD, Dahan O, Pilpel Y (2013) tRNA genes rapidly change in evolution to meet novel translational demands. *eLife* **2**: e01339.

Yusupov MM, Yusupova GZ, Baucom A, Lieberman K, Earnest TN, Cate JHD, Noller HF (2001) Crystal structure of the ribosome at 5.5 A resolution. *Science* **292**: 883–896.

Zhao T, Li G, Mi S, Li S, Hannon GJ, Wang XJ, Qi Y (2007) A complex system of small RNAs in the unicellular green alga Chlamydomonas reinhardtii. *Genes Dev* **21**: 1190–1203.

Zhu Y, Pulukkunat DK, Li Y (2007) Deciphering RNA structural diversity and systematic phylogeny from microbial metagenomes. *Nucleic Acids Res* **35**: 2283–2294.

Zuckerkandl E, Derancourt J, Vogel H (1971) Mutational trends and random processes in the evolution of informational macromolecules. *J Mol Biol* **59**: 473–490.

It takes a well rooted trunk to bear a tree — reductionist models of divergent evolution and further aspects of reality

Richard Egel[1]

Summary. In this visionary and rather personal essay I try to untangle certain ambiguities about tentative origins of molecular biodiversity. A "leitmotif" in my considerations will be the tree-like dendrograms encountered in various evolutionary settings from Darwinian species and replicating genes to the intricate folding patterns of protein domains. These patterns are powerful aids to comprehending present biological diversity, but looking back in evolutionary time may run into difficulties of inconclusive generalization. Extrapolating backward toward the bottom node of tree-like bifurcations, one inevitably ends up in some singularity, where the comparison-based algorithm to construct the tree is virtually stripped for its basic rationale. This is a serious conceptual problem. The following aspects are scrutinized and discussed in this regard for their likely evolutionary beginnings: Darwinian speciation,

[1]*Department of Biology, University of Copenhagen Biocenter, Copenhagen, Denmark*

cells, genomes, fold families of protein domains, and replicating RNA molecules. In some of these examples, the principle that generates the branching pattern loses its predictive power for looking deeper yet, i.e. below the earliest bifurcation events that started to define the diversifying tree. Other and perhaps unthought-of principles may allow escape from such a conceptual impasse. To really comprehend those evolutionary phase transitions we need plausible links to thermodynamically sound reactions in large numbers, and collective coevolution of multiple components should also be considered. To clearly define a common frame of reference, I do not start with Darwin, who only had prescient yet limited access to the basic mechanisms beneath the macroscopic appearance of biodiversity at large, but rather from the Woesean perspective of three fundamentally different superkingdoms (domains or phylodomains) in the universal organismal tree of — presently apparent — life.

1. Introduction

For conciseness and clarity of reading, references to this chapter for further reading are kept to minimum. The list contains previous writings of mine on the subject (Egel 2009, 2011, 2012 a,b, 2014, 2017), all of which are fully referenced, as followed by key articles and recent reviews of other authors. A substantial update with additional references has been incorporated at the final revision stage — especially addressing the still controversial debate whether our understanding of eukaryogenesis should be constrained by assuming two or three organismal domains in the canonical Tree of Life. This concerns whether Eukarya represent a third domain, along with Archaea and Bacteria, or rather should be nested within the archaeal domain.

In terms of energy procurement, to enable the emergence of self-maintaining proto-life, this essay presupposes ultraviolet sunlight as the primary basis of photo-autotrophic origins (Guzman, 2011), but not in the ocean. Rather, terrestrial settings at geothermal fields (Mulkidjanian *et al.*, 2012) appear more appropriate, especially in combination with frequent dehydration cycles (Damer and Deamer, 2015). In metabolic terms, both non-oxygenic phototrophy and methanogenesis are considered amongst the oldest pathways — already of relevance to a complex common source of all the organismal lineages in modern life (Glansdorff *et al.*, 2008), to be accompanied by heterotrophic turnover of accumulating organic matter and chemo-autotrophic adaptations as a derived or secondary mode. When looking at the very origins of life as a special phase state transition of organic matter, it is worth considering the pumping effect of temporal oscillations in the environment as a very important or even essential prerequisite for life's emergence (Kompanichenko, 2008, 2012). As mentioned further below, the photo-activating short-wave input of ultraviolet light and/or the much longer phases of Earth's spinning about its axis while orbiting around the sun — resulting in wet/drying cycles — can substantiate this general claim.

2. The Woesean perspective on biodiversity in molecular phylogenies

Applying tree-building algorithms to gene sequence comparison, Carl Woese has presented evolutionary biology with an endowment nobody had asked for or expected, by splitting the multitude of bacteria-like microorganisms in two disparate lots of Bacteria and Archaea. And for the worse, these two categories of similar appearance were not even sister groups in the overall phylogeny but were separated by the nuclear genomes of the composite Eukarya, to which the microbial Archaea appeared peculiarly related (Woese, 2000) — as suggested by Woese's criterion of rRNA sequence comparison. While the stick-figure relationships in this basic three-domain phylogeny have widely been held as common scientific knowledge, Woese's more circumstantial inferences about the evolutionary rooting of the organismal Tree of Life (ToL), as conjectured in his later writings, have not fully come into their own by now. I can only notice a limited or reluctant comprehension of the deep and remarkable insights when I skim the pertinent literature in this regard. A competing two-domain phylogeny (the eocyte hypothesis) will be addressed further below.

In particular, the fundamental nature of the traditional "prokaryote"/eukaryote distinction has been seriously questioned by Woese's three-domain model. Yet, a corresponding evolutionary presupposition seems to be very much alive today, assuming that simple prokaryote organization must have preceded emergence of the much more diversified cytoplasmic infrastructure in eukaryotes. Together with a dissenting minority of other researchers I do not share this view, and for that matter, Bacteria and Archaea are collectively addressed as akaryotes in the remaining text, which is a more neutral term avoiding the outmoded implications of "prokaryotes" (Penny *et al.*, 2014). This caveat aside, the main results of Woese's evolutionary considerations at the organismal level are summarized as follows.

- *Woesean progenote:* hypothetical stage precursory to modern organismal life when virtually everything must have been different from what we know about cellular organisms today (Woese, 2005). The sequence-determining processes of gene replication and protein synthesis were necessarily rather inaccurate to begin with, resulting in many different, partly randomized and relatively short products. With many different genes but no integral genome or assorting mechanism in the beginning, there was no regular cell division cycle either and vertically stable organismal lineages could not possibly exist. Instead, early evolutionary change was mainly focused on the collective optimization of many system-stabilizing interactions, which very much depended on communal innovation sharing and the frequent exchange of interactive components on a system-wide scale. In the absence of vertically stable sub-lineages of Darwinian speciation, the unbranched character of this early phase can be described as universal trunk line evolution of a distinctly Woesean type.
- *Woesean–Darwinian transition:* passage from progenote communality to the emergence of vertically discrete lineages of genealogically related organisms, which subsequently diversified into ever more sublineages of a binary branching phylogram. Woese referred to this turning point as the crossing of a "Darwinian threshold" (Woese, 2002), which critically depended upon having consolidated a modular genome, not only capable of being replicated faithfully and distributed equally among the progeny, but also sufficiently large to support all vital functions of entire organisms. This pivotal step represents the earliest possible time for the emergence of modular cells.

- *Woesean asymmetry:* peculiar disparity between the branches at the earliest bifurcation(s) of the emergent ToL. This is an inevitable consequence of the foregoing considerations. It merely restates, in general terms, that the very first lineage of modular cells emerged from a larger population of non-modular progenote entities (further below I call them "paracells") and that all other members of the preexisting population were not necessarily outcompeted very fast by the emergent cellular newcomers. In consequence, the asymmetric branching away from the communal bulk population could continue by repeating itself with the generation of additional cellular lineages several times.
- *Addressing the residual progenote trunk line — after the genomic bacterial lineages had just branched off:* Eventually this still pre-genomic community gave rise to both Archaea and Eukarya. Tentatively, therefore, I call this progenote remnant the Archgenote trunk line, and it is worth emphasizing that the common ancestors of Archaea and Eukarya were neither archaea nor eukaryotes in any modern sense. To designate the latest trunk-line stage, after the genomic archaeal lineages too had branched off, the formerly suggested term of "Urkaryote" deserves to be revived.

As of recently (Egel, 2017), I have tried to substantiate some of the rather abstract and generally phrased inferences by several not widely considered yet rationally grounded possibilities involving endogenous subcellular compartments, which are incorporated into the following presentation. In short, I envision these transitions as a staggered evolution at multiple levels, all being superimposed on a communal protocytoplasm and gradually diversifying under the directional and regulatory influence from rather simple genetic elements to compartmentalized genomic agents. To the former category I count individual genes, introns and transposable elements; to the latter viruses, "paraorganellar" vesicles and eukaryote-like proto-nuclei.

To extend the basic progenote concept, I have conjectured that the Woesean asymmetry feature, in particular, must have changed the evolutionary frame work — not only in the formal terms of vertical lineage competition versus horizontal trunk-line communality but also in kinetic terms of adaption for selective speed and major evolutionary transitions. The latter argument is especially important at early stages of genome nucleation considered here specifically, such as for selfish viral lineages and self-replicative, potentially symbiotic plasmid-like minigenomes.

3. Origins of Darwinian speciation

It is important to emphasize that the Woesean transition that crossed a Darwinian threshold for the first time did not as such define the origin of biological evolution — generally understood as descent with modification — but only the origin of Darwinian speciation. This means that the dynamics of evolution shifted in character from a single collective trunk lineage to the tree-like branching mode, when separate sublineages began to specialize, so as to effectively compete against each other. Before reaching that critical threshold, therefore, the entire biosphere was formally comprised of a single Darwinian "meta-species" with a large communal gene exchange pool.

How then can we reasonably visualize or imagine the eventual material representations of Woese's abstract and rather intangible evolutionary concepts? Scientists in general are reluctant to substantiate potentially possible phenomena in concrete terms if multiple degrees of freedom remain unconstrained by rigorous testing or exact assumptions. First and foremost, of course, such mental sobriety is considered a virtue rather than a lack of insight, but it might also be a hindrance if some tacit yet unjustified assumptions stand in the way of deeper understanding. A prime candidate for a misleading presupposition of this kind, I suspect, is the virtually dogmatic view that "eukaryotes descended and evolved from prokaryotes," which still pervades the current literature. The following considerations argue for a more reasonable alternative, where both akaryotes and eukaryotes eventually diversified from a complex and inclusive, yet non-modular and polyphenotypic source.

It is a central tenet of Woesean reasoning that the emergence of modular cells required the establishment of modular genomes first. This begs several questions — not only of how non-modular proto-genomes and non-modular proto-organisms may have coevolved at the long-lasting polyphenotypic progenote stage, but also how this type of coevolution was embedded in a dependable and system-wide proto-metabolism. In the following I will try to disentangle some of these aspects under different headings, even though all the various features are intrinsically interrelated.

4. Origins of energy procurement, compartmentalization and genuine cells

Under the influence of the gene-centered paradigm, which had dominated molecular biology for many decades, the origins of genomes would probably have been given first priority, but I am not inclined to follow that pursuit. Such reluctance is based on serious doubts that gene-like molecules might have emerged directly from some "primordial soup" in free solution. Rather, a kind of "surface metabolism" in an energized reactor-type microenvironment appears more likely to have kick-started a hodge-podge of many organic synthesis reactions simultaneously, the most interactive products of which could then have started to reinforce their mutual regeneration in autocatalytic cycles and hypercycles. The metabolic aspects of this scenario, however, are not encompassed by tree-like hierarchical concepts and, therefore, go beyond the scope of this essay. Suffice it to say that the envisioned "georeactor" was driven by solar energy input and microcatalytic properties at mineral surfaces and nanoparticles to gradually accumulate soft organic matter — a kind of surface-lining "prebiotic goo" — which in turn took over various catalytic activities. The notion of "neighborhood selection" from emergent "surface protoplasm" (Baum, 2015) is a conceptual link to biologically relevant structural organization, and I have suggested "protobiofilms" with similar intentions (Egel, 2014).

In contrast to most other models of early pre-cellular evolution, the concepts of "protobiofilms" and "surface protoplasm" put less emphasis on outer membranes as self-defining or self-containing boundaries from the beginning. This follows the lead of early proponents of alternative concepts, such as Alexander Oparin and Sidney Fox, who favored border-less models based on proteinaceous self-cohesiveness. In fact, phase-separated colloidal droplets of predominantly proteinaceous character are able to secondarily assemble lipophilic peptides and other amphiphilic compounds at their outer surface in membrane-like associations.

Yet, if permanent external boundaries were not essential early on, why and how did composite and persistent biomembranes come into being in the first place? There is in fact another important feature tightly associated with membranes throughout all the organismal branches: procurement and transduction of free energy to keep life going all along. This notion may lead, perhaps, to evolutionary incentives for proteinaceous complexes to nucleate around energizable inorganic compounds and in turn expand into more and more membrane-like biopartitioning systems.

I made the point before that photochemical charge separation, as coupled to organic redox reactions, is worth considering a potential link to energy harvesting at membrane-bounded compartments, which may have evolved in several steps from simple beginnings (Egel, 2011, 2012a). The coupling of organic syntheses to inelastic absorption of UVB photons by MeS minerals is facilitated by the colloidal properties of photoactive particles (Hoffmann *et al.*, 1995; Guzman and Martin, 2009). Photon absorption then leads to momentary charge separation of a conductance-band electron (e^-) and a complementary valence-band hole (h^+), both of which can readily migrate through the crystalline lattice. At the colloidal nanoscale, both e^- and h^+ can diffuse to the particle surface — engaging in redox reactions with external scavenger compounds, rather than being annihilated by charge recombination within the mineral grains themselves.

Conceivably, emergent life had complementary means of optimizing various components to utilize this elementary effect for organic synthesis reactions: delivering suitable scavenger compounds to receive the separated charges and developing insulation coatings to suppress spontaneous charge recombination. Peptides and proteins were especially suited to serve both objectives, due to the variation of up to 20 different amino acid side chains. In due course, therefore, the primitive scavenger compounds may have evolved into protein-mediated electron transfer chains of various electron donor-acceptor complexes, and the initial insulation coatings made it into both membrane-spanning protein domains and hydrophobic cores in globular proteins, the former being assisted by phospholipids in building up to more extensive and topologically closed membrane configurations.

Similar to other composite systems, biomembranes need a critical nucleation process to be formed at new locations, and modern biomembranes can only grow from preexisting membrane structures in a cell. By this token, the biomembranes at large constitute one of the oldest analog (non-genetic) continuance systems in the living world. Notably, the membrane phospholipid components are chirally and chemically different in the bacterial and archaeal domains of akaryotes, consisting of glycerol-3-phosphate with two long-chain fatty acids and of glycerol-1-phosphate with two isoprenoid chains, respectively. This deep-branching dichotomy has led to the interpretation that cell boundaries at the common ancestor stage were about as leaky as simple fatty-acid vesicles (Wächtershäuser, 2003), whereafter the energy-related tightening of membranes was achieved by different means in each domain (Sojo *et al*., 2014).

Accordingly, I herein begin with discussing cytoplasmic continuity — in terms of catalytic functionality and structural organization — as the natural link between basic metabolism on the one hand and the superposition informational repositories — genes and genomes — on the other.

Not the least in the context of membrane-mediated energy procurement and transduction, considering that electron bifurcating charge separation by reduction of FeS-rich proteins, such as ferredoxin, seems to have particularly ancient roots, it appears advantageous to begin with tiny vesicles or yet more limited planar patches, since ion-tight membrane systems are easiest to set up from scratch at smaller scale.

Under the reign of the traditional prokaryotes-before-eukaryotes presupposition, this section would have to elaborate on how the bacteria-like type of cell organization may have been accomplished as smoothly as possible, perhaps by taking recourse to yet smaller lipid vesicles and certain organic inclusions as a potential starting point. The Woesean challenge against simplistic concepts of this kind, however, has severely shaken confidence in the classical scenario, and other arguments have also been raised against it before. The main point here is that the Woesean progenote phase required a large communicative system, at least throughout the time required for the optimization of the genetic coding and translation machinery and very likely for other complex tasks as well, such as highly reliable replication mechanisms appropriate for the faithful propagation of modular genomes in general.

Accordingly, I here advocate a rather different basic model, assuming "protoplasmic continuance" as the most basic principle to connect the earliest molecular ecosystem of surface protoplasm with all biotic life, as known today. Already well before the emergence of genome-directed organisms, this compositional connectivity may have led to on semi-fluid (colloidal) sessile layers of slowly growing surface protoplasm of potentially confluent protobiofilms. At the earliest stages presumably, the functional coherence of the emerging proto-cytoplasmic system was not yet dominated by linear or processive polymerization, such as template replication, transcription or step-by-step translation. Instead, some prereplicative tinkering prevailed in more or less stochastic processes of terminal extension, patch/paste ligation and mutual stabilization by affinity binding of certain macromolecular products.

Already early on, I think, the externally amorphous biofilm-like layers began to differentiate internally, well before they eventually attained more cell-like properties in their overall appearance. A recent model of cycling between micellar microsphere dispersions and dried-down multi-lamellar films (Van Kranendonk *et al.*, 2017) appears related to this point. In my opinion at least, this model should not only imply a redispersion into simple vesicles but also allow multiply connected lamellar spaces to engage in evolution by differential neighborhood selection. In consequence, each round of episodic film redispersion may have resulted in separate entities of unequal size: many smallish vesicles and fewer larger bodies — some of the latter also bearing vesicular inclusions inside a larger mass of protoplasm.

To give internally compartmented precellular associations a tentative name that has not been used before with different implications,

I have suggested pan-mictic and polyphenotypic populations of composite "paracells." Though not yet organized as modular cells — in the way it is known to us today — the paracells were already proto-organisms in a collective sense by being capable of sustaining and self-propelling life-like continuity throughout an interactive population.

The partly miscible micellar membrane systems — perhaps being peptide-dominated, as I presume — may have prevailed into the putative stage of paracell organization, where their inherent tendency to merge on contact supported potential confluence at a population-wide scale. The latter property was significant in allowing wide-spread material exchange, as required for the system-wide communality inferred by the Woesean progenote concept. The evolutionary importance of pre-cell fusion, complementary to pre-cell fission, has actually been pointed out before by Günter Wächtershäuser (2003), who is a dedicated chemist rather than a bacteriologist and, for that matter, may have approached the subject with a more open mind.

In contrast to most other models, I put particular emphasis on the — potentially endogenous — formation of internal compartments at the paracell stage of early evolution. To me at least, it appears surprising that this interesting possibility, which had been taken for granted before the adoption of mitochondrial symbiogenesis as leading theory, has not been given much attention for many years. Moreover, I find it harder than many others to believe that tiny vesicles with sufficiently tight membrane enclosures would have had any reasonable chance at all to come forth from a "primordial soup" directly, so as to attain life-like robustness and resilience early on. All this looks very different if large connected masses of organic proto-biomatter with protoplasmic functionality, continuance and confluence could have accumulated in sessile films, without necessarily being surrounded by tight membranes from the beginning. Under the structural and functional umbrella of some surface protoplasm of this kind, it should be much easier to imagine the generation and sustenance of internal vesicles, considering their adaptive value as energy-procuring compartments in productive coevolution with the surrounding cytoplasm.

As for evolutionary origins of mitochondrial organelles, a partial contribution of the now widely dismissed model of endogenous compartmentalization is not purely speculative. The recent comparative analysis of mitochondrial proteomics by Michael Gray (2014) has led to a pioneering Pre-Endosymbiont Hypothesis as follows. When the ancestral precursors to all the extant mitochondria and hydrogenosomes derived thereof were being internalized from free-living α-proteobacteria by an ancient pre-eukaryotic ancestor, this putative host line may already have carried other organelles beforehand, including an elaborate protein import system to support their propagation and functionality. This conjecture is based on the observation that the majority of mitochondrial proteins did not converge back into the α-proteobacterial component, as it would have been expected by the classical endosymbiont paradigm, but apparently has come from other sources. Gray denoted these putative precursors as premitochondria, which may have been of endogenous origin and eventually were replaced by the current mitochondrial lineage of proteobacterial descent.

These notions correlate with my reinterpretation of Woesean conjectures in what I have dubbed the extended progenote hypothesis, involving internally diversifying paracells. In a peculiar way, that is, preceding proposals to explain the emergence of complex eukaryotes by endosymbiosis — resulting from various fusion theories — are turned upside down, deriving somewhat equivalent states of endobiosis-like coevolution from within the common protoplasm — by endogenous symbiosis, so to speak. The genomic implications of this scenario are taken up in the next section, considering especially how different kinds of genuine cells may have been derived from the putative paracells.

5. Origins of modular genomes

5.1. *The rise of akaryotic bacterial genomes*

Arguably, I think, Woese's most interesting and important insight in his later evolutionary writings is the conceptual coupling between the consolidation of reliably inheritable genomes and the emergence of vertically stable lineages of modular cells. This begs the mechanistic questions of how the numerous short individual genes became lined up in large and self-sufficient genomes and what kind of evolutionary driving forces could have acted in favor of such directional changes. Woese himself was rather reluctant to speculate further on these crucial issues, other than that he considered operon-like clustering on mini-chromosomes and some tentative similarity with the highly redundant association of many genes in the macronuclei of ciliates. Neither did he pay particular attention to the potential contribution of viruses in this regard.

In general, I think that the initial clustering of functionally related genes in the progenote era must have occurred at subcellular levels. It has commonly been conjectured that the communal populations of progenote entities were very diverse in genes and phenotypes, i.e. highly polyphenotypic. This means that the local distribution of various genes was not the same throughout the entire population of progenote paracells but varied by stochastic fluctuation and adaptive clustering according to locally variable conditions. It also means that reversible adjustment to temporal fluctuations — in cyclic manners, as well as by more chaotic turbulence — was a very early objective for emergent proto-life. Accordingly, epigenetic modification pathways and regulatory feedback from poly-disperse RNA populations may be about as old in evolutionary terms as genetic repository mechanisms in general. In addition, the communal optimization of replicative, transcriptional and translational capabilities inadvertently rendered the progenote paracell system vulnerable to over-exploitation by selfish genetic elements and virulent genomic agents. Yet such potentially "rogue" genetic entities, in turn, could also contribute productively to further optimization in the communal host systems itself, so as to approach sustainable degrees of genomic and organismal modularization at various levels.

The progressive integration of individual genes into self-perpetuating genomes was likely initiated by higher evolutionary speed in vertical lineages of specializing smaller replicative entities, as compared to the conservative buffering capacity of the much larger, communally organized progenote gene exchange pool. Conceivably, the first and smallest genomic agents to exploit this evolutionary strategy were RNA viruses, which could afford to specialize on the specificity and speed of replication as such by still relying on the metabolic potential of the protoplasm in progenote paracells at large. Next in line presumably, the chemical modifications that resulted in more stable DNA instead of RNA allowed a novel class of viruses to assemble larger genomes (Forterre, 2006; Forterre and Prangishvili, 2013), potentially draining the more sluggish RNA-based paracells for much of their precious biomass and metabolic capacity. By adopting the novel invention of DNA for metabolic genes as well, however, the progenote paracells were able to strike a new balance, and also cellular evolution was channeled in new directions.

In hindsight, for that matter, the superposition of genomic DNA upon the metabolically functional protein/RNA-based protoplasmic system comprised the most important transition toward the generation of cellular organisms. It was probably too much of a task, however, to quickly assemble all the necessary genes of the collective progenote system in a unified genome in a single round. The pre-endosymbiont hypothesis mentioned above can pave the way for a more gradual evolutionary approach to modern life. The vesicular mediation of electron transfer reactions in particular, in combination with self-replicative plasmids inside such vesicles, could allow relatively small initial proto-genomes to grow bit by bit over an extended period, whilst these "paraorganellar" vesicles were generally nourished and supported by the functional cytoplasm of the surrounding paracells. By specializing and evolving faster than the communal progenote host system and the corresponding extra-organellar "paragenome," however, the modular paraorganelle lineages with their autonomously replicating plasmid genomes would increasingly compete with one another for diminishing dependence on the host. Gaining independence under these presumptions may eventually have resulted in "cellular escape" of bacterial lineages first and of archaeal lineages in later rounds, so as to partly explain the general pattern of the Woesean ToL.

5.2. *Asgard archaea: "Jokers" of the game*

Woese's three-domain phylogeny, in fact, is severely challenged by new data from large-scale environmental DNA analyses (Saw *et al.*, 2015; Hug *et al.*, 2016), suggesting a two-domain model by rooting the eukaryotic organisms within the archaeal domain. This started when genome reconstruction from submarine sediments pointed at Lokiarchaeota as a novel lineage, which since has widened to Asgard archaea as a separate branch of related — still unculturable — archaeal microorganisms. Collectively, this superphylum contains the largest set of homologs that previously had been considered specific of the eukaryotic signature (Zaremba-Niedzwiedzka *et al.*, 2017).

Although most experts seem convinced by the new paradigm of eukaryotes originating deep within the Asgard archaea, consensus about appropriate interpretations is not fully unanimous (Da Cunha *et al.*, 2018;

Gaia *et al.*, 2018; for a summary of continuing debates, see Watson, 2019). The crux of the matter boils down to several persistent ambiguities: Does it take two or three principal components of gene flow from the communal progenote stage to best account for the different system properties that characterize Bacteria and Archaea on the one hand and Eukarya on the other? Is it reasonable to presuppose that evolution goes from simpler to more complex system properties — also in crossing the Darwinian threshold between very different phase/state parameters? And is it indeed warranted to consider multiple gene losses as being less plausible than single gain events in modeling the deepest branches of the ToL?

Somewhat disturbingly, many different genes do not necessarily tell the same story: only some supporting the two-domain eocyte hypothesis, whereas a subset addressed as "Woese proteins" are more in line with Woese's three-domain topology (Da Cunha *et al.*, 2017; Gaia *et al.*, 2018). With the paracell model and the protobiofilm-to-biofilm concept presented herein I intend to provide additional arguments in support of the Woesean view, allowing multiple gene flows from a system dynamics perspective.

5.3. *Protobiofilm to biofilm transition: a mixed model to bridge deep knowledge gaps*

In looking at recent representations of the formal ToL (Hug *et al.*, 2016; Castelle and Banfield, 2018), it is particularly striking that various groups — newly added but deeply rooted all the same — comprise of still unculturable organisms, such as the archaeal DPANN cluster and Asgard archaea on one side and the bacterial "Candidate Phyla Radiation" (CPR) of Microgenomates and Parcubacteria on the other. It is predicted that symbiotic lifestyles underlie various forms of microbial unculturability but peculiar nutritional preferences may also play a role (Carini, 2019). It is also worth noting that by far the longest interval on the scaled, unrooted tree keeps the last common ancestor of all archaea and eukaryotes apart from the branching point between CPR and other bacteria.

More likely than not, the common root of the entire ToL is buried somewhere inside this central spacer region — or is it? This seems quite valid in a strictly formal sense, which would contract the base to just a single point of bifurcation, but Woese's communal progenote concept puts this question in a different perspective. The Woesean quest, in fact, is more concerned about how flat the common basis of communal system optimizing really was, or how wide and spacious the common trunk had become before it got strong enough to bear the entire treetop of the modern biosphere.

Localizing the formal root — most likely equidistal to the averaged species end points on either side — would bisect the unrooted spacer line and convert it into paired arms of Y-shaped branching in a rooted tree. While the usual interpretation would have us believe that either arm already corresponded to genuine cells, respectively resembling the bacterial and archaeal types as known today, the Woesean point of view is very different. The collective progenote concept (Woese, 2002, 2005) — and/or the related precell model of Kandler (1993) — may actually have covered most of the Y-shaped connector regions to the common trunk. To be sure, Woese did not explicitly address the possibility of splitting the trunk line without one of the branches crossing the Darwinian threshold of genome consolidation already at the bifurcation point. Kandler (1993), however, was well aware that physical barriers of geo-/topographical isolation — the classical mediator of "allopatric" Darwinian speciation in the modern world as well — may have let two subpopulations of the trunk-line community evolve into very different directions, before either branch gave rise to genuine cells of different types later on.

In presenting the aforementioned paracell model (Egel, 2017), I have argued for more open discussion on how the principal implications of Woese's general insights into genome–lineage interdependence — in terms of vertical persistence over evolutionary time — may gradually have materialized in terms of mechanistic and organizational possibilities in the living world. This particular progenote model is based on the possibility of internally compartmentalized paracells and fully independent lineages arising by "cellullar escape," but it is also open to further elaboration. Partly independent cell lines, too — expected to arise from early escape events — would likewise continue to evolve by staying together in mixed, cooperative biofilms. The clustered emergence of deep-branching unculturable microorganisms (Hug *et al*., 2016; Castelle and Banfield, 2018) indicates that primordial-biofilm hypotheses deserve further consideration.

5.4. *Founding the organismal ToL: the first dichotomy between Archaea and Bacteria*

As mentioned before, the Woesean concepts couple the emergence of modular cells to the consolidation of modular organismal genomes. The original nucleation of the corresponding genomes, however, has been given rather little attention. In summary of additional considerations in this regard, I am suggesting that the bacterial genome precursor was first to nucleate and that it did so as a self-replicating DNA plasmid — possibly related to an early DNA virus (Forterre, 2002, 2006). Together with emergent, still rather leaky means of membrane-bounded compartmentation — internally at first, as I presume, and externally as cellular partners in biofilm-like consortia later on — these smallish genome initials began their evolutionary journey in partial separation, but not completely isolated from the much larger communal gene pool of the progenote trunk-line population.

Before this evolutionary turning point, the mutual cooperation of RNA and proteins — as manifested in ribosomal mRNA translation by way of the genetic code, together with RNP/protein-mediated replication and processing of appropriate RNA components — was sophisticated enough to support a collectively evolving "progenote" state of life, but RNA as such was not well suited as genetic material at the scale required for vertically stable organismal genomes. The founding of a diversifying ToL may, therefore, have been tightly coupled to the invention of chemically more stable DNA — putatively by one or more DNA viruses, some of which converted from the pure strategy of virulent parasitism to a more temperate — potentially symbiotic — mode by turning latent in a non-infective plasmid form (Forterre, 2006).

In view of puzzling indications that the proteins for cellular DNA replication are homologous in Archaea and Eukarya but non-homologous in Bacteria (Leipe *et al*., 1999), I find it reasonable to assume that the DNA plasmids just mentioned may represent the first — and most profound — example of Woese's asymmetry principle discussed further above. On the one hand, these plasmids established the genomic lineage that eventually led up to the free-living bacteria abounding in the modern biosphere. On the other hand, however, the initial DNA plasmids and corresponding proto-bacterial compartments depended on the residual, still RNA-dominated progenote community for a good time to come.

Accepting the non-homology of bacterial replication factors as evidence for a bacteria-specific genome nucleation mechanism, it is quite reasonable to assume that archaeal lineages emerged from the residual progenote community by different means of genome nucleation — perhaps initiated by a different kind of DNA virus — which in turn may have led up to at least partly different escape routes for free-living archaeal cells to gain independence from symbiotic interactions with the residual progenote population.

As a corollary to the assumption that bacterial and archaeal cells may have arisen by separate and partly independent "escape" events from the communal progenote state, two features indicate that cytoplasmic membranes in archaeal and bacterial cells became multifunctional by convergent evolution — the dissimilar chemistry of their membrane lipids and the remote phylogeny of the membrane-associated ATPases. In the light of the extended progenote hypothesis proposed herein, and the related pre-endosymbiont hypothesis, these fundamental differences are suggestive of, or at least compatible with, the following interpretation.

If the pre-mitochondrial (bacteria-related) lineages of paraorganelles were founded at a time when early proto-membranes, perhaps, were dominated by lipophilic peptides but long-chain lipids in general and triple-chain lipids in particular did not yet exist, these pioneering vesicular/genomic "power houses" began to tighten their surrounding membranes from inside and were first in making ester-linked lipids based on long-chain fatty acids for that purpose. There is another ancient source for making lipophilic compounds in all three domains of life, however, as represented by isoprenyl anchors in many membrane-associated proteins. This may have become the basis for making ether-linked isoprenoid lipids for membrane assembly in other compartments of the composite progenote paracells. Lipids of the latter type are still abounding in the archaeal domain (Jain *et al.*, 2014), and may even represent the oldest pathway (Coleman *et al.*, 2019), whereas all eukaryotic membrane systems have been taken over by the bacterial type.

Membrane-associated ATPases, which are deeply involved in ion transport and/or osmotic energy conversion, may tell a similar story. Of two ancient and related families, one is characteristic of bacteria and mitochondria, as well as chloroplasts, whereas the other is represented on archaea and eukaryotes outside these organelles — preferentially in vacuoles and lysosomes (Mulkidjanian *et al.*,

2007). Tellingly, vacuoles and lysosomes are internal membrane systems, which under the extended progenote hypothesis may represent the ancestral or primitive state as to where this particular protein family has first appeared. The placement in the outer boundaries of bacterial and archaeal cells later on may thus find a natural explanation by way of transitional membrane continuity during the founding events of "cellular escape" when these akaryotes emerged as vertical lineages of free-living organisms from the prior population of collective progenote paracells.

Under these assumptions, the very first branch point in the canonical ToL would only mark the primordial split between proto-bacterial genome initials and all the rest of the collective progenote community. The second branchpoint, again, would only mark a subsequent split between proto-archaeal genome initials and the residual trunk line, which also then was still remaining at the communal progenote stage. The proto-cellular organization would, however, reach far back into the communal state of trunk-line evolution, originally with micellar boundary systems of substantial potential for membrane coalescence and widespread gene pool exchange. Accordingly, most of — if not even more than — the lengths of both arms resulting from first Y-shaped branching may still have been part of a collectively optimizing trunk-line population.

5.5. *Consolidation of eukaryotic genomes*

What still remains now is to rationalize the emergence of the more complex eukaryote organisms as well. These may have evolved from the residual progenote community (the "Archgenote" trunk line) by different means than the akaryotic cell types of Bacteria and Archaea. It is worth noting that none of the minimalistic akaryote lineages had necessarily gathered the whole range of various genes in the communal progenote gene pool on a single circular molecule of DNA. They had just to fit the needs of a tiny microcell, in order to become more or less independent in the beginning and to optimize further akaryote evolution for rapid growth and reproduction at a minimum cell size. It is reasonable to assume, therefore, that the residual trunk line continued to make a living mainly based on the additional set of genes that were not really essential in akaryotic cells. The legacy of this extra lot of genes is still apparent as eukaryotic signature proteins (Han *et al.*, 2013). Ultimately, some of these additional genes were also instrumental in organizing the multichromosomal genomes of eukaryotes in novel ways, which showed little resemblance to common features of akaryote cell division cycles. Many others were involved in cytoplasmic activities and infrastructure, such as cytoskeleton components, membrane trafficking and lysosomal protein degradation, as well as spliceosomal proteins.

The consolidation of uniformly propagative genomes is intimately related to two other characteristics of the biosphere at large, the generation of vertically stable organismal lineages — or species — on the one hand and the adaptive specialization to different ecological niches on the other. Before there were genomes at the organismal level, there were no different biological species either, but locally different environmental possibilities did already exist before. As shown by the ancient oriental game of rock–paper–scissors or "roshambo," where "A beats B beats C beats A," the mutual entanglement of three or more categories as free parameters allows for certain configurations where uniquely definable winning strategies no longer exist. An indeterminate yet cyclic relation of this kind can set in train a complex and never ending path of diversification and mutually adaptive coevolution — wandering on the edge of eternity yet always at the risk of irreversible extinction of individual sublines. Of special note, the environmental input of various ecological niches is never fixed but is likewise subject to evolutionary changes caused by preexisting biological activity.

Up to this day there is no eukaryotic species that has gathered its entire genome on a single chromosome and presumably, therefore, such unification of eukaryotic genomes on single molecules of DNA has never been a major evolutionary objective in the distant past. Instead, the endogenous inclusion of multiple chromosomes in proto-nuclear compartments already at the communal progenote stage represents a conceivable, and more realistic, alternative hypothesis. Under this presumption, however, it is necessary to generate and optimize a reliable system for distributive segregation by different means than those being sufficient to segregate single plasmid-like chromosomes in the case of akaryotic cells. The most conspicuous differences are the separate compartmentation of multiple linear chromosomes in eukaryotic nuclei and the elaborate cytoskeleton machinery involved in reliable chromosome partitioning.

Just as remarkable as either feature may be in its own right, when viewed in a larger perspective both facets fall in line with many other aspects of spatial and temporal differentiation within the confines of a polyphenotypic common whole. It is also obvious that akaryotes and eukaryotes have settled on very different survival strategies concerning size, complexity and trophic interactions with the environment. These fundamentally different tendencies trace back to the level at the last common ancestors of Bacteria, Archaea and Eukarya, respectively. The Woesean progenote scenario, extended by the pre-endosymbiont hypothesis, merges the divergent evolutionary tendencies into a single polyphenotypic "meta-species" early on, where different strategies were being pursued in different parts of larger polymorphic "paracells."

Evolutionary melting pots suited to model a primordial breeding chamber of this kind do actually exist in anaerobic ciliates and giant amoebae of the modern world today. For one thing, these eukaryotic protists represent two different modes of locomotion — pseudopodial protrusion and retraction on a solid surface and flagellar swimming in a liquid habitat — which both trace back to the ancestry of all eukaryotes. For another, such anaerobic protists often harbor a "syntrophic" association of akaryotic endosymbionts and/or much reduced organelles derived thereof. And furthermore, by means of phagocytosis, these grazing macrocells primarily feed on free-living akaryote microcells for a living. The intracellular syntrophy in extant eukaryotes is based on the transfer of molecular hydrogen from bacterial-type, mitochondria-derived hydrogenosomes to endosymbiotic archaeal methanogens — comparable to syntrophy prevailing in anaerobic biofilms composed of genuine archaeal and bacterial cells (Sieber *et al.*, 2012).

Arguably, therefore, tentative primordial hydrogenosomes and methanogenosomes are good models for complementary paraorganelles already in the anaerobic progenote paracells, from which bacterial and archaeal lineages, respectively, may have originated and "escaped" as the free-living akaryotes we know today. At intermediate stages of mixed biofilms, this kind of endosymbiotic complementation could have continued as a network of epi-symbiotic relationships, until major lineages of free-living akaryote cells in turn managed to gain full nutritional and organizational autonomy by directional gene flow from the dispersed communal pool into the unimolecular plasmid genome.

Later on, after cyanobacteria had perfected photosynthesis into the oxygen-generating kind, free-living bacteria were probably first to deal with the toxic effects of molecular oxygen. Some of these oxygen-utilizing bacteria, in turn, were then taken up by the proto-eukaryotic trunk lineage as exogenous endosymbionts, from which all modern mitochondria and related organelles were then derived by strongly reductive evolution (Martin and Russell, 2013). This eventual de novo acquisition from free-living bacteria is now considered uncontested knowledge. On the other hand, the still ongoing debate concerns the likely state of the receptive host, which herein is perceived as replacing pre-existing, endogenous "paraorganelles" with novel endosymbionts from the outside world.

The management of energetics by semi-autonomously evolving monopartite minigenomes enclosed in vesicular compartments of the "akaryotic" type is one thing; the eventual consolidation of multipartite eukaryotic genomes from the multitude of other genes and minichromosomes originally present in the progenote paracells is quite a different matter. To this end I suppose that the many other chromosomes of the progenote "meta-genomic" gene pool may likewise have become enwrapped in separate membrane-enveloped compartments, as long proposed in various models of autogenous karyogenesis, before endosymbiosis became the reigning paradigm (Lake and Rivera, 1994). In a generally complex and diversifying cytoplasmic system this is a very natural idea. Distinct chromosome-bearing entities would subsequently begin to optimize the properties now common to genuine eukaryotic nuclei, such as the double-layered and pore-gated nuclear envelopes and the involvement of membrane-independent cytoskeleton components in the equal or proportionate partitioning of multi-chromosomal genomes. Two alternative karyogenesis models are briefly addressed further below. In putting focus on particular detail, each of these models may well bear some grain of thruth, but other aspects of the intricate nuclear–cytoplasmic relationship in eukaryotic cells remain to be fully understood in evolutionary terms.

In general, these considerations also beg the question how the complementary interplay of mitotic and meiotic nuclear divisions of modern eukaryotes may have arisen from semi-stochastic origins at the communal progenote stage of pre-genomic paracells. I tend to favor the idea that this complementary pair of related but also peculiar systems of chromosome shuffling and reshuffling have emerged together, in mutual coevolution from more indeterminate beginnings, where neither mitosis nor meiosis had come into existence first (Egel and Penny, 2008). This interpretation is not very common yet, since most observers still favor the traditional view that meiosis "merely" is a set of secondary modifications of the more basic mitotic mechanisms, but it allows to envision a more intimate and direct connection to the primordial sharing of a large communal gene exchange pool that is intrinsic to the Woesean progenote scenario.

Moreover, it is worth noting that the peculiar meiotic mechanisms of organized recombination and genome reduction cannot stand alone. They have to go hand in hand with complementary systems of cytoplasmic fusion (fertilization) and organized karyogamy (nuclear fusion) — once per generation — so as to complete a full cycle of sexual reproduction. Once again, this functional web of many interactive components is difficult to explain if it had to be derived from an isolated fusion event between two different akaryote cells and their streamlined unipartite genomes, as these are known to us today. If viewed from the Woesean perspective, however — implying a collective and inherently confluent population of polymorphic, polygenic and polyphenotypic paracells — the gradual consolidation of sexual propagation in the eukaryotic stem line appears more natural in evolutionary terms.

On this presumption, the combined emergence of mitosis, sexual fusion and meiosis becomes a matter of gradual optimization of population resilience and sustainability, promoting polychromosomal genome consolidation and stability against several counteracting factors. This formidable and progressive course could never have been embarked on in the first place, had the still pre-genomic paracells not found a means of living that made the best of being large. The tiny and more efficient akaryotes, with their monomolecular streamlined genomes, were probably fast in outcompeting the residual progenote communities at the main sites of energy procurement from the environment. The photosynthesizing proto-biofilms, therefore, were overgrown by genuine biofilms of akaryotic cells, and the larger yet sluggish paracells were driven aside into the darker zones where sunlight could not reach. Their best chance thereafter was to assimilate the steadily accumulating bio-matter produced by others and to adopt a phago-heterotrophic lifestyle.

In the *roshambo* hand game, the soft yet spread-out sheet of paper beats a hard piece of rock by enwrapping the pebble. By the same token, the tiny akaryotes became subject to engulfment by the larger — "protoeukaryotic" — paracells. The crucial instrument for subsequent digestion in intracellular vacuolar/lysosomal compartments may have derived from some precursory state, so as to recycle excessive amounts of the increasingly proliferative proto-organelles, which in the primordial progenote system had been in charge of primary energy procurement and many synthetic activities. The sustainability of being large at the cellular level very much hinges upon malleable infrastructure and intracellular diversification, much of which can be organized by combinatorial interactions of multi-purpose tools composed of versatile components. This overall strategy is much more apparent in eukaryotic than in akaryotic cells, and it is highly dependent on the characteristic set of eukaryotic signature proteins. The cytoskeleton components of this set, in particular, are deeply involved in virtually all the eukaryotic core activities, such as membrane trafficking, nuclear divisions, cellular mobility, phagocytosis and the like.

With rather few exceptions, the various eukaryotic organisms belong to sexual species, which are organized quite differently from the clonal structure of vertical lineages in akaryotes. Once per life cycle, that is, individual eukaryote genomes are resampled by recombination and reassortment from a common gene pool, which allows the entire population of a species to utilize a considerably broader ecological niche than what clonal progenies of a single organism are capable of doing in the long run. Moreover, it also acts as buffer and important safeguard against extinction by fluctuating detrimental factors of many kinds, especially if overall population size is constrained to low values. This provision, therefore, is especially important in eukaryotic species for which effective population size is much smaller on average than for tiny akaryotic cells. A major driving force for perfecting and retaining the multifaceted meiotic system is seen in the eliminating power of meiotic recombination and chromosome reassortment, which strongly facilitates selection against marginally deleterious mutations by stochastically gathering multiple genes of this kind in just a few meiotic progeny (Kondrashov, 1988). This factor gains particular importance in small populations, as generally prevail in large-bodied eukaryotes versus tiny akaryotic cells.

In the present context it is most significant to note that sexual propagation in eukaryotes may have direct links to pre-genomic, primordial confluence as posited here for a communal gene pool in the Woesean progenote state. I am strongly inclined to think that the co-emergence of mitosis and meiosis has been the eukaryotic way of preserving the best of the ancient progenote scenario by carrying over the early principle of optimization by communal innovation sharing into the modern world of genomic competition — and more than that. This coevolutionary coupling depended on yet another aspect of concurrent refinement, by which the primordial tendency of fusing indiscriminately at outer membranes was confined into narrow, manageable limits, rather than being abolished altogether and then reinvented de novo later on.

Both meiotic genome reduction and sexually motivated membrane fusion can, in fact, be understood as rarely occurring processes that serve specific adaptive functions which are distinctly complementary to the much more frequent mitotic nuclear division cycle.

Whilst accurate replication and mitosis serve the general purpose of clonal propagation in vertical lineages of individual genomes, the complementary set of sexual fusion, molecular recombination, chromosome reshuffling and meiotic genome reduction serve the overarching principle of saving the species at large from piecemeal extinction by accidental causes bit by bit (Goodenough and Heitman, 2014). Moreover, the supplementary action of appropriate switches between one or the other kind of cell cycle and the distinction between self and non-self at the outer boundary are some of the oldest examples of temporal differentiation in eukaryotes on the one hand and trans-membrane signaling mechanisms on the other.

In terms of a "Woesean transition," the full implementation of mitotic and meiotic mechanisms in the distributive partitioning of nuclear genomes marked the eukaryote way of crossing their "Darwinian threshold" eventually. This had major implications for the evolutionary dynamics of host–organelle relationships. It is my general conjecture that this transition marked a major turning point when the directional selective pressure shifted from driving vesicular paraorganelles toward gaining independence from the host system to favoring the newly established eukaryote host lines in competing for effective containment of their gradually expanding paraorganelle compartments. Later on, when the paracells eventually had passed their own Darwinian threshold as well, even newly acquired endosymbionts derived from free-living akaryotic cells became subject to severe reductive evolution inside their now stably propagable eukaryotic host cells.

Are all these deliberations still worth considering together with mounting evidence for a two-domain model of placing eukaryotes in the universal ToL? Before I return to this issue, I briefly address the largely unidirectional gene flow of viral genes that is notable for all cellular lineages. Contrary to common thought, it has been found that cellular genes were only rarely captured by viruses, whereas the gene flow from viruses to cells was very much larger (Forterre and Prangishvili, 2013). Arguably, the evolutionary impact of this phenomenon was most profound on major transitions early on, when innovative and especially successful components — which originally had given certain viral lineages a competitive edge on other viruses — proved valuable for widespread use in organismal populations as well.

The "three viruses, three domains" model (Forterre, 2006) is the most striking corollary to the general inference of more than two principal components in the ancestral gene flows parental to eukaryote emergence. There is just the additional caveat that such major viral contributions might have been most effective very early on, when the RNA-dominated progenote trunk-line population had not yet nucleated DNA-based genome initials with system-stabilizing potential but was being challenged by the first lineages of innovative DNA viruses. Also along this line, the high complexity of virus evolution in eukaryotes is more compatible with the proto-eukaryotic three-domain scenario than with the symbiogenic two-domain alternative (Koonin *et al.*, 2015). More specific models have been proposed for a *"viral origin"* of eukaryogenesis within the two-domain paradigm — as exemplified by the phylogenies of mRNA capping-related nucleotidyltransferases (Bell, 2001) or three well-conserved DNA polymerase domains (Takemura, 2001). Arguably, however, the overall complexity of DNA virus phylogenies at a global scale (Kazlauskas *et al.*, 2016) does not really suggest a one-to-one relationship between either two or three particular viruses and correspondingly initiated host domains.

At the morphogenetic level, the unorthodox proposal of an "*inside-out origin for the eukaryotic cell*" (Baum and Baum, 2014) deserves particular attention. It is inspired by a striking example of epibiosis amongst now living cells — giant archaea being surrounded by a sheath of much smaller γ-proteobacteria (Muller *et al.*, 2010).

Envisioning a similar relationship in the distant past, it is then hypothesized that ectosymbiosis evolved into endosymbiosis in a series of incremental steps: punching holes through the S-layer cell wall; extruding blebs of continuous membrane through the holes; reinforcing the rim at the holes by membrane-coating proteins; stabilizing the blebs by cytoskeleton proteins; surrounding the epibiotic bacteria by expanding the blebs; spreading the blebs outside the remaining S-layer by additional anchoring proteins, thus forming a perforated double-membrane proto-nuclear envelope; transporting ribosomal precursors and mRNA into the blebs through the proto-nuclear pores; developing a functional endoplasmatic reticulum (ER); and eventually closing the remaining gaps toward the outside world.

The model of Baum and Baum (2014) is thus returning to an earlier notion that the characteristic eukaryotic nucleus represents the dominant partner in an ancient endosymbiotic relationship — originally considered as entry into another cell (Lake and Rivera, 1994) — but focus is now shifted to initially ectosymbiotic interactions instead. Although the new model is tentatively based on the two-domain paradigm — and the paradigmatic coupling of eukaryogenesis as such to the emergence of mitochondrial endosymbiosis, for that matter — it might also be adaptable to slightly altered preconditions.

Notwithstanding the mounting evidence for a two-domain ToL, a range of apparent inconsistencies are still remaining, such as the clustering of potentially primitive RNA diversity in eukaryotes (Jeffares *et al.*, 1998; Penny *et al.*, 2009), the complexity of virus phylogenies and host range distributions (Koonin *et al.*, 2015; Kazlauskas *et al.*, 2016), and the presence of "Woese proteins" (compatible with a three-domain phylogeny) among universally conserved genes, together with a robust phylogenetic analysis of RNA polymerases (Da Cunha *et al.*, 2017). To naturally account for these particular correlations, a conceptually accessible shortcut for substantial gene flow from the collective progenote state to the proto-eukaryotic stem line should be highly desirable.

To this end, I suggest that several modifications to the "inside-out" model of Baum and Baum (2014) may provide for such a shortcut: As assumed here, three kinds of symbiotic lineages may have coexisted in mixed biofilms — semiautonomous bacterial and archaeal cells in close proximity to paracells of the residual progenote trunk line. The Asgard-related lineages, in particular, may have remained in such biofilms for considerably longer periods than other archaeal branches. At least one archaeal lineage may even have kept close membrane to membrane contact to trunk line paracells, and the dominant archaeal partners may have developed modular pores for molecular exchange between the cytoplasmic spaces. Eventually, the unidirectional transfer of archaeal mRNA and ribosome precursors may have paved the way for a functional take-over of the paracell protoplasm, also allowing substantial gene flow from the paracells into the archaeal partners. Such means would serve the needs of the archaeal master compartment primarily and allowed the chimaeric organisms to follow a novel common strategy of living in the long run. Subsequently, the two compartments evolved into the proto-eukaryotic nucleus and cytoplasm, respectively. The uptake of proto-mitochondrial precursors from a proteobacterial lineage, however, would only have occurred at a later stage (Pittis and Gabaldón, 2016).

Going more into molecular detail, the modular pores may also have been introduced by slightly different means than proposed in the Baum and Baum model. In modern nuclei, in fact, de novo nuclear pore assembly occurs by inside-out extrusion from the inner membrane of the nuclear envelope (Otsuka *et al.*, 2016), as critically mediated by ring-forming structural components first and membrane fusion factors later on. Not inconceivably, these kinds of critical perforation factors may first have been invented by viruses as part of their infection mechanism.

6. Origins of protein domains and fold families

Genes and proteins make a wonderful couple, without which life as we know it would not exist. More likely than not, this intimate teamwork is rooted in some coevolutionary coupling from very early, still rather indeterminate beginnings, where potential precursors for both proteins and nucleic acids already took part in functional networks of a proto-metabolic kind (Smith and Morowitz, 2010). To set in motion a self-reinforcing spiral of coordinated synthesis that eventually led to gene-encoded proteins, it is appropriate to consider some basic properties of amino acids, ribonucleotides and their respective oligomerization products. (i) A variety of different amino acids can be synthesized under early-Earth conditions as easily as various nucleotides or easier yet. (ii) Even small peptides among stochastic populations are suited better than oligonucleotides as catalysts in metabolically relevant reactions — albeit at still rather rudimentary levels. (iii) Polyphosphates, nucleotides and oligomers thereof are potentially capable of activating amino acids, which substantially accelerated the rate of prebiotic peptide synthesis. Subsequently, therefore, it was mainly a matter of gradual optimization to increase the size of coevolving oligomerization products — not only of peptides but oligonucleotides as well — and also to increase sequence specificity in functional products. To be sure, there is a long way to go from small stochastic peptides to the characteristic folding patterns of compact and conserved domains, which are the structural hallmark of modern enzymes and many other proteins.

Conceptually at least, this section is to bridge that knowledge gap by making reasonable assumptions, as motivated by basic questions about peptide–peptide cooperativity.

Question 1: Can numerous small peptides co-aggregate in domain-like heteromeric complexes? Most probably, I think, they can.

In modern proteins, a recognizable domain is defined as a 3-dimensional, relatively compact folding pattern that is covalently spanned by a single protein chain. The folding is stabilized in a secondary network of crosslinks or noncovalent contacts so as to minimize the direct exposure of non-polar side-chain residues to the polar influence of water. It is well known, in fact, that certain small peptides specifically adsorb to and thereby expand the characteristic folding pattern of certain protein domains, even if these peptides in free solution may appear as random coils — without any recognizable substructure on their own. It is not unreasonable to assume, therefore, that the same forces that stabilize the folding pattern in established protein domains cannot only attract individual peptides to this pattern on the scaffold of a covalently linked protein chain but may also act between several independent peptides, so as to initiate the loose assembly of some fold-like complexes. This should especially occur in response to gradual dehydration, since the nucleation of specific adsorption complexes is exceedingly difficult for soluble smaller peptides in a dilute and fully hydrated state. Model experiments bearing on this critical point are scarce but should be highly desirable.

Question 2: Can two peptides be ligated to form a longer chain, if they are held together by adsorption to a third one? Experimental modeling results support this assumption.

A proof-of-concept example — much celebrated as a "self-replicating peptide" — is based on strong binding affinity at so-called leucine zipper dimerization sites in many transcription factors (Ghadiri, 1998). The self-complementary motif comprises 30 amino acids in a perfect and contiguous α-helix (7–8 turns), forming homodimers at the hydrophobic interface in a repetitive pattern of "knobs-into-holes" packing. The full sequence can act as a template on which two partial fragments with reactive end groups are ligated in an autocatalytic amplification cascade. Such overly perfect conditions, of course, cannot be expected to prevail at conceivable early Earth scenarios, but the principle as such remains valid, even though the initial rates at any system-supportive reaction step must have been rather small in the beginning.

Question 3: Can numerous small peptides co-aggregate around composite non-peptide molecules of proto-metabolic potential? In spite of its obvious importance to prebiotic evolution, this reasonable possibility has not yet attracted particular attention for experimental studies.

A positive answer to this question would not only link peptides and cofactors to the emergence of a metabolically sustainable system but also, quite naturally, integrate mono- and oligonucleotides in the coevolutionary process at a very early stage. Further analyses of this question, therefore, rank high on a sensible priority list for future work. As a suitable candidate to test this particular conjecture, one may want to look for peptide arrangements resembling the ubiquitous nucleotide-binding Rossman fold (see below). Moreover, a yet more general problem is not just limited to the remote and hypothetical sphere of primordial peptide emergence but also concerns a somewhat enigmatic sector of modern evolution.

Question 4: How do new or novel proteins arise in the first place? Some cues can be taken from comparative genomics.

Large-scale genome sequencing, in fact, has not only connected molecular biology to evolutionary reasoning in terms of quantitative homology assessments but also revealed a "mystery of orphans," in that a remarkable lot of coding genes in any fully sequenced species are without noticeable homologs in other members of the biosphere. Such "taxonomically restricted genes" appeared as newcomers not so long ago, viewed on a larger evolutionary time scale (Tautz and Domazet-Lošo, 2011). The recency of their emergence notwithstanding, some of these new arrivals, in fact, are important or even essential for the well-being of the respective organisms. More often than not, orphan genes arise within previously untranslated intergenic regions, and the respective proteins do not in general conform to the folding pattern of common protein domains. Most likely, I think, novel orphan proteins emerged as relatively unstructured sequences which happened to have some binding affinity to one or more preexisting and functionally important interaction partners. The initially rather low affinity was subsequently strengthened and optimized by adaptive evolution to suit the ecological niche of the biological species, such as by increasing orphan length around the critical binding sites and allowing the extended proteins to fold up into supportive scaffold patterns of their own.

As for putatively ancient proteins, the most promising model studies pertaining to intrinsic affinity and early function address small metal- or phosphate-binding peptides and the apparent antiquity of metalloenzymes in general (Nitschke *et al.*, 2013). It has long been conjectured that iron–sulfur cluster proteins are particularly ancient among all the proteins known today. The most primitive ferredoxins, for example, are preferentially composed of simple, "primary" amino acids, which are most readily synthesized under abiotic conditions. Also, the binding of metal–sulfur clusters in these proteins occurs mostly at the backbone of the peptide chain, so that incidental sequence variations at the side chains were of lesser importance at the beginning.

Next in line are the numerous proteins that share the most "popular" folding patterns among super families, which carry the characteristic phosphate-binding P-loop and/or Rossmann-like configurations.

It has even been argued that these particular superfamilies represent the very oldest protein domains in a tree-like "phylogram" of structural evolution, since the members occur in highest numbers per fully sequenced genome throughout the entire organismal biosphere (Bukhari and Caetano-Anollés, 2013). There is, however, no general consensus on this matter and I am also inclined to question the conceptual basis for this line of reasoning.

To my understanding, there is no intrinsic coupling between the evolutionary age of a particular domain and its numerical abundance in modern genomes, and neither can I perceive any mechanistic rationale of universal phylogenetic significance behind a tree-building algorithm for all the protein structures in the first place, which a priori presupposes that there must be a particular "first" domain from which even structurally unrelated other domains were "derived" later on. Alternatively, I suppose, a variety of different domains emerged together — more or less independently of one another — from even smaller structural units, such as various kinds of hairpin, turn or corner motifs, mainly composed of α-helix, β-strand and flexible coil segments. Before the eventual consolidation of sequence-dependent replication, transcription and translation mechanisms, however, proteins as such had no deterministic means of complexifying in a bifurcating, tree-like manner from a single source. Instead, the synthesis, ligation and recombination of still uncoded peptide elements could only have occurred by more stochastic and agglutinative processes, which arguably cooperated in a collective phase of trunk line evolution where tree-like branching patterns had but little meaning and no dominating basis.

Question 5: Can suitable subsets of stochastic peptides be assembled around ADP/ATP so as to resemble the spatial arrangement of a Rossmann fold preferentially, especially if the congealing is seeded by a surplus of phosphate-binding peptides? This item relates to the more general query as to what extent the virtual monopoly on nucleotide binding by Rossmann-like configurations might have resulted from just a "frozen accident" of early evolutionary history, or whether it actually reflects the most prominent optimum amongst a rather limited number of possibilities in phosphate- and nucleotide-facilitated protein folding space at large.

The eventual answers to all these questions should shed some light on a central issue of biomolecular evolution, whether there were any basic rules that remained the same in the putative transition from uncoded to gene-directed protein synthesis. It is commonly assumed that the genetic code and the ribosomal translation system evolved together and that the selective driving force behind this grand achievement was to replace a preceding more primitive system of uncoded peptide synthesis by a dependable supply of particular and fully reproducible protein sequences. For this to work from the beginning, it was necessary that uncoded peptides, too, already served important functions in terms of system-supportive activity, and some of these basic functions need not have changed substantially as such. To start with, I think, mainly their quantitative aspects were adjusted for better performance in a collective and cooperative optimization process.

Towards the end of this transition, of course, the sequence-specific translation system could only reach perfection if RNA/DNA replication, too, was being perfected to a similar degree. The latter accomplishment itself — to my opinion — need not have come about beforehand independently, rather than resulting from the collective overall optimization process. In consequence, when both replication and transcription had become proficient at sequence-preserving fidelity, the laws of protein evolution were lifted to a new and partly different level after all. Notably, the selective evolution of proteins for functional utility, efficiency and specificity occurs by way of exploration in two different spaces of possibility and opportunity. At first there is the relatively limited space of topologically different folding patterns of protein chains in partly hydrated environments. This is contrasted with the enormous and virtually unlimited space of different amino acid sequences, even though the number of biologically relevant side chain variants is constrained by just twenty different constituents. The dimensional discrepancy between these disparate spaces reflects the factual redundancy that many different amino acid sequences can fold up into the same or very similar topological arrangements of secondary structure. As a conjecture, I would assume that an even more restricted subspace of protein folding patterns was particularly easy to access by early evolution, considering only those peptides and associated folds that had intrinsic affinity to certain preexisting and metabolically relevant non-protein scaffolds as initial nucleation centers.

The combinatorial differences mean that the sparsely populated spaces of scaffold-directed folding patterns first, and self-contained fold configurations later on, were considerably easier to explore by random sampling than the unlimited extent of sequence space. It is not inconceivable, therefore, that even uncoded — quasi-stochastic — peptides and small proteins were capable of congealing into simple and easily accessible modules of secondary structure, and some of these may have shown sufficient pairwise complementarity at their surface to interact with one another preferentially. Provided that activated and affinity-facilitated peptide ligation was feasible already at that early stage, also some composite protein folds were capable of being formed before the intervention of genetic template replication and transcription. All this may only have happened at rather low frequencies, of course, but evolutionary optimizing could reasonably start from there. Conceivably, this optimization process also had both stream-lining and diversifying aspects. A final question is to conclude this section.

Question 6: Can the emergence and gradual complexification of protein domains be rationalized by tree-like hierarchies that potentially remained invariant before and after the genetic/replicative transition? There are two different approaches in the literature that appear relevant to discussing this point. One of these is based on structural detail but its full evolutionary perspective remains to be assessed. The other proposal stresses phylogenetic relationships but is not primarily concerned about structural aspects of the protein folds in question. Naively speaking, I tend to see more evolutionarily relevant potential in the former case, which is therefore presented first.

The folding domains in many protein families and superfamilies, in fact, appear as being composed of much shorter sequence motifs in structural elements — often in a repetitive pattern. This can be seen as a result of step-wise addition by recombinational processes, be it at the level of interactive peptides early and/or RNA/DNA processing later on. Such patterns allow the construction of "structural trees" for individual protein superfamilies, which are seeded by a structural root motif and diverge at branch points where similar or different elements are added by incremental growth (Efimov, 2013). Overall, that is, early protein evolution can be comprehended by a multi-track system of separate structural trees, many of which are rooted independently in an ancient communal world of coevolving peptides and RNA. This system develops a natural hierarchy of combinatorial relatedness from simple elements and modular units to the complex folds of distinctive protein domains.

Taking a rather different approach, Caetano-Anollés and his productive bioinformatics group have enthusiastically propagated the vision of a unitary tree of universal protein fold emergence (Wang *et al.*, 2011). This polarized notion, however, is only based on a single and somewhat indirect conjecture. To be sure, their operational procedure may actually appear quite straightforward. By using phylogenetic analysis algorithms on large sets of protein-structural data, as represented in the growing lists of fully sequenced organismal genomes, they indeed arrived at impressive and well-defined cladistic trees of protein domain families and superfamilies. To this end the different fold categories were ranked according to their occurrence (abundance) in the various genomes — so far, so good for the operational soundness of their basic method. Somewhat alarmingly, however, their line of reasoning has not seriously been taken up or supported by any other group of protein experts.

The tricky part, in fact, begins when the resulting dendrograms are given apodictic validity in terms of phylogenetic descendence and relationships. After more than a decade's worth of publications dependent on the presumed validity of this relational succession, it becomes more and more cumbersome to check the underlying model for its conceptual roots and actual reliability. The key assumption, maintained throughout the entire series of this work, is that fold occurrence per current genome is quantitatively linked to the evolutionary age of the respective protein family. Moreover, it is also assumed that the age differences inferred are sufficient to establish direct ancestral relationships, even between structurally unrelated folds at the postulated, unitary root (Caetano-Anollés and Caetano-Anollés, 2003). Seen from a mechanistic perspective, however, "it ain't necessarily so." What I find most disturbing is the inbuilt tendency of this approach to overrule the natural hierarchy from short and simple to longer and more composite configurations.

Alternatively, and arguably by more natural means, there are other ways to rationalize the data. As I see it, the striking differences in genomic abundance between the various fold families may relate more to their functional versatility than to their evolutionary age as such. In other words, a high degree of functional versatility in association with a relatively large and composite fold is not necessarily a very ancient trait. Moreover, there is no conceivable mechanism that links the assembly of a particular domain-like peptide complex to the presence of preexisting molecules with a similar fold topology, which should be a necessary precondition for self-similar propagation with occasional variation in a tree-like branching pattern from a single common root. Instead, the duration of certain molecular complexes could be stabilized selectively by physical binding to other agents, such as external scaffolds, metabolic cofactors or substrate moieties attached to distinctive affinity handles. In consequence, the emergence of fold-like aggregates from uncoded peptides would be more naturally linked to their collective networking capacity — together with many unrelated potential partners simultaneously — than to a monophyletic dendrogram originating with and from a uniquely definable ancestral fold. Notably, the versatile Rossmann fold is especially remarkable in its handle-binding capacity; so its overall abundance is also correlated to the high number of substrates with similar affinity handles, in this case riboside derivatives.

When it comes to the intricate interplay between structure and function of peptides and proteins, a few general remarks are worth adding here. In enzymes, typically, rather few critical atoms of functional groups may participate directly in the chemical reaction being catalyzed, but a large structural scaffold embraces the reaction center from several sides, so as to constrain the functional groups in sterically favorable configurations — before and after the unstable transition stage. Much of the collective optimizing, in fact, has been directed toward growing chain length and better buttressed scaffolding — in individual proteins as well as in cooperative ribonucleoprotein (RNP) machines.

7. Origins of functional and replicating RNA molecules

Faithful replication of nucleic acid templates, with occasional mutation, is prototypical for generating tree-like patterns of reiterative duplication at the molecular level. Understandably, therefore, much attention has been focused on the presumption that the tentative beginnings of molecular replication must, and can, be solved as an isolated phenomenon before anything else. Since DNA, which chemically derives from RNA, can be conveniently set aside as a late-comer in biological context, this leaves us with the central assumption of the long cherished "RNA world" conception that a self-replicating ribozyme should be regarded the most likely starting point of prebiotic evolution. There are growing concerns, however, about the validity of this reductionist approach in its purest form, not the least from a systems-oriented perspective (Bernhardt, 2012). It probably needs some lateral thinking to break the conceptual deadlock.

A complex interactive system, such as life at large, is more than the sum of its constituent parts. Its comprehensive understanding also requires a full description of preferential interactions and dynamic interconversions amongst the various parts. The progressive emergence of life, therefore, cannot merely be reduced to singular events, such as a series of unique material inventions, but must also be understood as a continual improvement and optimization process for functional utility on many fronts. It is a basic fact of life that proteins and nucleic acids stand out as highly versatile constituents in these regards, and both kinds of compounds critically interdepend on one another in their respective ways of being made. RNA in particular is remarkable in its versatile potential for catalytic, archival and structural functionality. Not only can folded RNA ribozymes participate directly in certain chemical reactions, foremost concerned with the processing of RNA itself; RNA in general can also serve as a repository of sequence information for full recall upon unfolding and template utilization in the future; and perhaps most relevant in a primordial context, the fold-back looping of relatively short RNA sequences can have provided early structural scaffolds for certain uncoded peptides to adsorb for selective optimization to emerge.

It is mainly owing to the unique capacity to serve as self-complementary templates that RNA-centered theories of biogenesis have come to dominate the field. This said, however, it is also remarkable that, nowadays, functional replication of RNA or DNA depends on protein enzymes, and no obvious remnants of corresponding ribozyme activities have yet been found. The necessary helicase action, in particular, is probably hardest to achieve by protein-free ribozymes alone.

In terms of emergent evolution, therefore, I suppose that functional utility of small RNAs came first early on, together with uncoded peptides. This may reasonably have resulted in a functionally interactive "RNP world" scenario (Francis, 2011), presumably arising from simple origins in a peptide•RNA partnership (Carter and Wills, 2017). In this perspective, the more complex and indirect means of full template recall from archival or informational RNA/DNA only developed somewhat later on. More likely than not, the first generations of oligoribonucleotides were mainly polymerized by quasi-stochastic means, such as indiscriminate elongation at the end. At a certain length thereafter, folding back upon itself during synthesis could also result in local self-complementarity at base-paired segments, as a first round of deviating from fully random sequences.

A functional key role may have been played by simple stem–loop structures as rudimentary tRNA precursors, which supposedly featured some peptidyl transferase activity after amino-acylation at the terminal ribose moiety, and "minihelix-based peptide bond formation" is considered key to understanding further evolution (Tamura, 2011). This was still rather unspecific to begin with and thus could merely facilitate the synthesis of more or less stochastic peptides. A few such peptides, in turn, would also more than others associate with RNA and thus expand the repertoire of interactive agents in a Kauffman-type network of "catalytic closure." By joining forces in mixed complexes, these different types of chain molecules could assemble compact and rigid scaffold structures more frequently than either kind alone. In turn, the RNP complexes of highest networking potential engaged in an unbroken continuity of function and retained a recognizable identity up into the modern biosphere. Some of these even evolved into veritable RNP machines of remarkable sophistication, which commonly are unable to function without most of their proteins. The paramount example of this is the ribosome, which constitutes the central hub of gene-encoded protein synthesis in all the cellular organisms living today, and many theories have been proposed to rationalize its emergence and further evolution.

Characteristically, ribosomes consist of a small and a large subunit, which together contain 3–4 rRNA molecules and some 50 to >70 proteins, depending on the kind of organism. Initially, more primitive precursors of either subunit may likely have assembled from considerably more and smaller RNA pieces, held together non-covalently at conserved short segments of complementary base pairing. The subunits are functionally connected by aminoacylated tRNAs and rotate relative to one another back and forth during the repetitive protein elongation cycle, which has been referred to as a reciprocating ratchet mechanism. The heart of the large subunit bears the "peptidyl transferase center," from which the growing protein chain is leaving through a narrow exit tunnel. The small subunit, on the other hand, is organized around the "decoding center" where the mRNA sequence is being converted into amino acid code, and the ratchet mechanism for processive repetition is likewise based on the small subunit.

All three activities are crucial for overall performance, and the experts are not united as to which subunit is likely to have originated first. It has even been suggested that both subunits emerged independently of one another, having some unrelated functions to begin with. The most intriguing proposal to illustrate the latter notion assumes that the iterative decoding mechanism of the small subunit actually evolved as a ratcheting "triplicase" device — a composite RNA replicase that modular triplet bricks for transfer, rather than mononucleotides directly, to reproduce an entire sequence according to some RNA template (Penny, 2005). There are no obvious "relics" of a tentative triplicase activity in modern life, unfortunately, but more indirect hints point at a residual triplet pattern, RNY > RNR > YNY > YNR, in various nontranslated RNAs of ancient evolutionary origin (Lankenau, 2011) — where N includes any of the four bases, R only purins (A,G) and Y only pyrimidins (U,C).

The canonical cloverleaf structure of modern tRNAs could have arisen by recombinational processing, positioning a decoding loop and an aminoacylatable stem at different corners of a pseudo-knotted secondary structure. These composite units could subsequently interact with two other RNP machines — precursory to the two subunits of current ribosomes — so as to integrate and evolve the two centers for template decoding and peptidyl transferase activity into an effective protein assembly line. It is worth adding here that the various tRNAs, for most of the time, are not free-floating molecules in vivo but rather act as accessory RNP modules, together with aminoacyltransferases, translation initiation/elongation factors, and the like.

Finally, hosting the peptidyl transferase center, the large subunit also deserves special mentioning here, but with a peculiar twist. It happens to strain our general conception of what a ribozyme is expected or supposed to do. This is because the rRNA lining around the reaction center is very little involved in the peptidyl transfer reaction as such, whereas, for all we know, the crucial transition states are greatly facilitated by terminal ribose moieties of the participating tRNA substrates themselves. By that token, this is an intriguing example of "substrate-assisted catalysis," where a functional group that is critical for catalytic action is positioned on one of the reaction partners rather than provided by an external catalyst (Weinger *et al.*, 2004). It is very much a matter of semantics, therefore, which individual component of this complex system should be addressed as genuine, and potentially primordial, "peptidyl transferase."

For the purpose of this integrative essay, I here take the provocative position that terminally aminoacylated oligoribonucleotides sufficed to start uncoded peptide synthesis. These pioneering compounds were then assisted by an ensemble of other oligoribonucleotides, which were partly complementary to one another and together formed a narrow orifice just capable of surrounding two aminoacylated partners. The legacy of this initial pinhole is still apparent in the universally conserved central loop of the largest rRNA component. Its mechanical function is to serve as a one-way collar valve or nozzle, so as to assist tRNA-mediated peptide synthesis with some directional processivity by a structural scaffold, allowing cyclic repetition of an intrinsically stochastic reaction type. While it seems easy for an RNA-charged amino acid to slip into the receptive pinhole, the other tRNA is impeded by the drag imposed by the nascent peptide chain at the other side. Subsequently, this inbuilt directionality was reinforced by adding further RNA loops to ribosomal RNA and stabilizing proteins on either side of the central bottleneck, as well as by supplementing this initial scaffold with the GTP/GTPase-driven ratchet mechanism of the small subunit.

Furthermore, an additional function of the ribosomal reaction center as a protective casing should not be discounted. As the growing peptide chain represents a precious yet incomplete investment, it is of great advantage to keep spontaneous hydrolysis of the labile RNA–peptide bond at a minimum by restricting the accessibility of water in a narrow channel. Finally, as the RNA-lined exit tunnel can accommodate highly hydrophobic and other peptides alike, the ribosome is also a very effective appliance for injecting integral membrane proteins into membranes, in tight association with yet another RNP machine, the universally conserved signal recognition particle, which may have originated very early on (Smith *et al.*, 2008). Along these lines, distinguishing between hydrophobic and other amino acids may be one of the earliest bifurcations in the evolution of the genetic code, thus facilitating the early emergence of peptide-dominated membranes.

In addition to ribosomes, a variety of other RNP machines have survived into the modern, predominantly protein-catalyzed biosphere. Most of these, in fact, are involved in various aspects of RNA processing, such as cleaving, trimming, base modifying, recombining or splicing at particular motifs in RNA sequence. Some are widely conserved in all three domains of organismal life, but others are more restricted in their phylogenetic distribution. The most striking and conspicuous anomaly in this regard concerns the entire domain of extant eukaryotes at large. Eukarya, that is, are hosting the largest domain-specific variety of untranslated yet functional RNA species, including those participating in the arguably most complex systems of composite and cooperative RNP machines (Collins *et al.*, 2009). The two types of spliceosomes, in fact, are essential for recombinational removal (splicing) of two corresponding intron sets from eukaryotic mRNA precursors — some or many, depending on the organism. How much this variety has, or may have, direct relationship to preceding RNA/RNP world scenarios is still a matter of substantial debate.

8. Outlook and synopsis: A Tree is a Tree is a Tree — or is it?

This essay is meant to tell a story, one of the "EF" kind, in favor of a so-called "Eukaryotes First" scenario. Already many EF stories have been told before — too many to the taste of some. Why then add another variation on this, allegedly rather overworked, theme? And, yet more important for the topical book at hand: Why should bothering about this devious question be relevant anyway to the origins of biodiversity, molecular evolution and genomic biology in general?

In raising these questions, I hope to prompt considerable rethinking on various different front lines. To be sure, many of the seemingly unconventional suggestions mentioned here have been put forth before. Yet proposed one by one, they have hitherto been unable to gather momentum and gain enough support to change long-held preconceptions of what is primitive and what is derived when comparing cell organization in eukaryotic versus akaryotic organisms. In accordance with the unifying theme of this entire essay — the emphasis on trunk line evolution, well before the persistent establishment of competitiveness between different and diversifying organismal lineages — I here propose to integrate a unifying "Eukaryotes First" scenario in a primordial phase of organismal trunk line evolution of the Woesean kind.

Arguably, Carl Woese is the most progressive rethinker on early evolution from a biological perspective in two of his powerful claims: that genome consolidation must have preceded the emergence of organismal individuality, and that the long preceding phase was dominated by communal interactions of collective innovation sharing. But his great organism-minded ideas apart, Woese stopped short of analyzing the communal possibilities of internal compartmentalization as well. What is still missing is a conceptual umbrella to combine and rationalize the various EF proposals under a common and coherent framework. In this essay, therefore, I suggest that the internal compartmentalization of emerging minigenomes, being accompanied by gradual differentiation of the surrounding membranes — here also dubbed the extended progenote hypothesis — can do just that: provide a unifying concept under which the likely probabilities for the early emergence of both akaryote- and eukaryote-specific features can be reevaluated and, hopefully, combined into a plausible and coherent theory.

After all, are there any akaryote-specific traits that are certain to be homologous — inherited directly from a common ancestor — in archaea and bacteria in the first place? This critical question, in fact, has rarely been given serious attention. In a both lucid and critical overview of the various "eukaryotes first" scenarios proposed before, Mariscal and Doolittle (2015) provide a series of distinguishing akaryotic traits, tentatively graded "from possibly convergent to likely common descent." In addition, they also make a critical distinction between "Eukarya/Eukaryota" and "eukaryotes," respectively referring to a monophyletic or cladal category, "which comprises the last eukaryotic common ancestor (LECA) and all its descendants," and a phenological term, embracing the general type of cell organization and its possibly early origins. Notably, it is only in the latter sense that the various EF scenarios can have a meaningful interpretation.

I shall here concentrate only on the three highest ranking examples on the provisional scale toward tentatively common descent of archaea and bacteria. As seen in the perspective of the most passionate opponents against any EF scenario, these cases are particularly difficult to explain, as to how a transition to the akaryote condition could be effected twice, convergently: (1) multifunctional cytoplasmic membranes, (2) transcriptional–translational coupling, and (3) lack of spliceosomal introns.

Clearly, these akaryote-specific traits are in contrast with the eukaryotic counterparts. In eukaryotes, for that matter, the cytoplasmic membranes toward the outside world do not couple proton-motive forces to chemiosmotic processes, which instead are delegated internally to organellar membranes; the translation on cytoplasmic ribosomes occurs uncoupled from transcription in the nucleus; and the generally intron-carrying pre-mRNAs are subject to processing by one or two types of sophisticated spliceosomes. It should be revealing, therefore, to reassess the three akaryotic traits mentioned above in the light of the generalized EF scenario suggested by the extended progenote hypothesis, allowing for the internal compartmentalization of emerging minigenomes and gradual differentiation of the internal surrounding membranes.

1. *Multifunctional cell boundaries.* It is an important assumption of the generalized hypothesis that electron transfer reactions and related aspects of basic energy conversion have been coordinated at internal (proto)membranes from very early on and that this coordination was corroborated by the early inclusion of monopartite minigenomes inside the energy-converting vesicles, provided that some of these genes coded for energy-converting proteins. Next, it is a secondary assumption that functionally different, potentially complementary, nucleation events of such genome-containing vesicles occurred at different times and were partly independent of one another, thereby giving rise to the internal lineages of bacteria- and archaea-related (para)organelles. Under this hypothesis, therefore, it comes as no surprise that the cytoplasmic membranes of Bacteria and Archaea have carried different energy-related functions from the beginning and that boundary-related functions were added later on, in association with the eventual "escape" from the hosting paracells of the progenote community.
2. *Transcriptional–translational coupling.* Under the general assumptions of the extended progenote hypothesis, it is a matter of relative semantics whether the coupling of transcription and translation in akaryote cells is to be considered primitive or derived as an evolutionary trait. At any rate, the founding principle of nucleating a minimal genome in a semi-autonomously

evolving vesicle makes it important to translate the nascent mRNAs right away as they are being transcribed. This localized coupling incentive exists irrespective of whether the majority of progenote genes and minichromosomes outside the vesicular protoorganelles already are surrounded by protonuclear envelopes or not. Personally, I tend to the assumption that the organized inclusion of eukaryote-type chromosomes in such envelopes may well have occurred somewhat later as a secondary or derived trait, but this only happened outside the energy-converting paraorganelle compartments.

3. *Lack of spliceosomal introns.* By a similar token, the same founding principle — starting out with as few essential constituents as possible — favors the utilization of intron-less genes from the very beginning. Again, this localized incentive for exclusion exists irrespective of whether the abundance of intron-rich genes on progenote minichromosomes outside the vesicular protoorganelles is brought about by any "introns-early" or "introns-late" scenario. To the best of my judgement, intron splicing may have served the important role of RNA recombination and reshuffling of small motifs and emerging functional domains already at very early stages, as pointed out by the "introns first theory" (Penny *et al.*, 2009), which is readily compatible with the extended progenote hypothesis proposed herein.

Rethinking on many frontiers is also asked for at yet earlier stages of proto-metabolic beginnings, in parallel and simultaneously:

- *Coevolutionary coupling* of various mutually interactive components in a progressively self-organizing molecular ecosystem;
- *Substrate-assisted catalysis* among accumulating reaction partners, such as colloidal metal–sulfate complexes, organic cofactors, phosphorylated lipids and carbohydrates, as well as oligopeptides and oligoribonucleotides;

- *Sustainment of protoplasmic continuance* by sunlight-driven energetics first, organic heterotrophic reactions next and litho-autotrophic possibilities later on.

Altogether I am convinced that no single assumption or theory will ever be able to fully explain the emergence of life on Earth. As a non-reproducible historical event, which has not left any solid relics other than ever changing life itself, the numerous details of this transition will always remain unknowable. New kinds of relevant evidence are cropping up incessantly, with the discovery of Asgard archaea as a splendid example — here introduced as "*Jokers of the game.*" Their story, though, is not yet fully told either, for as long as not much is known about the biology of these newcomers or the physiological roles that are played by the spurious eukaryote-like proteins they happen to possess. After all, I can only hope that momentarily unconventional considerations, too, will be taken seriously in future attempts to integrate the most promising proposals into a rational, coherent and convincing story.

Besides, to modulate the leading "tree" motif as part of a grander whole just once again: The structural framework of deciduous trees is best seen at winter time — when all the leaves have fallen — but vibrant life has almost vanished by absence of the leafy greens.

References

Baum DA (2015) Selection and the origin of cells. *Bioscience* **65**: 678–684.

Baum DA, Baum B (2014) An inside-out origin for the eukaryotic cell. *BMC Biol* **12**: 76.

Bell PJL (2001) Viral eukaryogenesis: was the ancestor of the nucleus a complex DNA virus? *J Mol Evol* **53**: 251–256.

Bernhardt HS (2012) The RNA world hypothesis: the worst theory of the early evolution of life (except for all the others). *Biol Direct* **7**: 23.

Bukhari SA, Caetano-Anollés G (2013) Origin and evolution of protein fold designs inferred from phylogenomic analysis of CATH domain structures in proteomes. *PLoS Comput Biol* **9**: e1003009.

Caetano-Anollés G, Caetano-Anollés D (2003) An evolutionarily structured universe of protein architecture. *Genome Res* **13**: 1563–1571.

Carini P (2019) A "cultural" renaissance: genomics breathes new life into an old craft. *mSystems* **4**: e00092–19.

Carter CW, Wills PR (2017) Interdependence, reflexivity, fidelity, impedance matching, and the evolution of genetic coding. *Mol Biol Evol* **35**: 269–286.

Castelle CJ, Banfield JF (2018) Major new microbial groups expand diversity and alter our understanding of the tree of life. *Cell* **172**: 1181–1197.

Coleman GA, Pancost RD, Williams TA (2019) Investigating the origins of membrane phospholipid biosynthesis genes using outgroup-free rooting. *Genome Biol Evol* **11**: 883–898.

Collins LJ, Kurland CG, Biggs P, Penny D (2009) The modern RNP world of eukaryotes. *J Hered* **100**: 597–604.

Da Cunha V, Gaia M, Gadelle D, Nasir A, Forterre P (2017) Lokiarchaea are close relatives of Euryarchaeota, not bridging the gap between prokaryotes and eukaryotes. *PLoS Genet* **13**: e1006810.

Da Cunha V, Gaia M, Nasir A, Forterre P (2018) Asgard archaea do not close the debate about the universal tree of life topology. *PLoS Genet* **14**: e1007215.

Damer B, Deamer D (2015) Coupled phases and combinatorial selection in fluctuating hydrothermal pools: A scenario to guide experimental approaches to the origin of cellular life. *Life* **5**: 872–887.

Efimov AV (2013) Super-secondary structures and modeling of protein folds. *Meth Mol Biol* **932**: 177–189.

Egel R (2009) Peptide-dominated membranes preceding the genetic takeover by RNA: latest thinking on a classic controversy. *BioEssays* **31**: 1100–1109.

Egel R (2011) Integrative perspectives: in quest of a coherent framework for origins of life on earth. In *Origins of Life: The Primal Self-Organization*, Egel R, Lankenau D-H, Mulkidjanian AY (eds). Springer, Berlin Heidelberg, pp 289–360.

Egel R (2012a) Primal eukaryogenesis: on the communal nature of precellular states, ancestral to modern life. *Life* **2**: 170–212.

Egel R (2012b) Life's order, complexity, organization, and its thermodynamic–holistic imperatives. *Life* **2**: 323–363.

Egel R (2014) Origins and emergent evolution of life: the colloid microsphere hypothesis revisited. *Orig Life Evol Biosph* **44**: 87–110.

Egel R (2017) 'Parabiotic Evolution': from stochasticity in geochemical and subsequent processes to genes, genomes and modular cells. Preprints 2017100153 doi: 10.20944/preprints201710.0153.v1.

Egel R, Penny D (2008) On the origin of meiosis in eukaryotic evolution: coevolution of meiosis and mitosis from feeble beginnings. In: *Recombination and Meiosis — Models, Means and Evolution*, Egel R, Lankenau D-H (eds). Springer, Berlin Heidelberg, pp 249–288.

Forterre P (2002) The origin of DNA genomes and DNA replication. *Curr Opin Microbiol* **5**: 525e532.

Forterre P (2006) Three RNA cells for ribosomal lineages and three DNA viruses to replicate their genomes: a hypothesis for the origin of cellular domain. *Proc Natl Acad Sci USA* **103**: 3669–3674.

Forterre P, Prangishvili D (2013) The major role of viruses in cellular evolution: facts and hypotheses. *Curr Opin Virol* **3**: 558–565.

Francis BR (2011) An alternative to the RNA world hypothesis. *Trends Evol Biol* 3.**1**: 2.

Gaia M, Da Cunha V, Forterre P (2018) The Tree of Life. In: *Molecular Mechanisms of Microbial Evolution*, Rampelotto PH (ed). Springer, Cham, Switzerland, pp. 55–99.

Ghadiri MR (1998) Self-organized autocatalytic chemical networks and molecular ecosystems: Do they provide the experimental tools for modeling the transition from inanimate to animate chemistry? In: *Recent Trends in Molecular Recognition*, Diederich F, Künzer H (eds). Springer Berlin Heidelberg, pp 213–238.

Glansdorff N, Xu Y, Labedan B (2008) The last universal common ancestor: emergence, constitution and genetic legacy of an elusive forerunner. *Biol Direct* **3**: 29.

Goodenough U, Heitman J (2014) Origins of eukaryotic sexual reproduction. *Cold Spring Harbor Persp Biol* **6**: a016154.

Gray MW (2014) The pre-endosymbiont hypothesis: a new perspective on the origin and evolution of mitochondria. *Cold Spring Harbor Persp Biol* **6**: a016097.

Guzman MI (2011) Abiotic photosynthesis: from prebiotic chemistry to metabolism. In: *Origins of Life: The Primal Self-Organization*, Egel R, Lankenau D-H, Mulkidjanian AY (eds). Springer Berlin Heidelberg, pp 85–105.

Guzman MI, Martin ST (2009) Prebiotic metabolism: production by mineral photoelectrochemistry of α-ketocarboxylic acids in the reductive tricarboxylic acid cycle. *Astrobiology* **9**: 833–842.

Han J, Collins LJ, Biggs PJ, White WT, Penny D (2013) Are signature proteins the key to evaluating eukaryotic phylogeny? *Rev Bioinform Biometr (RBB)* **2**: 37–44.

Hoffmann MR, Martin ST, Choi W, Bahnemann DW (1995) Environmental applications of semiconductor photocatalysis. *Chem Rev* **95**: 69–96.

Hug LA, Baker BJ, Anantharaman K, Brown CT, Probst AJ, Castelle CJ, Butterfield CN, Hernsdorf AW, Amano Y, Ise K, Suzuki Y, Dudek N, Relman DA, Finstad KM, Amundson R, Thomas BC, Banfield JF (2016) A new view of the tree of life. *Nat Microbiol* **1**: 16048.

Jain S, Caforio A, Driessen AJ (2014) Biosynthesis of archaeal membrane ether lipids. *Front Microbiol* **5**: 641.

Jeffares DC, Poole AM, Penny D (1998) Relics from the RNA world. *J Mol Evol* **46**: 18–36.

Kandler O (1993) Cell wall biochemistry and three-domain concept of life. *Syst Appl Microbiol* **16**(4): 501–509.

Kazlauskas D, Krupovic M, Venclovas Č (2016) The logic of DNA replication in double-stranded DNA viruses: insights from global analysis of viral genomes. *Nucleic Acids Res* **44**: 4551–4564.

Kompanichenko VN (2008) Three stages of the origin of life process: bifurcation, stabilization and inversion. *Int J Astrobiol* **7**: 27–46.

Kompanichenko VN (2012) Inversion concept of the origin of life. *Orig Life Evol Biosph* **42**: 153–178.

Kondrashov AS (1988) Deleterious mutations and the evolution of sexual reproduction. *Nature* **336**: 435–440.

Koonin EV, Dolja VV, Krupovic M. (2015) Origins and evolution of viruses of eukaryotes: the ultimate modularity. *Virology* **479**: 2–25.

Lake JA, Rivera MC (1994) Was the nucleus the first endosymbiont? *Proc Natl Acad Sci USA* **91**: 2880.

Lankenau D-H (2011) Two RNA worlds: toward the origin of replication, genes, recombination, and repair. In: *Origins of Life: The Primal Self-Organization*, Egel R, Lankenau D-H, Mulkidjanian AY (eds). Springer Berlin Heidelberg, pp 225–286.

Leipe DD, Aravind L, Koonin EV (1999) Did DNA replication evolve twice independently? *Nucleic Acids Res* **27**: 3389–3401.

Mariscal C, Doolittle WF (2015) Eukaryotes first: how could that be? *Phil Trans R Soc B* **370**: 20140322.

Martin W, Russell MJ (2003) On the origins of cells: a hypothesis for the evolutionary transitions from abiotic geochemistry to chemoautotrophic prokaryotes, and from prokaryotes to nucleated cells. *Phil Trans R Soc B* **358**: 59–85.

Mulkidjanian AY, Bychkov AY, Dibrova DV, Galperin MY, Koonin EV (2012) Origin of first cells at terrestrial, anoxic geothermal fields. *Proc Natl Acad Sci USA* **109**: E821–E830.

Mulkidjanian AY, Makarova KS, Galperin MY, Koonin EV (2007) Inventing the dynamo machine: the evolution of the F-type and V-type ATPases. *Nat Rev Microbiol* **5**: 892–899.

Muller F, Brissac T, Le Bris N, Felbeck H, Gros O (2010) First description of giant Archaea (Thaumarchaeota) associated with putative bacterial ectosymbionts in a sulfidic marine habitat. *Environ Microbiol* **12**: 2371–2383.

Nitschke W, McGlynn S, Milner-White EJ, Russell MJ (2013) On the antiquity of metalloenzymes and their substrates in bioenergetics. *Biochim Biophys Acta — Bioenergetics* **1827**: 871–881.

Otsuka S, Bui KH, Schorb M, Hossain MJ, Politi AZ, Koch B, Eltsov M, Beck M, Ellenberg J (2016) Nuclear pore assembly proceeds by an inside-out extrusion of the nuclear envelope. *eLife* **5**: e19071.

Penny D (2005) An interpretive review of the origin of life research. *Biol Phil* **20**: 633–671.

Penny D, Hoeppner MP, Poole AM, Jeffares DC (2009) An overview of the introns-first theory. *J Mol Evol* **69**: 527–540.

Penny D, Collins LJ, Daly TK, Cox SJ (2014) The relative ages of eukaryotes and akaryotes. *J Mol Evol* **79**: 228–239.

Pittis AA, Gabaldón T (2016) Late acquisition of mitochondria by a host with chimaeric prokaryotic ancestry. *Nature* **531**: 101.

Saw JH, Spang A, Zaremba-Niedzwiedzka K, Juzokaite L, Dodsworth JA, Murugapiran SK, Colman DR, Takacs-Vesback=HC, Hedlund BP, Guy L, Ettema TJ (2015) Exploring microbial dark matter to resolve the deep archaeal ancestry of eukaryotes. *Phil Trans R Soc B* **370**: 20140328.

Sieber JR, McInerney MJ, Gunsalus RP (2012) Genomic insights into syntrophy: the paradigm for anaerobic metabolic cooperation. *Annu Rev Microbiol* **66**: 429–452.

Smith E, Morowitz HJ (2010) The autotrophic origins paradigm and small-molecule organocatalysis. *Orig Life Evol Biosph* **40**: 397–402.

Smith TF, Lee JC, Gutell RR, Hartman H (2008) The origin and evolution of the ribosome. *Biol Direct* **3**: 16.

Sojo V, Pomiankowski A, Lane N (2014) A bioenergetic basis for membrane divergence in Archaea and Bacteria. *PLoS Biol* **12**: e1001926.

Takemura M (2001) Poxviruses and the origin of the eukaryotic nucleus. *J Mol Evol* **52**: 419–425.

Tamura K (2011) Ribosome evolution: emergence of peptide synthesis machinery. *J Biosci* **36**: 921–928.

Tautz D, Domazet-Lošo T (2011) The evolutionary origin of orphan genes. *Nat Rev Genet* **12**: 692–702.

Van Kranendonk MJ, Deamer DW, Djokic T (2017) Life springs. *Sci Am* 317.**2**: 22–29.

Wächtershäuser G (2003) From pre-cells to Eukarya — a tale of two lipids. *Mol Microbiol* **47**: 13–22.

Wang M, Jiang YY, Kim KM, Qu G, Ji HF, Mittenthal JE, Zhang HH, Caetano-Anollés G. (2011) A universal molecular clock of protein folds and its power in tracing the early history of aerobic metabolism and planet oxygenation. *Mol Biol Evol* **28**: 567–582.

Watson T (2019) The trickster microbes that are shaking up the tree of life. *Nature* **569**: 322–324

Weinger JS, Parnell KM, Dorner S, Green R, Strobel SA (2004) Substrate-assisted catalysis of peptide bond formation by the ribosome. *Nat Struct Mol Biol* **11**: 1101–1106.

Woese CR (2000) Interpreting the universal phylogenetic tree. *Proc Natl Acad Sci USA* **97**: 8392–8396.

Woese CR (2002) On the evolution of cells. *Proc Natl Acad Sci USA* **99**: 8742–8747.

Woese CR (2005) Evolving biological organization. In: *Microbial Phylogeny and Evolution: Concepts and Controversies*, Sapp J (ed). Oxford University Press, New York, pp 99–117

Zaremba-Niedzwiedzka K, Caceres EF, Saw JH, Bäckström D, Juzokaite L, Vancaester E, Seitz KW, Ananthataman K, Starnawski P, Kjeldsen KU, Stott MB, Nunoura T, Banfield JF, Schramm A, Baker BJ, Spang A, Ettema TJ (2017) Asgard archaea illuminate the origin of eukaryotic cellular complexity. *Nature* **541**: 353–258.

The language of biomolecular communication

G. Caetano-Anollés[1]

Summary. Molecules communicate when they interact in the crowded cellular environment. This communication requires an evolving molecular language capable of adapting to the varying environments that are internal and external to the cells. Here I explore the existence of a growing vocabulary in the molecules and molecular functions of the cellular world. A genomic census of structural domains and ontological terms of molecular functions provides an opportunity to find meaningful patterns and processes driving biological language change. The census provides evidence supporting the existence of grammar dependencies and statistical laws of language in protein domain structure and organization. Language laws arise from principles of decreasing returns that are responsible for a number of scaling patterns. I illustrate these principles with the Menzerath-Altmann law of language and other language-related universals. Patterns of origin and diversification of repertoires of proteins (proteomes)

[1]*Department of Crop Sciences, University of Illinois, Urbana, Illinois, USA*

and functions (functionomes) in organisms reveal that their make up depends on trade-off solutions between economy, flexibility and robustness when optimizing for organismal and molecular persistence. Collectively, organisms of the three superkingdoms of life use different trade-off strategies. Archaea and Bacteria maximize economy, while Eukarya expand flexibility and robustness. Comparative genomic and phylogenomic data suggests that organisms engage in extreme semantic and pragmatic compression of their genetic messages. These tendencies are especially notable when organisms adapt to microbial lifestyles and harsh environments.

1. Introduction

Communication is the intentional act of conveying meaning through the flow of "messages." Messages are discrete communication units that make up languages. Every language involves creating messages and encoding them in semiotic entities or elementary units of lexical meaning (lexemes), the collective of which form vocabularies (lexicons). Using a more accurate definition, languages are formal symbolic systems of communication that use *signs* and *rules* drawn from some alphabet to convey a meaning of objects (things, beings or concepts). The signs of these systems are the lexical entities or units whose existence bear causal relations to their corresponding objects and act as symbol (likeness, token, index) of those

Table 1. Terms defining levels of complexity in language make up.

Term	Definition
Syntaxis	The systematic combination of the inventory of signs according to rules of grammar.
Semantics	Meaning of signs and their relationships at different hierarchical systems of the message.
Pragmatics	Context dependent relations of signs and sign users that follow pragmatic rules, i.e. rules that govern and discriminate successful from non-successful interactional motifs.

relationships. Examples of signs are words with syntactic, semantic and/or pragmatic content drawn from some alphabet (Table 1). In turn, rules are logical forms, abstractions or functions capable of producing a syntax governing the structure and meaning of signs. Examples of rules are the structural rules of grammar that govern the composition of words, clauses and phrases of a natural language.

Natural languages are languages that have evolved to enable communication between humans. While we use language in our everyday activities, its use remains mysterious. Languages have been the subject of intense study by philosophers, linguists, and biologists, ever since Plato, Gorgias and other sophists considered the link of words to experience and memory. However, communication is a central and undisputed activity of all living systems and should be considered a biological phenomenon that emerges from interpreting signs in complex structure and organization (Witzany, 2014, 2016). Natural languages for example are the initial product of animal brain activities. It is therefore unsurprising that biologists have used language paradigms to describe fundamental biological processes, including those that drive molecular biology (e.g. replication, transcription and translation to describe the flow of information from nucleic acids to proteins). For example, the genetic code has been considered a natural language and not simply a metaphor (Eigen, 1971). We note that this kind of communication cannot occur in absence of living agents (components of more complex living systems) capable of speaking a host of deep evolving languages. These languages pervade the structure of these agents at all levels of agent organization and somehow mediate the flow of information. They are at the center of thermodynamics, energy dissipation and information theory (Caetano-Anollés *et al.*, 2010).

Here I discuss the existence of a "meaningful" communication system that expresses at the molecular level during interaction of different macromolecules and molecular parts. This system produces coordinated outcomes. First, I showcase evidence to discuss that there is active communication between molecules, showing communication follows language rules and laws. In doing so, I explore the existence of a vocabulary in the molecular repertoires of proteins (proteomes) and their associated molecular functions (functionomes). Second, I discuss how these laws manifest within constraints imposed by the engineering of the emerging biological systems and trade-off solutions between economy, flexibility and robustness. Third, I focus on evolution of entire molecular repertoires of organisms and show that there are clear patterns of molecular origin and diversification that are consistent with these trade-offs.

2. Connecting linguistics and molecular biology at different levels of organization

There are clear correspondences between language and molecular biology (Searls, 2001, 2002). One notable correspondence is how the progression *Lexicon* ⇒ *Syntaxis* ⇒ *Semantics* ⇒ *Pragmatics* of language matches the progression *Structure* ⇒ *Function* ⇒ *Fitness* of biological paradigms that are commonly used to describe biological make up and change, which I have modified from Searls (2001).

2.1. *The lexicon and syntaxis of natural languages and their correspondence to the structure of macromolecules*

The lexicon of a language (the inventory of words or subwords) and its syntaxis (the systematic combination of the inventory according to grammar rules) can be seen as corresponding to all sorts of manifestations of molecular structure. This includes the entire spectrum of hierarchical structural organization, from molecular sequences, secondary and supersecondary structures, folded structural domains, and quaternary structures to the combination of these low and high-level structures in molecules and molecular complexes to make up cellular structure. The essence of the correspondence is the existence of "atomic" entities (units) that are capable of engaging in combinatorial games known to enhance the diversity of the biological system as a whole. These units should be regarded as *modules*, sets of integrated parts of the system that cooperate to perform a task, with parts in modules interacting more extensively with each other than with other parts and modules of the system. For example, low-level structural motifs are recurrent in proteins and can associate within different molecular or cellular contexts to form more complex 3-dimensional atomic structures (Söding and Lupas, 2003; Trifonov and Frenkel, 2009). Supersecondary structures such as αα-hairpins, ββ-hairpins and βαβ-elements combine in numerous folds to form complex and sometimes sophisticated topologies. One of the most sophisticated fold structures, which is confined to eukaryotes and viruses, is the β-propeller (Figure 1A). In this highly symmetrical structure, sets of well-positioned ββ-hairpins form the characteristic 4–8 blade-shaped β-sheets that are arranged toroidally around the central axis of the all-β protein architecture. The ββ-hairpins substructures also form the porin fold typical of membrane channels with β-barrel structural domains (Figure 1A). In all of these cases, ββ-hairpins represent substrings (subwords) of a string of secondary structures, which combine to form different folded structures in ways that

(A)

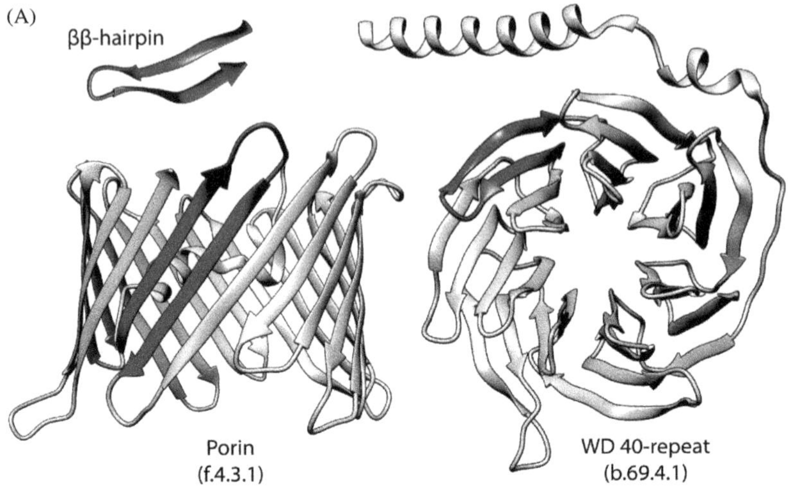

(B)

morph-{shape,form}: allo**morph**, a**morph**ous, ana**morphic**, anthropo**morphic**, di**morphic**, ecto**morph**, endo**morph**, geo**morph**ology, iso**morph**ic, meso**morph**, meta**morph**hosis, **morph**eme, **morph**ology, poly**morph**ism, zygo**morphic**
< From Gr morphē: form, shape

Figure 1. ββ-hairpins combine to form all-β protein domain structures as morphemes combine to form cognate words. (A) The ββ-hairpin (highlighted in red) is a minimal structural unit of the voltage-dependent anion channel 2 of zebrafish (*Danio rerio*; PDB entry 4bum), which holds the transmembrane β-fold structure typical of the porins with a (16,20)-barrel structure made of repeating ββ-hairpins. Similarly, the ββ-hairpin is the structural unit of the β1 subunit of the signal transducing G protein heterotrimer from Norwegian rat (*Rattus norvegicus*; 1gp2_B), which holds the 7-bladed β-propeller domain structure typical of the WD 40-repeat variant. Structures are labeled with SCOP family names using *concise classification strings* (*ccs*). (B) The morpheme "morph" is a lexical unit (highlighted in red) of cognate words related by common origin. The list of cognates includes the word "morpheme" and was retrieved from the Cognātarium (http://www.cognatarium.com/cognatarium/).

depend on the molecular context. Similarly, morphemes are word substrings that represent the smallest grammatical units of a language (Figure 1B). They can form words by themselves or when they combine with other morphemes.

Smaller motifs that form shorter loop structures (~25–30 amino acid residues long) and are stabilized by the formation of van der Waals locks can also combine to form structural domains (Berezovsky and Trifonov, 2001). Some of these structures are "elementary functional loops" that hold molecular functions, are evolutionarily highly conserved, and combine typically in groups of 2–3 to bind cofactors and create active, binding or regulatory sites (Goncearenco and Berezovsky, 2015). When analyzed with phylogenomic approaches, the loops that were most widely spread in organisms appeared associated with very old structural domains and recruited for central metabolic functions (Aziz *et al.*, 2016). Other loop structures have been proposed to represent a more primordial vocabulary responsible for the creation of the very first folded domains (Alva *et al.*, 2015). Remarkably, these structures are not combined but appear recurrently in different fold contexts. The difference between these two modes of recurrence has not yet been explained. Even lower hierarchical levels of structural organization reveal a language lexicon and an associated syntaxis. Patterns of dipeptide composition and enrichment characteristic of individual structural domains were found to be highly biased and characteristic of individual fold structures (Caetano-Anollés *et al.*, 2013). The evolutionary tracing of these patterns onto phylogenomic reconstructions of domain history revealed an ancient evolutionary process of dipeptide combination that contributes to the formation of higher levels protein structure. Thus, even the amino acid compositions of protein sequences carry a hidden molecular vocabulary.

The syntaxis of natural languages involves the combination of lexical units to form meaningful "expressions" such as phrases, idioms, clauses, sentences, texts and narratives. For example, a clause is a unit of grammatical organization, ranked next to the sentence itself. Clauses are proposition statements. Their make up is conditioned by grammar rules, the combinatorial landscape of the inventory of words, and ultimately the implied meaning. A common grammar must be identified in order to discover syntactic rules common to clauses and molecular biology. In this regard, the recent graph-theoretical "dependency" paradigm can be useful. A *dependency grammar* is a modern syntactic theory that focuses on dependency relations between linguistic units (morphemes, words) instead of the more standard "constituent" relations. Words are the emphasis of these grammars. Constituent relations are rewrite rules responsible for phrase structure grammar (Chomsky, 1957). In constituent grammar, a binary division splits the noun phrase (subject) from the verb phrase (predicate), which can be further atomized into their constituent parts in a binary tree. Instead, a dependency grammar implies a graph-theoretical view of language in which the linguistic units are connected by directed links (Tesnière, 1957). Figure 2A shows an example of a dependency grammar introduced by

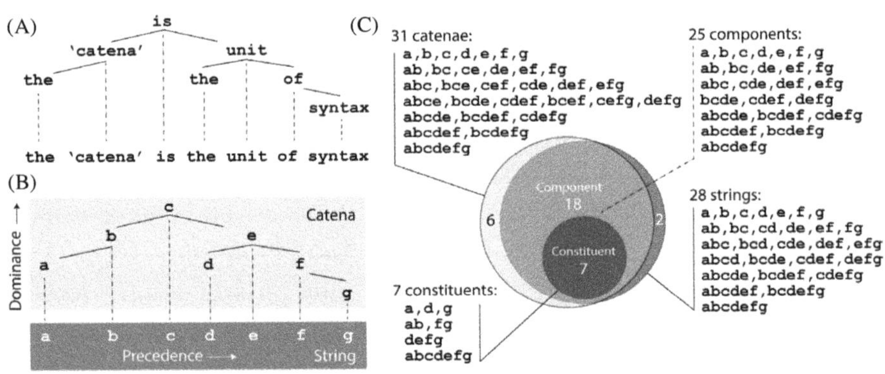

Figure 2. Dependency grammar and the linguistic value of catenae. (A) A clause is described with an ordered dependency tree, a "catenae" that reflects actual word order. The syntactic hypothesis is constructed by placing the verb "is" at the center of clause structure and building around it dependencies (one-to-one word correspondences) vertically, as the "dominant" parent node (word) "is" at the top connects to its children at the bottom. (B) The nodes of the catena are labeled with letters and correspondences to the string of words made explicit with dashed lines. Node *c* "dominates" node *e* because it is above node *e* and one can trace a path from *c* to *e* moving only downwards in the tree, i.e. *c* is a "parent" of *e* and *e* is "child" of *c*. (C) The combinatorial landscape of catenae and strings, and their two associated lexical units, components and constituents is described with a Venn diagram.

O'Grady (1998), the *catena* (Latin for "chain"), a unit of syntax and morphology, and its projection to the string of words. Given the expression *"the 'catena' is the unit of syntax,"* the verb *"is"* is placed at the center of the clause structure and its dependencies (one-to-one word correspondences) are built around it from top to bottom as the clause unfolds from left to right in the string of words (Osborne *et al.*, 2012). Within this structure, the *string* is a lexical unit (e.g., word, morph) or combination of units that is continuous on the horizontal dimension of precedence (Figure 2B). Precedence refers here to a "transitive" relationship that describes a unit preceding another that is adjacent in the string structure of units. Continuous refers to units immediately preceding other units in sequential manner as to define a "linear order." In turn, the *catena* is a lexical unit or combination of units that is continuous on the vertical dimension of dominance, with the most dominant (or primitive) on the top of the "hierarchical order" (Figure 2B). Dominance defines a dependence of some lexical unit on others. For example, the verb *"is"* dominates (licenses) its argument *"catena"* making it dependent on its presence. Similarly, the determinant *"the"* is an adjunct to *"catena"* and can be omitted without loss of meaning. Dependent lexical units are therefore placed below the dominant words (Figure 2) and a *dependency edge* drawn between the words to establish dominance relationships. Vertical edges connecting the lexical tree structure of the catena to the string are called *projection edges*. Note that a catena is any subgraph of the tree.

Two other lexical units are related to strings and catenae. The *component* is a lexical unit or combination of units that is both string and catena. In turn, the *constituent* is a component that is "complete," i.e. it includes all the units that are children of it (the node plus all nodes that the node dominates). Figure 2C describes the possible catenae, strings, components and constituents of the clause that was used as example. The figure shows a Venn diagram describing how these different lexical units are unique or relate to each other. Note that the dependency grammar of catenae differs from a constituent grammar in that it reflects a one-to-one correspondence between the units while the constituent grammar reflects a one-to-one-to-many correspondence (O'Grady, 1998; Osborne *et al.*, 2011, 2012). Also note that the dependency grammar of catenae offers the most informative of all units, since it encompasses most of the landscape or lexical unit combinations, which is more general.

As with phrases, idioms, clauses or sentences, protein macromolecules carry a syntactic organization of lexical units of higher structural organization, the *structural domains*. These units, which are evolutionarily conserved (i.e. they are somehow "frozen" in time), often combine with other similar units to form multidomain arrangements in proteins (Chothia *et al.*, 2003; Moore *et al.*, 2008; Wang and Caetano-Anollés, 2009). Structural domains are higher-level structural modules that fold into well-packed and compact atomic structures of the polypeptide chain (Wetlaufer, 1973; Richardson, 1981). Their folds carry independent molecular functions such as enzymatic activities, signaling and structural roles. Each domain contributes one or more specific functions to a protein, including a multiplicity of intramolecular interactions necessary for protein stability (Bhaskara and Srinivasan, 2011). Domains exhibit two important properties (Janin and Wodak, 1983; Murzin *et al.*, 1995; Riley and Labedan, 1997). First, they are folded units that are evolutionarily conserved. Domains are several orders of magnitude more resistant to the effect of mutation than their sequences (Illegård *et al.*, 2009). Second, domains are recurrent. Their evolutionary conservation implies a longer historical existence, which generally translates into numerous opportunities of protein reuse when domains are recruited into different molecular and functional contexts. Thus, domains are considered *structural*, *functional* and *evolutionary* units, i.e. modules with an existence that sometimes spans billions of years (Caetano-Anollés *et al.*, 2009; Wang *et al.*, 2011a), which rival lexical units of dependency grammars. To illustrate the combinatorial potential of domains in proteins, Figure 3 shows dependency grammars of domain

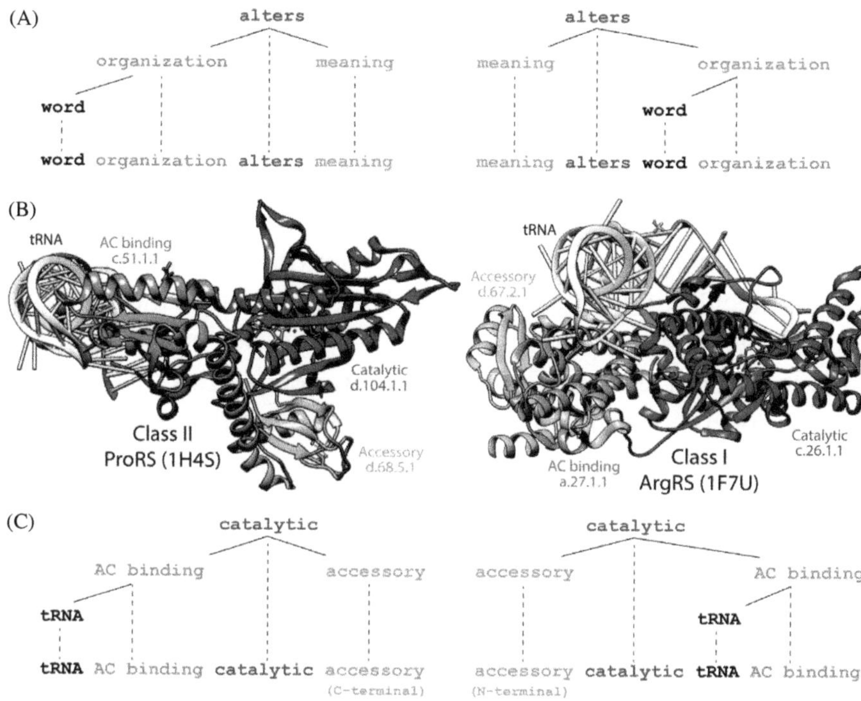

Figure 3. Protein domains combine to form protein structures as words combine to form language expressions. (A) The dependency grammar of two clauses with a same lexicon is described with ordered trees and catenae that reflect actual word order. Syntactic hypotheses are constructed by placing the verb "alters" at the center of clause structure and building around it dependencies (one-to-one word correspondences) to provide a meaning of the message. Note how the meaning is altered by word permutation, i.e. by the context of clause construction. (B) Domain organization of aminoacyl-tRNA synthetases (aaRSs) representatives of class I and II enzymes. The different domains of Class II proline-specific synthetase (ProRS; PDB entry 1h4s) and Class I arginine-specific synthetase (ArgRS; PDB entry 1f7u) are traced using different colors along the backbone of the cartoon depiction of their structures. Note that the red-orange-blue sequence follows the evolutionary age of domains. The tRNA is viewed coaxially from it anticodon-binding (AC) arm in both structures, serving as point of reference. (C) A dependency molecular grammar can be constructed for the aaRS enzymes by placing the oldest domain with the most central functional role (the catalytic aminoacylating function) as center of the multidomain structure and building dependencies with more evolutionarily derived domains to enhance the specificities of the initial catalytic core.

structures made explicit with lexical trees, strings and catenae. The order of words in an expression alters its meaning (Figure 3A). Similarly, the order of domains in a protein alters its structural and functional make up, as shown with two aminoacyl-tRNA synthetase enzymes harboring a catalytic, anticodon (AC) binding and accessory domain (Figure 3B, C). These synthetases attach specific amino acids by aminoacylation to the acceptor arm of their cognate tRNA molecules, which then act as adaptors to unfold the specificity of the genetic code during translation. The two enzymes in the example recognize their tRNAs from different angles, the Class I arginine-specific synthetase molecules do so from the minor groove of the nucleic acid molecule and the Class II proline-specific synthetase from the major groove. The examples show that the different modes of recognition and specificity involve different orders of domains in the protein chain as one travels from the amino-terminal (N-terminal) to the carboxyl-terminal (C-terminal) ends of the protein chains. These differences also place constraints on the 3-dimensional structure of the molecules and their interaction with the tRNA molecule. Lexical trees and catenae can be used to illustrate the physical and functional dependencies and dominance relations between the different domains in these two synthetases (Figure 3C). The catalytic aminoacylating domain defines the main catalytic activity of these synthetases and is placed at the most dominant position of the lexical tree. The domain is also the oldest, appearing much earlier than the other two (Caetano-Anollés *et al.*, 2013). Both the AC binding domain and the accessory domain provide specificity enhancements to the functions of these molecule and are of more recent evolutionary origin. They are therefore placed at lower positions of the y-axis. In particular, the AC binding domain interacts with specificity determinants of the tRNA to guarantee that the anticodon sequence of the tRNA matches the charged amino acid. Thus, it is dominant with respect to its cognate tRNA. The example illustrates how the syntaxis of natural languages have a correlate in the complex structure of multi-domain proteins.

2.2. *Semantics and its correspondence to components, events and activities of the cell*

The semantics of a language dissects the meaning of linguistic expressions and does so at different levels. In *lexical semantics* the focus is on the meaning of words or smaller lexical units while in *conceptual semantics* the focus is on the meaning of arguments, statements that convey messages. Some goals of semantics can be accomplished by developing theory that explains how the speakers use language to convey meaning and by constructing precise mathematical models with the help of formal logic (Devitt and Hanley, 2006). Lexical semantics explores the syntax-semantics interface, including how lexical units are classified, are decomposed, and are related to each other. A number of semantic relations for example make possible to build semantic networks that relate concepts to each other. These semantic relations include *meronymy* (part-of-whole relation), *holonymy* (whole-of-part), *hyponymy* (subordination), *hypernymy* (superordination), *synonymy* (sameness in meaning), *antonymy* (opposite) and *homonymy* (sameness with different meanings). For example, *hypernyms* are higher level terms that group lower level *hyponyms* in a transitive relation; the hypernym "color" has a more general meaning that its hyponym "purple," which is more specific. Semantic relations of these kinds are used by a number of ontology developers (see OBO foundry; http://www.obofoundry.org), including the seminal Gene Ontology (GO) database (Ashburner *et al.*, 2000). The controlled and species-neutral vocabulary developed by the GO consortium standardizes the functional annotation of gene products with a vocabulary of ontological terms describing the biological processes (*bp*), molecular functions (*mf*), and cellular components (*cc*) of the cell (Table 2).

These terms are often related to each other through hypernym-hyponym and "is a" relationships. GO *mf* terms define *activities*, actions characterizing agents (also known as actors) of the molecular realm. GO *bp* terms define *events*, objects and their properties manifesting

Table 2. Main ontological domains of the Gene Ontology (GO) database describing controlled vocabularies of the properties of gene products.

Term	Definition
Molecular function (mf)	GO terms of molecular functions describe *activities* that are performed by individual gene products or by assembled molecular complexes. These activities occur at the molecular level, such as "catalytic activity" or "binding activity". These terms represent activities (e.g. "catalytic activity", "binding activity") rather than the entities (molecules or complexes) conducting those activities themselves. They do not specify where, when, or in what context the action takes place.
Biological process (bp)	GO terms of biological process describe series of *events* arising from the activity of one or more molecular functions. In general, biological processes such as "cellular physiological process," "signal transduction," or "pyrimidine metabolic process," are collectives of distinct steps enabled by a repertoire of distinct molecular functions.
Cellular component (cc)	GO terms of cellular component describe *components* that are part of larger objects, such as groups of gene products associated with common functions of the cell (e.g. ribosome, proteasome) or an anatomical cellular structure (e.g. mitochondrion, plastic, nucleus).

in time. They are constrained by causes and effects and by the spatiotemporal dimension. Philosophically GO *mf* and *bp* terms are "occurrents," observables that are actively occurring (Hawley, 2015). In contrast, GO *cc* terms define *components*, parts of a biological whole and "places" where activities and events occur. Philosophically, they are "continuants," entities that continue or persist through time even though they can change (Hawley, 2015).

The relationship of GO terms has a complex tree-like network structure. Each of the root terms for separate *mf*, *bp* and *cc* networks develop into a directed acyclic graph (DAG) in which child terms can be connected with multiple parents. GO terms at lower levels of the DAG represent more specialized functional annotations (Harris *et al.*, 2004; Shegogue and Zheng, 2005). For example, "ATP binding" [GO:0005524] is a specific *mf* term that is child and establishes an "is a" relationship to a number of higher-level terms such as "purine ribonucleoside binding" [GO:0032550] or "binding" [GO:0005488], which represent broader functional categories (Figure 4A). GO annotations of structural domains provide an ontological description of the functions of these molecular structures. Thus, lexical trees of domains (Figure 3) can translate into lexical constructs indexed with biological meaning (Figure 4B). For example, the lexical and conceptual semantics dictated by domain organization of the Class I arginine-specific synthetase enzyme can be dissected into its component parts. The three structural domains are organized in a lexical tree structure that reveals a dependency molecular grammar. The x-axis unfolds the order of domains in the polypeptide chain, i.e. physical precedence. The y-axis unfolds both the evolutionary age of the domains, i.e. dominance by evolutionary precedence, and the semantic complexity of biological meaning as an indexing of *mf*, *bp* and *cc* GO terms in word clouds. This places the oldest domain with the most central functional role (the catalytic aminoacylating function) at the center of the multidomain structure, very much as the verb acts as focal point for assembling the tree structure of lexical relations. Dependencies are then built with more evolutionarily derived domains that enhance the specificities of the initial catalytic core. Figure 4C shows that the vast majority of *mf*, *bf* and *cc* GO terms (depicted as vertical, horizontal and oblique words, respectively) are associated with the central catalytic domain (in red), especially *bf* terms describing events such as "macromolecule metabolic process" and "biosynthetic process," while terms associated uniquely with the AC-binding domain (mostly *mf* and *cc* terms such as "RNA binding" and "plastid") are peripheral. The only *mf* term of the catalytic core, "ligase activity" describes the most fundamental property of aminoacylation of the acceptor arm

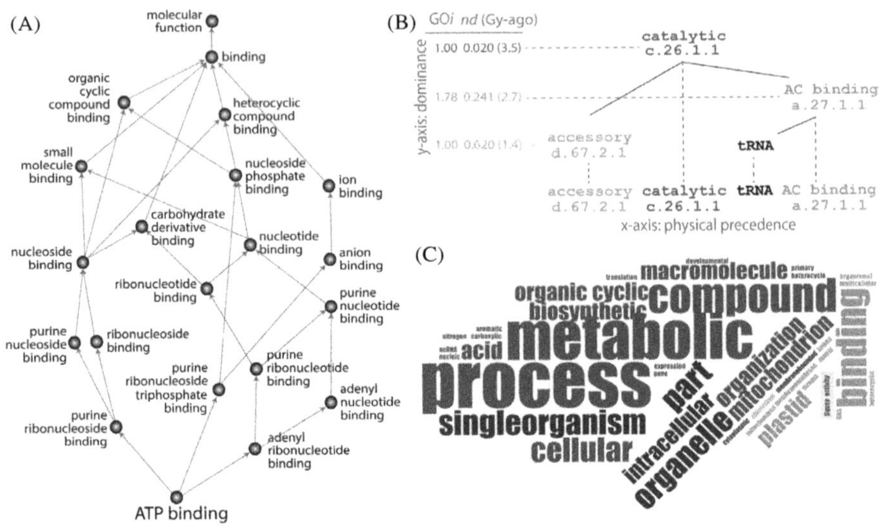

Figure 4. GO-based lexical and conceptual semantic statements of an arginyl-tRNA synthetase (argRS; PDB entry 1f7u). (A) Subgraph of the DAG of molecular functions derived from the child GO term "ATP binding" (GO:0005524). Each arrow describes an "is a" relationship. (B) A dependency grammar of ordered trees and catenae reflects structural domain order and syntaxis of the argRS protein molecule. Syntactic hypotheses are constructed by placing the oldest domain with significant GO term make up as the "verb" proxy of the functional statement and building dependencies according to those same variables and physical precedence. Domain age is given as node distance (*nd*) and in billions of years (Gy). *nd* ranges from 0 (origin of domains) to 1 (the present day). GO term make up was given as an GOi index, which divides the number of GO terms of a domain by the number of core GO terms that are common between domains. (C) Word cloud of GO terms describing the domains of argRS derived from the *superfamily* database, with *mf*, *bf* and cc terms depicted as vertical, horizontal and oblique words, respectively. Red terms describe properties of the catalytic domain (which are common to all domains) and orange terms describe properties unique to the AC-binding domain. Highlighted in yellow is the only *mf* terms describing the catalytic core, 'ligase activity'. Darker shades are terms that do not apply directly to the respective domains. Note how the accessory domain does not have unique terms since there are no words colored turquoise.

of the tRNA molecule, and the *mf* terms of the AC-binding domain are summarized by "RNA binding" recognition of tRNA identity elements. The accessory domain holds the same functions as the core catalytic domain, showing an adjunct role that is semantically non-consequential for the molecule. The example illustrates how the semantics of languages corresponds to components, events and activities of the cell.

2.3. *Pragmatics and its correspondence to fitness*

> *"I don't know what you mean by 'glory'," Alice said.*
> *Humpty Dumpty smiled contemptuously. "Of course you don't — till I tell you.*
> *I meant, 'There's a nice knock-down argument for you!'"*
> *"But 'glory' doesn't mean a 'a nice knock-down argument'," Alice objected.*
> *"When I use a word," Humpty Dumpty said in rather a scornful tone,*
> *"it means just what I choose it to mean — neither more nor less."*
> *"The question is," said Alice, "whether you can make words mean so many different things."*
> *"The question is," said Humpty Dumpty, "which is to be master — that's all."*
>
> — Through the Looking-Glass, Chapter 6, Humpty Dumpty. Lewis Carroll

In his literary piece, the "subversive" pragmaticist Lewis Carroll makes his Alice and Humpty Dumpty characters question the ambiguity of language (Lakoff, 1993). His poetical answer: utterances and their meaning are simply the intentional acts of speakers at time and place. The exploration of the meaning of utterances is the purview of pragmatics. This discipline of linguistics and philosophy studies how context influences meaning (Korta and Perry, 2015). In semantics, meaning is "encoded" in language. In pragmatics, meaning is conveyed by those involved in speech, by their intent, and by the context of their messages. Roman Jakobson (1960) proposed utterances as "speech events," acts of communication that can be deconstructed with a powerful analytical device, which I will summarize and illustrate following Hébert (2011). The messages of speech events are conveyed through six "constitutive factors," each determining a function of language that relates the message and the factor (Figure 5A). These factors include the sender and the receiver of the message, the context and message, and the contact and code of the speech event. All factors do not need to be present during speech. However, each results in a corresponding function. The sender, as a speaker, provides an "*emotive*" function of expression that conveys information in the form of an attitude or emotion. In turn, the receiver reacts by providing an appellative *"conative"* function of the vocative and imperative. The verbal or verbalizable context of speech invokes a *"referential"* function that describes the situation, the object and mental state of the participants. The message provides a *"rhetorical"* function that stresses the esthetic or poetic effect of the message ("the message for its own sake"). The contact provides a "*relational"* function that focuses on the interaction between the sender and the receiver (the channel of communication between them). The code provides a *"reflexive"* function that focuses on the language used and its meaning. Note that pairs of functions related to the communicating entities (emotive-conative), the message and its context (rhetorical-referential), and the communication channel of contact and code (relational-reflexive) are in trade-off relationship, i.e. when one functionality changes the other changes accordingly.

The analytical tool starts by enumerating the language functions of a speech event (e.g. text, sign, image, utterance, movie). It then builds a hierarchy of functions with an intention-based criterion that

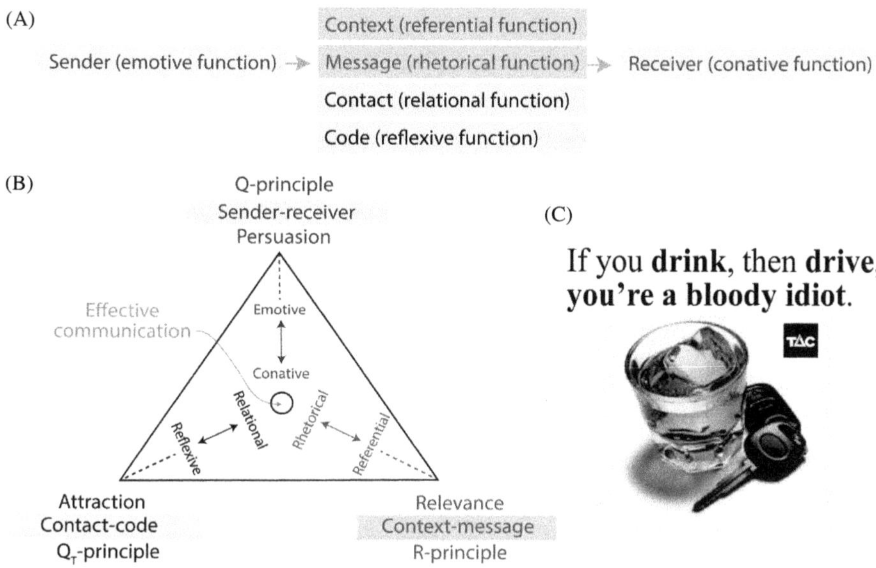

Figure 5. The pragmatics of language. (A) A diagram of the Jakobson's structuralist model of linguistic communication. Speech events are conveyed by six factors, each determining an associated function (in parenthesis): emotive (also known as expressive), conative (appellative, imperative, directive), referential (denotative, representative, informative), rhetorical (poetic, esthetic), relational (phatic, cognitive), and reflexive (metalingual, metasemiotic). (B) A triangle of effective communication shows trade-off solutions (Attraction, Persuasion and Relevance) imposed by associated linguistic functions. The Attraction, Persuasion and Relevance goals are in trade-off relationships, each defined by trade-offs between pairs of linguistic functions (inside triangle). Trade-offs in vertices correspond to neo-Gricean pragmatic principles of quality (Q_T), quantity (Q) and relation (R), which are also used to label the vertices. (C) A slogan of an advertising campaign against drinking-and-driving shows the main goals of a message and its associated functions: (1) *attract* attention (relational function), (2) *persuade* (conative function) by appealing to reason (referential functions) or emotion (emotive function), and (3) *impact* by triggering the conative and referential functions through relevance and faithfulness. See text for the structural interpretation of this message.

places dominant functions at the top of that hierarchy. Note that intention can manifest directly ("answer the call") or indirectly ("the phone rang") and that cause-and-effect relationships can help organize the hierarchy (e.g. cognitive cause can activate the rhetorical effect). The tool then utilizes an "energy-based model" that describes the dynamic relationship between opposing language forces and their inter-relationships (Hébert, 2011). To make the idea explicit I will use a triangle of "effective communication" as example (Figure 5B). The triangle places main goals at its vertices, each defining pairs of opposing forces: (1) in the *Persuasion* vertex, the sender must "persuade" the receiver through an emotive-conative dynamic relationship (note that persuasion is an umbrella term for "influence"); (2) in the *Relevance* vertex, context and message, and their respective referential and rhetorical functions, respectively, are used to generate an "action" that is relevant and faithful to the message, i.e. that is impactful; and (3) in the *Attraction* vertex, contact and language and their relational and reflexive functions are used to "attract" attention to the subject of the message. The three goals determine the influence, action and conduit of the message, respectively. Depending on the speech event, the opposing forces of the goals can push the functions of language towards one of the contending alternatives (towards each vertex of the triangle or towards its center) in a highly frustrated exercise of trade-offs or compromises. This places the speech event as an instance of effective communication within a particular location of the triangle, and does so depending on the language functions it unfolds. For example, in epic poetry (from Homer's *Iliad* and *Odyssey* to Dante's *Divine Comedy* or John Milton's *Paradise Lost*), narration and its referential function, together with the conative function, are intensified. This pushes communication away from the Persuasion vertex towards the Relevance vertex. In contrast, in lyric poetry (from the musical poetry of Sappho to the work of Rabindranath Tagore) the emotive and poetic (rhetorical) functions are intensified, pushing communication towards the Persuasion vertex and away from the

Relevance vertex. In contrast, scientific discourse (from science popularization to scientific writing) activates the message and its poetic rhetorical function. However, its communicative aspect is fundamentally conative and reflexive, pushing communication away from the Persuasion vertex towards the Attraction vertex.

Figure 5C illustrates how the structuralist model of linguistic communication can be used to analyze a slogan of an advertising campaign sponsored by the Transport Accident Commission (TAC) of Australia against drinking-and-driving. A description of the analysis can be found in Hébert (2011). The goal of the slogan is to stress the Relevance component of the triangle so that it can have a direct emotional impact on its audience that will induce action (refrain from drinking and driving). This is approached by stressing the poetic referential, conative and rhetorical functions. The slogan starts by making explicit the context of bad behavior (referential function of drinking-and-driving) of the receiver (the conative function of "you"), which is the worst of all three possibilities (you not drinking or driving, you drinking but not driving, you drinking and driving). Implied in the polysemic "you're a bloody idiot" rhetorical message, which draws attention to itself, is the negative moral value of having chosen a conative behavior that causes death (stressed by the word "bloody"), a pragmatic consequence or punishment not made explicit in words. Since being a bloody idiot presents death as an avoidable outcome, it persuades the receiver (you) towards rejection of the idiocy condition and pushes communication towards the Relevance (and action-driven) vertex of the triangle.

While Jakobson's communication model has been criticized as dated and incomplete, or that its premises are exceedingly "informational" (after all it is based on the Shannon-Weaver communication model), the tool offers insights into the complexities of language pragmatics and its possible links to biology. Pragmatics complements semantics by extending the study of the meaning of words and sentences to the intent of communicating entities. In this regard, Grice (1989) developed the influential theory of "implicature," which has been used in a number of more advanced theoretical propositions (Davis, 2014). Implicature is the act of meaning or implying one thing by saying something else, or the object of that act. For example, if I ask: "Are you going on vacation?," and you reply "I have to work," you imply that you are not going on vacation but do not state it directly. Under this concept, utterances implicate communication that is not made explicit in the message, embracing the concept that pragmatics adds an additional layer of complexity to language. For example, many sentences with explicit semantic meaning may have implications that cannot be derived from words. This includes figures of speech such as metaphor, irony, and understatement. When meaning goes beyond what is said in speech, including inferences drawn from non-linguistic (environmental) aspects of the utterance, Grice posits that communication must abide by a *Cooperative Principle*. The principle demands that in conversation, you as a speaker (sender) must contribute what is required by the accepted purpose of the conversation so that both sender and receiver engage in communicative cooperation. The principle also demands that four maxims be maximized for cooperation, intention, and rationality: (i) *Maxim of Quality* (Truth): Be truthful by avoiding falsehood or statements that lack adequate evidence; (ii) *Maxim of Quantity* (Information): Be as informative as required; (iii) *Maxim of Relation* (Relevance): Be relevant (efficient in communication); and (iv) *Maxim of Manner* (Clarity): Provide clarity by avoiding obscurity or ambiguity and fostering brevity and order. Horn (2004) improved Grice's formulation by replacing the last three maxims with two inter-related and counterbalancing principles that establish what he calls "division of pragmatic labor": (1) *Q[uantity]-principle*: A principle of sufficiency, in which you say as much as you can (given the R-principle); and (2) *R[elation]-principle*: A relation principle that seeks minimization, in which you say no more than you must (given the Q-principle).

The Q-principle unifies the maxims of quantity and manner (brevity) while the R-principle subsumes the maxim of relation by stressing the relevance of the conversation. Horn treats the maxim of quality as a separate principle, which I will name the *Q_T[uality]-principle*, a principle of truth or sincerity.

Grice's maxims have been considered circular and dependent on each other as "antinomic forces." Similarly, Horn's division of labor has been challenged by its incapacity to distinguish between semantically general or specific expressions (semantic minimization) or between shorter or longer expressions (expression minimization) (reviewed in Davis, 2014). Levinson for example proposed the re-definition of Horn's Q- and R-principles and the re-introduction of a version of the maxim of manner, though Horn contested his rationale. Instead of focusing on principle definition, I will instead focus on the fact that the Neo-Gricean Q_T-, Q- and R-principles can be considered the subjects of optimization. They can establish Pareto fronts in optimality theory or Nash equilibrium in decision or game theoretical formulations (van Rooy, 2004). This justifies placing maxims as trade-off solution in a *triangle of conversation*, with the Q_T-, Q- and R-principles as vertices. This triangle mimics the triangle of communication that I proposed in Figure 5 and the rationale of an energy-based model of optimization. The Q_T-principle vertex embodies the relational and reflexive functions of the Attraction vertex of the triangle of communication, especially because truthfulness requires a sender-receiver conduit that follows a faithful code of communication. Establishing the meaning of a message requires establishing the landscape in which it is true, which in turn requires benchmarking against a set of pragmatic implications (an "umwelt") in which the message is true. The driver of such an effort is *flexibility*, an ability to dissect meaning within the space of possible syntactic and pragmatic contexts. Similarly, the Q principle vertex embodies the emotive and conative functions of the Persuasion vertex and follows an *economy* of communication

Table 3. Trade-off solutions, their drivers, and their links to factors, functions, and Neo-Gricean principles of language.

Trade-off	Driver	Factors	Function	Principles
Attraction	Flexibility	Contact-Code	Relational-reflexive	Quality (Q_T)
Persuasion	Economy	Sender-Receiver	Emotive-Conative	Quantity (Q)
Relevance	Robustness	Context-Message	Referential-Rhetorical	Relation (R)

that fosters brevity, order and other minimalistic principles. This interpretation stresses the long-held idea that economy considerations impact language through a minimization of effort by both sender and receiver (Zipf, 1949). Finally, the R-principle seeks relevance, a property that embodies the context and message of the conversation and the referential and rhetorical functions of the Relevance vertex. The driver is *robustness,* the persistence of meaning in utterances in the presence of varied syntactic and pragmatic contexts. Table 3 summarizes the links between structuralist and Neo-Gricean views of language within a landscape of trade-off solutions that explain meaning in communication.

I will now explore how context influences the syntaxis and semantics of language in molecular biology using the structuralist language paradigm of Figure 5, which I have complemented with the Neo-Gricean pragmatics view. The word "pragmatics" (*Greek*, πραγματικός) means "fit for action." This definition is in itself remarkable. It describes how the context in which language contributes to meaning associates to "fitness." In this regard, the triangle of communication embodies a fitness function for the factors, functions, and principles of language. This fitness function is optimized when the three trade-off solutions of Attraction, Persuasion and

Relevance find a balance in the communication system that is being sought. In Figure 5 the sphere represents a single point in the trade-off space or a "quasispecies" collective that describes a global performance of a more complex communication system. These points or spheres locate in Pareto fronts, boundaries in multidimensional performance spaces that provide best-fitness solutions. The geometries for best-fitness solutions have been mathematically elaborated (Sheftel *et al.*, 2013). They correspond to planes in 3-dimensional performance spaces such as that of the triangle of communication. However, they can be simpler (line segments describing trade-offs between two alternative strategies) or more complex (polygons describing a multiplicity of strategies). In biology, maximizing fitness implies maximizing the potential of persistence of a system in different or changing environments. A triangle of trade-offs explaining communication between molecules that follows the principles and functions of language can explore the correspondence of biological fitness with the pragmatics of language.

The triangle paradigm can be tested with the most complex molecular machine of the cell, the ribosomal ribonucleoprotein complex that is responsible for protein synthesis (Figure 6A). The ribosome contains two main subunits. The small subunit (SSU) typically harbors one rRNA molecule (16S/18S rRNA) with ~50 universal helices folded into 3 domains and 30–40 ribosomal proteins. SSU holds both the "decoding" center that enables the reading of mRNA information by the anticodon loops of tRNA molecules and the central mechanism (ratchet) of a turnstile that moves the mRNA through a groove in the complex. The large subunit (LSU) typically contains 2–3 rRNA molecules (23S/28S and 5S/5.8S rRNA) with ~100 universal helices folded into six domains and 30–45 ribosomal proteins. The structure holds the peptidyl transferase center (PTC) that is responsible for protein biosynthesis and structures specialized in

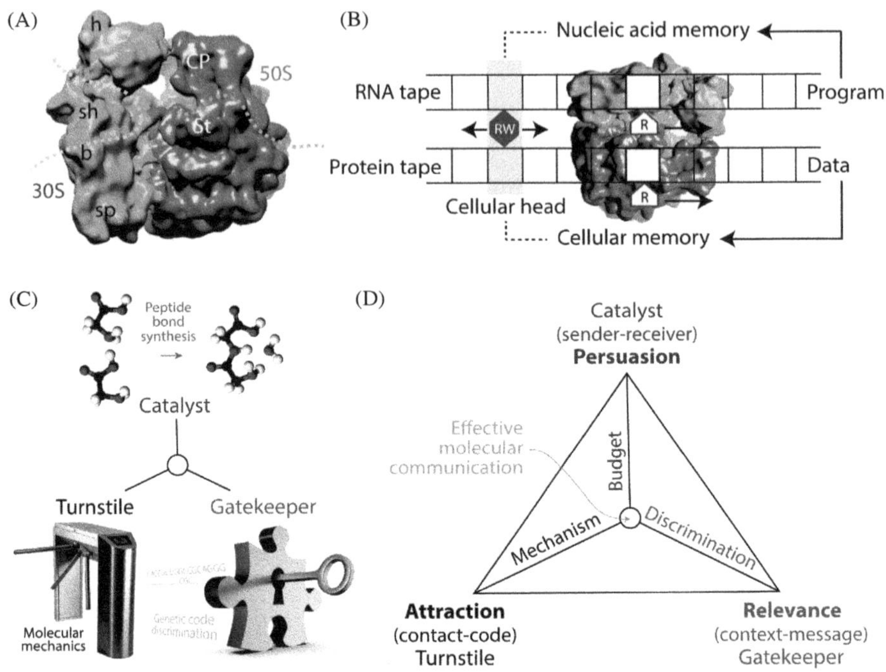

Figure 6. The pragmatics of molecular machines. (A) A cryogenic electron microscopy side view of the *Escherichia coli* 70S ribosome at ~10 Å resolution (entry EMD-2978), with its small (SSU) and large (LSU) subunits colored green and blue, respectively (Chen *et al.*, 2015). Some domains features are labeled: b, SSU body; h, SSU head; sh, SSU shoulder; sp, SSU spur; CP, LSU central protuberance; St, LSU stalk base. The trajectory of mRNA and the synthesized polypeptide is indicated with yellow and orange curves. The groove between the two subunits accommodates tRNA molecules carrying amino acids for polypeptide biosynthesis. (B) The ribosome is a universal Turing machine with two tapes. One stores the "program" for reading machines (mRNA tape) and the other stores the "data" of machines (polypeptide tape). In the universal ribosomal Turing machine the program tape defines the nucleic acid memory and the data tape stores the memory of polypeptide machines. Both move left during biosynthesis through read-only (R) turnstile heads (the ribosomal subunits) induced by a rotation of the SSU ribosomal head. The read-write (RW) "cellular head" operates at different timescales and in multiple cells through selective optimization (through mutation) at DNA and protein level, driven by fitness for cellular and molecular persistence. (C) The three main roles of the ribosome involve the synthesis of peptide bonds, the movement of RNA through mechanical gates and switches that resemble a turnstile,

mechanics and energetics that are needed to move the tRNA molecules through the central groove of the complex as they unload their amino acids.

Operationally, the ribosome resembles a two-tape universal Turing machine (Figure 6B), with one tape storing the program in mRNA and the other storing the data of machines in polypeptides (Caetano-Anollés and Caetano-Anollés, 2015). A Turing machine must have editable tapes and heads that move in both directions to read and write symbols on the the string of "placeholders" of the tape. In addition, a Turing machine is universal when it can be used to construct any other computing machine. In other words, a universal machine behaves as an automaton system capable of self-replication and universal computation. When exploring the molecular ribosomal complex, the heads of both tapes only move in one direction and do not have molecular editing capabilities. They are simply "read" mobile heads and not "read-write" editing heads. Thus, this double "turnstile" is a *finite state machine* program with limited computational power.

However, when the finite state machine of the ribosome is placed within its cellular context and when enough time is provided for cells to replicate and change by mutation, the proteins that are encoded in genes and are transcribed into the mRNA tape act as sensors of cellular fitness and are edited by mutation

(Figure 6 on facing page) and the ability to discriminate aminoacylated tRNA. (D) A triangle of effective molecular communication for the ribosome describes trade-off solutions that maximize molecular persistence. The Persuasion, Attraction and Relevance trade-off solutions of language at the vertices of the triangle correspond to the catalytic, turnstile and gatekeeper ribosomal roles, respectively. Their associated goals stress chemical budget (protein synthesis), molecular mechanism (dynamics of tape recognition and movement) and discrimination (genetic code specificities and cellular memory).

(Caetano-Anollés and Caetano-Anollés, 2015). Thus, these products of ribosomal protein biosynthesis produce a potentially unlimited editable tape, acting as an additional cellular "read-write" tape. This crucial capability turns the ribosome into a universal Turing machine.

This informational view of the ribosome therefore entails three clear elements of computation: (i) the capability to produce an infinite tape of proteins through biosynthesis, (ii) a program responsible for the editing of this infinite tape; and (iii) a set of mobile heads with read and write editing capabilities that are controlled by the program. Since the ribosome must satisfy these three elements of computation, it must perform three fundamental roles (Figure 6C) that are associated with matter-energy (metabolic) budget, molecular discrimination, and molecular mechanics, in that order (Caetano-Anollés and Caetano-Anollés, 2015). These three roles make the ribosome a *catalyst*, a *gatekeeper* and a *turnstile*. I will name these three fundamental roles with generic terms describing their function and will link these roles to corresponding goals of the triangle of communication to help in the discussion:

(1) *Budget (Persuasion):* From a metabolic perspective, the ribosome through its PTC acts as a *catalyst* for the synthesis of peptide bonds. Catalysis involves several steps (initiation, protein elongation, termination and ribosome recycling) that utilize hydrolysis of guanosine triphosphate (GTP) as the driving force of protein biosynthesis (Wallin *et al.*, 2012). These steps are mediated by translational GTPases, auxiliary protein factors that include tRNA-transporting translation factors that become activated upon binding to the ribosome. The catalytic function ensures that an optimal and functional matter-energy budget is maintained (within the confines of metabolism) by the molecular system. This carefully optimized budget enables, but also

defines, the biosynthetic production of the infinite “tape” of polypeptides that is needed for ribosomal computation, including the make up of its molecular components. The term “persuasion” is used here to describe the global “influence” that the matter-energy budget (economy) has on the information flow that is established by the communicating (sender-receiver) entities, i.e. the protein machines computed by the ribosome and the computing machine itself.

(2) *Discrimination (Relevance):* The ribosome is also a *gatekeeper*, policing genetic code discrimination during decoding by ensuring that the tRNAs are adequately transferred from their elongation factor transporters to the corresponding ribosomal active site and are matched to corresponding mRNA positions. This is accomplished through conformational proofreading mechanisms that are at least 100–1,000 times less stringent than that of aaRS enzymes and elongation factor binding (Reynolds *et al.*, 2010). The gatekeeper function, which is carried on by the ribosome and a host of other proteins (including the aaRSs), ensures that the memory of the genetic code is maintained in the molecular system. This function directs the editing of the infinite tape of proteins through a program written using a code for the string of “placeholders” of the inter-related nucleic acid and polypeptide tapes (the genetic code). The term “relevance” is used here to describe the discriminating specificity and stability that is needed to make an editing Turing program actionable and relevant, i.e. of practical value for the cell.

(3) *Mechanism (Attraction):* Finally, the ribosome machinery itself is a *turnstile* capable of molecular mechanics and information and energy transfer. The turnstile is enabled by the rotation of the SSU head (Figure 6A), a movement that involves flexing two crucial hinges in the neck between the head and body of SSU (Achenbach and Nierhaus, 2015). This mechanical movement of the turnstile is directly coupled to the decoding process,

which matches the codon-anticodon segments of the mRNA and tRNA molecules, and the movement of the anticodon arm of tRNA in SSU, which involves a gate that includes the central ratchet and prevents tape back-sliding. The turnstile is the mechanism that is necessary for the logical operation of the ribosome. It harbors at least three gates and one switch for universal biological computation. The ribosomal turnstile makes the moving head of the Turing machine and its moving parts and molecular contacts define a stable apparatus for displacing tapes and ensuring the budget and discrimination functions are preserved. The term "attraction" describes the accretion of molecular components that is needed for stable and reliable communication contacts.

The roles of budget, discrimination and mechanism can be considered determinants of the fundamental goals of Persuasion, Relevance and Attraction in the triangle of trade-offs describing any molecular communication system (Figure 6D). In this triangle, effective molecular communication (or computation) is optimized by establishing a balance between the Persuasion, Relevance and Attraction trade-off solutions. Persuasion stresses the molecular system's matter-energy budget while both Relevance and Attraction are more devoted to the transfer of information, one stressing the abstract (design) and the other the material (atomic) side of molecular communication. Note however that all three goals embody energy, matter and information. In the case of the ribosome, the Persuasion vertex defines the molecular machines that the ribosomal computation produces (their biosynthesis), the Relevance vertex defines their fitness within the confines of the cell (their discrimination), and the Attraction vertex defines the actual stable ribosomal operation (its mechanics).

The triangle of molecular communication does not apply only to the ribosome. Its validity extends to other biomolecular systems, including protein enzymes (e.g. GTPase switches), functional nucleic acids (e.g. tRNA), and multi-molecular complexes making up simple and advanced molecular machines (e.g. proteasome, ATPase, flagellum). For example, metabolic enzymes fold into conformations that reflect different functional states of the protein catalysts. These functional states have been carefully optimized in evolution to achieve both efficient function, stability and design, with the ultimate goal of increasing fitness (Mittenthal and Caetano-Anollés, 2010). Conformational changes are sometimes induced by ligands that act as allosteric activators or inhibitors, blocking or facilitating access of small molecules, metal ions, cofactors, proteins, and nucleic acids to pockets in their structures. Proteins can also toggle between "on" and "off" states when they behave as molecular switches. For example, canonical molecular switches involve conformational changes driven by nucleotide triphosphate binding and hydrolysis. Dependencies of molecular communication on mechanisms and designs that are enabled by functional states push protein enzymes towards the Attraction vertex. Dependencies that rely on the catalytic functionalities of the enzyme, including maintaining its associated matter-energy budget, push the system towards the Persuasion vertex. Finally, communication dependencies that rely on the ability to discriminate substrates and foster protein stability push the system towards the Relevance vertex. The molecular communication system is frustrated in its design, function, and stability, explaining why enzymatic functions are locked into defined structural roles (Kim *et al.*, 2013), hold catalytic functions that are evolutionarily non-optimal (Newton *et al.*, 2015), and are marginally stable (Mittenthal and Caetano-Anollés, 2010).

3. A Chomsky's hierarchy of formal language in biomolecules

A formal language is a set of *terminal* symbols (an "alphabet"), often presented as combinations in strings, organized by a set of rules specific to those symbols (Chomsky, 1956, 1957). Alphabet symbols and rules are here described with lower-case and upper-case letters, respectively. More formally, a rule-based grammar G requires a finite set of *nonterminal symbols* N that is disjoint with the strings formed from G (i.e. does not occur in the alphabet), a finite set of *terminal symbols* Σ of the alphabet that is disjoint from N, and a finite set of algorithmic re-writing rules termed *production rules* P, with each rule mapping one string of symbols ("head") to another ("body"): $(\Sigma \cup N)^* \; N \; (\Sigma \cup N)^* \rightarrow (\Sigma \cup N)^*$; where * is the Kleene star operator (unary operation on sets of strings, symbols or characters). The grammar also requires a non-terminal symbol S, $S \in N$, which starts the algorithmic implementation (the sentence symbol). Thus, a grammar is a rewriting system defined by the tuple (N, Σ, P, S).

To exemplify how a formal grammar can be applied to the generation of biopolymers, Searls (2002) generated a string representing an RNA molecule with a growing hairpin structure from nucleotides symbols, g, c, u and a, using the following 5 production rules of a context-free grammar:

$$S \rightarrow gSc; \; S \rightarrow cSg; \; S \rightarrow aSu; \; S \rightarrow uSa; \text{ and } S \rightarrow \varnothing$$

with the last rule forcing insertion of an empty string $\varnothing$ that erases S and terminates the computation. Derivations of this grammar produced nucleotide strings that grow from the nonterminal "sentence" symbol S and produces complementary base pairs towards the 5′ and 3′ ends, respectively. One example set of derivation is the following:

$$S \Rightarrow gSc \Rightarrow gaSuc \Rightarrow gauSauc \Rightarrow gaucSgauc \Rightarrow gaucgScgauc \Rightarrow gaucgcgauc$$

Note that each derivation that is indicated by double arrows ($\Rightarrow$) applies single arrow rules ($\rightarrow$) to "produce" or "expand" the string, and that the process generates nested dependencies. Adding the additional rule $S \rightarrow SS$, generates a bifurcation when used in a derivation. Each duplication of the generative non-terminal symbol results in a branching of the RNA structure. A single duplication generates 5S rRNA-like structures.

Chomsky organized formal languages and associated formal grammars into a hierarchy, in which higher-level language classes strictly include lower-level counterparts (Chomsky, 1956, 1957). In this hierarchy, generative linguistic algorithms produce increasingly complicated computation (Table 4):

(1) The *regular class* of languages is composed of languages that are the most restricted and the simplest of the hierarchy. They are generated by a regular grammar, a set of rules that start from an initial symbol and replace non-terminal symbols with terminal symbols using allowed production rules that grow strings in a single direction. These regular languages are recognized by deterministic or non-deterministic *finite-state automata* (DFA and NFS), machines that are built out of a finite number of states

Table 4. Chomsky's hierarchy of formal languages.

Language class	Grammar	Additional P*	Automaton	Example
Regular	Regular (right-linear) No long-range string dependencies	$A \to aA$ $A \to a$ $A \to \varnothing$	Finite-state	Enzymes
Context-free	Context-free Nested dependencies	$A \to X$	Pushdown	RNA, proteins
Context-sensitive	Context-sensitive Crossing dependencies	$X \to Y$ (Y as long as X)	Linear bounded	NRPS complexes
Unrestricted	Recursively enumerable	$X \to Y$ (X is not $\varnothing$)	Turing machine	Ribosomes

*Listed production rules P are additional to those listed in the preceding language class. a, terminal symbol; A, nonterminal symbol; X, Y, strings of terminal and nonterminal symbols, with Y being as long as X.

(typically one state per nonterminal symbol) interconnected by transitions, which emit symbols of the alphabet as the they are traversed. The machines, which resemble Hidden Markov Models (HMMs), transition between states based only on the current state, without any memory storage apart from the states themselves. Regular grammars can generate patterns of symbols in strings (e.g. alternating symbols). However, they cannot handle long-range dependencies in strings.

(2) The *context-free class* of languages is the second more restricted group of languages. Nonterminal symbols are replaced by strings of terminal and nonterminal symbols, including the empty ∅ string. They are used to specify computer language syntax, build compilers and model natural languages. These regular languages are recognized by *pushdown automata*, nondeterministic finite-state machines (DFAs) comprised of a finite number of transitioning states and a stack of unlimited size that provides an arbitrarily large amount of memory to the machine. Only the top of the stack is accessible for transitioning at any one time, acting as a "one-place" memory holder. The memory is used to count symbols and generate patterns typical of nonregular languages, including nested dependencies of the kind that exist in RNA.

(3) The *context-sensitive class* of languages have strings generated by context-sensitive grammars that are similar to context-free grammars. However, these languages are recognized by *linear bounded automata* (LBS), which are deterministic machines based on states defined by a finite tape (with size related to the input string) that contains the input string and a read-write head that moves left and write along that tape. The machine determines its next state by comparing its current state and the symbol on the tape at the head's position, using a finite number of rules, including symbol available for re-writing, and direction of head movement along the tape. The LBA recognizes any language that have strings growing linearly, including $\{a^n b^n c^n\}$ with

$n \geq 0$. Context-sensitive grammars can also generate crossing dependencies in the string, implemented thanks to the freedom of movement of nonterminal symbols during context-sensitive derivation.

(4) The *unrestricted class* of languages is generated by unrestricted grammars that are free and recursively enumerable, i.e. they hold recursive functions that can potentially generate all possible strings on them. These languages are recognized by *Turing machines* with properties identical to LBA. However, the tapes of these machines are infinite in length. This makes them able to emulate other machines (they become Universal Turing machines) with computational power equivalent to most programming languages run on a computer with infinite memory.

All language classes of the hierarchy are represented in molecular biology and may be interrelated. For example, enzymes act as nonterminal terms of a regular grammar with production stoichiometric rules that transform substrates into intermediates and products of metabolism. Terminal symbols describe metabolites (typically small molecules) and nonterminal symbols enzyme activities mediated by protein catalysts. Since metabolism represents a network of enzymatic pathways of metabolite transformation, regular grammars can be built that are capable of deriving metabolic network connectivity for metabolic network simulations (e.g., Pfeiffer *et al.*, 2005). Similarly, context-free grammars of the kind described above can be used to model RNA structure (reviewed in Searls, 2002). In turn, context-sensitive grammars are induced by nonribosomal peptide synthetase (NRPS) complexes that produce small functional peptides from over 500 amino acid components in the absence of a nucleic acid template. This combinatorial game requires enzymatic "assembly lines" responsible for the sequential formation of peptide bonds catalyzed by specialized enzymatic modules (Marahiel, 2016). NRPS complexes are Turing machines with linearly-bounded tapes embodied in their polypeptides tapes. While the complexes have read-only heads, the read-write editing "cellular" head used by the ribosome completes the NRPS modules of the complex and fulfills requirements of context-sensitive languages of a left-and-right movement of the head for re-writing and handling crossing dependencies. Recall that the cellular head operates at different time-scales and in multiple cells through selective optimization. Finally, and as mentioned earlier, ribosomes are universal Turing machines with unbounded tapes capable of implementing recursively enumerable languages (see Figure 6B).

Note that all languages in molecular biology are dependent on each other: no language class exists in isolation. The regular languages of enzymes require protein catalysts that are synthesized by the ribosome and are dependent on its unrestricted language. Similarly, the synthesis of functional RNA (e.g. tRNA, RNase P RNA, rRNA) with its embedded context-free grammars, requires decoding from DNA by the RNA polymerase complex and associated transcription enzymatic apparatus. The synthesis of these proteins, as well as those of the NRPS modules, also require of ribosomal protein synthesis to operate. Thus, regular, context-free and context-sensitive grammars are all dependent on the existence of the universal Turing machine automaton shared by the ribosome and the evolving cell. Note also that the Turing machine paradigm implies life follows a digital model of information. However, all processes that involve biomolecules have qualities that are far from digital when these fold to gain functionalities. They resemble analog systems.

4. Quantitative linguistics and universal stochastic hypotheses

Quantitative linguistic patterns support the existence of meaningful communication in messages and meaningful language in the structure of the entities that are communicating. These empirical patterns can also manifest in universal laws that describe statistical regularities in language (Altmann *et al.*, 2006). Within an hypothetico-deductive framework, these regularities are often considered stochastic hypotheses that cannot be violated. In sharp contrast, language rules such as syntactic, semantic and language-change rules can be violated and are therefore contextual and non-universal. There are three general groups of language laws, *probability distribution laws*, *functional type laws* and *developmental type laws* (Altmann and Gerlach, 2016), some of which I will now describe.

4.1. *Probability distribution laws*

Probability distribution laws are statistical laws that make predictions of possible outcomes. The typical example is the Zipf's law that describes a property of language in which the frequency of a word in a given text is inversely proportional to its rank ordering in a frequency table (Zipf, 1948). The simplest mathematical formulation for the frequency of the k-th most frequent word in a text, $f(k)$, is that it is inversely proportional to its rank k, $f(k) \sim 1/k$. A more advanced version of the law however is formulated in Equation (1)

$$f(k) = Ck^{-\gamma} \tag{1}$$

where C is a normalizing constant and γ is an exponential parameter, usually $\gamma \geq 1$. If $f(k)$ follows a fat-tailed distribution, this frequency reflects a probability $P(k)$ of a word with frequency k such that $P(k) \sim k^{-\gamma'}$ when $\gamma' = 1 + 1/\gamma$. Within this context, the Zipf's law is a special case of the scale-free distribution that it explains, which is characteristic of many networks, including biological networks such as metabolism (Strogatz, 2001). Scale-free networks follow power-law distributions $P(k) \sim k^{-\gamma}$, with k representing the number of links that connect nodes in the network and $P(k)$ the probability that the network has exactly k links. The power-law behavior typical of scale-free distributions can arise from "preferential attachment" or "Yule" processes of network growth, explaining the *"rich get richer"* motto summarizing the intuition that nodes with many connections are prone to get more connected (Barabási, 2009).

4.2. *Functional type laws*

These laws operate thanks to mathematical functions that link two kinds of variables, including word frequency and word length or the frequency of syntactic constructions and their complexity. The typical example of these types is the Menzerath-Altmann (MA) law. Menzerath (1928) found that longer syllables contained shorter articulated units of sounds (phonemes), and later, that words with more syllables were phonetically shorter. He summarized these qualitative statements of language with the motto "*the greater the whole, the smaller its constituents*" (Menzerath, 1954). Later, Altmann (1980) provided mathematical elaborations that were later supported by statistical analyses of many languages, and linguistic and phonetic relationships of many types, including vocal communications outside humans (gelada primates; Gustison *et al.*, 2016). Recently, the MA law has been extended to genomes (e.g., Ferrer-i-Cancho and Forns, 2010; Baixeries *et al.*, 2012) and the organization of protein structural domains in proteomes (Shahzad *et al.*, 2015), showing that the principle behind its functional link is general and not restricted to language. One mathematical formulation of the MA law that is very general (Altmann, 1980) is described in Equation (2)

$$y(x) = Ax^{b}e^{-cx} \tag{2}$$

with $y(x)$ being the length of the parts, x representing the number of parts that make up the system (or constructs of parts) or its length, and A, b and c fitting parameters. Note that this general formulation adds the effect of the hierarchical make up of the parts of the system that is being studied (the multi-level structure that is typical of language) and that $y(x)$ is often measured by counting parts defined at a deeper level of the system's organization. A more general formulation also adds dependences on additional variables (Strauss and Altmann, 2006).

4.3. *Developmental type laws*

Developmental type laws relate language properties to time. These laws generally describe processes of development, growth and evolution of vocabularies, including language variation at different levels, from lexical and phonetic to semantic and syntactic. In some formulations, time can also be expressed as accumulating change or innovation in the presence of a process of growth. A typical example of these laws is the Heaps' law, which describes how the number of different words V in a document or corpora (the word types or vocabulary) scales with the database of words N measured in number of words (the length of text or number of word tokens) (Herdan, 1964; Heaps, 1978). A typical formulation of the Heaps' law is described in Equation (3)

$$V(n) = KN^{\beta} \tag{3}$$

with K and β being empirically determined parameters. Note that the Heaps' law can be interpreted in the context of a Zipf's distribution when $\beta = 1/\gamma$, and that constituents of vocabularies are constant over centuries if they represent "core" words that appear with high frequency (Gerlach and Altmann, 2013). Thus, vocabulary growth (a dynamic process) can be described in terms of database size (a static view). The link between the Heaps' and Zipf's laws has been made explicit in a study of a large body of English text (Ferrer i Cancho and Solé, 2001). In this study, the probability density function showed two scaling regimes, a steepest regime that followed a Zipf's law characterizing a "kernel" lexicon of frequently used words and a second regime characterizing an "unlimited" lexicon of growing words of less frequent use. This two-regime Zipf's distribution translated into a two-regime Heaps' law with β exponents close to 1 for the kernel and 0.4–0.7 for the unlimited lexicon, with exponent variation reflecting differences in language organization. Multi-regime Heaps' laws describe a decreasing marginal need for new words and an evolutionary slowdown (cooling) of vocabularies (Petersen *et al.*, 2012; Lü *et al.*, 2013), as shown for Indo-European, Chinese, Japanese and Korean languages.

5. Laws of language in the structure and organization of macromolecules

Is there an exact science of language in molecular biology? If that is the case, then there must be stochastic hypotheses in the form of laws that are universal for the world of biological macromolecules at different level of structural organization. In other words, biomolecules are expected to comply with statistical laws that exist in language (Altmann *et al.*, 2006). Indeed, the probability distribution of the Zipf's law explains patterns of occurrence of Pfam domains in proteins (Searls, 2002) as well as protein structural domains in proteomes (Qian *et al.*, 2001, Caetano-Anollés and Caetano-Anollés, 2003). The γ power law decay values averages ~2 for the proteomes of Bacteria and Archaea and ~1.4 for the proteomes of Eukarya (Caetano-Anollés and Caetano-Anollés, 2003). These values are similar to those for the English and Chinese languages, respectively (Li *et al.*, 2016). Thus, the vocabulary of protein structural domains and natural languages share a same probability distribution but with different scaling regimes for akaryotic microbes and eukaryotic organisms (Nasir *et al.*, 2017). Similarly, a number of scaling relationships exist in the vocabulary of genomes that are consistent with the Heaps' law, a developmental type law, including the scaling of the number of genes in different functional categories as a function of genome size (Molina and van Nimwegen, 2009). The scaling appears universal for most functional categories. This is expected since the Heaps' and Zipf's laws are mathematically related. An analysis of the proteomes of thousands of organisms and viruses has shown that the vocabulary of protein structural domains (defined at superfamily level of SCOP classification) scales with the total number of domains (Nasir *et al.*, 2017). This scaling follows a Heaps' law. Remarkably, results showed not two but four distinct power law patterns suggestive of four regimes of slowdown of domain vocabulary growth, each corresponding to the proteomes of supergroups, viruses, Archaea, Bacteria and Eukarya, in that order. Exponents showed sublinear growth ($\beta < 1$), a signature of the Heaps' law, with β approaching unity for the proteomes of viruses. Average values ($\beta = 0.81$) paraphrased the very ancient kernel-like vocabulary of natural languages but were not far from the second regime of languages with limited dictionary sizes such as Chinese, Japanese and Korean ($\beta = 0.7$–0.77, Lü *et al.*, 2013; Petersen *et al.*, 2012). Values of β for the proteomes of Archaea ($\beta = 0.36$–0.40) were not far away from those of English text corpora ($\beta = 0.4$–0.7), such as the Gutenberg Project e-book collection ($\beta = 0.45$, Tria *et al.*, 2014). The values of β for the proteomes of Bacteria ($\beta =$ 0.19–0.26) matched those of the third regime of Chinese language (Li *et al.*, 2016; Petersen *et al.*, 2012). Finally, exponents approached 0 for Eukarya, showing extreme evolutionary slowdown in the protein repertoires of higher organisms. It is noteworthy that a recent study of Chinese text revealed a 3-regime Heaps' law with β scaling exponents of 1, 0.7 and 0.3 for increasing text lengths, which is explained by a stochastic feedback model of vocabulary growth driven by two probabilities, one for the reuse of frequently used words and the other for the rise of word novelties (Li *et al.*, 2016). This matches scaling patterns in the microbial world of cellular organisms.

A functional-type law, the MA law, also governs the organization of protein structural domains in proteomes (Shahzad *et al.*, 2015). Structural domains are not only structural units of proteins folding.

They also behave as evolutionary modules they can combine to form multidomain proteins (Chothia *et al.*, 2003; Moore *et al.*, 2008). Similar or different domains can fuse in evolution or excise from multidomain proteins by fission, driven mostly by the forces of genome rearrangement (Wang and Caetano-Anollés, 2009). Fusions often occur directionally in the protein chains (Bashton and Chothia, 2002; Ekman *et al.*, 2005; Toll-Riera *et al.*, 2013). The rise of *domain organization* appears massive, resulting in an arrangement of protein "architectures" that enhances the functional toolkit of the cell and enables a molecular vocabulary (Bashton and Chothia, 2007). This additional level or structural organization has been shown to impact evolutionary constraints imposed by folding speed and protein flexibility (Debès *et al.*, 2013) and length of domains and proteins (Wang *et al.*, 2011b). Shahzad *et al.* (2015) uncovered the MA law by studying a highly curated dataset of structural domains in ~3 million sequences of 745 proteomes (Wang *et al.*, 2011b) covering the main kingdoms of life (Archaea, Bacteria, Protista, Plants, Fungi and Metazoa), which was carefully annotated with domain and linker length. The lengths of structural domains in this dataset decreased linearly with increasing domain number when data was fitted in log-log plots for all kingdoms (Table 5). Figure 7 shows an example of linear fitting for the proteome of the haloalkaliphilic archaeon *Natronomonas pharaonis* that lives in extreme saline and alkaline conditions. The linear regression lines in the log-log plots showed significant negative correlations that are typical of a general two-parameter power law form of the MA law with high coefficients of determination (R^2 ranging 0.85–1.00) and significant *F* test-derived correlations ($p <$ 0.0001–0.133). We also observed an MA law governing the relationship between the length of certain secondary structures and their number in structural domains of proteins (Minhas, Mittenthal and Caetano-Anollés, ms. in preparation). There are also linguistic patterns at even lower level of protein structural abstractions. Quantitative linguistic distribution models of decreasing returns

Table 5. Correlation data supporting a Menzerath-Altmann (MA) law driving the combination of structural domains in proteins at proteome level. Linear regressions and F-test correlations were examined for sets of 10 proteomes per kingdom of life.

Kingdom	Slope ($-b$)	R^2	F-value	p-value
Archaea	0.180 – 0.374	0.91 – 0.98	78.9 – 251.5	<0.0001 – 0.0047
Bacteria	0.246 – 0.331	0.90 – 0.90	78.9 – 251.5	<0.0001 – 0.0464
Protista	0.184 – 0.238	0.87 – 1.00	19.9 – 271.4	<0.0012 – 0.1327
Plants	0.045 – 0.256	0.83 – 0.99	11.5 – 159.0	<0.0002 – 0.0773
Fungi	0.164 – 0.288	0.90 – 0.98	34.6 – 137.8	<0.0013 – 0.0365
Metazoa	0.252 – 0.404	0.91 – 0.95	78.9 – 251.5	<0.0001

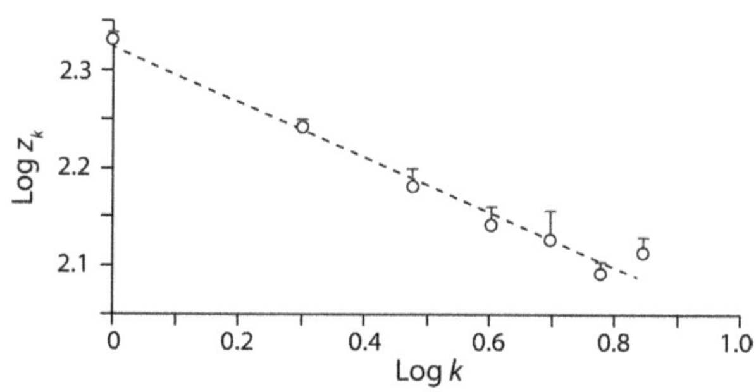

Figure 7. The Menzerath-Altmann (MA) law of language in the structural domains of an archaeal proteome. A log-log plot describes how the average length (z_k) of structural domains in a protein decreases with the number (k) of structural domains it contains for the entire proteome of *Natronomonas pharaonis*. The vertical bars on the circles represent the standard error of the means. The $y(x) = Ax^b$ two-parameter power-law formulation of the MA law facilitates fitting parameters in log-log plots to a straight line. The hatched regression line indicates the weighted linear fitting ($b = -0.291$; $R^2 = 0.97$; $p < 0.0001$). Data from Shazhad *et al.* (2015).

have been shown to accurately describe the distribution of the length of proteins in proteomes (Eroglu, 2014). Thus, language-like behavior exists in protein structure and its significance can be explained.

6. Communication and the persistence triangle

"Tis but thy name that is my enemy."
—Romeo and Juliet, Act 2, Scene 2. W. Shakespeare

When computers communicate across the internet, let us say to send and receive an e-mail message, they must connect and transmit data by delivering information without error to the right computer address. If that would not be the case communications would not be persistent. They would break down or fail to maintain "identity over time." When computer network participants of such a digital communication system transmit information they must be named. These participating entities must abide by a system of rules, a "*communication protocol*," a set of standards defining the syntax, semantics and synchronization (pragmatics) of their naming. The most important internet protocol system is the TCP/IP maintained by the Internet Engineering Task Force. The goal is to exchange information as efficiently and reliably as possible using a set of naming rules that is universal, so that all computers can establish communications that are human-meaningful, secure, and decentralized. These three crucial properties define what is known as Zooko's triangle, a paradigm of compromises during naming, which often involve open-ended data structures (key-values) that would make communications persistent (Wilcox-O'Hearn, 2001). Figure 6A describes Zooko's triangle.

Names must be human-meaningful, i.e. *memorable*, so that users that are transmitting the data can remember the name without errors (Ferdous *et al*., 2009). Names need to be *secure* to ensure that there is only one specific entity to which the name maps during name validation, even in the presence of random failure from ambiguity, noise or malicious activity. Names must be *decentralized*, readily available and distributed, so that there is absence of a central authority that controls the naming system and can collapse following targeted attack. While Wilcox-O'Hearn (2001) suggested only two of the three properties could be achieved by the communication system at one time, all three properties have been shown to be in tradeoff relationships and all three contribute to the persistence of the naming system (reviewed in Ferdous *et al.*, 2009).

Zooko's *memorable-secure-decentralized* persistence triangle for naming of computer agents is reminiscent of the *quantity-relation-quality* triangle of conversation (Figure 8B) and the *budget-discrimination-mechanism* triangle of molecular persistence (Figure 8C). Memorability in language usually involves semantic compression, an expression of economy of information (Chomsky, 1995; Sayood, 1996). Securing a naming system to avoid ambiguities or failure usually involves optimization to enhance communication robustness. Finally, a decentralized system allows versatility which is an endowment of flexibility. Remarkably, Zooko's triangle and all triangles I have previously introduced are compatible with a general *economy-robustness-flexibility* persistence triangle that has been

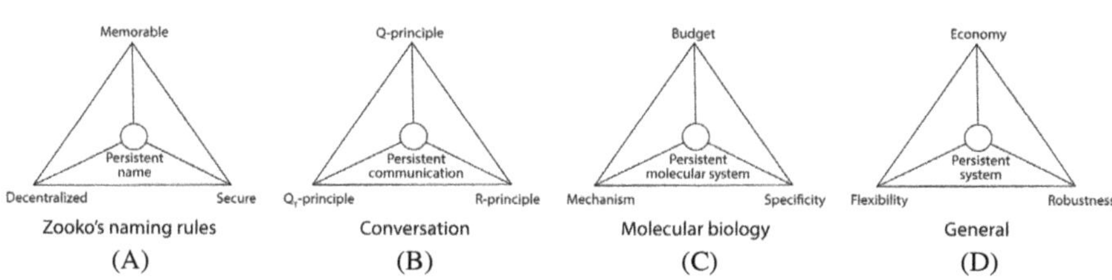

Figure 8. Triangles of persistence. (A) Zooko's *memorable-secure-decentralized* triangle explaining strategies of persistence of names in a gobal namespace that allows to navigate the internet through negotiations of authority and legitimacy. (B) The *quantity-relation-quality* triangle of conversation that explains persistence of communication. (C) The *budget-specificity-mechanism* triangle of molecular persistence explains trade-off solutions for a molecular system in biology. (D) The "general" *economy-robustness-flexibility* triangle of biological persistence explains trade-off solutions for the engineering of the components or the whole of a biological system. This triangle subsumes all previous alternatives.

proposed for biology (Figure 8D) and is supported by ecological and molecular data (Yafremava *et al.*, 2013). The three general principles that are invoked in this general triangle of persistence are defined in Table 6 and show clear correlates to trade-off solutions for naming, language and molecules. In this persistence triangle for biology, a set of solutions to the tradeoff problem of endowing a biological system with properties of economy, flexibility and robustness provides alternative outcomes. I will now discuss how this general space of trade-offs can be applied to biological language in general, and the world of biomolecules in particular.

There is a size-dependent minimization principle in the MA law that is responsible for reducing the size of the components of a larger system. Altmann (1980) suspected that it was *"somehow connected with the principle of least effort or with some not yet known*

Table 6. Three principles driving the persistence of a system.

Term	Definition
Economy	The budget of the system in terms of matter-energy costs to maintain its functioning mechanisms.
Flexibility	The structural and functional mechanisms responding to *internal* and *external* changes imposed on the molecular system. These mechanisms require processing of information and delimit the potential of the system to adapt to environmental change. More flexible systems are generally larger, harbor more complex functionalities, and are more diverse in finding trade-off solutions.
Robustness	The mechanisms that use information to maintain structure and function despite external influence and protect the system from malfunction. These mechanisms provide and ability to resist damage and change. Robustness includes stability but refers to broader processes that are passive from an information point of view.

principle of balance recompensating lengthening on one hand with shortening on the other." The minimization principle was subsequently re-interpreted as an instance of "compression of information" or self-organization (Köhler, 1987; Cramer, 2005; Gustison *et al.*, 2016). We recently reinterpreted it differently by explaining minimization with the triangle of persistence (Shahzad *et al.*, 2015). Using the general framework of persistence (Yafremava *et al.*, 2013), we illustrated how economy, flexibility and robustness of biological systems (defined in Table 6) impose constraints on the engineering of the macromolecular system of a protein. Figure 9A shows the central concept. First, a molecular matter-energy budget defines a central transect of a persistence triangle with vertices representing the three main strategies. Second, the triangle unfolds fully once a new transect of information (suitably defined in Yafremava *et al.*, 2013) is projected perpendicular to the matter-energy budget. This new transect contributes to the trade-offs and the design of the proteins. In fact, the persistence triangle can be made mathematically explicit. We discovered that the MA law represents a power law relationship when molecular persistence (P) is made a function of domain accretion, measured as number of domains k in proteins (Shahzad *et al.*, 2015). P holds two terms, one reflecting the matter-energy cost of adding domains and extending their length in proteins (P_{ME}). Every new domain added to a protein during evolution taxes the biological system as a whole. Longer proteins and corresponding genes must be synthesized and kept functional or appropriately stored. This requires matter in the form of chemical backbones and energy for the synthesis of the amino acid and nucleic acid constituents. It also requires maintenance of a number of biological processes, including metabolism, translation, replication and other housekeeping roles (e.g. chaperone activities) needed to produce the longer protein molecules and maintain their activity in the crowded cellular environment. The second term describes the influence of information present in domain length and

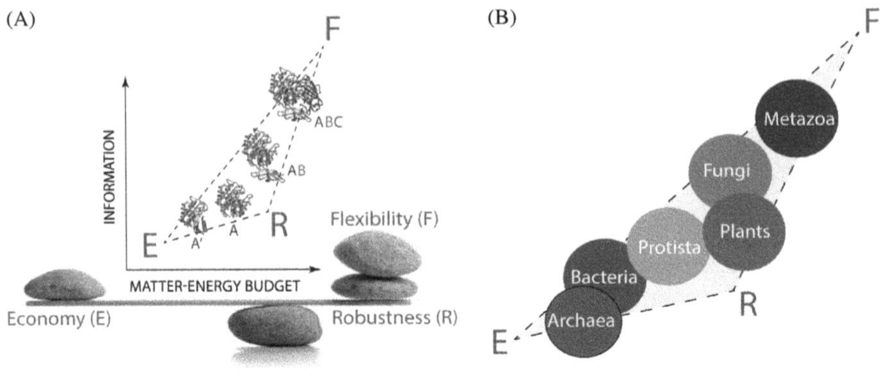

Figure 9. The persistence triangle and its impact on the proteomic vocabulary of organismal kingdoms. (A) Diagram showing how domain size and organization in proteins affects molecular persistence. Domain A′ extends its length in A, and A combines with B and C in evolution to form multi-domain proteins with different propensities towards economy (E), flexibility (F) and robustness (R). Within this landscape, molecules find optimal trade-off solutions given a matter-energy budget and information. (B) Patterns of trade-off solutions for entire complements of protein domains in the proteomes of organisms grouped within the six major kingdoms of life.

number on the flexibility and robustness of the molecular system (P_{FR}). Information is here stored in intramolecular and intermolecular interactions necessary for molecular function and stability. For example and as I have discussed above, information in the combinations of the domains of multi-domain proteins generates an additional vocabulary that can be equated with language information (Bashton and Chothia, 2007). Note that P depends only on the length and number of domains. Thus, the derivative of P with respect to k, when set equal to zero, gives the power law version of the MA law of Equation (2), $z_k = Ak^b$, where z_k is domain length, and b and A are the slope and intercept of log-log plot linear fittings, respectively. P appropriately distills trade-off relationships of the persistence triangle, with one component reflecting an axis of *matter-energy budget* and another reflecting an axis of information. Thus, plots of b versus A, and b and A versus effective protein length (sum of length of domains in a protein without considering linkers and terminal sequences) showed distinct trade-off solutions for the 60 proteomes of the six kingdoms (Table 3) that were examined (Shahzad *et al.*, 2015). Archaea, Fungi and to a lesser degree Plants fostered molecular economy, each at their own economy stratum, as evidenced by the number of domains they harbor. Fungi increased domain size in single domain proteins while reinforcing the MA pattern of diminishing returns. In contrast, Metazoa, and to lesser degrees Protista and Bacteria, relaxed economy, each at their own economy stratum. Metazoa achieved maximum flexibility and robustness by harboring compact molecules and complex domain organization. Figure 9B summarizes the findings and reveals a continuum of kingdom clouds of proteomes that matches a similar continuum derived from physiological and ecological data (Yafremava *et al.*, 2013).

7. The semantic and pragmatic compression of "omic" vocabularies

The triangle of persistence is also useful when studying entire repertoires of proteins or molecular functions. I will now illustrate how the trade-off dynamics materializes in the domain make up of proteomes (the genomic collective of proteins) and functionomes (the genomic collective of biological functions) when these are studied following a census of domains and molecular functions in genomes (Nasir *et al.*, 2014a,b). The diversity (use) and abundance (reuse) of structural domains defined at the fold family (FF) level of a "gold standard" protein classification (SCOP; Murzin *et al.*, 1995) were used as proxies to study the economy and flexibility of proteomes, respectively. These proxies were then plotted against a reuse-to-use ratio of FFs (corresponding to the scaling coefficient β of the Heaps law), which was previously shown to represent an index of robustness (Yafremava *et al.*, 2013). The proteomes of superkingdom Eukarya traded economy for maximal flexibility and robustness. In contrast, the proteomes of microbial superkingdoms Archaea and Bacteria relaxed flexibility and robustness in favor of economy, with Archaea showing the greatest push towards an economy of scale (Figure 10A). A Venn diagram describing the structural diversity and distribution of FF domains in the three superkingdoms of life confirms that the proteomes of superkingdoms have repertoires that are distinct in terms of their number and distribution in superkingdoms (Figure 10A). Archaea harbored the smallest vocabulary

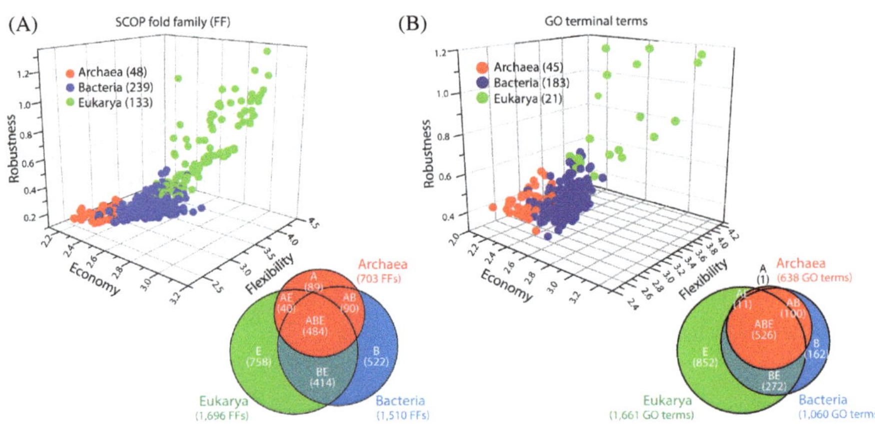

Figure 10. The persistence triangle and its impact on the proteomic and functionomic vocabulary of superkingdoms. (A) Scatter plot describing the persistence strategies of the organisms of superkingdoms measured by the occurrence and abundance of SCOP family (FF) domains. Axes are in logarithmic scales and the numbers in parentheses indicate the numbers of proteomes that were studied in each supergroup. The Venn diagram describes the diversity and distribution of 2,397 FFs in 420 proteomes of superkingdoms Archaea (A), Bacteria (B) and Eukarya (E) describing seven possible taxonomic groups of FF sharing (Kim and Caetano-Anollés, 2012). Data from Nasir *et al.* (2014a). (B) Scatter plot describing the persistence strategies of organisms measured by the occurrence and abundance of their associated GO terms defined at terminal level. Axes are in logarithmic scales and number in parentheses indicate number of proteomes that were examined in the study. The Venn diagram describes the diversity and distribution of 1,924 terminal GO terms in superkingdoms. Data from Nasir *et al.* (2014b).

(703 FFs) and the smallest number of superkingdom-specific FFs (89 FFs). In contrast Eukarya contained both the largest vocabulary (1,696 FFs) and the largest number of superkingdom-specific FFs (758 FFs). The vocabulary of Bacteria was somehow in between. While a substantial number of FFs (20%) is common to all three superkingdoms (the ABE Venn group), over two thirds of the archaeal vocabulary (69%) was part to this group. Thus, the push towards economy has made the domain vocabularies of the proteomes of Archaea the most "compressed" of the cellular world. Their highly universal nature suggest they are also the most ancient. In sharp contrast, the FF repertoire that is unique to Eukarya is almost two thirds larger than the vocabulary that is common to all superkingdoms, showing a remarkable push towards flexibility and robustness in this superkingdom.

A genomic survey of GO annotations of genes at GO terminal level shows similar trade-off dynamics between economy, flexibility and robustness and similar patterns of GO diversity and distribution (Figure 10B). Diversity (use), abundance (reuse) and reuse-to-use ratios of GO terminal terms for a set of 249 functionomes were again used as proxies for functionome economy, flexibility and robustness, respectively, and were displayed in a 3-dimensional plot describing a persistence triangle (Nasir *et al.*, 2014b). The patterns resembled those observed with FF domains (Figure 10A). The functionomes of Archaea and Bacteria favored economy. In turn, the functionomes of Eukarya favored flexibility and robustness. Again, Archaea showed the highest levels of economy-driven compression but their levels of robustness were not far away from those of Bacteria. A Venn diagram that compares the diversity of GO terminal terms in the three superkingdoms confirms that the functionomes of Archaea are the most compressed and least diverse at this lowest level of ontological classification. The 638 terminal GO terms of Archaea were substantially lower than those of Bacteria (1,060) and Eukarya (1,661). Most terms (82%) were universal (shared by the three superkingdoms) in Archaea. In contrast, only 49% and 32% of terms were universal in Bacteria and Eukarya, respectively. The number of superkingdom-specific terms increased in the order Archaea, Bacteria and Eukarya. Remarkably, only one GO terminal term was Archaea-specific, "methylenetetrahydromethanopterin dehydrogenase activity" [GO:0030268]. Thus, the vocabularies of molecular functions of Archaea should be considered the most evolutionarily compressed and homogeneous of the cellular world.

Taking it at face value, the analysis of domain structures in proteomes shows that the microbial push towards economy of structures can be equated to a process of *semantic compression* that is characteristic of evolved natural languages. Semantic compression can be defined as the concurrent reduction of both the vocabulary and language heterogeneity, while at the same time maintaining its semantics (Chomsky, 1995; Sayood, 1996). In other words, a same message is expressed with a smaller number of words. This semantic compression of the structural domain make up of proteins and proteomes is also apparent in its semantics (Kim and Caetano-Anollés, 2012; Yafremava *et al.*, 2013; Nasir *et al.*, 2014a,b; Shahzad *et al.*, 2015). The analysis of molecular functions defined with GO terminal terms reveals that the semantic compression occurring in proteomes also resulted in highly reduced vocabularies of molecular functions. This evolutionarily reduced repertoires describe a process of *pragmatic compression* that reduces the semantics of molecular function without sacrificing its pragmatics. This new compression conveys all possible messages with a smaller set of meanings. In other words, the compressed messages use fewer meanings while preserving the context of the vocabulary. Under this paradigm, the overarching goal is to increase the fitness of an organism in interaction with its environment by pushing the persistence of the molecular language towards the Persuasion vertex of the persistence triangle (Figure 6).

8. The evolution of semantic and pragmatic compression in molecules

Over the past two decades we have used phylogenomic methods to study the history of the protein world and their associated molecular functions. We generated phylogenomic trees of structural domains defined at different hierarchical levels of the SCOP and CATH gold standards of protein classification (e.g., Caetano-Anollés and Caetano-Anollés, 2003; Wang and Caetano-Anollés, 2009; Caetano-Anollés *et al.*, 2012; Kim and Caetano-Anollés, 2012; Bukhari and Caetano-Anollés, 2013; Nasir *et al.*, 2014a; Nasir and Caetano-Anollés, 2015) and trees of molecular functions at different levels of the GO classification (e.g. Nasir *et al.*, 2014b; Koç and Caetano-Anollés, 2017). Tables of evolutionary ages for the origin of each structural domain group and GO *mf* terms are available. These chronologies, which are derived directly from the reconstructed trees, have been benchmarked to the geological record (Wang *et al.*, 2011a). They provide considerable insight into the origin and diversification of the molecular and organismal world and have also considerable predictive power. For example, phylogenomic analysis of GO terms produced a picture of organismal diversification that is congruent with that obtained from structural domains (Kim *et al.*, 2014) and other evidence. In all of these studies, trees of life with leaves describing proteomes and functionomes showed the appearance of superkingdom Archaea at its base (reviewed in Caetano-Anollés *et al.*, 2014). Thus, Archaea represents the first domain of diversified life to appear in evolution.

The possibility of indexing an evolutionary age to each structural domain or GO term has uncovered data-driven hypotheses for the rise of complexity in biochemistry and the piecemeal evolution of proteomes and functionomes. The endeavor is evolutionary. It entails three crucial starting of point of reason [modified from

Wiley (1976)]: (i) *Historical continuity:* Change entails spatiotemporal continuity; (ii) *Historical singularity:* Only one account of history exists, which is not statistical; and (iii) *Historical memory:* Biological attributes (phylogenetic characters) are transmitted from one generation to the next (modified or unmodified) and can be mined for information about the past. The evolutionary timelines of domains and functions comply both with language laws and the principle of continuity. For example, while domain gains override losses in proteomes (Nasir *et al.*, 2014a), new structural domains and their associated molecular functions accumulate in time according to a multi-regime Heaps' law (Nasir *et al.*, 2017). This law resembles the law for languages with limited vocabulary sizes (e.g. Korean, Chinese). Similarly, the accumulation of domains and their associated GO terms of molecular functions can inform about the natural history of semantic and pragmatic compression in molecular evolution. Figure 11 for example describes the evolution of level 1 GO *mf* terms with a heat diagram of genomic abundances of terminal GO terms mapped to that level for a set of 249 organisms (Koç and Caetano-Anollés, 2017). Columns of the heat diagram are organized according to the age of GP *mf* terms at this top level of the ontological hierarchy. As expected, the first three GO terms (the oldest) were the most abundant and widely distributed in functionomes. They were linked to catalysis, binding, and transport, functions that are metabolic and truly universal, i.e. present in all 249 genomes examined. The next five were also universal (they were part of the ABE Venn category of GO terms shared by the three superkingdoms of life) but were not present in all functionomes. They were specially absent in some Archaea. The rest were present in Eukarya and very few lineages of Bacteria or were Eukarya-specific, all of which appeared more recently. One conclusion that can be derived from the heat diagram and timeline is that the functional vocabulary of Archaea was probably the oldest and most compressed. The most basal functions (binding, metabolism and transport) are also highly

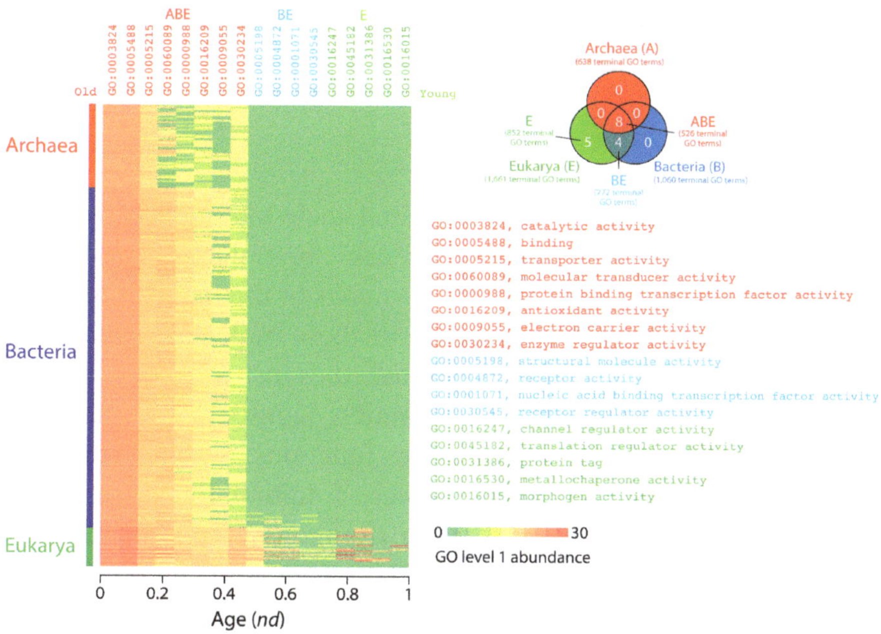

Figure 11. The evolutionary history of level 1 GO terms and their abundance in genomes reveals the early origin of semantic and pragmatic compression and supports the early origin of Archaea. The heat map shows the distribution of genomic abundances of 17 level 1 GO terms mapped to terminal GO terms in 249 organisms and places them along an evolutionary timeline of molecular function innovation of level 1 GO terms derived from a phylogenomic reconstruction. The age of GO terms is expressed in a relative scale of node distance (*nd*) values, from 0 (origin of molecular functions) to 1 (the present). The Venn diagram describes how GO terms distribute between the the superkingdoms of life. The definition of GO terms is provided as a list. Data from Koç and Caetano-Anollés (2017).

homogeneous in terms of abundance levels (Figure 11). This same global patterns manifests at other levels of the GO hierarchy, suggesting that the diversification of the metabolic vocabulary of Archaea arose from strong evolutionary reductive constraints (Koç and Caetano-Anollés, 2017). These constraints were likely imposed by the microbial lifestyle of early organisms and the harsh environments that are expected of early Earth, which modern archaeal organisms have learnt to master.

9. Conclusions

The structural domains of proteins and their molecular functions can be considered the signs of an ancient molecular language. I have argued that the combinatorial syntaxis of these signs produces multidomain proteins. The make up of these molecular phrases is likely driven by sets of hidden grammar rules that define the molecular functions of the cell. In semantics, signs convey meaning through *predicates*, i.e. relations or functions imposed through grammar or first order logic rationales. Traditionally, these predicates are verb phrases that modify or act upon *arguments*, subjects such as noun phrases that complete the meaning of the predicates. Since predicates reflect truth about its signs, I extend these linguistic concepts to the meaning of proteins using dependency grammars and catenae. Dominance relations derived for example from the evolutionary age of domains and their GO annotations are used to define molecular predicates, which are then combined with physical precedence relationships to gain insight into molecular syntaxis. I also prompted analysis of the pragmatics of biomolecular language by introducing triangles of molecular persistence that describe the high levels of complexity that pervades molecular meaning, its links to molecular function, and its intimate relation to fitness in performance spaces. Note that both biomolecules and language are highly structured and hold multiple layers of organization, which in very simplistic ways can be distilled by Chomsky's hierarchy of formal languages. Both comply with laws that describe their diversity, cohesion and growth. As with language, the natural history of biomolecules impacts their syntaxis, semantics and pragmatics. I showcase this fact with the proteomes and functionomes of cellular life, showing that Archaea holds the smallest, oldest and most homogeneous vocabulary of all superkingdoms. I propose that this molecular vocabulary has been compressed in evolution with semantic and pragmatic rationales. In contrast, Eukarya showcases the most diverse and complex molecular vocabulary.

"The place where I come from is a small town,
They think so small, they use small words
But not me, I'm smarter than that, I worked it out
I've been stretching my mouth, to let those big words come right out"
— Peter Gabriel, "Big time" from his 1986 album "So"

References

Achenbach J, Nierhaus KH (2015) The mechanics of ribosomal translocation. *Biochimie* **114**: 80–9.

Altmann EG, Gerlach M (2016) Statistical laws of linguistics. In: *Creativity and Universality in Language, Lecture Notes in Morphogenesis*, Degli Esposti M, Altmann EG, Pachet F (eds). Springer, Berlin, pp. 7–26.

Altmann G (1980) Prolegomena to Menzerath's law. *Glottometrika* **2**: 1–10.

Altmann G, Köhler R, Vulanović R (2006) Encyclopedia of linguistic laws. http://lql.uni-trier.de/index.php/Main_Page (Accessed January 2, 2017.)

Ashburner M, Ball CA, Blake JA, Botstein D, Butler H, Cherry JM, Davis AP, Dolinski K, Dwight SS, Eppig JT (2000) Gene Ontology: Tool for the unification of biology. *Nat Genet* **25**: 25–29.

Alva V, Söding J, Lupas AN (2015) A vocabulary of ancient peptides at the origin of folded proteins. *eLife* **4**: e09410.

Aziz MF, Caetano-Anollés K, Caetano-Anollés G (2016) The early history and emergence of molecular functions and modular scale-free network behavior. *Sci Rep* **6**: 25058.

Baixeries J, Hernández-Fernández A, Ferrer-i-Cancho R (2012) Random models of Menzerath-Altmann law in genomes. *Biosystems* **107**: 167–173.

Barabási A-L (2009) Scale-free networks: A decade and beyond. *Science* **325**: 412–413.

Bashton M, Chothia C (2002) The geometry of domain combinations in proteins. *J Mol Biol* **315**: 927–939.

Bashton M, Chothia C (2007) The generation of new protein functions by the combination of domains. *Structure* **15**: 85–99

Berezovsky IN, Trifonov EN (2001) Van der Waals locks: Loop-n-lock structure of globular proteins. *J Mol Biol* **307**: 1419–1426.

Bhaskara RM, Srinivasan N (2011) Stability of domain structures in multi-domain proteins. *Sci Rep* **1**: 40.

Bukhari SA, Caetano-Anollés G (2013) Origin and evolution of protein fold designs inferred from phylogenomic analysis of CATH domain structures in proteomes. *PLoS Comp Biol* **3**: e1003009.

Caetano-Anollés G, Caetano-Anollés D (2003) An evolutionary structured universe of protein architecture. *Genome Res* **13**: 1563–1571.

Caetano-Anollés G, Caetano-Anollés D (2015) Computing the origin and evolution of the ribosome from its structure — Uncovering processes of macromolecular accretion benefiting synthetic biology. *Comput Struct Biotech J* **13**: 427–447.

Caetano-Anollés G, Wang M, Caetano-Anollés D, Mittenthal J (2009) The origin, evolution and structure of the protein world. *Biochem J* **417**: 621–637.

Caetano-Anollés G, Yafremava L, Mittenthal JE (2010) Modularity and dissipation in evolution of macromolecular structures, functions, and networks. In: *Evolutionary Bioinformatics and Systems Biology*, Caetano-Anollés G (ed). Wiley-Blackwell, Hoboken, NJ, USA, pp. 443–449.

Caetano-Anollés D, Kim KM, Mittenthal JE, Caetano-Anollés G (2011) Proteome evolution and the metabolic origins of translation and cellular life. *J Mol Evol* **72**: 14–33.

Caetano-Anollés G, Wang M, Caetano-Anollés D (2013) Structural phylogenomics retrodicts the origin of the genetic code and uncovers the evolutionary impact of protein flexibility. *PLoS One* **8**: e72225.

Caetano-Anollés G, Nasir A, Zhou K, Caetano-Anollés D, Mittenthal JE, Sun F-J, Kim KM (2014) Archaea: The first domain of diversified life. *Archaea* **2014**: 590214.

Chen B, Kaledhonkar S, Sun M, Shen B, Lu Z, Barnard D, Lu T, Gonzalez Jr R, Frank J (2015) Structural dynamics of ribosome subunit association studied by mixing-spraying time-resolved cryogenic electron microscopy. *Structure* **23**: 1097–1105.

Chomsky N (1956) Three models for the description of language. *IRE Trans Information Theory* **2**(3): 113–124.

Chomsky N (1957) *Syntactic Structures*. Mouton, The Hague/Paris.

Chomsky N (1995) *The Minimalist Program*. MIT Press, Cambridge, MA.

Chothia C, Gough J, Vogel C, Teichmann SA (2003) Evolution of the protein repertoire. *Science* **300**: 1701–1703.

Coin, L, Bateman A, Durbin R (2003) Enhanced protein domain discovery by using language modeling techniques from speech recognition. *Proc Natl Acad Sci USA* **100**(8): 4516–20.

Cramer I (2005) The parameters of the Altmann-Menzerath law. *J Quant Linguist* **12**(1): 41–52.

Davis W (2014) Implicature. In: *The Stanford Encyclopedia of Philosophy*, Zalta EN (ed). Fall Edition, URL = <https://plato.stanford.edu/archives/fall2014/entries/implicature/>

Debès C, Wang M, Caetano-Anollés G, Gräter F (2013) Evolutionary optimization of protein folding. *PLoS Comput Biol* **9**: e1002861.

Devitt M, Hanley R (2006) *The Blackwell Guide to the Philosophy of Language*. Blackwell Publishing, Oxford.

Eigen M (1971) Selforganization of matter and the evolution of biological macromolecules. *Naturwissenschaften* **58**: 465–523.

Ekman D, Bjorklund AK, Frey-Skott J, Elofsson A (2005) Multi-domain proteins in the three kingdoms of life: Orphan domains and other unassigned regions. *J Mol Biol* **348**: 231–243.

Eroglu S (2014) Language-like behavior of protein length distribution in proteomes. *Complexity* **20**: 12–21.

Ferdous SMd, Jøsang A, Singh K, Borgaonkar R (2009) Security usability of petname systems. In: *Identity and Privacy in the Internet Age, NordSec 2009. Lecture Notes in Computer Science*, Vol 5838, Jøsang A, Maseng T, Knapskog SJ (eds). Springer, Berlin, Heidelberg, Germany, pp. 44–59.

Ferrer-i-Cancho R, Forns N (2010) The self-organization of genomes. *Complexity* **15**: 34–36.

Ferrer i Cancho R, Solé RV (2001) Two regimes in the frequency of words and the origins of complex lexicons: Zipf's law revisited. *J Quant Linguist* **8**: 165–173.

Gerlach M, Altmann EG (2013) Stochastic model for the vocabulary growth in natural languages. *Phys Rev* **3**: 021006

Gimona M (2006) Protein linguistics — a grammar for modular protein assembly? *Nature Rev Mol Cell Biol* **7**: 68–73.

Goncearenco A, Berezovsky IN (2015) Protein function from its emergence to diversity in contemporary proteins. *Phys Biol* **12**: 045002.

Grice HP (1989) *Studies in the Way of Words*. Harvard University Press.

Gustison ML, Semple S, Ferrer-i-Cancho R, Bergman TJ (2016) Gelada vocal sequences follow Menzerath's linguistic law. *Proc Natl Acad Sci USA* **113**(19): E2750–E2758.

Harris M, Clark J, Ireland A, Lomax J, Ashburner M, Foulger R, *et al.* (2004) The Gene Ontology (GO) data-base and informatics resource. *Nucleic Acids Res* **32**: D258–D261.

Hawley K (2015) Temporal Parts. In: *The Stanford Encyclopedia of Philosophy*, Winter edition, Zalta EN (ed). https://plato.stanford.edu/archives/win2015/entries/temporal-parts/

Heaps HS (1978) Information Retrieval: Computational and Theoretical Aspects. Academic Press, New York.

Hébert L (2011) The functions of language. In: Hébert L (dir.), Signo [online], Rimouski, Quebec, http://www.signosemio.com/jakobson/functions-of-language.asp.

Herdan G (1964) *Quantitative Linguistics*. Butterworth Press, Oxford.

Horn L (1984) Towards a new taxonomy of pragmatic inference: Q-based and R-based implicature. In: *Meaning, Form, and Use in Context: Linguistic Applications, GURT84*, Schiffrin D (ed). Georgetown University Press, Washington, pp. 11–42.

Illegård K, Ardell DH, Elofsson A (2009) Structure is three to ten times more conserved than sequence — A study of structural response in protein cores. *Proteins* **77**: 499–508.

Jakobson R (1960) Linguistics and poetics. In: *Style in Language*, Sebeok T (ed.). MIT Press, Cambridge, MA, pp. 350–377.

Janin J, Wodak SJ (1983) Structural domains in proteins and their role in the dynamics of protein function. *Prog Biophys Mol Biol* **42**: 21–78.

Kim HS, Mittenthal JE, Caetano-Anollés G (2013) Widespread recruitment of ancient do- main structures in modern enzymes during metabolic evolution. *J Integr Bioinform* **10**: 214.

Kim KM, Caetano-Anollés G (2012) The evolutionary history of protein fold families and proteomes confirms that the archaeal ancestor is more ancient than the ancestors of other superkingdoms. *BMC Evol Biol* **12**: 13.

Kim KM, Nasir A, Caetano-Anollés G (2014) A tree of cellular life inferred from a genomic census of molecular functions. *J Mol Evol* **79**: 240–262.

Koç I, Caetano-Anollés G (2017) The natural history of molecular functions inferred from an extensive phylogenomic analysis of Gene Ontology data. *PLoS One* **12**(5): e0176129.

Köhler R (1987) System theoretical linguistics. *Theor Linguist* **14**(2–3): 241–258.

Korta K, Perry J (2015) Pragmatics. In: The Stanford Encyclopedia of Philosophy, Winter edition, Zalta EN (ed). https://plato.stanford.edu/archives/win2015/entries/pragmatics/

Lakoff RT (1993) Lewis Carroll: Subversive pragmaticist. *Pragmatics* **3**: 367–385.
Li S, Lin R, Bian C, Ma QDY, Ivanov PC (2016) Model of the dynamic construction process of texts and scaling laws of words organization in language systems. *PLoS One* **11**(12): e01688971.
Lü L, Zhang Z-K, Zhou T (2013) Deviation of Zipf's and Heaps' laws in human languages with limited dictionary sizes. *Sci Rep* **3**: 8028–8033.
Marahiel MA (2016) A structural model for multimodular NRPS assembly lines. *Nat Prod Rep* **33**: 136–140.
Menzerath, P (1928) Uber einige phonetische probleme. In: Actes du Premier Congrès International de Linguists. Sijthhof, Leiden, pp. 104–105.
Menzerath P (1954) *Die Architektonik des Deutschen Wortschatzes*. Dümmler, Bonn, Germany.
Mittenthal JE, Caetano-Anollés G (2010) Exploring the interplay of stability and function in protein evolution. *Bioessays* **32**(8): 655–658.
Molina N, van Nimwegen E (2009) Scaling laws in functional genome content across prokaryotic clades and lifestyle. *Trends Genet* **25**(6): 243–247.
Moore AD, Bjorklund AK, Ekman D, Bornberg-Bauer E, Elofsson A (2008) Arrangements in the modular evolution of proteins. *Trends Biochem Sci* **33**: 444–451.
Murzin A, Brenner SE, Hubbard T, Clothia C (1995) SCOP: A structural classification of proteins for the investigation of sequences and structures. *J Mol Biol* **247**: 536–540.
Nasir A, Caetano-Anollés G (2015) A phylogenomic data-driven exploration of viral origins and evolution. *Sci Adv* **1**: e1500527.
Nasir A, Kim KM, Caetano-Anollés G (2014a) Global patterns of domain gain and loss in superkingdoms. *PLoS Comput Biol* **10**: e1003452.
Nasir A, Kim KM, Caetano-Anollés G (2014b) A phylogenomic census of molecular functions identifies modern thermophilic Archaea as the most ancient form of cellular life. *Archaea* **2014**: 706468.
Nasir A, Kim KM, Caetano-Anollés G (2017) Phylogenetic tracings of proteome size support the gradual accretion of protein structural domains and the early origin of viruses from primordial cells. *Front Microbiol* **8**: 1178.
Newton MS, Arcus VL, Patrick WM (2015) Rapid bursts and slow declines: On the possible evolutionary trajectory of enzymes. *J R Soc Interface* **12**: 20150036.
O'Grady W (1998) The syntax of idioms. *Nat Lang Linguist Theory* **16**: 279–312.
Osborne T, Putnam M, Groß T (2011) Bare phrase structure, label-less trees, and specifier-less syntax: Is Minimalism becoming a dependency grammar? *Linguistic Rev* **28**: 315–364.

Osborne T, Putnam M, Groß T (2012) Catenae: Introducing a novel unit of syntactic analysis. *Syntax* 15(4):354–396.

Petersen AM, Tenenbaum JN, Havlin S, Stanley HE, Perc M (2012) Languages cool as they expand: Allometric scaling and the decreasing need for new words. *Sci Rep* **2**: 721–725.

Pfeiffer T, Soyer OS, Bonhoeffer S (2005) The evolution of connectivity in metabolic networks. *PLoS Biology* **3**(7): e228.

Qian J, Luscombe NM, Gersten M (2001) Protein family and fold occurrence in genomes: Power-law behavior and evolutionary model. *J Mol Biol* **313**: 673–681.

Reynolds NM, Lazazzera BA, Ibba M (2010) Cellular mechanisms that control mistranslation. *Nat Rev Microbiol* **8**: 849–56.

Richardson JS (1981) The anatomy and taxonomy of protein structure. *Adv Protein Chem* **34**: 167–339.

Riley M, Labedan B (1997) Protein evolution viewed through *Escherichia coli* protein sequences: Introducing the notion of a structural segment of homology, the module. *J Mol Biol* **268**: 857–868.

Sayood K (1996) *Introduction to Data Compression.* Morgan Kauffman, San Francisco.

Searls DB (2001) Reading the book of life. *Bioinformatics* **17**(7): 579–580.

Searls DB (2002) The language of genes. *Nature* **420**: 211–217.

Shahzad K, Mittenthal KE, Caetano-Anollés G (2015) The organization of domains in proteins obeys Menzerath-Altmann's law of language. *BMC Syst Biol* **9**: 44.

Sheftel H, Shoval O, Mayo A, Alon U (2013) The geometry of the Pareto front in biological phenotype space. *Ecol Evol* **3**: 1471–1483.

Shegogue D, Zheng WJ (2005) Integration of the Gene Ontology into an object-oriented architecture. *BMC Bioinformatics* **6**(1): 113.

Söding J, Lupas AN (2003) More than the sum of their parts: On the evolution of proteins from peptides. *BioEssays* **25**: 837–846.

Strauss S, Altmann G (2006) Hierarchic relations. In: *Encyclopedia of Linguistic Laws*, Altmann G, Köhler R, Vulanović R (eds). http://lql.uni-trier.de/index.php/Main_Page

Strogatz SH (2001) Exploring complex networks. *Nature* **410**: 268–276.

Tesnière L (1959) *Éléments de Syntaxe Structurale.* Klincksieck, Paris.

Tria F, Loreto V, Servedio VDP, Strogatz SH (2014) The dynamics of correlated novelties. *Sci Rep* **4**: 5890.

Trifonov EN, Frenkel ZM (2009) Evolution of protein modularity. *Curr Opin Struct Biol* **18**: 335–340.

Toll-Riera M, Alba MM (2013) Emergence of novel domains in proteins. *BMC Evol Biol* **13**: 47.

Van Rooy R (2004) Signaling games select Horn strategies. *Linguist Philos* **27**: 493–527.

Wang M, Caetano-Anollés G (2009) The evolutionary mechanics of domain organization in proteomes and the rise of modularity in the protein world. *Structure* **17**: 66–78.

Wang M, Jiang YY, Kim KM, Qu G, Ji HF, Mittenthal JE, Zhang HY, Caetano-Anolles G (2011a) Universal molecular clock of protein folds and its power in tracing the early history of aerobic metabolism and planet oxygenation. *Mol Biol Evol* **28**(1): 567–582.

Wang M, Kurland CG, Caetano-Anollés G (2011b) Reductive evolution of proteomes and protein structures. *Proc Natl Acad Sci USA* **108**: 11954–11958.

Wallin G, Kamerlin SCL, Åqvist J (2012) Energetics of activation of GTP hydrolysis on the ribosome. *Nature Comm* **4**: 1733.

Wetlaufer DB (1973) Nucleation, rapid folding, and globular intrachain regions in proteins. *Proc Natl Acad Sci USA* **70**: 697–701.

Wiley EO (1975) Karl R. Popper, systematics, and classification: A reply to Walter Bock and other evolutionary taxonomists. *Syst Zool* **24**: 233–243.

Wilcox-O'Hearn Z (2001) Names: Decentralized, secure, human-meaningful: Choose two. https://web.archive.org/web/20011020191610/http://zooko.com/distnames.html

Witzany G (2014) Language and Communication as Universal Requirements for Life. In: *Astrobiology: An Evolutionary Approach*, Kolb V (ed). CRC Press, pp. 349–370.

Witzany G (2016) The biocommunication method: On the road to an integrative biology. *Comm Integr Biol* **9**(2): e1164374

Yafremava LS, Wielgos M, Thomas S, Nasir A, Wang M, Mittenthal JE, Caetano-Anollés G (2013) A general framework of persistence strategies for biological systems helps explain domains of life. *Front Genet* **4**: 16.

Zipf GK (1948) *Human Behavior and the Principle of Least Effort*. Addison-Wesley, Boston, MA.

Growth

Molecular accretion, growth and innovation

D. Caetano-Anollés[1]
G. Caetano-Anollés[2]

Summary. Accretion brings together disparate parts to form bigger wholes. In doing so, it provides opportunities for growth and innovation. This process occurs pervasively in nature at widely different timeframes. In biological evolution, component parts are added to growing macromolecules, which also interact and merge with other growing molecules to form molecular complexes and higher levels of molecular and cellular organization. Here we study accretion by focusing on protein structural domains that make up proteomes. Phylogenomic reconstruction generates phylogenetic trees and chronologies of structural domains, which can be used to describe proteome growth. We find that the interface between evolutionary models, trees, and data supports Leibniz's principle of spatiotemporal continuity. The growth of macromolecular systems complies with mathematical universals,

[1]*Broad Institute of MIT and Harvard, Cambridge, Massachusetts, USA*
[2]*Department of Crop Sciences, University of Illinois, Urbana, Illinois, USA*

such as the Zipf and Heaps laws. Finally, sequential S-shaped logistic wavelets that model growth and diffusion of innovations describe processes of diversification. Remarkably, pairs of wavelets induce biphasic patterns that help model protein evolution. It appears fine-grained structural elaborations in growing funnels of diversification give rise to coarse-grained innovations that tailor the generation of structural modules and biological complexity.

1. Introduction

> *"Any change from small to large, or vice versa, passes through something which is, in respect of degrees as well as of parts, in between; as well as that no motion ever springs immediately from a state of rest, or passes into one except through a lesser motion; just as one could never traverse a certain line or distance without first traversing a shorter one."*
>
> — Sämtliche Schriften und Briefe, A, VI, page 56, Gottfried Wilhelm Leibniz

The existence of growth in our universe cannot be contested, from the overarching cosmos to the invisible world of molecules. Galaxies for example form by a process of gradual accumulation of matter. They arise from gravitational collapse of vast clouds of dust and gas, clumping of gas and dust lumps, and dry and wet mergers of stars (Fraix-Burnet *et al.*, 2012). These processes of accretion occur at different levels of granularity. Similarly, stars form by gravitational collapse of giant molecular clouds and dense interstellar media known as "stellar nurseries." These regions of matter collapse under their own gravitational pull forming circumstellar disks of orbiting matter that spiral inward towards the growing central bodies (Beckwith and Sargent, 1996; Jofré *et al.*, 2017). Planets arise by a similar process. They form by accretion from solar nebula, proto-planetary disks of gas and solids, and N-body interactions of dust grains when these orbit around the central proto-star (Mordasini *et al.*, 2015). Figure 1A offers a pictorial representation of the outcome of some of these cosmological accretion processes. Unsurprisingly and at the other end of the spectrum, macromolecules in the biological world arise from similar growth processes. They add component parts to growing molecules, which also interact and merge with other growing molecules to form molecular complexes and higher levels of molecular and cellular structure (Caetano-Anollés *et al.*, 2009). For example, Figure 1B describes a phylogenomic analysis of the origin and evolution of the ribosome, the molecular complex responsible for protein synthesis in the cell (Caetano-Anollés, 2002; Harish and Caetano-Anollés, 2012; Caetano-Anollés and Caetano-Anollés, 2015). Similarly, one could also envision accretion processes being responsible for organizing more complex cellular structure by generating cellular consortia, multicellular organisms and populations of diversifying individuals. For example, one can envision regarding a multicellular organism as the result of evolutionary accretion of cells. Even cities, the products of humanity, are often subject to inexorable processes of growth triggered by innovation, wealth creation, and population (Bettencourt *et al.*, 2007; Bettencourt, 2013). Urban organization and dynamics appear linked to knowledge creation and economies of scale that guarantee sustainable development. Figure 1C summarizes a 5,700-year timeline of global urbanization in our planet (Reba *et al.*, 2016). Thus, the development of higher and higher levels of organization result in more complex structure that is

(A)

(B)

(C)

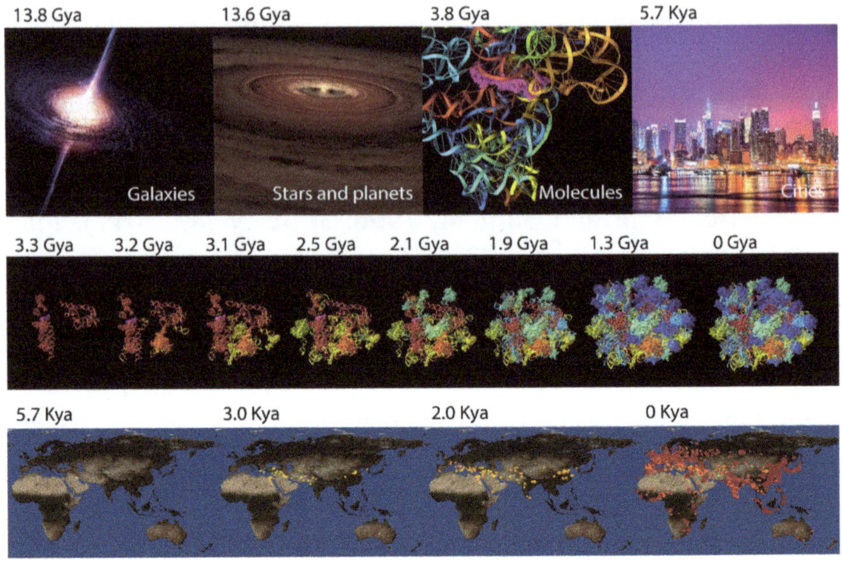

Figure 1. Accretion appears universal. (A) The origin and evolution of galaxies, stars, planets, macromolecules and cities involve processes of growth occurring at different timeframes, from billions of years to hours of a day. The origin of accretion is indicated in billions (Gya) or thousands (Kya) of years ago. (B) The natural history of the ribosome inferred from the phylogenetic analysis of protein structural domains and ribosomal RNA structures. RNA helical structures and structural domains of ribosomal proteins are colored according to their age in a scale from red (very ancient) to blue (very recent). Data from Harish and Caetano-Anollés (2012). (C) The origin and evolution of the cities of Afro-Eurasia and Oceania. Cities with >20,000 inhabitants were mapped according to Reba *et al.* (2016); city concentrations are described with size and color hues. The panel shows excerpts from an animation of city growth hosted by Metrocosm (http://metrocosm.com/history-of-cities/).

permanently growing and diversifying. Here we focus on the processes of molecular growth, structural nucleation, and innovation that exist in molecular biology. We unravel accretion with cladistics approaches. These methodologies have been used to successfully study the evolution of galaxies (Fraix-Burnet *et al.*, 2012), star systems (Jofré *et al.*, 2017) and macromolecules (Caetano-Anollés and

Caetano-Anollés, 2015). The ultimate goal is to provide a framework that would explain the rise of molecular complexity within the confines of Leibniz's spatiotemporal continuity axiom that is fundamental for biological evolution.

2. Macromolecular accretion

Accretion of macromolecules can manifest at different levels of complexity. A molecule can grow by an addition of molecular parts, molecules can accrete to form molecular complexes, or molecules can accrete within the confines of the cell and in doing so increase the complexity of the molecular repertoire. Molecular structures and their molecular functions are encoded in genetic information. The dynamics of macromolecular accretion depends on how evolution shapes genomic make up and its impact on 3-dimensional atomic structure and the associated functions of the biological macromolecules. Agents of genetic change include point mutations, insertions, deletions, rearrangements, fusions, and fissions. Many of these changes will provide benefits to the complex molecular interactions needed to enhance organismal persistence, promoting the multiplicity of inter- and intramolecular interactions that prompt nucleating foci for growing molecules. These will be retained in evolution. Some of these changes will represent molecular innovations. A small group of them will be of planetary significance (e.g. the origin of RubisCO and its role in carbon fixation; Caetano-Anollés, 2017). While the rates and distributions of these processes are not well understood for nucleic acids and proteins, they can materialize relatively quickly in organismal lineages or take millions of years (My) or billions of years (Gy) to unfold.

Recent computational and structural biology studies of evolutionary accretion have unraveled the origin and evolution of macromolecules with the approaches of evolutionary genomics (reviewed in

Caetano-Anollés and Caetano-Anollés, 2015). We will not be discussing here the accretion of molecular parts that make up molecules or molecular complexes. If interested in the origin and evolution of the ribosome, for example, explore Caetano-Anollés (2002) and Harish and Caetano-Anollés (2012). Instead, we illustrate our understanding of molecular growth, accretion and innovation by focusing on protein evolution and the accretion of domain structures in organisms. Proteins structures are organized in a nested hierarchy of structural modules (Caetano-Anollés *et al.*, 2009). Modules are sets of integrated parts that cooperate to perform a task and interact more extensively with each other than with other parts and modules of the system (Hartwell *et al.*, 1999). These modules appear recurrently in different molecular contexts at both protein and proteome levels. A proteome is here defined as both the set of all genes encoding the proteins of an organism or the set of the encoded proteins themselves. One important protein module is the *structural domain*, a structural folded 3-dimensional (3D) atomic core stabilized by maximizing low energy conformation of individual residues and formation of well packed structures through van der Waals forces and hydrophobic collapse. These cores act as independent units of structure (Wetlaufer, 1973; Richardson, 1981), evolution (Janin and Wodak, 1983; Murzin *et al.*, 1995; Riley and Labedan, 1997) and function (Han *et al.*, 2007), mainly because of their high evolutionary conservation and their ability to establish a multitude of stabilizing intramolecular interactions (Bhaskara and Srinivasan, 2011). A number of classifications schemes use the structural similarity and common evolutionary origin of structural domains to classify proteins. Two of them, the Structural Classification of Proteins (SCOP) (Murzin *et al.*, 1995) and the CATH database (Orengo *et al.*, 1997) are considered gold standards. They group proteins into a similar hierarchy of structures. For example, SCOP groups domains in fold families (FF), fold superfamilies (FSF), folds and protein classes. FFs unify domains with pairwise amino acid sequence identities of more than 30%. FSFs unify FFs that share similar structures and functions, which most likely have common evolutionary origins. Folds unify FSFs that share common core topologies, i.e. similar arrangements of secondary structures in 3D space. Finally, protein classes unify folds that share similar overall designs or properties (e.g. mostly helical or mostly strand). SCOPe, an extended version of SCOP, currently classifies 87,224 Protein Data Bank (PDB) structural entries with 276,231 domains into 1,232 folds, 2,026 FSFs and 4,919 FFs (release 2.07, March 2018). The numbers of domain structures of the highest hierarchical SCOP levels should be considered finite, since they have not increased despite significant increases in PDB entry numbers. This suggest that these protein structures represent the majority of those that exist nature.

Emile Zuckerkandl, one of the founders of the molecular evolutionary field, anticipated the need to describe protein diversity with an evolutionary rationale (Zuckerkandl, 1975). He did so despite patchy structural views that could be derived from the study of the 13 crystallographic entries that were available at that time. Four decades later and despite acquisition of ~167,000 PDB structural models (as of July 2020), the systematic classification of protein structure has been unable to unify the widely different structural fold topologies. Since a general metric for global pairwise (or multiple) comparisons of structures does not yet exist (Taylor, 2006), the problem has not been resolved. Zuckerkandl also realized that *"neither gene duplicates nor dormant genes represent*[ed] *promising materials for a de novo generation of proteins"*, and that with exceptions, *"new structural genes must nearly consistently derive from preexisting structural genes."* By the same token he thought that *"new functions can be evolved only on the basis of old proteins"* and that a *"principle of concomitant ascent"* enabled proteins to *"grow up together"* and be *"rejected if introduced at a period when other proteins reached adulthood."* These concepts were revolutionary for his time but were quickly forgotten once genomics and evolutionary bioinformatics advances focused almost exclusively on the sequence of molecules instead of their structures.

In order to go back in time and reconstruct the past, one needs to analyze data with an evolutionary model capable of producing a statement of history, i.e. a phylogeny. Standard phylogenies have the form of a phylogenetic tree, but can also be represented with a phylogenetic network. One turning point for the understanding of the origin and evolution of proteins was the recognition that a census of protein structural domains in proteomes carried deep phylogenetic signal. Gerstein (1998) was the first to build phylogenetic *trees of proteomes* (ToPs) from the occurrence of structural domains in proteomes. These ToPs are "trees of life" with leaves (taxa) representing organisms. Gerstein used distance methods and then maximum parsimony to build his trees. However, he did not root the trees he reconstructed, i.e. he did not identify and "pull down" a branch holding the ancestor of the unrooted tree to its base to reorient change in the direction of time. Thus, origins of proteomes could not be identified. Later on, however, we used standard cladistic methods of maximum parsimony to study the occurrence and abundance of structural domains in proteomes and generate rooted trees (Caetano-Anollés and Caetano-Anollés, 2003). Because ToPs were rooted, both genomic occurrence and abundance information was used to track the origin and evolution of proteomes and their corresponding proteomic repertoires. Following these initial studies, rooted and unrooted ToPs have been reconstructed from surveys of domain and domain organization in proteomes with different approaches (e.g., Yang *et al.*, 2005; Wang and Caetano-Anollés, 2006, 2009; Wang *et al.*, 2007; Fang *et al.*, 2013; Nasir and Caetano-Anollés, 2015), confirming that domain structure carries significant evolutionary information.

We also generated a novel kind of phylogenomic tree (Caetano-Anollés and Caetano-Anollés, 2003). We reconstructed rooted *trees*

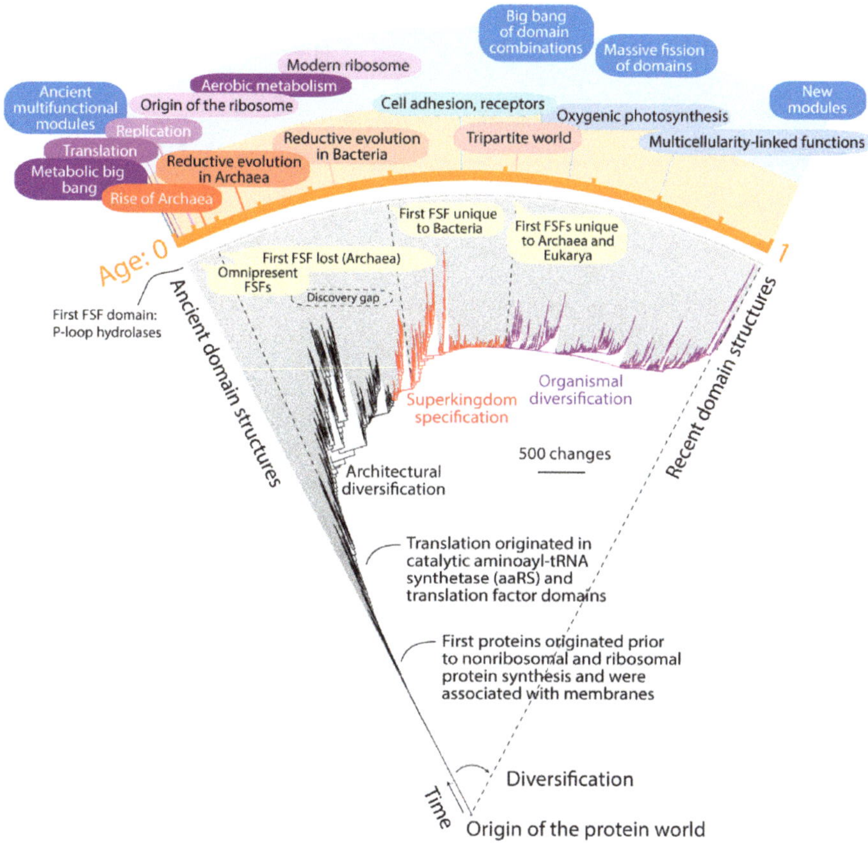

Figure 2. Phylogenomic analyses of protein structural domains reveal an evolutionary timeline of accumulating biochemical innovations. A *tree of domains* (ToD) describes the evolution of 1,453 FSF domains. The tree was reconstructed from a census of domain abundance in proteomes using SCOP 1.67 definitions (Wang *et al*., 2006). Tree branches are colored according to the three evolutionary epochs of the protein world (Wang *et al*., 2007). An indexed protein timeline in the top provides de relative age of individual domain structures (*nd*) in a scale from 0 (the most ancient) to 1 (the present). The figure was modified from Caetano-Anollés and Wang (2008).

of domains (ToDs), which described the structural diversification of the protein world and allowed to generate timelines or chronologies of domains structures. The work that ensued was initially reviewed (Caetano-Anollés *et al*., 2008, 2009) and is still developing. Figure 2

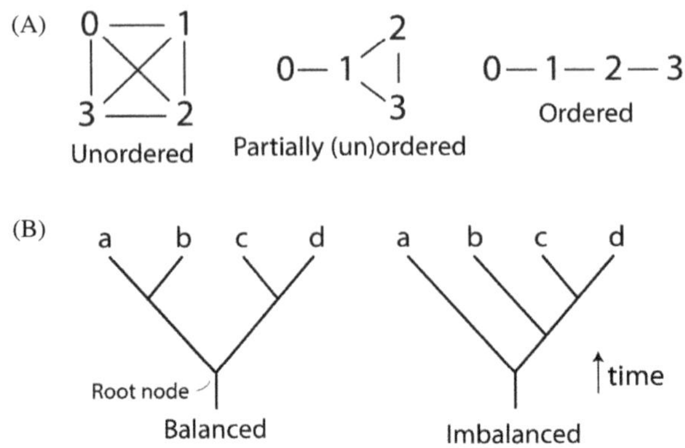

Figure 3. Evolutionary models in parsimony analysis. (A) Character state graphs (CSGs) for equally weighted undirected characters with 4-character states (0, 1, 2, and 3). The CSG in the left is a typical maximally connected character, an "unordered" character while the CSG in the right is a minimally connected character embodied in a "fully ordered" character. The CSG in the middle is a partially ordered CSG containing a reticulation. To illustrate with an example, nucleotides A, G, C and T can be considered states of an unordered character describing mutational substitutions in a sequence site of DNA. (B) Rooted 4-taxon tree statements with leaves labeled a, b, c and d showing different levels of balance. Tree balance usually refers to the topological structure of a tree and describes a frequency distribution of taxa in clades. The two trees have 3 clades. However, the more balanced tree in the left has 2, 2 and 4 taxa per clade, while the tree in the right has 2,3 and 4 taxa per clade.

shows a ToD built from SCOP FSFs indexed with major biochemical and functional innovations and other information. Note that a molecular clock of domain fold structures calibrated with biomarkers and geomarkers defines the relative age of structural domains in Gy and within the context of the geological record (Wang *et al.*, 2011). This places the origin and evolution of modern biochemistry within a framework of planetary history, and for example allows to assign geological ages to stages of molecular accretion (such as in Figure 1B). The taxa of rooted ToDs are structural domains defined at some level of abstraction (e.g. SCOP folds, FSFs or FFs or CATH architectures, topologies or homologous superfamilies) while phylogenetic characters represent corresponding domain occurrences or abundances in the proteomes that are being sampled. We now discuss how ToDs are evolutionary statements that contain crucial information about the accretion processes of the protein world.

3. Genomic abundance and the shape of phylogenetic trees

In our initial study, we recognized that increasing abundance of structural domains in genomes is costly (Caetano-Anollés and Caetano-Anollés, 2003). Each and every new domain variant of a same fold structure that arises for example from the duplication, subfunctionalization and neofunctionalization of genes takes millions of years to evolve. Once accumulated, these fold structures are almost impossible to lose in evolution as they spread in lineages and cellular networks and are canalized by a multiplicity of inter- and intramolecular interactions. Thus, the probability of losing a fold structure as time progresses and the structure spreads in lineages by vertical (e.g., duplication and divergence) or horizontal (e.g., recruitment) mechanisms becomes vanishingly low. This property of abundance extrapolates to an extreme when measuring abundance as an occurrence. Since domain occurrence represents a coarse-grained view of abundance of a genomic feature, the presence-absence bias of the fold structure has the consequence of decreasing even further the probability of loss.

Phylogenetic characters are useful biological features. They carry history when they spread in evolution. We have shown that the occurrence and abundance of structural domains are excellent characters because the structures of domains are highly conserved (Caetano-Anollés and Nasir, 2012). This evolutionary conservation ensures that ancestral structural features are stably retained in descendants as the features spread in the nested patterns of the tree of life. In other words, structures behave as heritable traits. In the case of the genomic abundance of structural domains, characters describing their number in a genome are particularly useful. They constitute "serial homologies" that transform from one particular character state to another when their numbers increase or decrease in a proteome. Note that domain gain and loss establish proteome and domain history when changes in domain abundance (character state transformations) are reconstructed by optimization along the branches of trees and when independent origins (homoplasy) are minimized (Nasir *et al.*, 2014).

The phylogenetic analysis of data requires postulating a *character state transformation model* (also known as a "transformation series") capable of describing character change along the branches of phylogenetic trees. There are several difficulties related to the evolutionary model that must be considered:

(i) *Character state transformations.* The way that character states transform into each other must be defined. This is generally done with character state graphs (CSGs). CSGs connect *n* vertices representing character states with edges representing allowed state transformations (Figure 3A). Many transformation arrangements are possible in CSGs. At one end of the spectrum, maximally connected CSGs are "complete graphs" that allow all possible transformations between states. Characters of this kind were first proposed by Kluge and Farris (1969) but their phylogenetic optimization in trees was elaborated by Fitch (1971). They are commonly known as *unordered characters*, or Fitch characters, and are very popular when change involves stochastic processes such as change in molecular sequences of proteins and nucleic acids (e.g., nucleotides A, G, C and T transforming to any other nucleotide in a DNA sequence). At the other end of the spectrum, minimally connected CSGs are non-reticulated graphs with *n*–1 edge connections. The simplest of these CSGs are linear CSGs with no vertices having three or more connecting edges. They can be represented with alphanumeric strings that are amenable for simple computation. Linearly connected characters were originally formalized by Kluge and Farris (1969) and optimized by Farris (1970). They are known as linearly *ordered characters* or Wagner characters. They are widely used with quantitative characters that have states describing gradients or clines (Grand *et al.*, 2013). One example of these serial homologies is the genomic abundance counts of domains. Thus, an appropriate CSG must be chosen that is compatible with the phylogenetic data.

(ii) *Transformation costs.* Character state models must also define the individual costs of transformations between characters. Costs for all possible transformations are usually measured in steps and are given in square matrices. In these stepmatrices, transformations requiring many steps are costlier from a parsimony point of view and are therefore less probable from an evolutionary point of view. Thus, cost and probability are inversely related. When characters are *undirected*, the number of steps between any two-character states in one direction must be the same as those in the opposite direction. The stepmatrices of these characters are symmetric and their phylogenetic optimization produces unrooted phylogenetic trees. For example, change between any two states of a linearly ordered Wagner character a cost equal to the number of edges separating the states, regardless of the direction of change. In sharp contrast, when characters are *directed*, the number of steps in one or the other direction can be different and the resulting stepmatrices become asymmetric. This imposes a directionality (a "polarization") to character transformation, which defines a flow of change in the direction of time and roots the optimized trees. Nelson (1973) clarified the difficulties of direct methods of character polarity, including the fatal flaw of *a priori* rooting methods (including the use of outgroups) of leading to infinite regress. In addition, models that use asymmetric stepmatrices violate the "triangle inequality," a fundamental property of distances that makes them metric. In Euclidean space, the length of one side of a triangle is always less than the sum of the lengths of the other two sides. Violation of this triangle inequality makes the shortest distance between two points not a straight line. Thus, *"non-metric distances can have unforeseen and sometimes bizarre effects"* (Wheeler, 2012), such as creating a "wormhole" in space-time that allows to reach all locations at zero cost. During tree searches the use of non-metric distances can even produce branches with negative distances. Thus, any phylogenetic methodology that proposes a new distance must evaluate the consequences of a triangle inequality violation.

(iii) *Direction of change.* Evolution is time-dependent and is therefore a "directed" process. Thus, establishing the direction of character transformation along branches of a CSG and tree is a necessary step that must be defined at some point in the course of phylogenetic analysis. Operationally, this "polarization" can be done at two different stages of phylogenetic analysis. *A priori* polarization is made by establishing direction of character change in the CSG before searching for optimal rooted trees. *A posteriori* polarization is done once optimal unrooted trees have been identified. In other words, *a priori* polarization defines assumptions of polarity prior to tree optimization while *a posteriori* polarization defines assumptions once trees have been already recovered. Both approaches will produce the same outcome in the absence of character conflict in phylogenetic data, i.e. in the absence of *ad hoc* hypotheses of independent origins (homoplasy). However, data is never conflict-free and the purpose of phylogenetic analysis is to resolve it (Farris, 1983). For that reason, Meacham (1984) suggested that characters be treated first as undirected and that the resulting unrooted trees be rooted with exploratory polarization methods and explicit parsimony arguments. This has been the popular view of modern phylogenetic analysis and solves the problem of infinite regress of the direct methods.

(iv) *Tree shapes.* The topology and shape of a rooted binary tree determines how change distributes in its branches and tells a lot about the evolutionary processes of diversification that the tree describes, as well as the potential of rooting a tree given an evolutionary model (Steel, 2012). One particularly important property that describes the shape of a tree is its imbalance. The property can be viewed as a distribution of taxa in subgroups of taxa that have a common origin. These subgroups are known as clades. At one end of the spectrum rooted trees can be perfectly balanced when taxa are evenly distributed in clades (Figure 3B). At the other end, trees can be completely imbalanced when the distribution of taxa in clades is highly biased. The comb-like (pectinate) appearance of imbalanced trees can be maximal when splits start to occur recurrently close to the crown. While trees of life (including ToPs) have been shown to depart from a strictly balanced configuration (e.g. Blum and Francois, 2006; Jones, 2011), they are far away from completely imbalanced configurations. In fact, two kinds of models of speciation-extinction, which have been used as priors in popular Bayesian analysis implementations, bracket the balance of empirical trees (Jones, 2011). At one end the *constant rate birth-death model* (CRDB) tends to generate trees with increased degrees of balance while at the other end the *proportional to distinguishable arrangement* (PDA) model tends towards imbalance. A number of other models of speciation-extinction that model topology, diversification, and age-dependencies lie in between. In contrast, ToDs are unique in that they are highly imbalanced (e.g. Wang *et al.*, 2011). The timing of domain discovery is largely defined by molecular speciation and not by changes in domain abundance, i.e. by the

> shape of trees and not the length of branches. Thus, the shape of trees can be explained as the consequence of the accumulation of an evolving heritable trait (Heard, 1996). Global dynamical models for the evolution of domains indeed suggest that structural explorations of fold space occur through coarse-grained discoveries that undergo fine-grained elaboration (Tal *et al*., 2016).

We have adopted the standard and well-tested methodology of using undirected characters of the ordered type and Wagner optimization to generate both unrooted ToDs and ToPs, which are then rooted *a posteriori* with Weston's generality criterion (Weston 1998). This rooting methodology uses the Lundberg (1972) implementation to reorient and root a tree by pulling down a branch that yields a minimum increase in tree length and maximizing the nesting of character states (serial homologies), from the most general at its base to the most lineage-specific at its crown. The "standard" implementation of Lundberg complies with Weston's rule (Bryant, 1997) by setting all character states of an ancestor as "missing" (ancstate=?) and proceeding to optimize attachment of all possible ancestors to optimal trees. Alternatively, arbitrarily defined ancestors enabling compatible Lundberg polarization schemes can be optimally attached to the most parsimonious tree reconstructions. These implementations can be compared to determine if some schemes are more parsimonious than others and are less affected by homoplasy. We have observed for more than a decade that ToPs obtained using the "missing" and "all-0" ancestors were topologically isomorphic and optimal in terms of tree length and character-to-tree fits such as the retention index (RI). Figure 4A plots ensemble RI against the length of the most parsimonious ToPs when these were rooted with ancestors holding the same ancestral state of 32 possible states in the ordered series (labeled in alphanumeric format from 0 to 9 and A to V) for every character. The "standard" and "all-0" ancestor

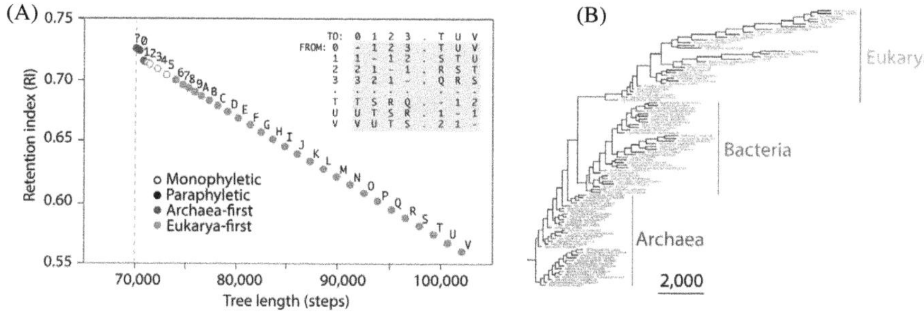

Figure 4. Rooting *trees of proteomes* (ToPs) with the Lundberg method. Trees of life describing the origin and evolution of proteomes were generated using ordered characters from a genomic census of 1,420 fold superfamilies (FSFs) of protein structural domains in 102 proteomes using the dataset of Kim *et al.* (2014). Proteomes were from organisms with free-living lifestyles belonging to superkingdoms Archaea, Bacteria and Eukarya. (A) The relationship between the retention index (RI) and the length of most parsimonious ToPs rooted *a posteriori* using Lundberg and alternative ancestors. An RI value of 1 implies perfect fit and absence of instances of independent origin (homoplasy). An RI value of 0 implies the tree fits data as poorly as possible. Topologies with paraphyletic or monophyletic basal superkingdoms were labeled with open and close circles, respectively, and colors describing support for an Archaea-first or Eukarya-first evolutionary scenario of origin. The inset shows the stepmatrix describing ordered character state transformations. RI scaled monotonically with length with a perfect fit ($y = -5.10{-}6x + 1.07$; $R^2 = 1.000$; $P < 0.001$). A control using unordered maximally connected characters also showed that the Lundberg standard "missing" and "all-0" ancestors were the most parsimonious and produced trees rooted paraphyletically in Archaea, there was no monotonic decrease pattern. (B) The single most parsimonious rooted ToP obtained with the all "missing" ancestor (ancstates = ?) has a length of 69,628 steps and an RI value of 0.725.

implementations of Lundberg produced the most parsimonious and conflict-free ToPs with identical topologies that supported an Archaea-first scenario for the origin of diversified life (Figure 4B). One remarkable finding is that both RI and tree length decrease monotonically when character states of Lundberg ancestors were defined with increasing values (from ancstate = 0 to ancstate = V). This regular decrease is not trivial. Instead, it strongly supports indications that proteome abundance data significantly fits the model of ordered characters and evolutionary growth in the nested lineages of the ToP. This pattern suggests that the nesting of serial homologies is equally affected by increasing suboptimality of the rooting variants. Thus, the rooting scheme of ToPs complies with Weston's rule and the generality criterion. The optimal character polarization with "all-0" ancestors also shows there is a tendency of growth of structural domains in proteomes, and not global tendencies of reductive evolution. This is even made evident by reconstructing character state change in ToPs, deriving frequencies of character change, and plotting them in "bubble" diagrams (Figure 5). While changes occurred most frequently in single steps, gains were favored over losses throughout the stepmatrix of character state transformations. This again reveals a clear pattern of growth.

We found similar patterns of growth in ToDs. Lundberg implementations with "missing" and "all-*X*" ancestors made rooted ToDs in all of our phylogenomic reconstructions also isomorphic and optimal (*X* represents the maximum valued character state). As with ToPs, they also complied with Weston's rule. ToDs are highly imbalanced trees (see Figure 2). This property enables to build chronologies describing the evolutionary appearance of structural domains directly from the trees by calculating a "node distance" (*nd*), the relative number of internal nodes from the root to a leaf of the tree (Figure 2). Since much of protein history has occurred in cells, even prior to episodes of cellular diversification, chronologies capture the history of protein domains in proteomes from the origin of modern proteins ($nd = 0$) to the present ($nd = 1$). Effectively, ToDs describe the growth of the protein world. They resemble archeological digs supporting chronologies describing the settlement and growth of cities that are illustrated in Figure 1B. We have studied the chronology of structural domains for over a decade and half. Considerable background knowledge confirms its realism.

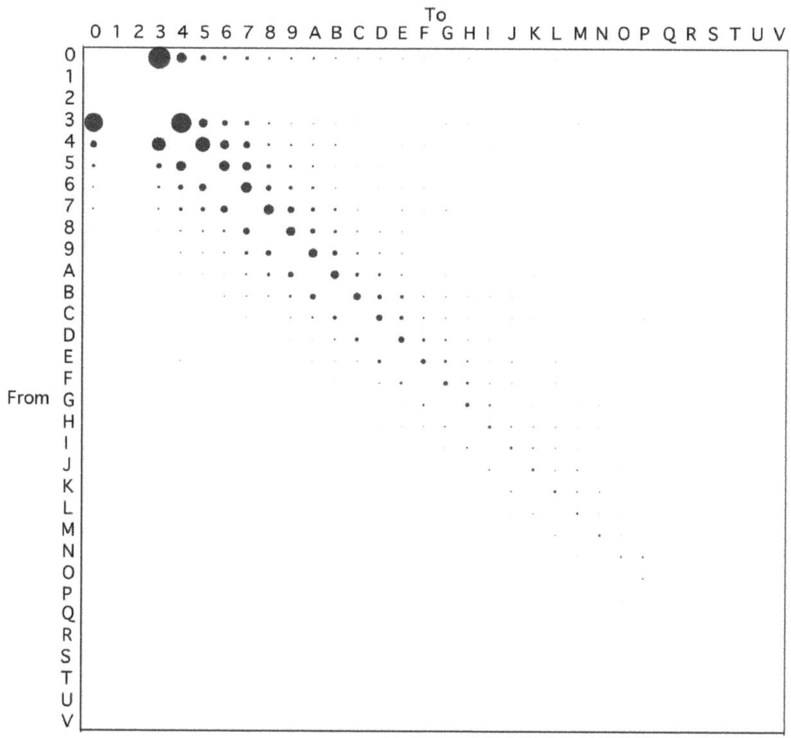

Figure 5. Deriving frequencies of character state changes from character state reconstructions. Phylogenetic trees were reconstructed treating coded domain abundance values as undirected ordered (Wagner) characters and producing the rooted ToP described in Figure 4B. Bubble charts that describe frequencies of changes between character states were derived from character state reconstruction of changes along the branches of the optimal tree.

Major functional breakthroughs appeared at expected timeframes, including the inception of molecules mediating cell adhesion, multicellularity and oxygenic photosynthesis (summarized with speech bubbles in Figure 2).

4. Macromolecular growth and the Heaps law

The interface between evolutionary model, data and tree shape add support to a long list of evidence backing the evolutionary principle of spatiotemporal continuity. This principle manifests in evolutionary growth (Nasir *et al.*, 2017). We now explore evidence supporting macromolecular accretion as an evolutionary process of innovation that tailors the generation of biological modules and complexity.

We have argued that there is a meaningful language in the structure of proteins (Caetano-Anollés *et al.*, 2017). This language follows quantitative linguistic patterns that can be described for example with statistical laws, including probability distributions and functional and developmental type laws (Altmann and Gerlach, 2016). These distributions and laws appear universal. For example, the Zipf law of language is a special case of the scale free distribution, which is a pervasive property of many biological networks, including metabolic networks and networks of protein–protein interactions. Scale-free networks follow power-law distributions $P(k) \sim k^{-\gamma}$, with k representing the number of links of the network and $P(k)$ the probability that the network has exactly k links. The power-law behavior of such scale-free distributions can be explained by a *"rich get richer"* model of network growth in which new links "attach preferentially" to nodes harboring many connections. For example, the Zipf law is followed by protein structural domains at fold and FSF levels (Qian *et al.*, 2001, Caetano-Anollés and Caetano-Anollés, 2003), with γ power law decay values of ~2 for Bacteria and Archaea and ~1.4 for Eukarya (Caetano-Anollés and Caetano-Anollés, 2003). These decay values match values for the English and Chinese languages, respectively (Li *et al.*, 2016). We have also shown that protein structural domains follow developmental type laws of language (Nasir *et al.*, 2017). These laws describe the growth and evolution of vocabularies through correlations of language properties with either time or accumulating innovation. The Heaps law for example describes how the number of different words V in a vocabulary scales with the database of words

N measured in number of words (the length of text) (Heaps, 1978). The law follows Equation (1),

$$V(n) = KN^{\beta} \tag{1}$$

with K and β being empirically determined parameters. Note that the Heaps law and Zipf distribution are related through $\beta = 1/\gamma$, a relationship that has been empirically confirmed under asymptotic conditions (Gerlach and Altmann, 2013). For human languages, "kernel" words that appear with high frequency in vocabularies are constant over centuries and their growth can be described in terms of database size (Petersen *et al.*, 2012; Gerlach and Altmann, 2013). Thus, a dynamic view of vocabulary innovation can be described using a static view. The signature of the law is sublinear growth ($\beta < 1$), a property that is typical of *"economies of scale"* that show increasingly weaker returns for new vocabulary innovations. The deviations in tail distributions linked to the Heap law regression can be studied. For example, a study of a large body of English text involved offsetting these tail distribution deviations (Ferrer i Cancho and Solé, 2001). The exercise showed that the probability density function had two scaling regimes. The steepest regime followed a Zipf law characterizing a "kernel" lexicon of frequently used words. The other regime characterized an "unlimited" lexicon of growing words of less frequent use. Note that the two-regime Zipf distribution translates into a two-regime Heaps law with β exponents close to 1 for the kernel and 0.4–0.7 for the unlimited lexicon of a number of Indo-European languages, with exponent variation reflecting differences in language organization. These different regimes describe the decreasing marginal need for new words and a slowdown (cooling) of linguistic evolution (Petersen *et al.*, 2012). Recent studies of languages with limited dictionary sizes such as Chinese, Japanese, and Korean (Petersen *et al.*, 2012; Lü *et al.*, 2013) have shown multi-regime Heaps laws. A recent study shows Chinese text follows a 3-regime Heaps law with β scaling exponents of 1, 0.7, and 0.3 for increasing text lengths, which is explained by a stochastic feedback model of vocabulary growth driven by two probabilities, one for the reuse of frequently used words and the other for the rise of word novelties (Li *et al.*, 2016).

When considering the vocabulary of genomes there are a number of scaling relationships, some of which reflect the relationship between the number of genes in different functional categories and genome size (Molina and van Nimwegen, 2009). The scaling relationship for many of these functional categories appears universal. Linguistic theory has been used to explain the modular make up of proteins (Gimona, 2006). For example, the combination of atomic units of language (morphemes) to form higher-level units such as words and phrases (lexemes) resemble the combination of structural domains to form multidomain proteins. Recently, we have shown that the use and reuse of structural domains defined at FSF level complies with the Heaps law (Nasir *et al.*, 2017). FSF use and FSF reuse were used as proxies for the number of words and the database of words that the Heaps law describes for protein domain structures. Remarkably, log–log plots of FSF use versus FSF reuse showed four distinct power law patterns (Figure 6). These patterns were suggestive of four regimes of slowdown of domain vocabulary growth, each corresponding to the proteomes of viruses, Archaea, Bacteria and Eukarya, in that order. The vocabulary of total FSF domains scales with corresponding proteomic datasets with decreasing β. Viral proteomes use a very ancient kernel-like vocabulary with β exponents of 0.81 approaching unity but not far from the second regime of languages with limited vocabularies ($\beta = 0.7$–0.77, Petersen *et al.*, 2012; Lü *et al.*, 2013). This ancestral kernel is then expanded successively by growing vocabularies with slowdowns in the proteomes of Archaea and Bacteria and to an extreme in the proteomes of Eukarya, as these gradually appeared in evolution. The values of β for the proteomes of Archaea ($\beta = 0.36$–0.40) are not far away from those of English text corpora ($\beta = 0.4$–0.7), such as

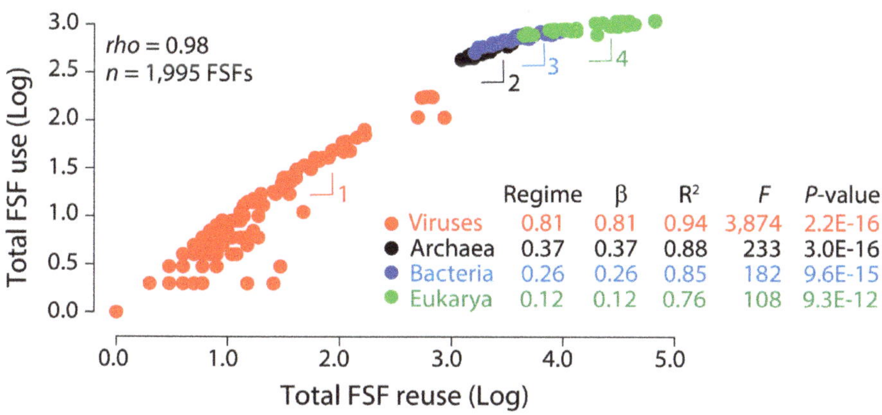

Figure 6. A four-regime Heaps law of vocabulary growth for protein structural domains. The plot describes strong correlations of use (occurrence) and reuse (abundance) of total fold superfamilies (FSFs) sets for the proteomes of viruses (266), Archaea (34), Bacteria (34) and Eukarya (34), which are colored red, black, blue and green, respectively. Scaling exponents summarizing the Heaps law $V \sim N^{\beta}$ for the four distinct regimes that correspond to viruses and the cellular superkingdoms are describe in table format together with their fits to linear regression models. Data from Nasir *et al.* (2017).

the Gutenberg Project e-book collection ($\beta = 0.45$, Tria *et al.*, 2014). The values of β for the proteomes of Bacteria ($\beta = 0.19$–0.26) match those of the third regime of Chinese language (Petersen *et al.*, 2012; Li *et al.*, 2016).

Both vocabulary and database size of the Heaps law of language are expected to be proportional to time. We recently unfolded this temporal connection for the Heaps law of proteome growth (Figure 6) by linking proteome size and information from phylogenomic ToDs (Nasir *et al.*, 2017) with the help of a new multidimensional scaling method we call *evolutionary principal coordinate* (evoPCO) analysis (Nasir and Caetano-Anollés, 2015). evoPCO analysis combines the power of cladistics and phenetics to produce a multidimensional view of evolutionary relationships among a set of proteomes. The strategy, which is described in Figure 7A, uses a simple principle:

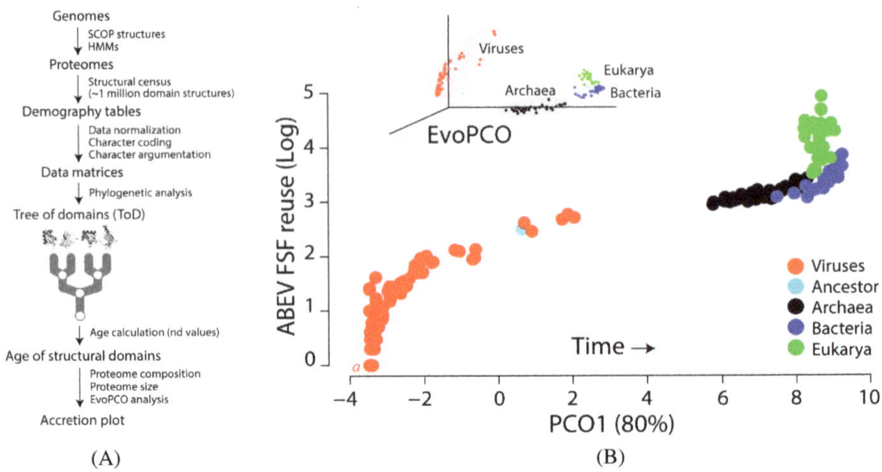

Figure 7. Revealing global evolutionary tendencies of growth in proteomes. (A) Flow diagram describing the study of the growth of structural domains in proteomes. The mining of genomic information with hidden Markov models (HMMs) of structural recognition produces a census of protein structural domains defined at SCOP FSF level. This census is used to reconstruct a phylogenomic tree of domains (ToD) with which to assign ages to known domain structures. These ages are then used to build a multidimensional view of the evolutionary relationships of proteomes. An evolutionary principal coordinate (evoPCO) analysis plot shows in its first three axes (85% variability explained) the evolutionary distances between cellular and viral proteomes. (B) Accretion plot describing proteome growth. The most important evoPCO component was plotted against the reuse of FSFs that are universally present in the four supergroups of life (ABEV) in logarithm scale. The reconstructed proteome of the last common ancestor of modern cells (Kim and Caetano-Anollés, 2011) was added as reference to confirm the direction of evolutionary change, which proceeds from left to right in the plot. Data from Nasir *et al.* (2017). The plot also supports results from many phylogenomic studies that show the early rise of viruses, followed by Archaea and a group of Bacteria and Eukarya, in that order. This matches evolutionary patterns of the rooted ToP (Figure 4B). It also indicates that the topology of the ToP is not due to an artifact induced by genome size because the evoPCO plot relies exclusively on the individual ages of the universal FSFs of proteome, which are derived from a ToD, a phylogenomic tree that describes the evolution of individual structural domains. Data from Nasir *et al.* (2017).

the overall age of a proteome is determined by the ages of its individual component parts, the structural domains. First, the ages of FSF domains are obtained directly from a ToD by calculating corresponding *nd* values. Second, the collective of all *nd* ages of a proteome are calculated for a set of proteomes. Third, an FSF occurrence (FSF use) matrix summarizing abundance information in the proteomes is then transformed into an FSF occurrence*(1–*nd*) matrix that summarizes evolutionary information. This makes it possible to study a multidimensional space of "reverse" evolutionary ages of domains without losing information of FSFs of very ancient origin or introducing biases from FSF absences. Fourth, the Euclidean distances describing dissimilarities between proteomes are summarized in a distance matrix, which is then used to calculate the first three principal coordinates describing maximum variability in the data. These three most significant loadings describe how FSF domain parts contribute to proteome history. The inset of Figure 7B shows an evoPCO analysis plot with its first three major axes, which explain 85% variability in evolutionary distances between 368 cellular and viral proteomes. The plot dissects four distinct temporal clouds of proteomes for viruses, Archaea, Bacteria and Eukarya. Figure 7B plots the most important evoPCO loading (responsible for 80% of variation) against the reuse of FSFs that are universally present in the proteomes of the four supergroups of life (the ABEV FSFs). We found a gradual increase of the genome size proxy as one travels in time through the main axis of the temporal cloud visiting the proteomes of viruses, Archaea, Bacteria and Eukarya, in that order. This confirms that the global tendencies of proteome growth arise from the evolutionary accretion of novel structural domains in the protein world. The result is significant because of two main advantages of the evoPCO method: (i) there is no genome size-related variable in the data matrix as FSF abundances are replaced by their relative reverse ages, and (ii) the method ensures that the fundamental assumption of character independence in phylogenetic tree reconstruction remains intact (Huelsenbeck and Nielsen, 1999; Nasir and Caetano-Anollés, 2015). Proteome growth tendencies are in line with the prevalence of domain gains over domain losses derived from character state reconstructions along the branches of a ToP that describes proteome evolution (Nasir *et al.*, 2014), as we will soon discuss.

5. Modeling processes of macromolecular growth and innovation

The processes of growth and innovation are associated and determine the behavior of a dynamic system. This behavior often takes the form of a series of logistic wavelets ("loglets"), S-shaped logistic components of growth and diffusion in systems that model time-series data (Meyer *et al.*, 1999). The centrality of logistic behavior has been long recognized, from linguistics to biology. The Piotrowski law for example is a typical developmental law of historical linguistics. It describes a process of accumulation of change, which starts with an exponential form and is followed by an S-shaped (sigmoid curve) form. This results in rates of accumulation first increasing and then decreasing. When developmental processes continue for very long time, the accumulation can sometimes end with a reverse S-shape. The elaboration of Beöthy and Altmann (1982) of the Piotrowski law, with assumptions borrowed from Weinreich *et al.* (1968), arrives at a logistic expression that describes the growth of the proportion p of new forms developed in time. This is described in Equation (2),

$$p = \frac{1}{1 + ae^{-bt}} \quad (2)$$

where a in an integration constant and b a constant parameter. Initially, the logistic expression has been used to explain aggregate growth of biological systems unfolding in time when these systems are non-linearly constrained by environments. The logistic model has been used to describe the growth of populations in biology ever since Verhulst (1838) modeled population dynamics in a constrained environment with Equation (3),

$$\frac{dN(t)}{dt} rN(t) - \alpha N(t)^2 \tag{3}$$

where $N(t)$ represents the number of individuals of a population at time t, r is the intrinsic growth rate, and α is a competition parameter that describes the density of individuals. Population equilibrium is established when the environmental carrying capacity of the population K becomes r/α. The ideas of cumulative growth of a biological system were later taken forward by Lotka–Volterra equations that model competition in predator-prey relationships.

Sigmoid curves of growth have been used to describe processes of innovation. The simple symmetrical logistic model has been complemented with non-symmetrical or skewed growth models such as the Gompertz or generalized logistic functions (Meyer *et al.*, 1999). Series of sequential logistic S-curves have been also modeled and subjected to Fisher-Pry decompositions, i.e. normalization of the logistic curves to straight lines. For example, double S-curves are typical of paths of high performance in diffusion of innovation models (Rogers, 1962). These bi-logistic curves (wavelets), illustrated in Figure 8, often overlap in time with different magnitude and with different speeds generating sequential patterns of wavelet superposition, convergence and divergence (Meyer *et al.*, 1999). In particular, the innovation diffusion model is inspired by the spread of cultural innovations from one society to another in socio-economic systems. Innovation is communicated among entities as time progresses, some of which are prone to adopt the innovation (Rogers, 1962). The innovation lifecycle involves a 5-step decision-making process that involves "innovators," "early adopters" (trendsetters), an "early

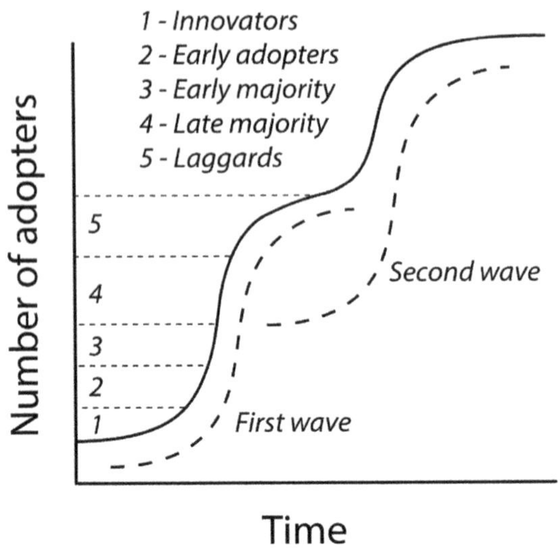

Figure 8. The discovery and growth of innovations described as a series of wavelets. The double S-curve is typical of paths of high performance in diffusion of innovation models. These bi-logistic curves (wavelets) often overlap in time with different magnitude and with different speeds generating sequential patterns of superposition, convergence and divergence (Meyer *et al.*, 1999).

majority," a "late majority," and "laggards," in that order (Figure 8). One of the most general and popular S-curve models of the "communication" process of adoption of innovations is the Bass method (Bass, 1969). The method considers a homogeneous population with a fraction of innovation adopters (ρ) increasing when they meet non-adopters ($1-\rho$) under intrinsic forces with rate q (coefficient of *imitation*) and under external forces with rate p (coefficient of *innovation*). The adoption process follows Equation (5),

$$\frac{d\rho(t)}{dt} = (p + q\rho(t))(1 - \rho(\mathrm{t})) \tag{4}$$

and its derivation leads to the differential Equation (6),

$$d\rho(t) = p + (q - p)\rho(t) - q[\rho(t)]^2 \tag{5}$$

which corresponds to a Ricatti form with constant coefficient, a first order ordinary differential equation that is quadratic in the unknown function. Guseo (2016) has shown that the Bass diffusion model becomes a logistic distribution when $p = 0$, i.e. when only imitation by non-adopters governs the diffusion process. In other words, the logistic and Bass models become isomorphic (but not identical). Conversely, the model reduces to the exponential distribution when $q = 0$, i.e. when the emergence of innovation drives change.

Under the Bass model, S-curves of vocabulary growth and replacement can be modeled to show differential effects of intrinsic imitation and exogenous innovation factors on the replacement process (Ghanbarnejad *et al.*, 2014). The Bass model can also be used to explain successive generations (waves) of innovation (Norton and Bass, 1987). These wavelets of innovations can compete with each other sometimes leading to substitutions of innovations (logistic substitution). Note however that when innovations occur in parts of more complex wholes their substitution enhances the entire system, which successfully continues adoption. This generally results in double S-curves or a multiplicity of them. The examples studied by Norton and Bass (1987) involve sequential innovations that result in technological substitutions in industries that undergo rapid change, such as that of the semiconductor industry. In their landmark study, estimation, fitting and prediction of data to demand-growth-decline models involved two basic types of integrated circuits for computers: static and dynamic random-access memory (SRAM and DRAM) and microprocessor and microcontroller units (MPU and MCU). Successive S-curves for the use of integrated circuits in computers explained the rise of four generation of DRAMs (4K, 16K, 64K and 256K), three generations of SRAMs, and two generations of logic circuits (MPUs and MCUs). The patterns and forecasts for DRAMs have been confirmed and extended a decade and a half later (Victor and Ausubel, 2002). These logistic substitution models are in line with evolution being the result of a sequence of "new-for-old" replacements or substitutions in social, industrial or biological systems (Montroll, 1978). However, evolution of life is much more complex and is not only driven by recruitments or other replacement strategies.

Innovation diffusion models apply to empirical molecular biology data and can help understand protein evolution. For example, innovations in the structural domains of proteomes accumulate as logistic wavelets. A phylogenomic analysis of the evolutionary accumulation of structural domains defined at SCOP FSF level corresponding to major categories of molecular functions showed complex logistic behavior (Figure 9A). FSFs were classified into seven general functional categories and 50 subcategories based on SCOP, Interpro, Swiss-Prot and literature sources (Vogel 2005) and their accumulation expressed as a cumulative frequency distribution (Caetano-Anollés *et al.*, 2011). FSF accumulation in all seven major categories produced S-like growth curves suggesting the existence of diffusion of innovation mechanisms. The growth of FSFs typical of early metabolism in the "Metabolism" and "General" categories, which includes the "Small molecule binding" subcategory, occurred before all others (Figure 9A). These initial FSFs were followed closely by FSFs in "Information," "Regulation," and "Intracellular processes." Molecular functions that are more evolutionarily derived grew later in evolution. This included functions typical of multicellular organisms ("Extracellular processes," with cell adhesion, immune response, toxins and defense, and blood clotting functions) and viruses ("Viral proteins" in category "Other"). Note that the more ancient categories showed clear double S-curves that are typical of paths of high performance in diffusion of innovation models (Rogers, 1962).

We also reconstructed ancestral character state vectors in the branches of ToDs and ToPs at FF structural domain level (Nasir

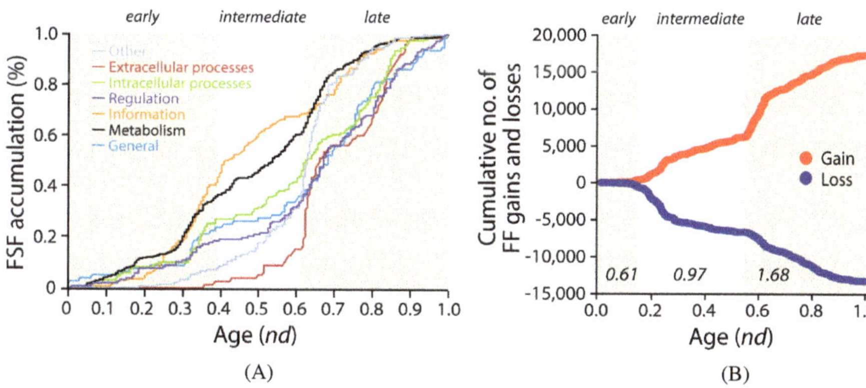

Figure 9. The growth of structural domains in proteomes. (A) S-shaped accumulation of SCOP FSFs associated with Vogel's seven broad categories of molecular functions along a chronology) derived from a ToD reconstructed from 184 proteomes representing all three superkingdoms of life. The age of domains is given as node distance (*nd*) values, in a scale from 0 (origin of proteins) to 1 (the present). Data from Caetano-Anollés *et al.* (2011). (B) Cumulative numbers of gains and losses in the occurrence of SCOP FFs along the branches of the tree of life. Scatter plots reveal growth trends in the accumulation of FF gains and losses in the proteomes of 420 organisms belonging to Archaea (48), Bacteria (239) and Eukarya (133). Gains are identified in red while losses are identified in blue. The three evolutionary epochs of the protein world (early, intermediate, late) are marked with corresponding average gain-to-loss ratios in italics. A similar pattern is obtained when studying abundance of FFs. Data from Nasir *et al.* (2014).

et al., 2014). This allowed to calculate the number of gains and losses of FFs for every time step of the chronology of domains. Figure 9B shows chains of continued adoptions and substitutions of innovations that are occurring in the branches of a tree of life describing the evolution of the proteomes of cellular organisms. Scatter plots reveal clear double S-curves of accumulation of both FF gains and losses. Gains however override losses at increasing ratios of 0.61, 0.97 and 1.68 for the early (architectural diversification), intermediate (superkingdom specification) and late (organismal diversification) epochs of domain evolution, respectively, which are defined for example in Caetano-Anollés *et al.* (2011). The first S-curve covers exactly the early and intermediate epochs, while the beginning of the second wave starts at the boundary of the intermediate and late epochs. Note that the boundary is defined by a burst of structural domains that establish domain combinations and are specific to lineages in a diversifying world of proteomes (Wang and Caetano-Anollés, 2009). The boundary is quickly followed in evolutionary time by a massive combinatorial "big bang" in domain organization. It is clear that the first wave of innovation and its associated S-curve involves the discovery of new domains while the second wave involves domains that have a capacity to combine with others to enhance the functional repertoire of the cell. Thus, two innovation modes appear in time as protein structural domains adopt new fold structures and molecular functions. The first mode involves domain make up while the second involves domain organization in proteins.

6. Innovation and the diffusion of mutations in sequence and structure space

Natural selection explains the evolutionary culling of phenotypes arising from random processes of mutation that are responsible for genetic change. Its initial consequence is to facilitate the persistence of a biological system given a changing environment. However, both natural selection and mutational change are not sufficient to explain the rise of innovations at structural, functional or semiotic levels. Natural selection acts on innovations but does not explain their origin. Similarly, genetic change is stochastic and does not necessarily result or explain innovations. A system's change can often end with less fit phenotypes, for example when these are impacted by changing environments. Since change is constrained by a restricted set of mutations out of an unfathomable world of mutational possibilities, its effects on persistence are only local and the fate of mutations constrained. Add to these limitations of understanding that genetic change and natural selection are separated by multiple levels of biological organization, from string representation of genotypes to fitness-defining persistence strategies, all of which involve vast evolutionary timeframes. Thus, new paradigms are needed to explain innovation and their spread in evolution.

One widely embraced metaphor is Sewall Wright's *fitness landscape.* This landscape portrays a multidimensional space of genotypes, which must be evolutionarily negotiated by a population of organisms with the goal of increasing their fitness (Wright, 1932). A visual and often mythical representation of such a landscape invokes a two-dimensional space of genotypes and a third dimension describing the fitness of the mutating population. Given a certain environment, genotype fitness levels give rise to peaks and valleys that make up smooth or rugged landscapes, each of which impact the likelihood of achieving evolutionary optimality.

While the fitness landscape metaphor is attractive, it is simplistic. Fitness is a relative concept. It depends on the phenotype's level of selection, the definition of the population, and other constraints acting on the fate of mutations. For example, macromolecular innovation involves finding novel structures, functions and interactions, such as novel metabolic or signaling pathways, which must be stable from a thermodynamic or energetic point of view and at the same time functionally versatile and compatible with the biophysical realities of the macromolecules. Biophysically grounded macromolecular modeling and empirical analyses must therefore support the workings and constraints of the process of innovation, including those that use the fitness landscape paradigm (Fontana, 2003; Wagner, 2011; Sikosek and Chan, 2014; Bershtein *et al*., 2016). Three important characteristics are particularly relevant for macromolecular innovation. First, numerous molecular nucleic acid or amino acid sequences (string representations of genotypes) map to a core set of fold structures, i.e. there is a sequence-to-structure map that is highly degenerate. Second, there are few, but many rare structures materializing in an incredibly vast space of possible structures. Third, extensive *neutral networks* percolate sequence space (Lipman and Wilbur, 1991; Schuster *et al.*, 1994). Within these networks, structure is impervious to change at sequence level. This last property, which was anticipated by Maynard Smith (1970), appears to be both random and "shape space" covering, i.e. most structures can materialize within relatively few mutational changes in sequence. In other words, it does not take much effort to access any structure in sequence space despite its vastness — the sequence landscape is percolated. Figure 10A illustrates neutral networks as highways of mutational diffusion in sequence space. The black, white and grey nodes of the networks represent genotypes that map to distinct fold structures, which can be attained by an evolving population of organisms. Suppose mutations cause population individuals to drift in sequence space, starting from the subnetwork of black nodes that folds into one triosephosphate isomerase (TIM) β/α barrel FSF structure (bottom left of Figure 10A). At some point, the population can attain a new FSF structural innovation by diffusing from the black subnetwork into the white subnetwork. Note however that the population cannot directly access the grey subnetwork without considerable mutational drift through the white subnetwork. The black and grey subnetworks are not neighbors. They cannot access each other.

The neutral network paradigm describes mappings between levels of organization. It does not explain the dynamics of how those mapping unfold in evolution. In other words, there is need to explain the innovation process, including the percolation phenomenon in neutral networks. One alternative is to couple modeling with prediction and retrodiction efforts. We have recently modeled the origin and evolution of structural domains defined at fold and FSF levels with "global" birth-death differential equations and proteomic

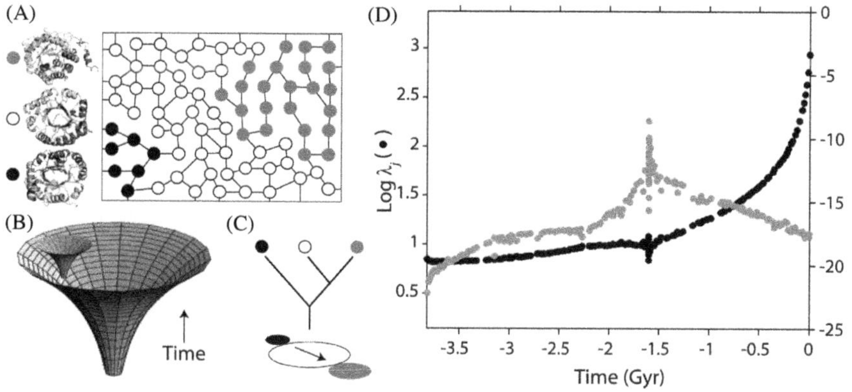

Figure 10. Modeling evolution of protein structural domains. (A) Neutral networks in sequence space. A two-dimensional view of a multidimensional space of domain sequences is described as a network, with colored nodes representing sequences and links representing allowed mutational changes between them. Nodes colored black, gray and white are sequences that fold into a same domain structure defined at some level of a structural hierarchy (here portrayed as variants of a same TIM β/α barrel fold). Each group represents a neutral network. (B) Funnel paradigm of the diversification of a domain structure defined at some level of a hierarchy. The breath of the funnel describes evolutionary increases in abundance. (C) A rooted phylogenomic tree of domains (ToD) explains the rise of domain innovations by a diffusion of mutations through walks (in the direction of the arrow) in the neutral networks (circles) of panel A. (D) Change of parameters specifying rates of generation of new variants of an FSF domain (λ_j) and rates of generation of new FSFs (a_{ij}) along the chronology of structural domains, with time expressed in billions of years (Gyr) ago. λ_j starts at 7 Gyr^{-1}, ranges 7–12 Gyr^{-1} at the transition occurring at 1.5 Gyr, and grows super-exponentially to 867 Gyr^{-1} due to the combinatorial explosion of domains in multidomain proteins. In contrast, a_{ij} increases towards the transition and then decreases, showing typical values of 10^{-17} Gyr^{-1}, which matches the 10^{-11}–10^{-77} fraction of sequence space that the structure occupies. Data from Tal *et al.* (2016).

abundances of domains as state variables (Tal *et al.*, 2016). The evolutionary model, which is global and independent from horizontal exchange between proteomes, is supported by three reasonable assumptions:

(i) *Growth*: Each fold or FSF domain diversifies into accumulating variants, which produce a growing "funnel" (Figure 10B). The funnel's mouth widens as the number of variants of the fold structure grows in time. This funnel metaphor is supported by considerable background knowledge. While structural domains have fold topologies that are highly conserved, the sites of their amino acid sequences can diverge >70% in identity and their structures can have a multiplicity of secondary structure decorations that still keep the structure and its associated functions intact (Caetano-Anollés *et al.*, 2009). These variants populate proteomes and are entrenched throughout the workings of the cell, from metabolism to signaling networks. Thus, accumulation of sequence and structural variants of a fold in an evolving proteome will necessarily increase that fold's abundance.

(ii) *Diffusion*: The chances that one of the structural variants will produce a new fold or FSF increases with the growing number of variants. Thus, a new growing funnel describing the diversification of a new fold or FSF will arise within the original funnel given enough time and enough exploration of sequence space (Figure 10B). This assumption holds good ground. Structures have been shown to be highly conserved in evolution when compared to sequences (Chothia and Lesk, 1986; Illegård *et al.*, 2009; Caetano-Anollés and Nasir, 2012; Lundin *et al.*, 2012). In fact, sequences can sometimes diverge >95% and preserve the fold structure (Rost, 1999). These observations support the presence of neutral networks preserving functions and structures, which facilitate diffusion in the percolated space of sequences. In a landmark bench experiment, in vitro evolution experiments revealed that a walk of 40 mutations was needed to engineer an intersection ribozyme with two catalytic tasks (Schultes and Bartel, 2000). This provided experimental confirmation that diffusion in neutral networks has the "shape space" covering property of sequence space. This facilitates innovation.

(iii) *Access*: Not all structures are easily accessible in sequence space. This is a reasonable assumption that is supported by studies of fitness landscapes, neutral networks and the percolation phenomenon, all of which are mediated by exploratory random walks in sequence space. For example, Ferrada and Wagner (2010) studied ~30,000 proteins with known structures and function and found that different regions of genotype space contained different sets of structures and functions. They found that sequences that diverge >25% in amino acid identity shared less than 30% functions, i.e. most functions were located in different structural neighborhoods. Since most functions were carried by a single structure but few were carried by many structures, large fractions of unique functions in a neighborhood likely resulted from a same function being carried out by different structures. Thus, the percolation of sequence space by neutral networks is limited and most structures and functions are not accessible to diffusion by random walks. This important property, which is illustrated by the neutral networks of Figure 10A, can be described with trees (Figure 10C). When genotypes drift in evolution from left to right in the figure, they must access structural novelties of subnetworks of black, white and grey nodes, in that defined order. The first time a subnetwork boundary is reached, a split occurs in the tree. That event initiates an evolutionary accumulation of structural phenotypes in that new subnetwork and the birth of a new growing funnel. We call this model the "irreversible tree-hugging" model of abundance growth of domain structures (Tal *et al.*, 2016). Thus, only "forward" transitions from ancestral domain structures to neighboring descendants along branches of a ToD are assumed possible, i.e. realized structures are unique to each structural network.

The global dynamic model of protein evolution is governed by Equation (7),

$$\frac{dN_j}{dt} = (\lambda_j - \sum_{i \neq j} a_{ij})N_j + \sum_{i \neq j} a_{ij} N_j \qquad (7)$$

N_j are global genomic abundances of domain structures of type j, λ_j is the rate (birth minus death) of generating new variants of structure j, and a_{ij} is the rate of "forward" transition from structure j to structure i. In the model, each fold or FSF diversifies by producing domain variants in approximately exponential manner with rate λ. At some point, given rate a, the growing funnel of accumulating variants generates a new fold or FSF structure. This new structure is a unique and rare innovation (since $a << \lambda$) that creates a new growing funnel of variants. The growth of protein domains in evolution is thus governed by two parameter classes, λ and a, which are analogous to imitation and innovation parameters q and p of the Bass model, respectively. We note that structural domains are building blocks of protein molecules with structures that can be coarse-grained into levels of a hierarchy, such as folds or FSFs. Thus, the rise of innovations at these levels must be separately explained.

The global dynamic models were fit to real ToDs and phylogenomic data to evaluate the distributions of model parameters along the branches of the tree. These distributions specify rates of generation of structural domain variants and rates of generation of new folds or FSFs. The equations were solved by simplifying the trees to a comb-like topology and treating branches as emerging directly from a trunk. While certain regions of the ToD deviated from a completely imbalanced topology there was no bias in the distribution of these non-comb topologies. Figure 10D plots rate constants a and λ against time of origin of corresponding domains, with age measured on a Gy scale. The plot describes growth and innovation of FSFs. However, we found that the rate constants for folds and FSFs evolved similarly. Remarkably, the plot revealed trends that are generative of a biphasic pattern. First, a clear pattern of increasing λ was observed, in which funnel growth appears to accelerate in evolution. This acceleration started to increase ~1.5 Gya (*nd* ~ 0.55), a time that coincides with the rise of the second wave of innovations of Figure 9. Second, *a* increased and then decreased in evolutionary time. The inflection point again occurred ~1.5 Gya. The sharp transient change coincides with a "big bang" of domain combinations that occurred during the rise of Eukarya and multicellularity (Figure 2), a period in which domains massively combined to form multidomain proteins and their arrangements distributed in novel lineages during the rise of organismal diversification (Wang and Caetano-Anollés, 2009). The modeling exercise revealed the same two innovation modes of the diffusion of innovation models that are illustrated in the proteomic growth curves of Figure 9. These modes became emergent properties of the dynamic model of protein evolution.

7. Conclusions and prospects

The world of biological macromolecules is complex and diverse. Such complexity and diversity could have not arisen through punctuation. Instead, it is much more likely that the process was gradual and followed Leibniz's principle of spatiotemporal continuity. Here we have proposed that accretion brings together parts to form bigger wholes in evolution, which ultimately provides opportunities for growth and innovation. Aggregation involves matter, energy, space, time and information. In all cases, the entities that are accreted include modules, sets of parts that interact more strongly with each other than with other parts of systems. These parts are tightly linked. Our study focused on the structural domains of proteins. They represent structural, functional and evolutionary modules and excellent models to study the processes of accretion, growth and innovation. However, we could have used other modules that exist in biology to illustrate our discussion, including modules of matter-energy (e.g. reaction networks, metabolic fluxes), time (e.g. clocks, circadian rhythms) or information (e.g. signals, languages, semantics).

Because parts exist at different hierarchical levels of biological organization (Simon, 1962), the accretion process is necessarily complex. For example, structural domains are one of many types of structural modules that exist in protein structures, which include amino acids, dipeptides, sequence motifs, secondary structures, and elementary loops. We have shown that computational mining of data-intensive information drawn from the genomic revolution uncover explicit timelines of accretion of structural parts in molecular repertoires and molecules. The exercise confirmed that growth of the protein world complied with the principle of continuity. We argue that there are mathematical universals that apply to the growth of macromolecular systems, such as the Zipf and Heaps laws. These laws are linked to processes of diversification encapsulated in sequential S-shaped loglets that model growth and diffusion in systems. The processes can be explained with diffusion of innovation models of the Bass type. Remarkably, pairs of loglets sometimes induce biphasic patterns (e.g. Figure 9). Global dynamical models of protein evolution explain these biphasic patterns (Tal *et al.*, 2016). Similarly, a general model of evolution of modules (Mittenthal *et al.*, 2012) describes how modules emerge through a biphasic "hourglass" process of diversification. In the first phase (wave), parts are at first weakly linked and associate variously as evolution explores alternative structural and functional capacities. As parts diversify, they compete with each other and are selected for performance to increase the persistence of systems. The emerging interactions increase linkage and therefore constrain their structures and functions. This causes parts to self-organize into modules with tight linkages, which increase system's organization. In the second phase (wave), variants of the modules evolve and become new parts for a new generative cycle of higher-level organization. The module evolution model is compatible with a protein evolution model (Tal *et al*., 2016). Fine-grained structural elaborations in growing funnels of diversification result in coarse-grained innovations.

References

Altmann EG, Gerlach M (2016) Statistical laws of linguistics. In: *Creativity and Universality in Language. Lecture Notes in Morphogenesis*, Degli Esposti M, Altmann EG, Pachet F (eds). Springer, Berlin, pp. 7–26.

Bass F (1969) A new product growth for model consumer durables. *Management Sci* **15(5)**: 215–227.

Beckwith SVW, Sargent AI (1996) Circumstellar disks and the search for neighbouring planetary systems. *Nature* **383**: 139–144.

Beöthy E, Altmann G (1982) Das piotrowski-gesetz und der lehnwortschatz. Zs Sprachwissenschaft **1**: 171–178.

Bershtein S, Serohijos AWR, Shakhnovich EI (2017) Bridging the physical scales in evolutionary biology: From protein sequence space to fitness of organisms and populations. *Curr Op Struct Biol* **42**: 31–40.

Bettencourt LMA (2013) The origin of scaling in cities. *Science* **340**: 1438–1441.

Bettencourt LMA, Lobo J, Helbing D, Kühnert C, West GB (2007) Growth, innovation, scaling, and the pace of life in cities. *Proc Natl Acad Sci USA* **104**: 7301–7306.

Bhaskara RM, Srinivasan N (2011) Stability of domain structures in multidomain proteins. *Sci Rep* **1**: 40.

Blum M, Francois O (2006) Which random processes describe the tree of life? A large-scale study of phylogenetic tree imbalance. *Syst Biol* **55**: 685–691.

Bryant HN (1997) Hypothetical ancestors and rooting in cladistic analysis. *Cladistics* **13**: 337–348.

Caetano-Anollés G (2002) Tracing the evolution of RNA structure in ribosomes. *Nucleic Acids Res* **30**: 2575–87.

Caetano-Anollés G (2017) RubisCO and the search for biomolecular culprits of planetary change. *Bioessays* **38(11)**: 1700174.

Caetano-Anollés G, Caetano-Anollés D (2003) An evolutionary structured universe of protein architecture. *Genome Res* **13**: 1563–1571.

Caetano-Anollés G, Caetano-Anollés D (2015) Computing the origin and evolution of the ribosome from its structure — Uncovering processes of macromolecular accretion benefiting synthetic biology. *Comput Struct Biotech J* **13**: 427–447.

Caetano-Anollés G, Wang M (2008) The protein world. *Biochemist* **30**: 4–8.

Caetano-Anollés G, Sun F-J, Wang M, Yafremava LS, Harish A, Kim HS, Knudsen V, Caetano-Anollés D, Mittenthal JE (2008) Origin and evolution of modern biochemistry: Insights from genomes and molecular structure. *Front Biosci* **13**: 5212–5240.

Caetano-Anollés D, Kim KM, Mittenthal JE, Caetano-Anollés G (2011) Proteome evolution and the metabolic origins of translation and cellular life. *J Mol Evol* **72**: 14–33.

Caetano-Anollés G, Minhas BF, Aziz F, Mughal F, Shahzad K, Tal G, Mittenthal JE, Caetano-Anollés D, Koç I, Nasir A, Caetano-Anollés K, Kim K.M (2017) The compressed vocabulary of the proteins of Archaea. In: *Biocommunication of Archaea*, Witzany G (ed). Springer, Dordretch, The Netherlands, pp. 147–264.

Caetano-Anollés G, Nasir A (2012) Benefits of using molecular structure and abundance in phylogenomic analysis. *Front Genet* **3**: 172.

Caetano-Anollés G, Wang M, Caetano-Anollés D, Mittenthal JE (2009) The origin, evolution and structure of the protein world. *Biochem J* **417**: 621–637.

Chothia C., Lesk AM (1986) The relation between the divergence of sequence and structure in proteins. *EMBO J* **5**: 823–826.

Fang H, Oates ME, Pethica RB, Greenwood JM, Sardar AJ, Rackham OJL, Donoghue PCJ, Stamatakis A, de Lima Morais DA, Gough J (2013) A daily-updated tree of (sequenced) life as a reference for genome research. *Sci Rep* **3**: 2015.

Farris JS (1971) Toward defining the course of evolution: Minimal change for a specific tree topology. *Syst Zool* **20**: 406–416.

Farris JS (1983) The logical basis of phylogenetic analysis. In: *Advances in Cladistics, Proceedings of the Second Meeting of the Willi Hennig Society*, Vol. 2, Platnick NI, Funk VA (eds), Columbia University Press, New York, USA, pp. 7–36.

Ferrada E, Wagner A (2010) Evolutionary innovation and the organization of protein functions in sequence space. *PLoS ONE* **5**: 14172.

Ferrer i Cancho R, Solé RV (2001) Two regimes in the frequency of words and the origins of complex lexicons: Zipf's law revisited. *J Quant Linguist* **8**: 165–173.

Fitch WM (1971) Toward defining the course of evolution: Minimal change for a specific tree topology. *Syst Zool* **20**: 406–416.

Fontana W (2003) Modelling 'evo-devo' with RNA. *BioEssays* **24**: 1164–1177.

Fraix-Burnet D, Chattopadhyay T, Chattopadhyay AK, Davoust E, Thuillard M (2012) A six-parameter space to describe galaxy diversification. Astronomy Astrophysics **545**: A80.

Gerlach M, Altmann EG (2013) Stochastic model for the vocabulary growth in natural languages. *Phys Rev* **3**: 021006.

Gerstein M (1998) Patterns of protein-fold usage in eight microbial genomes: A comprehensive structural census. *Proteins* **33**: 518–534.

Ghanbarnejad F, Gerlack M, Miotto JM, Altmann EG (2014) Extracting information from S-curves of language change. *J R Soc Interface* **11**: 20141044.

Gimona M (2006). Protein linguistics — a grammar for modular protein assembly? *Nat Rev Mol Cell Biol* **7**: 68–73.

Grand A, Corvez A, Duque Velez LM, Laurin M (2013) Phylogenetic inference using discrete characters: Performance of ordered and unordered parsimony and three-item statements. *Biol J Linn Soc* **110**: 914–930.

Guseo R (2016) Diffusion of innovations dynamics, biological growth and catenary function. *Phys A* **464**: 1–10.

Han J-H, Batey S, Nickson AA, Teichmann SA, Clarke J (2007) The folding and evolution of multidomain proteins. *Nature Rev Mol Cell Biol* **8**: 319–330.

Harish A, Caetano-Anollés G (2012) Ribosomal history reveals origins of modern protein synthesis. *PLoS ONE* **7(3)**: e32776.

Hartwell LH, Hopfield JJ, Leibler S, Murray AW (1999) From molecular to modular cell biology. *Nature* **401**: c47–c52.

Heaps HS (1978) *Information Retrieval: Computational and Theoretical Aspects*. Academic Press, New York.

Heard SB (1996) Patterns in phylogenetic tree balance with variable and evolving speciation rates. *Evolution* **50**: 2141–2148.

Huelsenbeck JP, Nielsen R (1999) Effect of nonindependent substitution on phylogenetic accuracy. *Syst Biol* **48**: 317–328.

Illergård K, Ardell DH, Elofsson A (2009) Structure is three to ten times more conserved than sequence — a study of structural response in protein cores. *Proteins* **77**: 499–508.

Janin J, Wodak SJ (1983) Structural domains in proteins and their role in the dynamics of protein function. *Prog Biophys Mol Biol* **42**: 21–78.

Jofré P, Das P, Bertrandpetit J, Foley R (2017) Cosmic phylogeny: Reconstructing the chmical history of the solar neighborhood with an evolutionary tree. *Monthly Notices Roy Astron Soc* **467**: 1140–1153.

Jones GR (2011) Three models for macroevolution and phylogenetic analysis. *Syst Biol* **60**: 735–746.

Kim K, Caetano-Anollés G (2011) The proteomic complexity and rise of the primordial ancestor of diversified life. *BMC Evol Biol* **11**: 140.

Kim KM, Nasir A, Caetano-Anollés G (2014) The importance of using realistic evolutionary models for retrodicting proteomes. *Biochimie* **99**: 129–137.

Kluge AG, Farris JS (1969) Quantitative phyletics and the evolution of anurans. *Syst Zool* **18**: 1–32.

Li S, Lin R, Bian C, Ma QDY, Ivanov PC (2016) Model of the dynamic construction process of texts and scaling laws of words organization in language systems. *PLoS ONE* **11(12)**: e0168971.

Lipman D, Wilbur W (1991) Modeling neutral and selective evolution of protein folding. *Proc R Soc Lond Ser B* **245**: 7–11.

Lü L, Zhang Z-K, Zhou T (2013) Deviation of Zipf's and Heaps' laws in human languages with limited dictionary sizes. *Sci Rep* **3**: 8028–8033.

Lundberg J (1972) Wagner networks and ancestor. *Syst Zool* **18**: 1–32.

Lundin D, Poole AM, Sjoberg B-M, Hogbom M (2012). Use of structural phylogenetic networks for classification of the ferritin-like superfamily. *J Biol Chem* **287**: 20565–20575.

Maynard Smith J (1970) Natural selection and the concept of protein space. *Nature* **225**: 563–564.

Meacham CA (1984) The role of hypothesized direction of characters in the estimation of evolutionary history. *Taxon* **3**: 26–38.

Meyer PS, Yung JW, Ausubel JH (1999) A primer on logistic growth and substitution. The mathematics of the Loglet lab software. *Tech Forecasting Social Change* **61**: 247–271.

Mittenthal J, Caetano-Anollés D, Caetano-Anollés G (2012) Biphasic patterns of diversification and the emergence of modules. *Front Genet* **3**: 147.

Molina N, van Nimwegen E (2009) Scaling laws in functional genome content across prokaryotic clades and lifestyle. *Trends Genet* **25(6)**: 243–247.

Montroll EW (1978) Social dynamics and the quantifying of social forces. *Proc Natl Acad Sci USA* **75(10)**: 4633–4637.

Mordasini C, Molliere P, Dittkrist K-M, Jin S, Alibert Y (2015) Global models of planet formation and evolution. *Intl J Astrobiol* **14**: 01–232.

Murzin A, Brenner SE, Hubbard T, Clothia C (1995) SCOP: A structural classification of proteins for the investigation of sequences and structures. *J Mol Biol* **247**: 536–540.

Nasir A, Caetano-Anollés G (2015) A phylogenomic data-driven exploration of viral origins and evolution. *Sci Adv* **1**: e1500527.

Nasir A, Kim KM, Caetano-Anollés G (2014) Global patterns of protein domain gain and loss in superkingdoms. *PLoS Comput Biol* **10**: e1003452.

Nasir A, Kim KM, Caetano-Anollés G (2017) Phylogenetic tracings of proteome size support the gradual accretion of protein structural domains and the early origin of viruses from primordial cells. *Front Microbiol* **8**: 1178.

Nelson GJ (1973) The higher-level phylogeny of vertebrates. *Syst Zool* **22**: 87–91.

Norton JA, Bass FM (1987) A diffusion theory model of adoption and substitution for successive generations of high-technology products. *Management Sci* **33(9)**: 1069–1086.

Orengo CA, Michie A, Jones S, Jones DT, Swindells M, Thornton JM (1997) CATH — a hierarchic classification of protein domain structures. *Structure* **5**: 1093–1109.

Petersen AM, Tenenbaum JN, Havlin S, Stanley HE, Perc M (2012) Languages cool as they expand: Allometric scaling and the decreasing need for new words. *Sci Rep* **2**: 721–725.

Qian J, Luscombe NM, Gersten M (2001) Protein family and fold occurrence in genomes: Power-law behavior and evolutionary model. *J Mol Biol* **313**: 673–681.

Reba M, Reistma F, Seto KC (2016) Spatializing 6,000 years of global urbanization from 3700 BC to AD 2000. Sci Data **3**: 160034.

Richardson JS (1981) The anatomy and taxonomy of protein structure. *Adv Protein Chem* **34**: 167–339.

Riley M, Labedan B (1997) Protein evolution viewed through *Escherichia Coli* protein sequences: Introducing the notion of a structural segment of homology, the module. *J Mol Biol* **268**: 857–868.

Rogers EM (1962) *Diffusion of Innovations*. Free Press, New York.

Rost B (1999) Twilight zone of protein sequence alignments. *Protein Eng* **12**: 85–94.

Schultes EA, Bartel DP (2000) One sequence, two ribozymes: Implications for the emergence of new ribozyme folds. *Science* **289**: 448–452.

Schuster P, Fontana W, Stadler P, Hofacker I (1994) From sequences to shapes and back: A case study in RNA secondary structures. *Proc R Soc London Ser B* **255**: 279–284.

Sikosek T, Chan HS (2014) Biophysics of protein evolution and evolutionary protein biophysics. *J R Soc Interface* **11**: 20140419.

Simon H (1962). The architecture of complexity. *Proc Am Phil Soc* **106**: 467–482.

Steel M (2012) Root location in random trees: A polarity property of all sampling consistent phylogenetic models except one. *Mol Phyl Evol* **65**: 345–348.

Tal G, Boca SM, Mittenthal J, Caetano-Anollés G (2016) A dynamic model for the evolution of protein structure. *J Mol Evol* **82**: 230–243.

Taylor WR (2006) Topological accessibility shows a distinct asymmetry in the folds of $\alpha\beta$ proteins. *FEBS Lett* **580**: 5263–5267.

Tria F, Loreto V, Servedio VDP, Strogatz SH (2014) The dynamics of correlated novelties. *Sci Rep* **4**: 721–723.

Verhulst PF (1838) Notice sur la loi que la population suit dans son accroissement. *Correspondance Math Phys* **10**: 113–121.

Victor NM, Ausubel JH (2002) DRAMs as model organisms for study of technological evolution. *Tech Forecasting Social Change* **69**: 243–262.

Vogel C (2005) Function annotation of SCOP domain superfamilies. In: Superfamily database, HMM library and genome assignments server. http://supfam.org/SUPERFAMILY/function.html.

Wagner A (2011) The molecular origins of evolutionary innovations. *Trends Genet* **27**: 397–410.

Wang M, Caetano-Anollés G (2006) Global phylogeny determined by the combination of protein domains in proteomes. *Mol Biol Evol* **23**: 2444–2454.

Wang M, Caetano-Anollés G (2009) The evolutionary mechanics of domain organization in proteomes and the rise of modularity in the protein world. *Structure* **17**: 66–78.

Wang M, Jiang Y-Y, Kim KM, Qu G, Ji H-F, Mittenthal JE, Zhang H-Y, Caetano-Anollés, G (2011) A universal molecular clock of protein folds and its power in tracing the early history of aerobic metabolism and planet oxygenation. *Mol Biol Evol* **28**: 567–582.

Wang M, Yafremava LS, Caetano-Anollés D, Mittenthal JE, Caetano- Anollés G (2007) Reductive evolution of architectural repertoires in proteomes and the birth of the tripartite world. *Genome Res* **17**: 1572–1585.

Weinreich U, Labov W, Herzog M (1968) *Empirical Foundations for a Theory of Language Change*. University of Texas Press, Austin, Texas.

Weston PH (1988) Indirect and direct methods in systematics. In: *Ontogeny and Systematics*, Humphries CJ (ed). Columbia University Press, New York, pp. 27–56.

Wetlaufer DB (1973) Nucleation, rapid folding, and globular intrachain regions in proteins. *Proc Natl Acad Sci USA* **70**: 697–701.

Wheeler WC (2012) *Systematics: A Course of Lectures*. John Wiley & Sons, Hoboken, NY, USA.

Wright S (1932) The role of mutation, inbreeding, crossbreeding, and selection in evolution. *Proc 6th Cong Genetics* **1**: 356–366.
Yang S, Doolittle RF, Bourne PE (2005) Phylogeny determined by protein domain content. *Proc Natl Acad Sci USA* **102**: 373–378.
Zuckerkandl E (1975) The appearance of new structures and functions in proteins during evolution. *J Mol Evol* **7**: 1–57.

A model of lineage evolution based on the persistence strategy hypothesis

L.S. Mainzer[1,2]
J.E. Mittenthal[3]
G. Caetano-Anollés[2]

Summary. We propose a model of lineage evolution to follow the emergence of trade-offs between three strategies of persistence: economy, flexibility and robustness. The objective properties of the environment, such as temperature, salinity, illumination, are linked with the subjective experience of the organism that is exploring the environment, via probability functions expressing the likelihood of being exposed to each environmental signal. Organisms themselves are modeled according to Miller's theory of living systems, each represented with subsystems that process matter, energy, and information, such as boundaries (membranes), sensors, and internal transducers. The mechanisms of flexibility and robustness are expressed via the properties of these critical subsystems, many well-defined in Miller's theory. The energetic costs of manufacturing and maintaining the critical subsystems

[1]*National Center for Supercomputer Applications (NCSA),*
[2]*Department of Crop Sciences,*
[3]*Department of Cell and Developmental Biology,*
University of Illinois, Urbana, Illinois, USA

factor into the organism's strategy of economy and help define the trade-offs between maximizing flexibility and robustness while minimizing their costs. These trade-offs are collectively depicted on a "persistence triangle," which encompasses the entire landscape of trade-off solutions that lineages can explore during their evolution. We discuss trajectories of movement on the triangle, both for lineages in the course of their evolution (via emergence of new mechanisms of flexibility and robustness, and bifurcation into new lineages), as well as individual organisms during their lifetime (via learning, dormancy and life stages, such as egg, larva, and pupa). This model can serve as a recipe for creating a simulation that visualizes how organisms explore persistence strategy trade-offs during evolution.

1. Introduction

The evolution of physiologies under the influence of the environment has long received attention from experimental, theoretical, and computational biologists. It is understood that the physical properties and signal content of the environment drive evolution of adaptations that improve persistence. The constraints of this evolutionary process, effect of the initial and boundary conditions, and the actual relationships between the environmental properties and resultant physiologies, including intermediary evolutionary states, can be studied systematically in a simulation. In this chapter, we describe a general model of environment-driven evolution of physiologies, based on the framework of persistence strategies we introduced in Yafremava *et al.* (2013).

The crux of the framework consists of two conceptual trios: scope/umwelt/gap and economy/flexibility/robustness. The first trio stems from the work of von Uexkull (1909), who defined "umwelt" as the totality of an organism's perception: the set of environmental signals that undergo sensory processing. We recognized that organisms do not process all of the signals they are exposed to. Those signals that do not undergo sensory processing, fall into the "gap" between umwelt and the entire "scope" of an organism' experience with the environment.

The scope/umwelt/gap trio and Miller's theory of living systems (Miller, 1995) help define three persistence strategies: flexibility, robustness, and economy. The main tenet of Miller's theory is that living systems are molecular machines that process matter-energy and information. For a signal to undergo sensory processing means that its information content is captured and transformed by the organism's internal systems, resulting in an internal change or outward behavior. When such a change routinely results in a suitable adaptation to the environment and ultimate survival, it becomes evolutionarily fixed in the lineage, promoting its persistence. Thus, we associate the extent and content of an organism's umwelt with a persistence strategy of "flexibility" — ability to flexibly adapt to environmental changes. In contrast, signals within the gap are not processed by the organism, even though they could be damaging. Organisms evolve mechanisms of withstanding the effect of those signals, or become "robust" against them. Thus, we associate the extent and content of the gap with the persistence strategy of "robustness." The mechanisms of flexibility and robustness compete for matter-energy, which an organism must constantly replenish. Due to the metabolic costs associated with these two mechanisms, we introduced "economy" as a third competing strategy of persistence — the strategy of maintaining reduced flexibility and robustness, where the loss in survivability of individuals is compensated by the gain in the speed of reproduction and the number of offspring. The trade-offs between the three competing persistence strategies are displayed as locations on the "persistence triangle" (Figure 1A). Proximity to a corner of the triangle conveys the prominence of that persistence strategy in a particular organismal lineage. Previous work has established that flexibility mechanisms are more metabolically expensive and more variegated, than properties conferring robustness, which makes the triangle asymmetric: the flexibility leg is more extended, and occupied by Metazoa. The robustness leg is shorter and occupied by Plants.

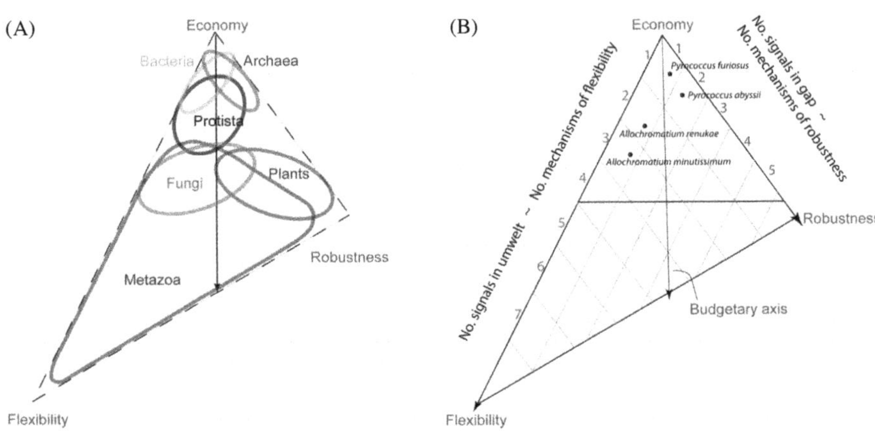

Figure 1. Interpretation of the persistence triangle. The persistence triangle depicts the space of evolutionary trade-off solutions among economy flexibility and robustness. (A) Previously proposed locations of the six kingdoms on the triangle. (B) The triangle is bounded by the respective measures of the three strategies. Economy is measured as matter-energy budget. Flexibility of an organism is measured with the number of signals it processes in its umwelt, and the number of flexibility mechanisms it uses to respond to them. Robustness is measured with the number of signals in an organism's gap, and the number of physical properties and mechanisms that confer robustness against those signals.

Fungi are more flexible than plants, but also have smaller budget, so they occupy the middle of the triangle. Unicellular organisms have the smallest budgets and occupy the Economy corner, with Protista separating akarya from multicellular eukarya. This fairly crude separation of the kingdoms of life on the triangle helps understand their persistence strategies, but needs to be refined and elaborated.

As the genotypes of individuals in a lineage evolve, their interaction with the environment changes. Over the course of evolution, individuals become exposed to different signals, acquire and lose mechanisms of flexibility and robustness, and develop new "economies" of matter-energy for maintaining their physiology. Here we map these changes as movements inside the persistence triangle that occur within an individual life time, across individuals in a lineage, and in form of bifurcations into new lineages. We describe an analytical model for converting objective environmental descriptors to the subjective experience of an individual organism, establishing a deeper correspondence between the two trios. This model helps explain how a lineage adopts a location on the persistence triangle in the course of evolution by following examples of extant lineages inside the triangle.

2. Interpretation of the persistence triangle

The persistence triangle can be a convenient and useful way of comparing lineages, once its axes and locations within it are correctly defined. Yafremava *et al.* (2013) introduced the triangle as the phase space of the trade-off solutions between flexibility, robustness and economy. An organism can be viewed as comparatively more flexible if it is able to respond to a greater number of environmental signals (i.e. bigger umwelt) with a greater number of inner changes and outward behaviors that lead to its adaptation to the environment. Thus, the flexibility axis of the triangle measures the number of flexibility mechanisms an organism has, which correlates with the number of signals in its umwelt (Figure 1B). For example, species of the purple sulfur bacterium *Allochromatium* differ in their abilities to use organic substrates for photo-heterotrophic growth (Kumar et al. 2008). *Allochromatium renukae* can utilize eight substrates (acetate, butyrate, pyruvate, furamate, succinate, malate, glucose and Casamino acids), but *A. minutissimum* can also utilize five more (formate, propionate, glycolate, crotonate, and valerate), as well as at least four inorganic substrates (hydrogen, thiosulfate, sulfur, and sulfite for chemolithotrophic growth). Thus, *A. minutissimum* is a more flexible organism, able to invoke a wider variety of metabolic reactions that sustain life in response to the changing availability of nutrients in the environment.

Likewise, an organism is comparatively more robust if it is able to withstand without change a greater number of environmental influences (signals). Physiologically such organisms have more mechanisms of robustness. The robustness axis of the triangle measures the number of robustness mechanisms an organism has, which correlates with the number of signals in its gap. For example, both *Pyrococcus abyssii* and *Pyrococcus furiosus* are thermophiles, robust against high temperatures (optimum at ~102°C; Erauso *et al.*, 1993; Fiala and Stetter, 1986). In addition, *P. abyssii* is also highly barophilic (can withstand living at ocean depths down to 2,000 m) and therefore its internal systems have the additional properties conferring robustness against high pressure (Kumar and Nussinov, 2001; Berezovsky and Shakhnovich, 2005; Oger and Jebbar, 2010).

According to these definitions, an organism can be placed on our persistence triangle in a similar way as Grime's plants on his well-known CSR triangle, which captures the continuum of trade-offs between the strategies of Competitiveness, Stress-tolerance, and the Ruderal habit of growing in disturbed habitats (Grime, 1977). The lines parallel to the flexibility and robustness axes section the triangle according to the respective values on those axes (Figure 1B), just as Grime's CSR triangle is sectioned according to the values of the variables that measure the competitiveness and stress tolerance. In our case, an organism is placed where its respective values of flexibility and robustness intersect. The resultant distance from the Economy vertex is proportional to its matter-energy budget. The flexibility/robustness space is not the standard Cartesian coordinate system, as the mechanisms of flexibility and robustness are not always independent. Some adaptations, such as sporulation or repair, contribute to both flexibility and robustness (Yafremava *et al.* 2013). This is why we do not draw the axes of flexibility and robustness perpendicular to each other. However, it is useful to define a "budget axis" — locations where flexibility is equal to robustness for a particular matter-energy budget. Deviation of an organism from this axis indicates the direction of the trade-off between flexibility and robustness.

As organisms evolve, they probe the trade-offs among the three persistence strategies in ways that depend on the properties of the environment, and the methods of persistence (flexibility mechanisms, robustness properties, and the associated economies) that the organisms already possess. It would be quite difficult to trace this process entirely even for a single lineage, due to the difficulty in measuring the exact content of scope/umwelt/gap and enumerating all of the mechanisms of persistence. However, this is possible in a simulation, and the following section presents a formalism for doing so.

3. A method of modeling evolutionary movements of organisms on the persistence triangle

In order to effectively simulate organisms on the triangle, one would need to establish the correspondence between objective properties of the environment (e.g. temperature, pressure,

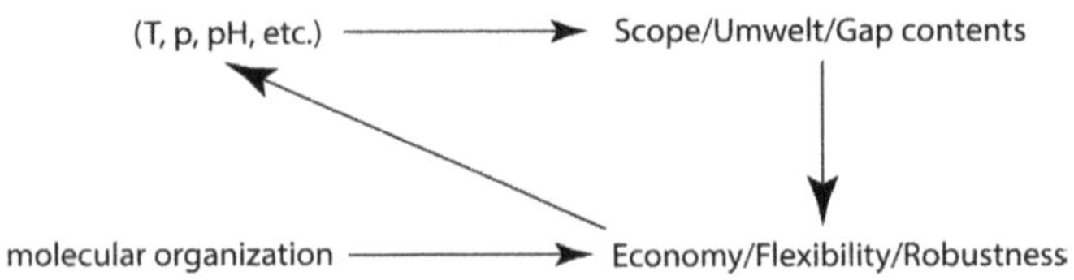

Figure 2. Schematic of mutual inter-relations between the objective environmental parameters, the subjective scope/umwelt/gap trio and the persistence strategies.

illumination, etc.), and the subjective experience of an individual (actual content of scope/umwelt/gap). Further, one would need to establish a correspondence between the elements of the organism's internal molecular organization and the resultant behaviors (mechanisms of flexibility) and properties (robustness). These behaviors and properties in turn constrain how the organism explores its environment, creating a feed-back loop between environment and individual (Figure 2).

3.1. *Simulating scope/umwelt/gap*

An individual is born with a certain set of evolved behaviors, such as ability to move in the environment and manipulate it in certain ways. As the individual performs those behaviors, its scope is filled with signal values according to their distribution in the environment at the moment when the individual visits any given location. For example, temperature differs in space and time due to the presence of overshadowing objects, movement of the air, daily and seasonal oscillations in illumination, proximity to geothermal heat, etc. This can be modeled with a probability density $p(T_{\bar{r},t})$, which defines the probability $p_0 dT$ that temperature T at a particular location $\bar{r}_0$, at any time t_0, is between T_0 and $T_0 + dT$ (e.g. Andronova and Schlesinger, 2001; Perkins *et al.*, 2007; Pitman and Perkins, 2009; Figure 3, panel A). As an individual travels through space-time, it becomes exposed to different temperatures according to this probability

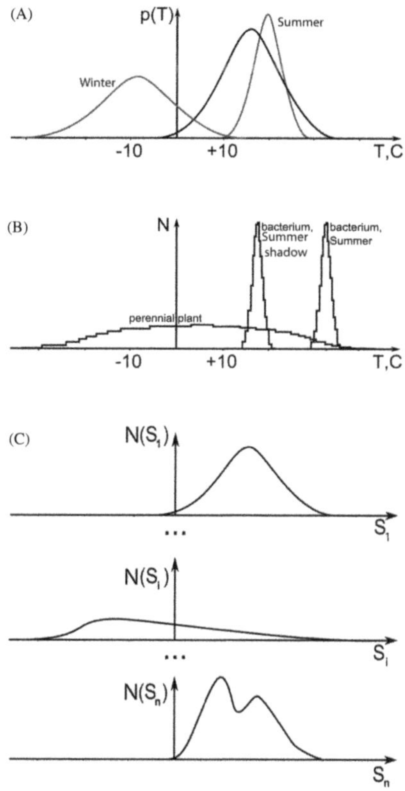

Figure 3. Scope content is expressed using probability functions of environmental signals. The likelihood for an individual organism to be exposed to a particular signal, such as temperature, can be modeled with a probability density that is a function of location and time. (A) Probability density function for temperature is different on an exposed surface in the summer and winter, or in a shadowed location in the summer (unlabeled curve). (B) Every individual organism arrives at its own scope content, having uniquely sampled its environment over the lifetime. Two bacteria that lived in the same time period but in slightly different locations may have sampled from two different signal probability distributions (e.g. summer and shadowed location in the summer in Panel A), and therefore arrive at drastically different scope contents despite similarity in physiology and behavior. A plant will have integrated across the entire range of temperatures in Panel A throughout the years of its lifetime. (C) The complete set of expected probability distributions for all environmental signals to which an individual from a given lineage was exposed during its lifetime. Each may have its own shape, depending on the signal and how the organism tends to explore it.

density function, modified by the individual's behavior, such as its movement path. The lifetime record of all temperature values to which an individual has been exposed, sums up its unique experience in a frequency distribution function $N(\mathrm{T})$ (Figure 3, panel B). Generalizing to other signals $\{\mathrm{S_i}\}$, such as illumination, pressure, salinity, etc., we can express the individual's post-mortem scope content as the set of all such frequency distribution functions $\{N(\mathrm{S_i})\}$ (Figure 3, panel C).

These "scope content" functions differ between individuals, reflecting their unique experiences due to their genetic and epigenetic variation, unique behavior, or local changes in the environment stemming from natural geological, climatic and other variations. However, individuals from the same lineage are genetically related, and similar to one another in terms of their flexibility/robustness mechanisms, which results in similar post-mortem scope contents. As the lineage evolves, subsequent generations inherit genotypes that predispose them toward experiencing similar scope content, with modifications due to mutations and epigenetic changes. The entire set of the scope content functions $\{N(\mathrm{S_i})\}$ across all individuals in a generation can be used to characterize the lineage, compare it to other lineages, and trace how the individual experience of the environment changes through evolution. This can be easily done in a simulation by recording the signal values encountered by the simulated individuals, and comparing the distributions themselves postmortem. However, when the number of individuals and signals is appropriately large to make a realistic simulation, visualizing and interpreting these data would be difficult. Instead, a good a-posteriori model of the lineage's "expected" scope content $\{P(S_i)\}$ can be constructed by averaging the individual scope contents $\{N(\mathrm{S_i})\}$ across generations. Over a large number of generations, the result will approach the integral of $p(S_i, r, t)$ for each signal S_i over the characteristic path taken by individuals through the spatial range R

and lifetime T, which is why we call it the "integral" scope content of the lineage:

$$P(S_i) = \frac{\iint_{R,T} p(\mathrm{S}_i, r, t) dr dt}{\rho \cdot \tau} \tag{1}$$

Here $\rho \cdot \tau$ symbolizes normalization over the spatial (ρ) and temporal (τ) scope size. The parameter ρ can be the volume (m^3) or area (m^2) that tend to be explored by the individuals, and τ is the longevity characteristic of individuals in the lineage. The same operation can be performed to obtain the integral umwelt content $\{U(S_i)\}$ and integral gap content $\{G(S_i)\}$ of the lineage. In those cases, only the signals that fall within umwelt or gap are integrated. By definition, $P(S_i) = U(S_i) + G(S_i)$.

The complete set of functions $\{P(S_i)\}$ reflects all environmental signals to which the individuals have been, and can potentially be exposed over the course of their lifetimes. These functions can be useful for comparing lineages and making inferences about effects of the environmental niche on the evolution of flexibility and robustness. Variability of signals in the inhabited niche has a particularly strong impact on the integral scope content. Individual experiences $N(S_i)$ in different generations will be substantially different if signals vary at a time scale greater than the temporal scope size of individuals (Yafremava *et al.* 2013). This will create a broad profile for the lineage's scope content $P(S_i)$, integrated over much narrower individual scopes ($N_i << P_i$, Figure 4). For example, individual bacteria from the same species, one of which came into being on a hot summer day, another on a cool autumn evening, experience very different temperatures. The post-mortem frequency distribution function for temperature is tall and narrow for both individuals, due to the narrow temporal scope. However, the two distributions are shifted significantly relative to one another (Figure 3B). The

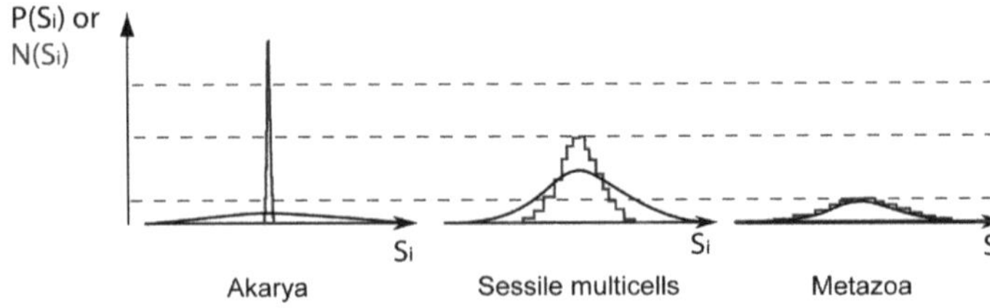

Figure 4. Relationship between integral scope content of a lineage and scope content of an individual. Smooth black curves depict integral scope content probability functions P, whereas the step-like blue functions depict the post-mortem frequency distributions N for the same signal. Akaryal individuals are exposed to fairly narrow ranges of environmental signals, but many akaryal lineages persist in a wide variety of conditions. Sessile multicells (fungi and plants) include all experiences of their constituent cells, and so have greater scope content than akarya. They tend to be dispersed across large territories with differing conditions, but individuals are localized, so have smaller scope content than lineage at large. Metazoa have the richest scope content among all kingdoms, with fairly small discrepancy between the individual and lineage scope content due to motility and other flexible behaviors.

integral scope content of their lineage encompasses the experiences of both these individuals, and many others who had very different lives. In contrast, the scope contents of two perennial plant individuals from the same species tend to be fairly similar to each other, and to the integral scope content of their lineage (Figure 3B). The relationship between the individual scope content $\{N(S_i)\}$ and scope content of the lineage $\{P(S_i)\}$ is important in evolution. In the sections below we show that the mechanisms of flexibility and robustness are affected by the shape of these functions and the extent of disparity between them. However, the probability functions $P(S_{i0})$ alone are not sufficient to characterize scope content, because they do not include the dynamic component of the signals. Two organisms with the same scope size and integral content may be exposed to different amount of information, depending on how quickly their environmental signals change. For example, plants may be exposed to a similar range of temperatures in a dry desert and in a temperate climate, with the total amplitude of variation being tens of degrees centigrade. However, the former experience this variation diurnally, whereas the latter experience it seasonally. Both frequency and amplitude of signal variation are important for defining density. Temperature oscillations of 1°C per minute with amplitude of 1°C are less relevant for life than oscillations of 10°C per 10 minutes with amplitude of 10°C, even though rate of change of the two signals is the same. An organism's scope content is "dense" when individuals are exposed to environmental signals that change rapidly at the individual's temporal scale, with amplitude that encompasses signal values that have different effects on the physiology. This may be due to the environment being highly inhomogeneous in space-time, or due to the individuals exploring the environment at high speed. Either way, this rate of signal variation is likely to affect evolution of flexible responses.

Meanwhile, to complete description of the model, Miller's theory of living systems will give us tools for describing the molecular mechanisms of flexibility and robustness.

3.2. *Simulating the mechanisms of flexibility and robustness*

Miller's theory permits us to relate the nature of an organism's environment to its internal processes, represented through the functions of the so-called "critical subsystems" (Miller, 1995). The critical subsystems (marked here in italics) process matter-energy and information within every living system, mostly through conversion and production (Figure 5). Material particles are taken into an organism by ingestors, then transformed by converters into forms used by producers to make necessary structures. Information is taken in by input transducers and is converted into the action of producing new material structures and new forms of information. Both matter-energy and information can be transported/transduced, stored, used for movement of the organism, somatic growth and

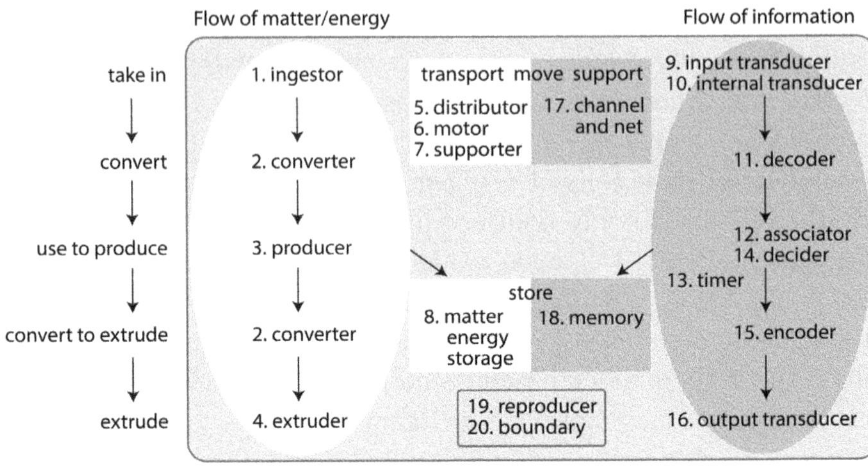

Figure 5. Organisms as molecular machines that process the vital fluxes of matter-energy and information. This figure summarizes the view of organisms in Miller's theory of living systems. The large rounded rectangle delineates a single organism. The pink and cyan shaded areas depict critical subsystems which process fluxes of matter-energy and information, respectively. Ellipses encompass critical subsystems dedicated exclusively to processing of one vital flux. Within them the flux follows the basic progression of intake, conversion, production, extrusion, as depicted by the flowchart on the left (see text). The rectangles encompass the critical systems, which are necessary, but tangential to that flow: they deal with transport, support and storage of matter-energy and information. Reproducer and boundary process both vital fluxes.

reproduction (forms of production), and extruded as waste or products usable by other organisms. For example, in mammals, food particles are ingested by mouths, converted into more easily digestible products by teeth, used to produce the necessary biomolecules in the various cells of the body, and waste is converted in the gut for extrusion through the anus. Visual information is taken in by the eye, transduced via axons to the visual cortex, where it is decoded, and used to produce a behavioral decision in the frontal lobes. Once a decision has been reached, it can be encoded in form of commands for muscle fibers of e.g. the larynx and mouth, which together serve as the output transducer for speech.

Miller's theory gives an exhaustive description of properties of each critical subsystem that is uniform across all levels of complexity. This takes away idiosyncrasies of individual body parts and organs and makes it easy to simulate all internal subsystems in the same way, with transducers and converters that carry fluxes of matter-energy and/or information with varying efficiency.

Among the properties assigned to the subsystems are: threshold of operation, minimum and maximum sustained flux, flux efficiency (loss or dissipation of matter-energy or information as they are carried through the subsystem), temporal variation in performance (e.g. circadian rhythms), etc. These properties can be subject to a genetic algorithm to simulate their evolution in different environments. To enable this, the model must connect the properties of critical subsystems to lineage persistence via some sort of "objective function" that links the costs and benefits of the subsystems to survival.

Flexibility is entirely implemented via the interaction of the critical subsystems that perform information processing (e.g. example of visual processing resulting in speech above). Their functioning is costly, as they frequently have to be manufactured and maintained even in the absence of the signals they have evolved to process. For example, acetylcholine receptors in a post-synaptic neuron incur metabolic costs whether they are idle or active due to excitation of the pre-synaptic neuron.

Robustness is conferred via properties of critical subsystems that make the organism able to withstand the effects of environment. Some of them, such as the bacterial capsule and cell wall, the bark and thorns of plants, incur costs of manufacture and maintenance that have to be satisfied even when the adverse physical impact is not present. Others, such as the thermophilic properties of proteins, restrict the organism to functioning optimally only under a limited set of conditions (Razvi and Scholtz, 2006).

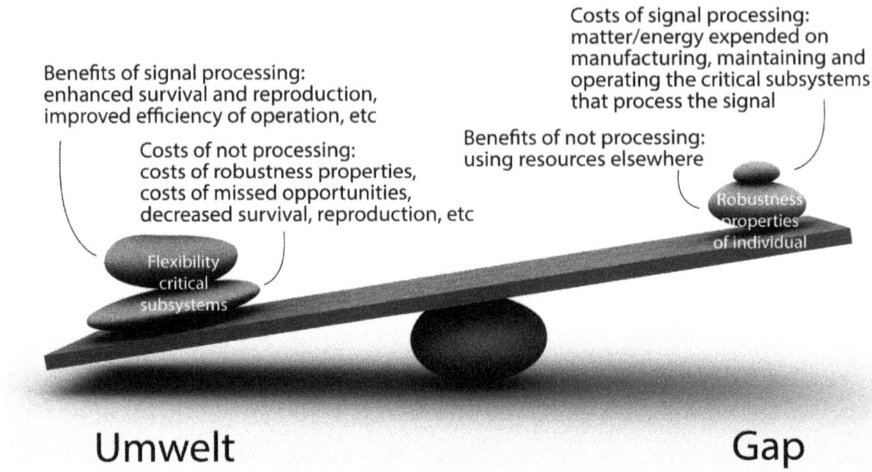

Figure 6. The evolutionary cost-benefit analysis for the Umwelt/Gap division. The evolution of critical subsystems that process umwelt signals, and physiological properties that confer robustness against gap signals, bring forth the division of scope into umwelt and gap. During this evolution the costs of processing or withstanding the signals are balanced against the benefits. The balance is influenced by the probability of being exposed to the signals and the extent to which the signals tend to affect the organism.

All those energetic costs are balanced against persistence benefits as the organisms evolve (Figure 6). Critical subsystems that process any particular signal persist through selection when the benefits brought forth by their functioning, and the costs of withstanding the signal without processing it or being robust against it, outweigh the cost of their operation. Conversely, the properties that confer robustness evolve when the costs of processing the signal outweigh the cost of maintaining robustness against it. Costs of robustness include the opportunities missed whenever the information contained in the signal is ignored. Benefits and costs can be measured in a simulation very easily by keeping track of mortality, fecundity, and offspring survival, as well as the metabolic costs: those of manufacturing, maintaining and operating the molecular components of the critical subsystems.

The cost-benefit analysis in the evolution of critical subsystems feeds back to the signal space in scope/umwelt/gap: evolution of flexibility or robustness with respect to a signal depends on the likelihood of encountering that signal. Indeed, the costs of maintaining the ability to deal with a low-probability signal may outweigh the benefits of doing so, because the signal's effects are rarely experienced. Such signals could remain in the gap, without any flexibility or robustness mechanisms evolving to deal with them. Therefore, a special mechanism of flexibility or robustness against a signal is more likely to evolve if encounters with that signal are frequent, and have substantial impact on survival of individuals.

In the model this means that flexibility and robustness mechanisms will tend to evolve with respect to signal values close to the peak of the distribution function N(Si). For example, mesophilic bacteria tend to experience mild temperatures with higher probability, and do not evolve the thermo-stabilizing adaptations of their thermophilic cousins, whose thermal scope peaks close to boiling temperatures.

Scope content and the nature of signals in it shape evolutionary paths on the persistence triangle by predisposing organisms toward evolving different kinds of flexibility and robustness mechanisms. Further text explores the different kinds of movement on the triangle.

4. Movements within the persistence triangle

4.1. *Evolutionary movements*

Since individuals from the same lineage are genetically related, they start life with similar sets of flexibility/robustness mechanisms, and are located close to each other on the persistence triangle (Figure 7A). That location generally does not change throughout lifetime, with a few exceptions discussed below. Nonetheless, every individual also

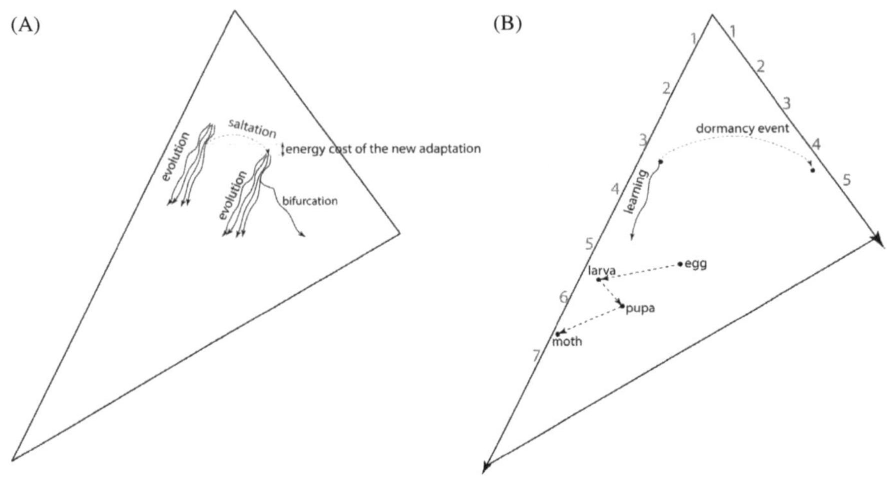

Figure 7. Movements on the persistence triangle. (A) Organisms in the same lineage tend to evolve roughly the same way, and roughly from the same locus on the triangle, due to their genetic similarity. An invention of an adaptation that significantly affects flexibility or robustness can result in a saltation, or evolutionary jump from one location on the triangle, to another. Invasion of novel environments leads to bifurcation of paths on the persistence triangle. (B) Generally an individual is stationary on the triangle during its lifetime. However, dormancy, learning and transitions between life stages can lead to legitimate movements on the triangle during an individual's lifetime.

has unique genetic features, which could enable it to have unique experiences and survive in slightly different conditions. At the end of life, each surviving individual adds to the integral scope content *{P(Si)}* of the lineage all the unique data points from its scope *{N(Si)}* that it had acquired by the time of reproduction. Each surviving individual also passes onto the next generation its own set of flexibility and robustness mechanisms that may have been modified via de novo mutation, genetic duplication with modification, deletion or recruitment, all of which are easily simulated using genetic algorithms. Thus, from generation to generation, mechanisms of flexibility and robustness are changed, added or lost, resulting in movement on the triangle (Figure 7A). The resultant traces are expected to bunch closely together, as the different "family histories" within the same lineage explore the balance of costs and benefits for each adaptation, and evolve more economical implementations of flexibility and robustness.

This process is expected to happen mostly around the peaks of integral scope content, evolving special persistence mechanisms for each of the signal values that have important meaning to the organism. However, when disparity exists between scope content of an individual and integral scope content of its lineage ($N_i << P_i$, Figure 4), we expect emergence of "blanket" mechanisms, which (however imperfectly) work for a wide variety of signal Si values (Yafremava *et al*, 2013). For example, bacterial sporulation turns the cell into a state that is robust against a number of adverse signals, such as desiccation, heat, freezing, and radiation (Stephens 1998). It has evolved in many of the cosmopolitan akarya (Sommaruga and Casamayor, 2009) that can be found throughout the globe in a wide variety of conditions. The integral scope content of a cosmopolitan species is rich with signals. However, the bacterial individuals themselves have excruciatingly tiny spatio-temporal scope size compared to the lineage at large, and thus experience only a miniscule fraction of the integral scope content. Rather than encoding all possible useful mechanisms of flexibility and robustness that could be appropriate for the lineage, sporulation evolves as a much cheaper solution, which allows individuals to "fall asleep" when conditions are not quite nice, and "wake up" again when things get better.

Introduction of a blanket mechanism of persistence can propel the lineage far along the axis of flexibility or robustness, since it buys persistence against a number of different signals at a relatively small cost.

4.2. *Lifetime movements*

Some organisms undergo, over the course of their lifetime, transitions between states of being that have very different complements of flexibility and robustness. In the above example of sporulation, as well as hibernation or any other state of dormancy, the dormant state is qualitatively different from the active state. It confers robustness against many influences, but also puts the organism into the state of stasis, where flexibility is limited. This is equivalent to a reversible "jump" on the persistence triangle, from a more "flexible" location on the triangle to a distant, more "robust" location, skipping some locations in-between (Figure 7B). Invoking this blanket robustness seems to be a meaningful way of dealing with highly destructive signals, which either occur rarely, or which could be costly to deal with over prolonged period.

Learning — another blanket mechanism of persistence — allows individuals to move smoothly within the persistence triangle during their lifetime. Unlike sporulation, which helps an individual converts into a more robust state and back, learning is a process by which an individual "invents" new mechanisms of flexibility. Each newly learned behavioral response moves the individual farther along the flexibility axis. Many of these learned flexibility mechanisms die with the individual that developed them, but they can also be spread between individuals within the same generation (e,g, conspecifics in social species copying each other's behavior) and passed down to the next generations in some organisms (via epigenetic mechanisms or further learning, e.g. historical memory in humans). Thus, learning can result in both individual lifetime movements and evolutionary lineage movements on the persistence triangle.

Finally, organisms with distinct life stages will appear to jump across the triangle as they transition from one state to another. For example, moths move through the developmental stages of an egg, larva, pupa and adult winged moth. At each stage they may consume different foods, move over different territories and be exposed to different environmental conditions. The scope content and density is vastly different at each of those stages, and therefore different flexibility and robustness mechanisms must evolve under the influence of the N(Si) prevalent during each life stage. During the lifetime of a single individual, N(Si) shifts within P(Si) during each transition between life stages.

4.3. *Bifurcations into new lineages*

Once an organism had evolved a mechanism of flexibility or robustness to deal with signals that have been sampled by the lineage so far, new environments may become accessible to its individuals. Through the process of adoptation, individuals can move into (adopt) a different environment depending on the coping mechanisms they already possess (Dover, 1986). There they can potentially discover new signals, adding those to the fringes of the integral scope content of their lineage. Such "fringe" individuals can be sufficiently distinct from the mother lineage to go on to found new lineages in their adopted niches, resulting in bifurcating traces on the persistence triangle (Figure 7A). We expect this to show up most often when simulating lineages that have two features together: one — blanket mechanisms of persistence, permitting the individuals to invade and survive in novel niches; two — highly redundant and quickly mutating genomes, permitting rapid divergence of the redundant gene copies to adapt to the new niche. Using learning in combination with proxy signals [both are blanket flexibility mechanisms, Yafremava *et al.* (2013)] is expected to be most successful in adopting new environments.

Proxies are the harmless signals that can be used to predict the arrival of other, significant and harmful signals. This is possible because physical signals that are generated by the same source tend to be correlated. For example, leaf surface temperature and illuminance are correlated. Temperature and pH in the same location of a geothermal pond are correlated. Because different signals have different properties (Dusenbery, 1992), some are better at being proxies than others.

- Light and electrical signals travel fastest, whereas chemical signals travel slowest. Thus, light and electrical signals are better proxies of spatially remote events: they bring information about the events sooner.
- It is easiest to localize the source of light and gravitational signals, but very difficult to localize the source of chemical signals. Light and gravity are therefore good proxies for spatial location.
- Light and sound are more easily obstructed than chemical and electrical signals, which is why chemical and electrical signals are better proxies in object-filled environments.

When an organism moves into (adopts) a new environment, some of the "important" signals (e.g. temperature) can still be clamped in their optimum range, by using proxy signals to generate the required flexible responses. For example, migratory animal species span a wide range of latitudes and could potentially be exposed to a wide range of temperatures, as expected from the local seasonal variations captured by the $p(T)$ functions. However, those animals do not wait for the extreme temperatures to arrive before they migrate away from the affected area. They use proxy signals, such as the changes in diurnal illumination patterns, to detect the imminent arrival of the climatic change, and move out before it happens. In terms of our formalism, a lineage's integral $P(T)$ evolves to be a narrower distribution than $p(T)$, and excludes the extreme temperatures. In turn, the lineages integral $P(S_i)$ for proxy signals, such as illuminance, evolve to be very broad and encompass as many values as possible.

Evolution of the critical subsystems to process signals by proxy can reduce the cost of flexibility as more scope signals become replaced with their proxies. Sounds and light patterns especially can pack a lot of information. If that information content can be discerned correctly, then a single set of critical subsystems (input transducer/decoder/associator/decider in the visual pathway) can be used to process a number of widely varying signals, resulting in many flexible responses at the cost of one sensory organ.

5. Conclusions and outlook

This chapter outlines elements of a model that could be used to demonstrate the evolutionary movements of simulated organisms on the persistence triangle. The simulation begins by seeding an environment, i.e. specifying the physical parameters, such as temperature, moisture and nutrient abundance as functions of space and time $p(S_{\bar{r},t})$ (Figure 8). Likewise, the population is seeded by

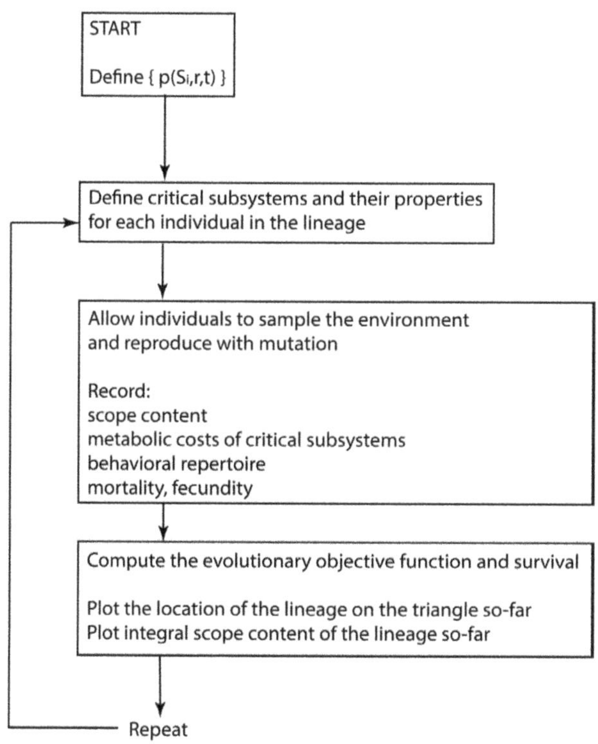

Figure 8. Summary of the proposed model and simulation pointers.

specifying the critical subsystems for a matter-energy processing pathway for at least one nutrient, and by specifying critical subsystems for an information-processing pathway for at least one physical parameter, ending with an effector to instantiate the first flexibility mechanism. Other critical subsystems are specified with some initial conditions. These initial organisms are allowed to explore the environment, reproduce with mutation, feed or die. The simulation records:

- the signals organisms encounter, in order to construct the $\{N(S_i)\}$,
- behavioral response repertoire these organisms managed over their lifetime, to segregate umwelt from gap in each generation,
- metabolic costs of all critical subsystems in each organism and
- mortality, fecundity and offspring survival, to compute the evolutionary objective function that keeps track of costs and benefits for each critical subsystem.

During reproduction, critical subsystems are generated by replication and duplication with mutation (changed parameters), or deleted. A new round of the simulation proceeds making similar measurements. Based on the recorded responses of the simulated organisms to environmental stimuli, their location on the triangle would be established.

Once this simulation becomes sufficiently advanced and successfully incorporates a substantial number of signals, flexibility mechanisms and robustness properties, it can be used to test and explore the applicability of the persistence triangle and the theory of persistence strategies to living organisms:

- Effects of scope density on evolution of critical subsystems.

Do organisms evolve rapidly acting information processing in dense, rapidly changing environments? For example, evolution of learning and behavior in animals is supported by the concomitant evolution of the system of rapid internal, communication. Neural nets supporting electric (as opposed to chemical) transduction accommodate dense mammalian scopes, in contrast to the sparser scopes of plants or fungi that use largely chemical signal transduction mechanisms in their bodies. Myelination further increases the transduction speed, allowing vertebrates to process more information per unit time than any other class of organisms, further improving their ability to use proxies, predict and respond to environmental signals and events, and invade novel environments. Comparisons of the evolution of internal signal transduction in organisms with dense scope vs. sparse scope can help confirm these elements of metazoan make up.

- Effect of the initial conditions on evolutionary paths within the triangle.

Does starting from an extremophilic environment reduce the propensity for evolving rich flexibility mechanisms? Does the initial body size (minimum scope size) constrain the number of bifurcations during reductive and expansive evolution on the triangle (Yafremava *et al.*, 2013)? Is there a minimum number of subsystems in an organism for effectively maintaining expansive evolution into the flexibility corner of the triangle?

- What is the full set of conditions that promote formation of blanket mechanisms of flexibility and robustness?
- What are the environmental factors and elements of internal organization of organisms that promote or impede the various patterns of movement on the persistence triangle: saltations, bifurcations, lifetime movements, etc?

Answering these questions in a simulation will enhance our understanding of the mechanisms of evolution and persistence.

References

Andronova NG, Schlesinger ME (2011) Objective estimation of the probability density function for climate sensitivity. *J Geophysical Res* **106(D19)**: 22605–22611.

Erauso G, Reysenbach A-L, Godfroy A, Meunier J-R, Crump B, Partensky F, Baross JA, Marteinsson V, Barbier G, Pace NR, Prieur D (1993) *Pyrococcus abyssi sp. nov*., a new hyperthermophilic archaeon isolated from a deep-sea hydrothermal vent. *Arch Microbiol* **160(5)**: 338–349.

Fiala G, Stetter KO (1986) *Pyrococcus furiosus sp. nov.* represents a novel genus of marine heterotrophic archaebacteria growing optimally at 100°C. *Arch Microbiol* **145(1)**: 56–61.

Dover GA (1986) Molecular drive in multigene families: how biological novelties arise, spread and are assimilated. *Trends Genet* **2**: 159–165.

Dusenbery DB (1992) *Sensory Ecology*. Freeman and Company, New York.

Grime JP (1977). Evidence for the existence of three primary strategies in plants and its rel-evance to ecological and evolutionary theory. *Am Nat* **111**: 1169–1194.

Kumar PA, Srinivas TNR, Sasikala C, Ramana CV (2008) *Allochromatium renukae sp. nov. Intl J Syst Evol Microbiol* 58: 404–407.

Miller JG (1995) *Living Systems*. University Press of Colorado, Niwot.

Oger PM, Jebbar M (2010) The many ways of coping with pressure. *Res Microbiol* **161(10)**: 799–809.

Perkins SE, Pitman AJ, Holbrook NJ, McAneney J (2007) Evaluation of the AR4 climate model's simulated daily maximum temperature, minimum temperature, and precipitation over Australia using probability density functions. *J Climate* **20**: 4356–4376.

Pitman AJ, Perkins SE (2009) Global and regional comparison of daily 2-m and 100-hPa maximum and minimum temperatures in three global reanalyses. *J Climate* **22**: 4667–4681.

Razvi A, Scholtz JM (2006) Lessons in stability from thermophilic proteins. *Protein Sci* **15(7)**: 1569–1578.

Sommaruga R, Casamayor EO (2009) Bacterial 'cosmopolitanism' and importance of local environmental factors for community composition in remote high-altitude lakes. *Freshw Biol* **55(5)**: 994–1005

Stephens C (1998) Bacterial sporulation: A question of commitment? *Curr Biol* **8(2)**: R45–R48.

von Uexküll J (1909) *Umwelt und Innenwelt der Tiere*. Springer, Berlin.

Yafremava LS, Wielgos M, Thomas S, Nasir A, Wang M, Mittenthal JE, Caetano-Anollés G (2013) A general framework of persistence strategies for biological systems helps explain domains of life. *Front Genet* **4**: 16.

An early cellular origin of viruses

A. Nasir[1,2]
G. Caetano-Anollés[3]

Summary. The data-driven studies of origin of viruses are challenging. Traditional alignment-dependent methods may be ill suited to study virus origins. Over long evolutionary timescales, DNA substitutions may completely erase the evolutionary signal, especially in the case of fast-evolving RNA and retro-transcribing viruses. Other complications such as limited choice of orthologous gene markers conserved across a broad range of organisms further complicate the recovery of a reliable alignment to use for evolutionary inferences. Under such circumstances, it can be better to focus on characters that are better conserved in evolution (such as protein structures) or to explore other alignment-independent alternatives. Here we share our experience of using alternative molecular and morphological characters to study virus origins and evolution. We discuss how our phylogenomic data-driven insights agree

[1]*Theoretical Biology and Biophysics Group, Los Alamos National Laboratory, Los Alamos, New Mexico, USA*
[2]*Department of Biosciences, COMSATS University Islamabad, Islamabad, Pakistan*
[3]*Evolutionary Bioinformatics Laboratory, Department of Crop Sciences, University of Illinois, Urbana, Illinois, USA*

with an early cellular origin of viruses and strongly refute the classical perceptions about viruses as mere byproducts of cellular evolution. We also propose a phylogenomic-data inspired model of virus origination that transforms primordial vesicles of horizontal genetic exchange into full-fledge viruses capable of establishing complex virus-host life cycles.

1. Introduction

Three general hypothetical frameworks have been used to explain the origin of viruses (Figure 1). The "*virus-first*" framework considers that the origin of viruses predated the origin of cells and that viruses contributed to the rise of cellular life (Koonin *et al.*, 2006; Koonin and Dolja, 2014). This framework is supported by the presence of unique viral genes lacking cellular homologs and a disjoined and patchy relationship of their sequences. The "*reductive evolution*" framework considers that cells originated first and that viruses, specifically large DNA viruses, are simply highly reduced forms of parasitic or symbiotic organismal forms (Bandea, 1983; Claverie, 2006). This hypothesis is supported by the discovery of "giant viruses" with genomic and physical features that resemble cells (La Scola *et al.*, 2003; Raoult *et al.*, 2004; Arslan *et al.*, 2011; Philippe *et al.*, 2013; Legendre *et al.*, 2015; Abrahão *et al.*, 2018) and by several phylogenomic analyses (Nasir *et al.*, 2012a; Colson *et al.*, 2018). Finally, the "*escape*" framework suggests viruses are rogue genetic entities that escaped control of diversified cells and later pickpocketed genes using horizontal genetic transfer (HGT) mechanisms, which ensured their

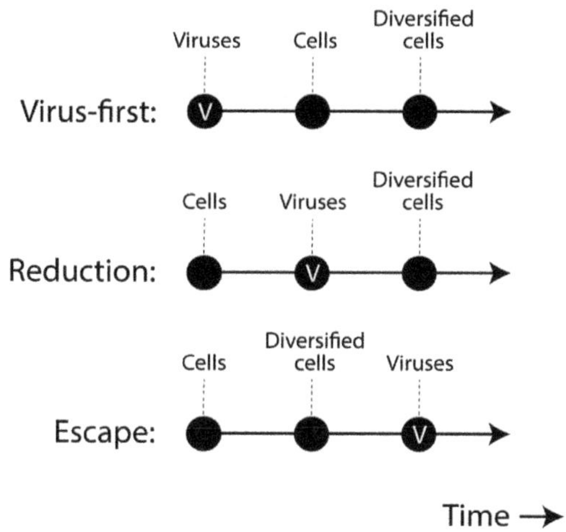

Figure 1. Three general frameworks that explain virus origins. The order in which the origin of viruses, cells and diversified cells of organisms appear in evolution establish three hypothetical frameworks to explain the origin and evolution of viruses. These frameworks can be elaborated into a multiplicity of hypotheses of origin, which must then be subjected to hypothetico-deductive tests.

existence (Moreira and Brochier-Armanet, 2008; Moreira and Lopez-Garcia, 2009). HGT is indeed a widespread process in the evolution of microorganisms (Jeong *et al.*, 2019) with enormous potential to shape the genomes of viruses (Malik *et al.*, 2017). Note that many alternative scenarios are possible within each framework (Forterre and Krupovic, 2012) including the hybrid models combining elements from the virus-first and escape models (Krupovic *et al.*, 2019) and those differentiating between the cell-like and parasite-like stages in virus evolution (Nasir *et al.*, 2012b). Similarly, all frameworks involve the ability of viruses to establish "virospheres," environments external to cells where they impact cellular life. Thus, viruses embody *"panspermia"* (from Ancient Greek πᾶν (pan), meaning "all," and σπέρμα (sperma), meaning "seed"), the ability to spread genetic agents that impact cellular life. Here, we do not use the term with a connotation of extraterrestrial origin, which is highly controversial, but rather as an appropriate term for genetic spread at planetary level.

2. Basic science questions are often the hardest

The pace of scientific research has accelerated significantly in the 21st century. The availability of online resources, technological advances, and increased international collaborations mean we are now advancing scientific discovery at unprecedented scales. For example, ~1.6 million research papers were published in 2018 that were listed in the Web of Science (2017–2018) database, a 5% increase from the previous year and the highest ever (Makri, 2018). This amounts to >4,300 papers published every day! Increased scientific output can be directly linked to knowledge creation, innovations, economic growth, and the welfare of our society. For example, 2018 saw significant advances in restoring voluntary movement in paralyzed individuals (Wagner *et al.*, 2018), reversing Alzheimer's in mice (Hu *et al.*, 2018), and exploring new horizons in space (Hand, 2018). Ironically, humans seem amazingly capable of solving complex scientific problems but remain divided over fundamental questions related to our evolution and classification. For example, there is no agreement about how to define "life" (Trifonov, 2011). Even if "life" is restricted to cellular organisms, it is undecided how many distinct domains or superkingdoms of life (Woese *et al.*, 1990) should be recognized and what are their evolutionary relationships (Woese *et al.*, 1990; Williams *et al.*, 2013; Nasir *et al.*, 2015a, 2016; Spang and Ettema, 2016; Da Cunha *et al.*, 2017; Spang *et al.*, 2018; Staley and Caetano-Anollés, 2018). Viruses pose additional challenges (Claverie and Abergel, 2016). Their status as living or non-living, definitions, classifications, and origins, create more confusion in the minds of scientists and public alike (Forterre, 2016).

3. Standard methods are not suited for Tree of Life research

The major obstacle of Tree of Life (ToL) research lies in our inability to accurately reconstruct evolutionary events that happened billions of years ago. Recent attempts have focused on answering these questions using genomic data that has accumulated rapidly over the past many years. For example, the NCBI Genome Browser provides genomic information for ~250,000 assemblies from prokaryotes, eukaryotes, viruses, and organelles (April 2019). However, "gold standard" approaches towards resolving deep phylogeny that include comparative genomics and gene phylogenies do not work well for the most ancient molecules (Hoeppner *et al.*, 2012) and lead to mixed results when the datasets are changed causing great confusion and controversies. One illustrative example is the exchange between Spang *et al.* (2015) and Da Cunha *et al.* (2017) regarding the position of eukaryotes in the ToL (read below).

In our view, three major limitations of the standard phylogenetic approaches can prohibit reliable reconstruction of the deep evolutionary splits.

(i) The standard methods rely, by default, on a good sequence alignment. These methods can therefore be referred to as alignment-dependent methods. These kinds of alignments refer to similarities between sets of atoms (nucleotide or amino acid residues) in macromolecules with known correspondences but restricted to their order. These correspondences induce alignments of parts of sequences. Unfortunately, alignment can sometimes become the "weakest" link in evolutionary studies (Morrison, 2009). Due to the relatively high rates of mutations and substitutions that alter the genomic record of an organism, comparing gene and protein sequences from distantly related organisms can misguide evolutionary interpretations

(Caetano-Anollés and Nasir, 2012). In general, sequence similarity fades over evolutionary time (Sober and Steel, 2002). Over long timescales, evolutionary signal can sometimes be completely lost, as shown recently in the case of RNA viruses by Holmes and Duchêne (2019). Thus, great caution needs to be taken in generating a robust and balanced alignment dataset for ToL experiments (Nasir *et al.*, 2016).

(ii) Another complicating factor is the underlying process of how genes and proteins evolve. Proteins are composed of domains that are independent structural, functional, and evolutionary units (Chothia *et al.*, 2003; Caetano-Anollés *et al.*, 2009). Protein domains can be gained, lost, rearranged, or duplicated at unequal rates in the genomes of different lineages (Nasir *et al.*, 2014b). These evolutionary events make it challenging to reliably juxtapose or align full-length protein sequences from diverse groups of organisms and viruses. Often, such alignments require significant manual inspections and curations (e.g. deletion of sites harboring >50% gaps), practices that impact reproducibility of results (Baurain *et al.*, 2017). We emphasize these problems with two recent examples. Spang *et al.* (2015) recently proposed an archaeal origin for the eukaryotic cell. Their conclusion relied on a phylogenetic tree reconstructed from the concatenation of 36 "conserved" proteins in 104 organisms (from 84 Archaea, 10 Bacteria, and 10 Eukarya). Due to the diverse composition of taxa, a large number of positions in the alignment had to be discarded in their analysis (i.e. sites with >50% gaps) (Spang *et al.*, 2015). In fact, out of the 36 concatenated proteins, 15 (42%) indicated alignment-related problems (Figure 2A). We illustrate this problem using the alignment of phenylalanyl-tRNA synthetase (PheRS) protein, one of the members of their concatenated dataset (Figure 2B). PheRs length varies naturally across Archaea, Bacteria, and Eukarya due to the differential

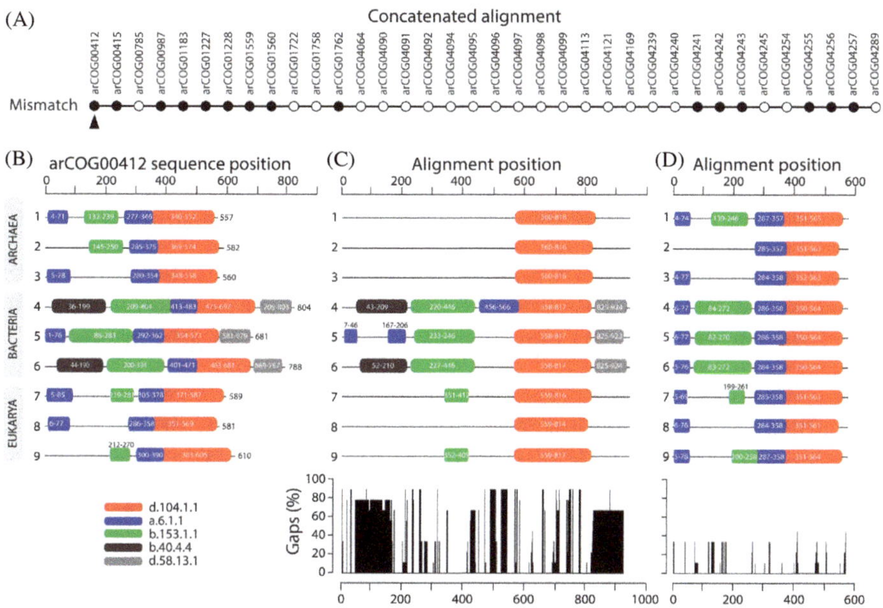

Figure 2. Alignments can distort the underlying protein domain structure and generate uncertainties in data analysis. (A) Out of the 36 archaeal clusters of orthologous groups (arCOGs) in Spang *et al.* (2015), 15 indicated alignment-related problems (filled circles). (B) Protein domains were detected in the raw unaligned sequences of PheRS protein (arCOG00412, see arrowhead) using SUPERFAMILY HMMs (Gough *et al.*, 2001; Gough and Chothia, 2002) (ver. 1.75). For demonstration purposes, results are shown only for 9 organisms (3 each from Archaea, Bacteria, and Eukarya). These organisms included 1, *Desulfurococcus kamchatkensis*; 2, uncultured Marine Group II euryarchaeote; 3, *Lokiarchaeum*; 4, *Bacillus subtilis* 168; 5, *Rhodopirellula baltica* SH1; 6, *Thermotoga maritima* MSB8; 7, *Homo sapiens*; 8, *Trichomonas vaginalis*; 9; *Tetrahymena thermophila* PT. Domains are defined using SCOP concise classification strings (*ccs*) (Andreeva *et al.*, 2008). (C) Aligning these sequences using MAFFT (Nakamura *et al.*, 2018) distorts domain read-through in an attempt to balance the alignment length. (D) A trimmed alignment solves some problems but eliminates two crucial domains, the Myf domain (b.40.4.4) and the anticodon-binding domain of PheRS (d.58.13.1) and shortens or splits the others. The histograms below show the percentage of gaps in each column of the alignments. Figure modified from Caetano-Anollés *et al.* (2018).

composition of protein domains that make up the protein (Figure 2B). Hence, alignment is only possible via the introductions of several gaps (insertions/deletions) that are unaware of domain order in protein sequences. As a result, several gaps are introduced within sequences encoding crucial domains effectively altering the sequence beyond recognition even by the specialist software for remote homology detection such as hidden Markov models (HMMs) (Gough *et al.*, 2001; Gough and Chothia, 2002) (Figure 2C). Trimming the alignment only partially solves the problem (Figure 2D). For example, trimming sites harboring >50% gaps, as in Spang *et al.* (2015), leads to complete loss of the crucial anticodon-binding domain (d.58.13) and the reduction of original sites from 925 to 571 (38% loss of data) (Figure 2D). Thus, trimming can have undesirable and often unpredictable consequences and can potentially remove segments of protein structures that are crucial from function and evolutionary perspectives.

Another example is the recent study of RNA virus origins and evolution by Wolf *et al.* (2018). Their analysis, based on the alignment of a single gene, which encoded RNA-dependent-RNA-polymerase, originally included >12,000 sites across >4,000 viral genomes, of which, only 3.6% (441) remained after curation (Holmes and Duchêne, 2019). While both the Spang *et al.* (2015) and Wolf *et al.* (2018) studies suggested major paradigm shifts in evolutionary biology, the reconstructed trees explained only a minor portion of the original data. In fact, it is logical to think that given a diverse and distantly-related set of genomes, gene or protein sequences cannot be safely aligned without throwing away large chunks of data. The discarded data contain valuable information about past evolutionary events (mutations, substitutions, gene duplications, transfers, and domain rearrangements) that should however be part of the scenarios describing the evolution of genomes being studied. These problems therefore highlight the need to model domain heterogeneities or insertion/deletions in proteins prior to alignment, practices that are rarely followed. Moreover, concatenated alignments can become worse than individual alignments since such problems can be amplified over the entire dataset. Unsurprisingly, therefore, the archaeal-ancestor scenario for the origin of eukaryotes was not supported by a large number of individual protein trees (see Da Cunha *et al.*, 2017 for a detailed reanalysis).

(iii) A third major problem associated with alignment-dependent or standard methods is the number and choice of "universal" genes that can be aligned for a diverse group of organisms. When sequences have evolved over billions of years, BLAST-based and even profile-based methods (e.g. HMMs) may perform poorly in recognizing distant homologies. Thus, many orthologous genes are simply missed from the analysis. In fact, only a few genes may be considered truly "universal" among cells (~60) (Dagan and Martin, 2006) and this number declines further when DNA viruses are included (7 in Raoult *et al.*, 2004). The solution in some of such situations is to settle for imbalanced taxon sampling and focus only on groups for which a large number of orthologs could be identified. This was perhaps the reason for the low number of bacterial and eukaryal taxa sampled in the Spang *et al.* (2015) dataset (10 each versus 84 Archaea) since their objective was to place the newly discovered "Lokiarchaeota" genomes in the ToL (Spang *et al.*, 2015). We have shown previously how such imbalanced taxon datasets can lead to contrasting topologies when the taxa are changed (Nasir *et al.*, 2016; Da Cunha *et al.*, 2017; Gaia *et al.*, 2018).

In summary, identifying a broad range of orthologous genes conserved across an equal and large number of genomes from diverse organism and viral groups and then recovering a biologically meaningful and reproducible alignment for phylogeny reconstruction can become a daunting, if not an impossible, task. It may be wise to reduce alignment trimming or to use information from conserved "indels" to infer evolution (Da Cunha *et al.*, 2017). Collectively, these problems alone should encourage scientists to explore alternative solutions or characters when the objective is to resolve very deep evolutionary scenarios. Despite these limitations, alternative characters are rarely used or even discouraged in contemporary evolutionary biology studies.

4. Protein structures offer many more useful characters to study virus origins

We have previously proposed to study ancient evolutionary relationships using conserved molecular features such as the fold structures of proteins (Caetano-Anolles and Caetano-Anollés, 2003; Caetano-Anollés and Nasir, 2012) and molecular functions (Kim and Caetano-Anollés, 2010; Nasir and Caetano-Anollés, 2013). There is often a direct link between protein structure and function, which places strong evolutionary constraints on structure to remain conserved for a longer period of time. Protein structures are thus better preserved in evolution (Illergård *et al.*, 2009) and can offer insights that are probably invisible to sequence-based methods. Fortunately, structural data are starting to increase in databases, as genome data did in the last decade. With ~150,000 structures available in the RCSB Protein Data Bank (Rose *et al.*, 2015) (April 2019), we can now revisit evolutionary scenarios using new sets of molecular characters.

In a recent analysis, we assigned protein structures to more than 11 million viral and cellular proteins (Nasir and Caetano-Anollés, 2015). Assigned structures were grouped into protein fold superfamilies (FSF) that are collections of distantly related protein domain structures (Andreeva *et al.*, 2008). We learned four major lessons from focusing on protein structures for deep evolutionary studies.

(i) While sequence-based studies repeatedly confirmed that there is no single gene unifying all viruses (Moreira and Lopez-Garcia, 2009), we could immediately identify 442 FSFs that were conserved across diverse groups of viruses and the cells of three superkingdoms. These folds comprised roughly one-fourth of all the defined superfamilies (i.e. 442 out of 1,995) studied in Nasir and Caetano-Anollés (2015). Thus, protein structures offered many more characters to study viral evolution. On balance, however, these structures were patchily dispersed in the sampled viral proteomes (only two were present in more than half of all viruses). This could indicate past and ongoing reductive tendencies in the proteomes of viruses (Nasir and Caetano-Anollés, 2015).

(ii) Many of the conserved FSFs in viruses and cells were membrane proteins that originated very early in evolution (Nasir *et al.*, 2015b). This led us to propose that viruses had a cellular history in their past and that viruses evolved from cells that either co-existed or pre-dated the ancestors of modern cells of superkingdoms Archaea, Bacteria, and Eukarya (Forterre, 2005; Forterre and Krupovic, 2012; Nasir *et al.*, 2012a,b). This conclusion was also supported by a phylogenomic tree of proteomes (taxa) and FSF abundance (442 universal characters) that pushed viruses to the base of the "four-supergroup" ToL and identified segmented RNA viruses as the most ancient viruses (Figure 2 in Nasir and Caetano-Anollés, 2015).

(iii) We identified 66 FSFs that were uniquely detected in viruses (i.e. virus-specific protein structures) (Nasir and Caetano-Anollés, 2015). It is sometimes quoted that viruses evolve by capturing genes from their hosts or cellular organisms (i.e. via cell-to-virus gene transfer) (Moreira and Brochier-Armanet,

2008; Moreira and Lopez-Garcia, 2009). This paradigm, called the "virus pickpocket paradigm" has been rightfully criticized (Forterre, 2016) since it generalizes virus evolution. Viral genomes are best characterized by the presence and abundance of a large number of genes that lack any detectable resemblance to cellular genes either at the sequence and structure levels (called ORFans) (Ogata and Claverie, 2007; Yin and Fischer, 2008; Cortez *et al.*, 2009; Nasir and Caetano-Anollés, 2015; Legendre *et al.*, 2018). Their origin thus cannot be simply attributed to cells since these genes have no representation in cells. Claverie and coworkers recently presented compelling evidence that virus-specific genes are created directly in viral lineages, *de novo* from the intergenic regions (Legendre *et al.*, 2019). Thus, the virus pickpocket paradigm is based on a very small subset of virus genes for which a cellular homolog can be identified (typically <30% even after implementing advanced profile based methods of remote homology detection according to our calculations; Nasir and Caetano-Anollés 2015) and is misleading. The rest, which is the majority, includes an astonishing number of genes and protein structures that are virus-specific (e.g. >90% in the genome of *Pandoravirus salinus*) (Philippe *et al.*, 2013). These facts support the idea that viruses are gene creators, not gene robbers, and that virus-to-cell gene transfer may be greater than gene transfer in the direction from cells-to-viruses (Forterre, 2011; Forterre and Prangishvili, 2013).

(iv) The classic version of the escape hypothesis constrains virus evolution to their host cells. Thus, viruses that evolve from a particular host should share more genetic similarities with their hosts rather than other cells. To test, we classified viruses into archaeoviruses, bacterioviruses (also known as bacteriophages) and eukaryoviruses based on the primary taxonomy of their infected host. Surprisingly, we discovered that several bacteriovirus genes were shared with eukaryotic cells (and not

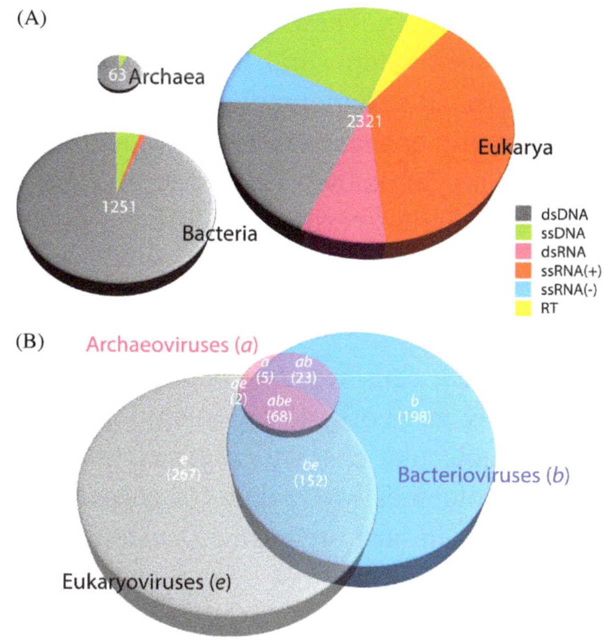

Figure 3. The replication modes and host preferences of viruses. (A) Pie-charts describing the abundance of viruses exhibiting different replication strategies in the three superkingdoms of life. Data and figure reproduced or modified from Nasir *et al*. (2014a). (B) Venn diagram showing the distribution of 715 FSFs detected in archaeoviruses (*a*), bacterioviruses (*b*) and eukaryoviruses (*e*). Venn groups represent FSFs specific to viral groups, *a* (archaeoviruses), *b* (bacterioviruses), *e* (eukaryoviruses) or present in two (*be*, *ae*, and *ab*) or in all three groups of viruses (*abe*). Data and figure reproduced or modified from Nasir and Caetano-Anollés (2015).

eukaryoviruses) and several eukaryoviral genes were shared with bacterial cells (and not bacterioviruses) (Malik *et al.*, 2017). We also identified 68 protein folds that were conserved among the three classes of viruses by host type (Nasir and Caetano-Anollés, 2015) (Figure 3). Previously, structural biologists have also revealed striking structural similarities in viruses infecting distantly related hosts (Bamford, 2003;

Abrescia *et al.*, 2012). These data are therefore compatible with the origin of modern viral lineages predating the emergence of modern cells (Forterre, 2005) and the intriguing possibility that genetic exchange is not restricted between viruses and their primary or known hosts but may even transcend known host boundaries (Nasir *et al.*, 2017).

5. Infection: An atypical way to study virus origins?

In addition to molecular data, we also focused on rather "atypical" viral features to study virus origins and evolution. Instead of looking for conserved molecular characters (genes, proteins, or protein structures) across proteomes, we looked for conserved virus infections (or lack thereof) in host organisms (Nasir *et al.*, 2014a). We hypothesized that over the course of evolutionary history, different cellular lineages have customized their virospheres (i.e. the ensemble of viruses that infect a particular group). This process has resulted in the gain-and-loss of viral lineages across host groups (Nasir *et al.*, 2015b). Tracing these gain and loss patterns can reveal how organisms and viruses have evolved over time.

We studied virus diversity based on types of genomic replicons and the methods of replication that are associated with each host superkingdom (Nasir *et al.*, 2014a). This diversity has been catalogued into seven classes, dsDNA, ssDNA, dsRNA, plus-ssRNA, minus-ssRNA, and retro-transcribing DNA and RNA replicons (Baltimore, 1971). Initially, we were surprised to note that there were no known RNA archaeoviruses (Figure 3A). In our phylogenies, RNA viruses occupied basal positions followed by retro-transcribing and DNA viruses (Nasir and Caetano-Anollés, 2015). They were thus the most ancient group among extant viruses. Similarly, Archaea formed a basal paraphyletic group in the cellular subtree separate from Bacteria and Eukarya (Nasir and Caetano-Anollés, 2015).

Archaea were thus identified as the most ancient cellular superkingdom (Caetano-Anollés *et al.*, 2014; Staley and Caetano-Anollés, 2018). The absence of RNA viruses in Archaea was therefore incompatible with our phylogeny. Since RNA viruses have been discovered in both Bacteria and Eukarya (Figure 3A), the best reconciliation was to invoke early loss of RNA viruses in ancestral archaeal cells because loss in one is more parsimonious than gain in two. This reconciliation is biologically well supported. Archaea are generally characterized by their preference to colonize extreme habitats (high or low temperatures, soil, etc.) and it is likely that the last common archaeal ancestor was a hyperthermophile (Brochier-Armanet *et al.*, 2011; Kim *et al.*, 2014; Nasir *et al.*, 2014c). Since RNA is relatively unstable at high temperatures (or extreme conditions), it could have been a survival strategy for ancestral archaeal cells to migrate to warmer habitats to get rid of ancient RNA viruses. Based on these observations, we supported the hypothesis that RNA viruses evolved first from ancient cells (Forterre, 2005). Adaptation of ancestral archaeal cells to warmer habitats marked the early rise of paraphyletic Archaea and ultimately loss of RNA viruses in Archaea. The last common ancestor of Bacteria and Eukarya continued to be infected by the ancestral RNA viruses. The method however obviously suffers from the huge ascertainment bias towards the discovery and sequencing of economically or health related viruses. For example, viruses of Archaea have been particularly understudied (Pina *et al.*, 2011; Prangishvili, 2013). Thus, future sequencing and discovery of novel viruses from diverse hosts and habitats will directly test our predictions. Similarly, one can wonder why ancient RNA viruses did not ease the pressure on early archaeal cells and thus prevented their migration to warmer habitats, a feature of antagonistic symbiosis. These are open questions. Nevertheless, atypical and novel solutions to hard and old problems should be welcomed in our quest to answer tough questions.

This gain-and-loss of virus lineages model for the evolution of cells and viruses supports an alternative version of the "Woesian" ToL (Woese *et al*., 1990) where Bacteria and Eukarya are sister groups and Archaea is the most ancient superkingdom (Di Giulio, 2007; Wang *et al*., 2007; Caetano-Anollés *et al*., 2014, 2017; Staley and Caetano-Anollés, 2018). Surprisingly, virus distributions in Archaea and Eukarya are markedly different. This contradicts evolutionary scenarios that suggest that the eukaryotic cell arose from an ancestral archaeon or a fusion between a bacterium and a host archaeon (Lake *et al*., 1984; Rivera and Lake, 2004). Under these scenarios (Spang and Ettema, 2016), one should expect to find a greater overlap of the genomes (or protein folds) of archaeoviruses and eukaryoviruses, which is however not the case. Remarkably, only two protein folds are unique to archaeoviruses and eukaryoviruses compared to 152 that are unique between bacterioviruses and eukaryoviruses and 23 between prokaryotic viruses (Figure 3B) (Nasir and Caetano-Anollés, 2015). There is therefore little or almost no overlap between the archaeoviruses and eukaryoviruses and a very significant overlap between bacterioviruses and eukaryoviruses, which further supports their origin from a common ancestor. In fact, Forterre (2015) provides additional arguments evaluating the biological plausibility of the archaeal ancestor scenario for the origin of eukaryotes.

6. Morphological characters may indicate a polyphyletic origin of virus

In our quest for atypical or *out-of-the-box* solutions to explain the origin and evolution of viruses, we also studied the emergence of virion morphotypes in the viruses of the three superkingdoms on the ToL. There are numerous distinct resolved virion shapes, especially in archaeoviruses (lemon-shaped, bottle-shaped, bullet-shaped) (Prangishvili, 2013) and eukaryoviruses (droplet-shaped, allantoid, flexuous, prolate) (see Nasir *et al*., 2014a for a distribution of virion morphyotypes). In contrast, the majority of bacterioviruses produce the classical head-tail virions (shared feature with some archaeoviruses) (Nasir *et al.*, 2017). Surprisingly, however, very few virion morphotypes are common to viruses infecting two or three superkingdoms. Only two virion morphotypes, filamentous and spherical, can be observed in archaeoviruses, bacterioviruses, and eukaryoviruses (Nasir *et al.*, 2015b), i.e. the viruses infecting the three superkingdoms of life. These virions are also the "simplest" shapes to assemble geometrically. In our phylogenies, viruses with filamentous and spherical virions occupied basal positions (Nasir and Caetano-Anollés, 2015). This led us to speculate that first virions were either spherical or filamentous (or both) that originated from one or multiple ancient cells (i.e. polyphyletic origins of viruses). These virions later became more decorated with additional features (e.g. double-capsid, spike) or perhaps joined together to form the classical head-tail structure in *Caudovirales* (Nasir *et al.*, 2014c).

Structural similarities among distantly related viruses however can also arise from convergent evolution. Hence, virion morphotype data and its power in explaining virus origins should also be taken with caution and is, at best, currently speculative. Note also that the origin of virions is not necessarily the origin of viruses. It is a big mistake to assimilate viruses to their virions, as now emphasized by several scientists (Bandea, 1983; Van Regenmortel, 2010; Forterre, 2016; Claverie and Abergel, 2016). The virion is the virus particle that disseminates virus genetic information. It is metabolically inert and is functionally analogous to bacterial spores or plant seeds (Claverie and Abergel, 2016). Viruses have thus been re-defined as capsid-encoding organisms (Raoult and Forterre, 2008) and later as virion-encoding organisms to emphasize on the intracellular stage of virus reproduction cycle and to distinguish them from their virions (Forterre, 2016). These definitions however also attempt to describe viruses based on the presence or absence of molecular or physical features and given the stark diversity among viruses (e.g.

capsid-less viruses) cannot be universally applied (Claverie and Abergel, 2016). Hence, we favor the alternative and more generic definition of viruses based on absolute parasitism (i.e. crucial dependency to reproduce using the macromolecular machinery of another biological entity) and generic modes of genome dissemination in any metabolically inert particle regardless of size or structure (Claverie and Abergel, 2016).

7. An "extrusion" model of viral panspermia: from vesicles to viruses

The protein structure-based census of the proteomes of cells and viruses reveal the existence of many ancient membrane-associated protein structures in virus genomes that are universal in cells (Nasir and Caetano-Anollés, 2015). This suggests a cellular history in the early evolution of modern viruses (Forterre, 2005; Nasir *et al.*, 2012b). Because a large number of virus genes involved in key virion assembly pathways and pathogenicity are shared by viruses infecting the three superkingdoms of life (Malik *et al.*, 2017), it is likely that the cellular ancestors of viruses either co-existed or pre-dated the ancestors of Archaea, Bacteria, and Eukarya (i.e. originated from ancient cells rather than modern cells (Forterre 2005). Viruses likely evolved from ancient cells via reductive evolution (Bandea, 1983; Claverie, 2006; Nasir *et al.*, 2012a; Claverie and Abergel, 2016), which is the dominant mode for the emergence of obligate parasites (McCutcheon and Moran, 2011) (see Nasir *et al.*, 2011 for a detailed analysis of organism lifestyles), although escape or multiple ways of emergence cannot be ruled out either (Forterre and Krupovic 2012). These ancient cells (or membrane-bound compartments) likely contained segmented RNA genomes and the earliest virus ancestors formed primitive or simple virions, as shown by the emergence of virus replicons and spherical and filamentous virions on the base of our ToL (Nasir and Caetano-Anollés, 2015),

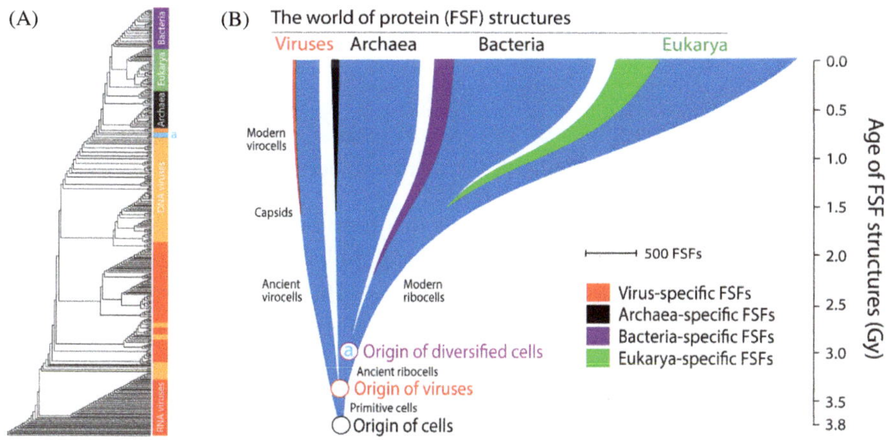

Figure 4. Phylogenomic analysis of the origin and evolution cells and viruses. (A) A tree of life (ToL) reconstructed from the occurrence of universal fold superfamilies (FSFs) defined by SCOP classification in 368 randomly sampled proteomes from superkingdoms Archaea, Bacteria and Eukarya and viruses. The proteome of the last common ancestor of modern cells (a) was added as an additional taxon to infer the evolution of evolutionary splits. Data and figure reproduced from Nasir and Caetano-Anollés (2015). (B) Cartoon of a universal tree of life inferred from a phylogeny of protein domains defined at FSF level (Nasir and Caetano-Anollés, 2015). Time unfolds from bottom to top according to the age of FSF protein domains in geological time (billions of years, Gy) according to a molecular clock of folds (Wang *et al.*, 2011). The horizontal axis is proportional to the number of FSFs. The FSFs that are unique to supergroups are highlighted with different colors in the phylogeny. The common ancestor of cells indicated with a purple circle (a) gives origin to modern diversified cells (ribocells) containing fully functional ribosomes. Data and figure modified from (Nasir *et al.*, 2012b).

and as previously argued (Forterre and Krupovic, 2012). RNA viruses evolved into retro-transcribing and later into DNA viruses (Forterre, 2005). Thus, according to our analysis, RNA viruses are the ancestors of DNA viruses. In fact, our ToL positions RNA viruses (especially filamentous viruses) at its base (Figure 4) (Nasir and Caetano-Anollés, 2015). Similarly, the most ancient protein structures that appear at the base of the tree are common to cellular

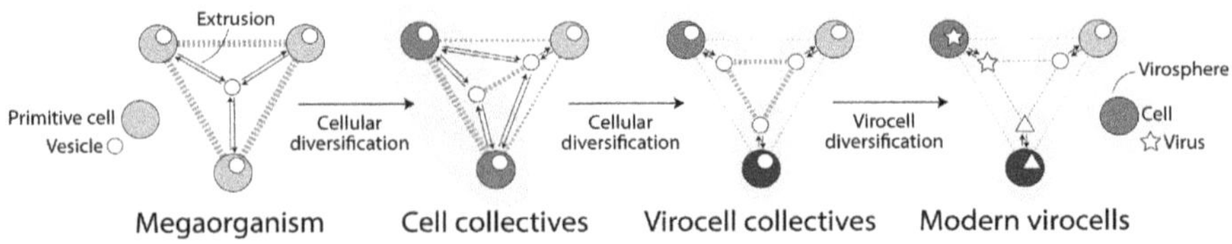

Figure 5. An extrusion model of virus origination: from vesicles to viruses. At the onset of life, primitive cells (big shaded circles) are free to exchange macromolecular and cellular make up, sometimes in the form of extruded vesicles. This in-out "extrusion" process fostered a single cell collective, a "megaorganism." Dashed lines connecting cells and vesicles describe copious biological interactions between them by free exchange of cellular and vesicle make up. With time, a cellular subset of the collective diversified (illustrated with grey-colored cells) by establishing preferential interactions between them. This is illustrated with weaker lines connecting the cellular subset with the rest. At a later stage, preferential interaction between cells of the subset resulted in preferential interaction of the extruded vesicles. This resulted in "virocell" collectives, in which extrusion occurred preferentially in each cellular subset. When cells stored memory and developed their own identities (rise of modern genetics) they diversified into lineages, carrying with them their more specific extruded vesicles. These modern virocells engaged in complex lifecycles. It is here when virions appear to enhance the panspermic process (illustrated with different shapes). Vesicles are now modern viruses and diversified cell lineages formed modern ribocells.

life and viruses. Their appearance is followed by structures that are common to cells. Thus, phylogenomic methods strongly support an early cellular origin of viruses, prior to the rise of diversified cellular lineages.

We previously developed an extrusion model to explain the cellular origin and ancient evolution of viruses (Nasir *et al.*, 2014b), which we here further elaborate (Figure 5). In absence of lineages, our model considers that the most primitive cells were free to exchange macromolecular and cellular make up, sometimes in the form of extruded vesicles. This "extrusion" process everted membrane and cellular contents and facilitated the formation of a single cell collective, a panorganism of planetary significance that we will call "megaorganism." The vesicles facilitated copious biological exchange between the primordial cells of this cellular entity. This exchange ensured the entity's planetary existence despite challenges from meteorite impacts, volcanism, and glaciations. Exchange was beneficious and parasitism was non-existent. With time, cellular subsets of the megaorganism diversified by adapting to the changing planetary environments. Isolation provided opportunities for innovation but curtailed the retention of cellular materials that vesicles widely exchanged. This resulted in a multiplicity of cellular collectives, which developed their own identities while widely exchanging materials with counterparts. At a later stage, preferential interaction between cells of the subsets also resulted in preferential interaction with their extruded vesicles. This limited the free exchange of cell and macromolecular components and resulted into a new kind of collective, a "virocell" collective, in which extrusion occurred preferentially in a cellular subset. We borrow here Forterre's virocell concept, a cell that upon virus infection produces viral particles rather than dividing into daughter cells (Forterre, 2011). We note, however, that the virocell collectives were still not "infected" by the vesicles, which were not pathogenic. Instead, vesicles freely distributed cellular wealth that was beneficial to their cellular counterparts, including metabolites, polypeptides, proteins, and nucleic acids. The rise of genetic "memory" and efficient ribosome-mediated translation helped cells develop their own cellular identities. The emergent "ribocells" gave rise to diversified lineages, which abandoned the copious horizontal exchange of their predecessors. However, and despite of having developed a "cellular self," the emerging lineages continued to reap the benefits of the vesicle-mediated panspermic process by establishing complex cell-vesicle "life cycles." These cycles fostered specific modes of virus propagation (e.g. virions, virus factories) and replication (e.g. viral replicons, replication strategies) that are typical of modern virocells. It is here when virions appear to enhance the panspermic process of the diversified cellular world, becoming true modern viruses. It is here also when viruses engaged in parasitic interactions. Consistent with these thoughts, key proteins that are components of modern viral capsids (e.g. the jelly-roll fold) appear late in evolution, a timepoint that strikingly coincides with the diversification of cells (Nasir and Caetano-Anollés, 2015). Interestingly, vesicle-mediated communication is now recognized as a universal feature of cellular life and a useful mean of transferring DNA (Barteneva *et al.*, 2013). Secreted vesicles share many biophysical properties with enveloped viruses and can even package viral proteins and RNA from virus-infected cells (Meckes Jr *et al.*, 2011). In fact, neuron cells secrete vesicles that resemble capsids from retroviruses (Ashley *et al.*, 2018; Pastuzyn *et al.*, 2018). Thus, there are striking similarities between extracellular vesicles and viruses and they could possibly be related (Nolte-'t Hoen *et al.*, 2016). They may share an ancient origin.

8. Synthesis

Objective reevaluations of virus origins and evolution scenarios are not possible, or very difficult, when using standard comparative and phylogenomic sequence-based methods. This is largely because of the fast mutation/substitution rates of DNA and the inability of existing algorithms to recognize the modular structure of gene and protein sequences. Our focus on alternative molecular characters (e.g. protein structure) and other atypical lines of investigation (virus-host mappings, virion morphotypes) overcome many of the technical limitations associated with standard methods and support the early cellular origin of viruses from primordial cells. These ancient cells co-existed or predated the ancestors of modern viruses, Archaea, Bacteria, and Eukarya. Figure 5 describes a co-evolutionary model explaining the origin of cells and viruses. According to our analysis, ancestors of RNA viruses evolved first from these ancient cells (likely via extreme forms of reductive evolution) and later evolved into retrotranscribing and DNA viruses. Thus, RNA viruses could be the ancestors of DNA viruses. Initial virions were likely not made up of sophisticated protein shells or capsids we observe in modern day viruses but rather had no restrictions on size or shape and simply were means to transmit cellular or genetic materials from cell to cell thus establishing an ancient mode of cellular communication.

In addition to elaborating virus origins, we also highlight recent findings that strongly challenge established or historical views regarding virus evolution. Viruses are capable of creating genes *de novo*, as shown by recent experimental and theoretical studies (Forterre, 2011; Legendre *et al.*, 2018). These newly created genes could be vastly different among even closely related strains of even the same pandoravirus family (the largest viruses known to date) (Philippe *et al.*, 2013; Legendre *et al.*, 2019). This suggests that the large size of many giant virus genomes is due to the very fast process of *de novo* gene creation and not due to HGT from host cells, as claimed by Moreira and Brochier-Armanet (2008). This also shows that new genes can originate directly in viral lineages (Forterre, 2011) without invoking the pre-cellular virus world scenario (Koonin and Dolja, 2014). In fact, viral genes can be frequently transferred to cellular genomes and these transfers can sometimes transcend known virus host boundaries (Malik *et al.*, 2017). The *de novo* gene creation in viruses and virus-to-cell HGT has been occurring over billions of years (Forterre and Prangishvili, 2009) and thus explain the large number of virus-specific genes without any cellular homologs and multiple insertions of virus genes and genetic elements into host genomes (Feschotte and Gilbert, 2012). Moreover, the concept of the "virus host" could be broader than what is generally believed.

There are now numerous examples of virus-cell interactions that are beneficial in nature (Roossinck, 2011, 2015; Nasir *et al.*, 2017). Genetic exchange can probably even occur among bacterioviruses and eukaryotes and eukaryoviruses and bacteria (Malik *et al.*, 2017). A stunning recent finding revealed that antibiotic resistant *Pseudomonas aeruginosa* bacterium produces pf phage (lysogenic to the host bacterium) that is endocytosed by the mammalian cells upon bacterial infection (Sweere *et al.*, 2019). Inside the mammalian cells, the pf phage produces its mRNA which prevents bacterial phagocytosis by the mammalian immune system (Sweere *et al.*, 2019). This is a striking bacteria-virus symbiosis providing preliminary proof that bacteriovirus RNA can be produced inside mammalian cells and involves multiple layers of virus-cell interactions that transcend known virus host boundaries (Sweere *et al.*, 2019). Another finding showed how segmented viruses can distribute reproduction in multiple host cells defining a multicellular way of "life" for viruses (Yvon *et al.*, 2019). Thus, defining viruses as mere pickpockets or robbers of cellular genes (Moreira and Lopez-Garcia, 2009) is an incorrect over-simplification and is incompatible with multiple lines of investigation. Viruses have played and continue to play major roles in the evolution of cellular organisms (Forterre, 2006) and there are big biases towards studying clinically or economically relevant viruses. Discovery of novel viruses from diverse hosts and habitats is therefore expected to unravel many more mysteries and are bound to change how we think about viruses and life itself.

Acknowledgements

GCA thanks the support of the National Science Foundation (OISE-1132791) and the National Institute of Food and Agriculture of the U.S. Department of Agriculture (ILLU-802-909 and ILLU-483-625). AN thanks the support of an Oppenheimer Fellowship and the

U.S. Department of Energy LDRD program at Los Alamos National Laboratory. GCA and AN thank Patrick Forterre for critical reading of the manuscript.

References

Abrahão J, Silva L, Silva LS, Khalil JYB, Rodrigues R, Arantes T, Assis F, Boratto P, Andrade M, Kroon EG, Ribeiro B, Bergier I, Seligmann H, Ghigo E, Colson P, Levasseur A, Kromere G, Raoult D, La Scola B (2018) Tailed giant Tupanvirus possesses the most complete translational apparatus of the known virosphere. *Nat Commun* **9**: 749.

Abrescia NGA, Bamford DH, Grimes JM, Stuart DI (2012) Structure unifies the viral universe. *Annu Rev Biochem* **81**: 795–822.

Andreeva A, Howorth D, Chandonia JM, Brenner SE, Hubbard TJP, Chothia C, Murzin AG (2008) Data growth and its impact on the SCOP database: new developments. *Nucleic Acids Res* **36**: D419–D425.

Arslan D, Legendre M, Seltzer V, Abergel C, Claverie J-M (2011) Distant Mimivirus relative with a larger genome highlights the fundamental features of Megaviridae. *Proc Natl Acad Sci USA* **108**: 17486–17491.

Ashley J, Cordy B, Lucia D, Fradkin LG, Budnik V, Thomson T (2018) Retrovirus-like gag protein arc1 binds RNA and traffics across synaptic boutons. *Cell* **172**: 262–274.e11.

Baltimore D (1971) Expression of animal virus genomes. *Bacteriol Rev* **35**: 235–41.

Bamford DH (2003) Do viruses form lineages across different domains of life? *Res Microbiol* **154**: 231–236.

Bandea CI (1983) A new theory on the origin and the nature of viruses. *J Theor Biol* **105**: 591–602.

Barteneva NS, Maltsev N, Vorobjev IA (2013) Microvesicles and intercellular communication in the context of parasitism. *Front Cell Infect Microbiol* **3**: 49.

Philippe H, M. deVienne D, Ranwez V, Roure B, Baurain D, Delsuc F (2017) Pitfalls in supermatrix phylogenomics. *Eur J Taxon* **283** doi: 10.5852/ejt.2017.283.

Brochier-Armanet C, Forterre P, Gribaldo S (2011) Phylogeny and evolution of the Archaea: one hundred genomes later. *Curr Opin Microbiol* **14**: 274–281.

Caetano-Anollés G, Caetano-Anollés D (2003) An evolutionarily structured universe of protein architecture. *Genome Res* **13**: 1563–1571.

Caetano-Anollés G, Minhas BFBF, Aziz F, Mughal F, Zhahzad K, Tal G, Mittenthal JE, Caetano-Anollés D, Koç I, Nasir A, Caetano-Anollés K, Kim KM (2017) The compressed vocabulary of the proteins of Archaea. In: *Biocommunication of Archaea*, Witzany G. (ed). Springer, Cham, pp. 147–174.

Caetano-Anollés G, Nasir A (2012) Benefits of using molecular structure and abundance in phylogenomic analysis. *Front Genet* **3**: 172.

Caetano-Anollés G, Nasir A, Kim KM, Caetano-Anollés D (2018) Rooting phylogenies and the tree of life while minimizing ad hoc and auxiliary assumptions. *Evol Bioinform Online* **14**: 1176934318805101.

Caetano-Anollés G, Nasir A, Zhou K, Caetano-Anollés D, Mittenthal JE, Sun F-J, Kim KM (2014) Archaea: The first domain of diversified life. *Archaea* **2014**: 590214.

Caetano-Anollés G, Wang M, Caetano-Anollés D, Mittenthal JE (2009) The origin, evolution and structure of the protein world. *Biochem J* **417**: 621–637.

Chothia C, Gough J, Vogel C, Teichmann SA (2003) Evolution of the protein repertoire. *Science* **300**: 1701–1703.

Claverie J-M (2006) Viruses take center stage in cellular evolution. *Genome Biol* **7**: 110.

Claverie J-M, Abergel C (2016) Giant viruses: The difficult breaking of multiple epistemological barriers. *Stud Hist Philos Biol Biomed Sci* **59**: 89–99.

Colson P, Levasseur A, La Scola B, Sharma V, Nasir A, Pontarotti P, Caetano-Anollés G, Raoult D (2018) Ancestrality and mosaicism of giant viruses supporting the definition of the fourth TRUC of microbes. *Front Microbiol* **9**: 2668.

Cortez D, Forterre P, Gribaldo S (2009) A hidden reservoir of integrative elements is the major source of recently acquired foreign genes and ORFans in archaeal and bacterial genomes. *Genome Biol* **10**: R65.

Da Cunha V, Gaia M, Gadelle D, Nasir A, Forterre P (2017) Lokiarchaea are close relatives of Euryarchaeota, not bridging the gap between prokaryotes and eukaryotes. *PLOS Genet* **13**: e1006810.

Dagan T, Martin W (2006) The tree of one percent. *Genome Biol* **7**: 118.

Di Giulio M (2007) The tree of life might be rooted in the branch leading to Nanoarchaeota. *Gene* **401**: 108–113.

Feschotte C, Gilbert C (2012) Endogenous viruses: insights into viral evolution and impact on host biology. *Nat Rev Genet* **13**: 283–296.

Forterre P (2006) The origin of viruses and their possible roles in major evolutionary transitions. *Virus Res* **117**: 5–16.

Forterre P (2016) To be or not to be alive: How recent discoveries challenge the traditional definitions of viruses and life. *Stud Hist Philos Biol Biomed Sci* **59**: 100–108.

Forterre P (2005) The two ages of the RNA world, and the transition to the DNA world: a story of viruses and cells. *Biochimie* **87**: 793–803.

Forterre P (2011) Manipulation of cellular syntheses and the nature of viruses: The virocell concept. *Comptes Rendus Chim* **14**: 392–399.

Forterre P (2015) The universal tree of life: an update. *Front Microbiol* **6**: 717.

Forterre P, Krupovic M (2012) The origin of virions and virocells: the escape hypothesis revisited. In: *Viruses: Essential Agents of Life*, Witzany G (ed). Springer, Dordrecht, pp. 43–60.

Forterre P, Prangishvili D (2013) The major role of viruses in cellular evolution: facts and hypotheses. *Curr Opin Virol* **3**: 558–565.

Forterre P, Prangishvili D (2009) The great billion-year war between ribosome- and capsid-encoding organisms (cells and viruses) as the major source of evolutionary novelties. *Ann NY Acad Sci* **1178**: 65–77.

Gaia M, Da Cunha V, Forterre P (2018) The tree of life. In: *Molecular Mechanisms of Microbial Evolution. Grand Challenges in Biology and Biotechnology*. Springer, Cham, Switzerland, pp. 55–99.

Gough J, Chothia C (2002) SUPERFAMILY: HMMs representing all proteins of known structure. *Nucleic Acids Res* **30**: 268–272.

Gough J, Karplus K, Hughey R, Chothia C (2001) Assignment of homology to genome sequences using a library of hidden Markov models that represent all proteins of known structure. *J Mol Biol* **313**: 903–919.

Hand E (2018) Interplanetary small satellites come of age. *Science* **361**: 736–737.

Hoeppner MP, Gardner PP, Poole AM (2012) Comparative analysis of RNA families reveals distinct repertoires for each domain of life. *PLoS Comput Biol* **8**: e1002752.

Holmes EC, Duchêne S (2019) Can sequence phylogenies safely infer the origin of the global virome? *mBio* **10**: e00289–19.

Hu X, Das B, Hou H, He W, Yan R (2018) BACE1 deletion in the adult mouse reverses preformed amyloid deposition and improves cognitive functions. *J Exp Med* **215**: 927.

Illergård K, Ardell DH, Elofsson A (2009) Structure is three to ten times more conserved than sequence — a study of structural response in protein cores. *Proteins* **77**: 499–508.

Jeong H, Arif B, Caetano-Anollés G, Kim KM, Nasir A (2019) Horizontal gene transfer in human-associated microorganisms inferred by phylogenetic reconstruction and reconciliation. *Sci Rep* **9**: 5953.

Kim KM, Caetano-Anollés G (2010) Emergence and evolution of modern molecular functions inferred from phylogenomic analysis of ontological data. *Mol Biol Evol* **27**: 1710–1733.

Kim KM, Nasir A, Hwang K, Caetano-Anollés G (2014) A tree of cellular life inferred from a genomic census of molecular functions. *J Mol Evol* **79**: 240–262.

Koonin EV, Dolja VV (2014) Virus world as an evolutionary network of viruses and capsidless selfish elements. *Microbiol Mol Biol Rev* **78**: 278–303.

Koonin EV, Senkevich TG, Dolja VV (2006) The ancient virus world and evolution of cells. *Biol Direct* **1**: 29.

Krupovic M, Dolja VV, Koonin EV (2019) Origin of viruses: primordial replicators recruiting capsids from hosts. *Nat Rev Microbiol* **17**: 449–458.

La Scola B, Audic S, Robert C, Jungang L, de Lamballerie X, Drancourt M, Birtles R, Claverie J-M, Raoult D (2003) A giant virus in amoebae. *Science* **299**: 2033.

Lake JA, Henderson E, Oakes M, Clark MW (1984) Eocytes: a new ribosome structure indicates a kingdom with a close relationship to eukaryotes. *Proc Natl Acad Sci USA* **81**: 3786–3790.

Legendre M, Alempic J, Philippe N, Lartigue A, Jeudy S, Poirot O, Ta NT, Nin S, Couté Y, Abergel C, Claverie J-M (2019) Pandoravirus celtis illustrates the microevolution processes at work in the giant pandoraviridae genomes. *Front Microbiol* **10**: 430.

Legendre M, Fabre E, Poirot O, Jeudy S, Lartigue A, Alempic J-M, Beucher L, Philippe N, Bertaux L, Christo-Foroux E, Labadie K, Couté Y, Abergel C, Claverie J-M (2018) Diversity and evolution of the emerging Pandoraviridae family. *Nat Commun* **9**: 2285.

Legendre M, Lartigue A, Bertaux L, Jeudy S, Bartoli J, Lescot M, Alempic J-M, Ramus C, Bruley C, Labadie K, Shmakova L, Rivkina E, Couté Y, Abergel C, Claverie J-M (2015) In-depth study of Mollivirus sibericum, a new 30,000-y-old giant virus infecting Acanthamoeba. *Proc Natl Acad Sci USA* **112**: E5327–35.

Makri A (2018) Pakistan and Egypt had highest rises in research output in 2018. *Nature doi*: 10.1038/d41586-018-07841-9

Malik SS, Azem-e-Zahra S, Kim KM, Caetano-Anollees G, Nasir A (2017) Do viruses exchange genes across superkingdoms of life? *Front Microbiol* **8**: 2110.

McCutcheon JP, Moran NA (2011) Extreme genome reduction in symbiotic bacteria. *Nat Rev Microbiol* **10**: 13–26.

Meckes Jr DG, Raab-Traub N (2011) Microvesicles and viral infection. *J Virol* **85**: 12844–12854.

Moreira D, Brochier-Armanet C (2008) Giant viruses, giant chimeras: the multiple evolutionary histories of Mimivirus genes. *BMC Evol Biol* **8**: 12.

Moreira D, Lopez-Garcia P (2009) Ten reasons to exclude viruses from the tree of life. *Nat Rev* **7**: 306–311.

Morrison DA (2009) Why would phylogeneticists ignore computerized sequence alignment? *Syst Biol* **58**: 150–158.

Nakamura T, Yamada KD, Tomii K, Katoh K (2018) Parallelization of MAFFT for large-scale multiple sequence alignments. *Bioinformatics* **34**: 2490-2492.

Nasir A, Caetano-Anollés G (2013) Comparative analysis of proteomes and functionomes provides insights into origins of cellular diversification. *Archaea* **2013**: 648746.

Nasir A, Caetano-Anollés G (2015) A phylogenomic data-driven exploration of viral origins and evolution. *Sci Adv* **1**: e1500527.

Nasir A, Forterre P, Kim KM, Caetano-Anollés G (2014a) The distribution and impact of viral lineages in domains of life. *Front Microbiol* **5**: 194.

Nasir A, Kim KM, Caetano-Anollés G (2012a) Giant viruses coexisted with the cellular ancestors and represent a distinct supergroup along with superkingdoms Archaea, Bacteria and Eukarya. *BMC Evol Biol* **12**: 156.

Nasir A, Kim KM, Caetano-Anollés G (2015a) Lokiarchaeota: Eukaryote-like missing links from microbial dark matter? *Trends Microbiol.* **23**: 448–450.

Nasir A, Kim KM, Caetano-Anollés G (2012b) Viral evolution: Primordial cellular origins and late adaptation to parasitism. *Mob Genet Elements* **2**: 247–252.

Nasir A, Kim KM, Caetano-Anollés G (2014b) Global patterns of protein domain gain and loss in superkingdoms. *PLoS Comput Biol* **10**: e1003452.

Nasir A, Kim KM, Caetano-Anollés G (2014c) A phylogenomic census of molecular functions identifies modern thermophilic archaea as the most ancient form of cellular life. *Archaea* **2014**: 706468.

Nasir A, Kim KM, Caetano-Anollés G (2017) Long-term evolution of viruses: A Janus-faced balance. *BioEssays* **39**: e201700026.

Nasir A, Kim KM, Da Cunha V, Caetano-Anollés G (2016) Arguments reinforcing the three-domain view of diversified cellular life. *Archaea* **2016**: 1851865.

Nasir A, Naeem A, Khan MJ, Lopez-Nicora HD, Caetano-Anollees G (2011) Annotation of protein domains reveals remarkable conservation in the functional make up of proteomes across superkingdoms. *Genes* **2**: 869–911.

Nasir A, Sun FJ, Kim KM, Caetano-Anollés G (2015b) Untangling the origin of viruses and their impact on cellular evolution. *Ann NY Acad Sci* **1341**: 61–74.

Nolte-'t Hoen E, Cremer T, Gallo RC, Margolis LB (2016) Extracellular vesicles and viruses: Are they close relatives? *Proc Natl Acad Sci USA* **113**: 9155–9161.

Ogata H, Claverie J-M (2007) Unique genes in giant viruses: regular substitution pattern and anomalously short size. *Genome Res* **17**: 1353–1361.

Pastuzyn ED, Day CE, Kearns RB, Kyrke-Smith M, Taibi Av, McCormick J, Yoder N, Belnap DM, Eriendsson S, Morado DR, Briggs JAG, Feschotte C, Shepherd JD (2018) The neuronal gene arc encodes a repurposed retrotransposon gag protein that mediates intercellular RNA transfer. *Cell* **172**: 275–288.e18.

Philippe N, Legendre M, Doutre G, Couté Y, Poirot O, Lescot M, Arslan D, Seltzer V, Bertaux L, Bruley C, Garin J, Claverie J-M, Abergel C (2013) Pandoraviruses: amoeba viruses with genomes up to 2.5 Mb reaching that of parasitic eukaryotes. *Science* **341**: 281–286.

Pina M, Bize A, Forterre P, Prangishvili D (2011) The archeoviruses. *FEMS Microbiol Rev* **35**: 1035–1054

Prangishvili D (2013) The wonderful world of archaeal viruses. *Annu Rev Microbiol* **67**: 565–585.

Raoult D, Audic S, Robert C, Abergel C, Renesto P, Ogata H, La Scola B, Suzan M, Claverie J-M (2004) The 1.2-megabase genome sequence of Mimivirus. *Science* **306**: 1344–1350.

Raoult D, Forterre P (2008) Redefining viruses: lessons from Mimivirus. *Nat Rev* **6**: 315–319.

Rivera MC, Lake JA (2004) The ring of life provides evidence for a genome fusion origin of eukaryotes. *Nature* **431**: 152–155.

Roossinck MJ (2011) The good viruses: viral mutualistic symbioses. *Nat Rev Microbiol* **9**: 99–108.

Roossinck MJ (2015) Move over, bacteria! viruses make their mark as mutualistic microbial symbionts. *J Virol* **89**: 6532–6535.

Rose PW, Prlic A, Bi C, Bluhm WF, Christie CH, Dutta S, Green RK, Goodsell DS, Westbrook JD, Woo J, Young J, Zardecki C, Berman HM, Bourne PE, Butley SK (2015) The RCSB protein data bank: views of structural biology for basic and applied research and education. *Nucleic Acids Res* **43**: D345–D356.

Sober E, Steel M (2002) Testing the hypothesis of common ancestry. *J Theor Biol* **218**: 395–408.

Spang A, Eme L, Saw JH, Caceres EF, Zaremba-Niedzwiedzka K, Lombard J, Guy L, Ettema TJG (2018) Asgard archaea are the closest prokaryotic relatives of eukaryotes. *PLOS Genet* **14**: e1007080.

Spang A, Ettema TJG (2016) Microbial diversity: The tree of life comes of age. *Nat Microbiol* **1**: 16056.

Spang A, Saw JH, Jørgensen SL, Zaremba-Niedzwiedzka K, Martijn J, Lind AE, van Eijk R, Schleper C, Guy L, Ettema TJG (2015) Complex archaea that bridge the gap between prokaryotes and eukaryotes. *Nature* **521**: 173–179.

Staley JT, Caetano-Anollés G (2018) Archaea-first and the co-evolutionary diversification of domains of life. *BioEssays* **40**: 1800036.

Sweere JM, Van Belleghem JD, Ishak H, bach MS, Popescu M, Sunkari V, Kaber G, Manasherob R, Suh GA, Cao X, de Vries CR, Lam DN, Marshal PL, Birukova M, Katznelson E, Lazzareschi DV, Balaji S, Keswani SG, Hawn TR, Secor PR, Bollyky PL (2019) Bacteriophage trigger anti-viral immunity and prevent clearance of bacterial infection. *Science* **363**: eaat9691.

Trifonov EN (2011) Vocabulary of definitions of life suggests a definition. *J Biomol Struct Dyn* **29**: 259–66.

Van Regenmortel MHV (2010) Logical puzzles and scientific controversies: The nature of species, viruses and living organisms. Syst. *Appl. Microbiol* **33**: 1–6.

Wagner FB, Mignardot JB, Le Goff-Mignardot CG, Demesmaeker R, Komi S, Capogrosso M, Rowald A, Seáñez I, Caban M, Pirondini E, Vat M, McCracken LA, Heimgartner R, Fodor I, Watrin A, Seguin P, Paoles E, Van Den Keybus K, Eberle G, Schuurch B, Pralong E, Becce F, Prior J, Buse N, Buschman R, Neufeld E, Kuster N, Carda S, von Zitsewitz J, Delattre V, Denison T, Lambert H, Minassian K, Bloch J, Vourtine G (2018) Targeted neurotechnology restores walking in humans with spinal cord injury. *Nature* **563**: 65–71.

Wang M, Jiang Y-Y, Kim KM, Qu G, Ji H-J, Mittenthal JE, Zhang H-Y, Caetano-Anollés G (2011) A universal molecular clock of protein folds and its power in tracing the early history of aerobic metabolism and planet oxygenation. *Mol Biol Evol* **28**: 567–582.

Wang M, Yafremava LS, Caetano-Anollés D, Mittenthal JE, Caetano-Anollees G (2007) Reductive evolution of architectural repertoires in proteomes and the birth of the tripartite world. *Genome Res* **17**: 1572–1585.

Williams TA, Foster PG, Cox CJ, Embley TM (2013) An archaeal origin of eukaryotes supports only two primary domains of life. *Nature* **504**: 231–236.

Woese CR, Kandler O, Wheelis ML (1990) Towards a natural system of organisms: proposal for the domains Archaea, Bacteria, and Eucarya. *Proc Natl Acad Sci USA* **87**: 4576–4579.

Wolf YI, Kazlauskas D, Iranzo J, Licia-Sanz A, Kuhn JH, Krupovic M, Dolja VV, Koonin EV (2018) Origins and evolution of the global RNA virome. *mBio* **9**: e02329–18.

Yin Y, Fischer D (2008) Identification and investigation of ORFans in the viral world. *BMC Genomics* **9**: 24

Yvon M, Urbino C, Peterschmitt M, Gutierrez S (2019) A multicellular way of life for a multipartite virus. *eLife* **8**: e43599.

Some considerations towards a predictive theory of life

S. Núñez-Corrales[1]

Summary. Current understanding of the massive complexity of life suggests the existence of an intricate world of interactions both below the ecosystem and above complex biochemistry. That particular cosmos of the organism has proven hard to untangle due to three key factors: the difficulty of reconstructing the past from traces left in the present, the dynamic nature of changes across multiple interacting scales and the contemporary, overall process of scientific discovery in the life sciences. By appealing to the integration of frontier research lines in the biosciences and a revision of the epistemology around the organism, some considerations are drawn towards the possible meaning — and form — of an integrated theory of life.

[1]*Illinois Informatics and National Center for Supercomputing Applications (NCSA), University of Illinois, Urbana IL, USA*

1. Introduction

Biology has been transformed in the last six decades thanks to three pivotal events: the consolidation of Darwinian evolution by natural selection as the overarching phenomenology behind the massive improbability of living organisms (Pigliucci, 2012; Dobzhansky, 2013), the rapid increase in the available quantities of information and its associated semantics about an ever-widening horizon of biological entities, from the molecular to the behavioral (Koonin, 2016), and the growing data processing infrastructure destined to cope both with the volume and the interrelatedness of biological data (Tang *et al.*, 2014). The organism sits in the middle of two extremes: at one side, biophysics determines the immediate dynamical background of molecules and structures critical to living systems; at the other side, the persistence of information across a wildly complex array of interactions that spans almost four billion years by now. Finding good explanations that transcend a traditionally mechanistic perspective on organisms, these very ontologically compact collections of molecules and history, seems critical towards a better picture of life in this century (Woese, 2004).

In the transition from the descriptive to the explanatory, and further to the predictive, theoretical excursions have become prevalent in the pursuit for an integrated theory of life despite known adoption difficulties often experienced by the experimentalist (Shou *et al.*, 2015). It is thus widely acknowledged that fundamental advances in experimental biology will not be possible without falsifiable theories and models that promote intuitions to system-range statements (Liepe *et al.*, 2013). Systems biology, starting from the rational decomposition of organisms into simpler units exhibiting emergence, robustness and modularity (Aderem, 2005) has become instrumental in providing a growing scaffold for theoreticians and experimentalists to express and test hypotheses, from description languages (Moodie *et al.*, 2015) to the characterization of information flows in regulatory networks (Tkačik *et al.*, 2008). In the language of category theory (Fong and Spivak, 2018), we are after generative effects that arise in organisms due to their compositionality. Such perspective has been explored in the past through the conceptualization of *memory evolutive systems,* for which Ehresmann and Vabremeersch (2007) constitutes a seminal reference.

The central challenge that system biology faces towards an integrated theory of biology resides on the classes of questions drawn at three recognizable interfaces: the genome (questions about phylogeny), the organism (questions about ontogeny) and the environment (questions about external evolutionary operators) (Losos *et al.*, 2013). Ferrada and Wagner (2010) brought to light one of the most critical unsolved issues in the field in connection with the three interfaces: evolution is multilevel, multivariate and overall, exceeds the conceptual simplicity of tree representations thanks to the concurrent nature of events, being more akin to how intermediate molecular entities traverse a highly dimensional — and at the same time — non-deterministic collection of potential surfaces (Bates *et al.*, 2009). Despite all difficulties, posing the problem of systemic evolution of organisms that share common ancestry but are exposed to different conditions in a more evolutionary neutral language provides a unified framework for interrogating nature about its apparent global two-component structure: the interplay between biochemistry (very short timescales) and heredity (very long timescales). Such a high-level view, at the same time, makes causality difficult to identify (Bialas *et al.*, 2003), in particular when events are non-linear and systems are not in equilibrium (Nicolis *et al.*, 1977).

2. Biology and the scientific method

Understanding[2] *life* involves addressing a dual problem: disentangling the complex relations to approximate a tree of life and, conversely, determining the accuracy in which trees are snapshots of the dynamic fabric in which evolution by natural selection operates (Wendel and Doyle, 1998; Samadi and Barberousse, 2006). In that sense, it is essential to take a step back into epistemological grounds with the goal of finding a more general (and possibly powerful) common language to describe evolutionary complexity (Edmonds, 1995). First, it must be stated that within biology — as an intellectually efficient interpretation of reality from the point of view of complex interacting entities — a reasonable goal is to center the attention in the organism as the topmost relevant hierarchical modular system that bears recognizable interactions with the environments, in the sense that any other higher level system often depends on rules that involve some form of *symbolic* information representation, processing and communication instead of purely biological and biochemical events (Arthur Woods *et al.*, 2015). Sets of physical symbols (i.e. organized matter and energy) are used to encode predicted future chains of events that have some particular value under known selective pressures. The prevalence of boundaries of individuality in both unicellular and multicellular organisms through selective membranes is likely driven by their evolutionary advantages as a general solution to both internal stability in the sense of maximum entropy production as described in Martyushev and Seleznev (2006) and resilience through anticipation of various

---

[2] *Understanding* in this context is to be defined precisely in terms of knowledge theory as the existence of models for all the possible (or rather, relevant) worlds that capture the phenomenology under scrutiny and later become a logic for reasoning about the satisfiability of new statements. Having only formulations of the models without proper specification of constraints is clearly insufficient for postulating decidable statements in the form of hypotheses for complex systems. See Fegin *et al.* (2003) for a complete account of reasoning in knowledge domains.

kinds (Michod, 2007). Second, general systems theory (GST) provides a substrate-agnostic framework for reasoning about biology from the vantage point of components and processes within the organism: it is an attempt to canonically describe sets of interacting elements for which a description of its dynamics can be elicited (Boulding, 1956).

Systems biology as a whole is not only inspired in GST, but is also heavily informed by its methods, theories and philosophical underpinnings (Kitano, 2002). As with any Popperian approach to science, GST allows postulating falsifiable arguments through a new dialectic: the complex entities are *queried* by asking questions in their language (e.g. biochemistry, genetics, epigenetics), then their answers are assessed and finally a conceptual ontology of theories — not just one theory — is then tested against further evidence (Zenko *et al.*, 2013). This interplay, or rather *conversation*, can proceed within the intellectually neutral grounds provided by novel arsenals in mathematics (Hinrichsen and Pritchard, 2011) capable of preserving original strengths of the hypothetico-deductive model while reaching more ambitious goals.

The hypothetico-deductive model of science is still the core of scientific discovery across the life sciences (Farji-Brener and Amador-Vargas, 2013). More precisely, this systematics of research pervades the existing conceptions about mechanisms used for investigating organisms within the broader horizon of the life sciences, and is assumed to spread into neighboring disciplines (Pouvreau, 2013). Following such path, biology has arrived to a point where the construction of theories, to a first approximation, spans a spectrum defined by two properties: explanatory power (from more specific to more general) and experimental accessibility (Figure 1). While at the moment any classification such as this one is somewhat arbitrary, one property holds for (almost) all theories: they depend upon "sloppy" models that focus on a single explanatory layer. These

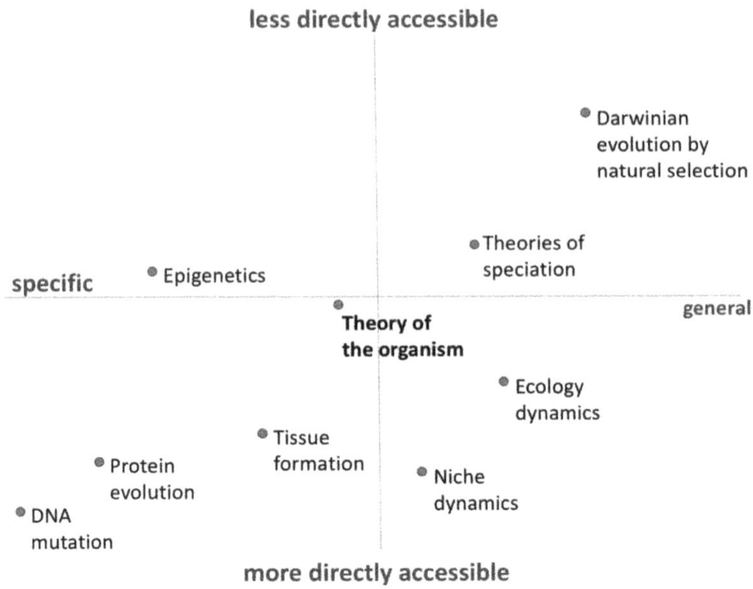

Figure 1. A conceptual, approximate classification of biological fields of enquiry in two parameters: explanatory power (horizontal axis) and experimental accessibility (vertical axis). A theory of the organism (center) stands in a pivotal position between the specific and more accessible molecular world and the more general and indirect level of explanation provided by evolution by natural selection.

models make drastic simplifications in order to be tractable (Tanstrum *et al.*, 2015). *Sloppiness* is a strong limitation, since model distinguishability becomes harder as more complex monolithic models arise (e.g. large number or parameters that require fitting), thereby the information that may be obtained the probability of its experimental falsification and its explanatory power are diminished. Assuming that increased model detail translates into high-accuracy predictions often backfires resulting in models that are increasingly hard to calibrate.

Part of the problem comes from the expectation that single layer models will suffice to meaningfully probe a multi-layered structure.

The formal language used most often for describing statistical hypotheses, the frequentist approach to probability (Efron, 2005), is to blame in many of the artificial limitations brought into the falsifiability process, such as well-known issues with *p*-values (Wagenmaker, 2007). These artifacts permeate even simple models (for instance, by fixing confidence intervals incorrectly when model aspects are related to otherwise complex and dynamical environmental conditions), and negatively impact the quality and soundness of the inferences. For instance, simplistic information-based models of the molecular clock leave plenty of room for interpretation, while models that also account for biophysics and biochemistry tend to place events at realistic evolutionary distances.

Why do these artifacts arise at all in the first place? One plausible explanation in the latter context is that parameters describing a variable landscape of microscopic states pertinent to organismal internal dynamics are crudely approximated by macroscale constants contained in deterministic continuum models. Many of these models the lack phenomenological accuracy required to describe responsive systems near thresholds, thereby producing bogus information (Hoover and Hoover, 2012). One solution may be the use of hierarchically nested, coupled systems of Fokker-Plank equations as a more realistic approximation that captures variations in the probability distributions governing such parameters within organisms, and then using such stochastic parameters to more fully describe dynamical couplings across scales (Núñez-Corrales and Jakobsson, 2011). It is likely for similar alternatives to arise as means of removing artifacts by acknowledging the inadequateness of continuum deterministic descriptions for certain classes of phenomena. At present, this remains an open research question.

Another part of the issue belongs to the nature of the hypothetico-deductive method itself. First, it is subsumed by a more general category known as *statistical relevance* in which multiple

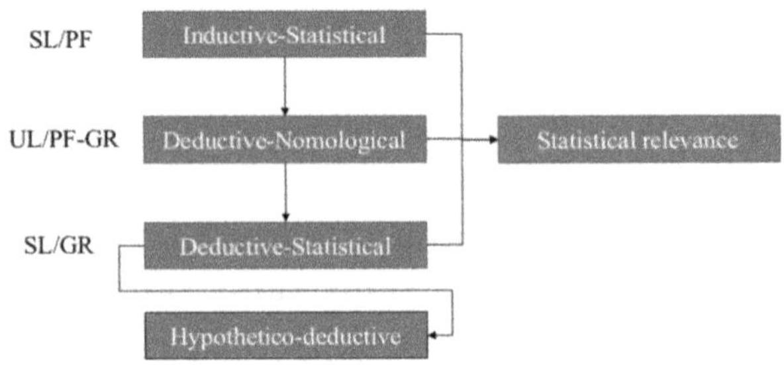

Figure 2. The hierarchy of scientific theories after Thomas Kuhn and their relation to the Statistical Relevance model of science. Nomenclature: SL — statistical laws, UL — universal laws, PF — particular facts, GR — general regularities.

propositions, with an assignment of probability related to their falsification, can be expressed and simultaneously attacked (Salmon, 1971); the model summarizes to a great extent the conception of scientific theories after Kuhn (Figure 2). Statistical relevance allows an arbitrary number of concomitant propositions to be tested simultaneously without sacrificing explanatory power. However, it remains limited due to a number of underlying assumptions about what it means for an explanation to be coherent (Thaggard, 1989), a notion largely constructed upon principles derived from single-scale systems. In addition, the place of computational simulation in the context of the scientific method -and in the hypothetico-deductive method itself- necessitates better demarcation (Thaggard, 1993), yet many of the contemporary assertions made about organisms depend crucially on computational renderings tied to hierarchies of hypothesis. Modeling and simulation have proven in practice to be fundamental tools for understanding the intricacies of organisms (and of other systems in general) by appealing to abstractions that can be empirically refined (Magnani *et al.*, 2012). Therefore, finding effective ways to obtain falsifiable and integrated descriptions of organisms will also contribute to expanding the scientific endeavor as a whole. Recent work compiled by

Jenschke and Heger (2018) on hierarchies of hypothesis (HoH) in ecology sheds light on how to tighten outcomes produced by the hypothetico-deductive method, and it is reasonable to expect HoH methods to be applicable to the organism itself. However, HoH only addresses some of the gaps in the hypothetico-deductive method without extending it.

2.1. *Probability in modern biology*

Let us consider another epistemological route. Conceptualizing the biology of organisms at multiple levels may benefit from the ideas and methods found in the study of Langevin dynamics (Paquet and Viktor, 2015) and stochastic methods in general. Instead of focusing on a single trajectory or state of an entity, stochastic methods attempt to find properties of collections of entities that are identical except in their initial conditions (Gibbs ensembles). Entities or phenomena are also described at two levels: a set of microstates governed by some dynamics, and a macrostate (i.e. a collection of average values for the ensemble) that can be measured for which some regularities can be expected. Since the number of microstates is often considered to be large, reasoning in terms of probability becomes inescapable. The term stochastic implies that the forces behind the dynamics contain both deterministic (e.g. gravity, chemical gradients) and random (e.g. photon emission, stochiometric variability) components. In structural biology, using Langevin dynamics has been successful in providing insights about polymer translocation through nanopores (Huopaniemi *et al*. 2006), the flexibility of cytoskeletal fibers (Nedelec and Foethke, 2007) and more recently modeling transition kinetics in DNA (Grazioli and Andricioaei, 2016).

What the former examples have in common is that *no individual outcome pertaining to states or trajectories in either entities or their constituents qualifies as an adequate representation for a given phenomenology.* This is particularly highlighted by the complexity and richness found in biological system, where compositionality takes precedence over single-scale phenomena and a collection of histories and interactions is needed as a starting point for explanation or prediction. Biological sciences may gain access to larger pieces of the intellectual land by finding global properties within ensembles describing organisms, stated as probability distributions tied to outcomes obtained either from first principles or estimated from experimental measurements. In this scientific view, understanding an organisms involves considering not one but many of its possible worlds.

Can a deeper link between the observable discrete diversity of life and chance be found? To a first approximation, let us consider organisms as responsive systems with complex hysteresis. Organisms evolve thanks to natural selection to converge to a finite set of preferential states, likely those granting stability and fitness (Friedman *et al.*, 2014). Some states correspond to internal representations of environmental variability and how to successfully cope with it (Adami *et al.*, 2000). Other representations encode useful information received from other organisms that, given the similarities in biological processes, point to triggers for various types of decisions, or proximal threats or opportunities; quorum sensing constitutes a prime example (Bassler, 2002). The more successful an organism becomes, the more reversible its hysteresis does too (i.e. the more likely it is to return to some steady state at the macroscale). Repeated success (i.e. survival and reproduction) leads to an embodied form of guessing relevant probability distributions of events at the organismal level, and energy constraints refine the resulting encodings to be economical in the long term. Natural evolution thus exemplifies a missing piece in information theory as described by Shannon (1948): the mechanisms by which communication infrastructure emerges such that it acquires its expected mathematical properties.

Organisms have evolved to function within a certain class of signal-to-noise ratios sufficiently large to produce new dynamical biasing components by harnessing thermodynamic forces (Zotin and Zotina, 1967; Fronczak *et al*., 2007), but at the same time sufficiently small not to completely disrupt their structure (Shlesinger, 1987), producing a form of self-organization with preference for stable states and threshold handling mechanisms (Nicolis and Prigogine, 1977) that increasingly shield structure and function from fluctuations. For instance, the biophysical context is extremely important at the onset of the protein folding process but becomes less and less significant as structures converge towards one stable form or another (Dave *et al*., 2016). The same mechanism appears to be present in other instances in the form of stochastic molecular switches that determine the fate of development in organisms (Norman *et al*., 2015; Weinberger, 2015). In particular, individuality may well be considered the hallmark of the stochastically guided architecture of life, optimized for surviving multilevel organismal selection (Michod, 1997).

While the role of probability is ample within biology, the ability to harness it towards better understanding of the organism remains somewhat unexplored. The patterns of biological connectivity in the organism as a network of interactions, the role of thermodynamic irreversibility across scales and the semantics behind what constitutes a prediction require much deeper excursions. Moreover, capturing the variation of probability distributions with time and using that information to refine predictions may constitute a promising road to uncovering the specifics during biological interactions.

3. Promising paths: joining visions of the organism

Once the theoretical landscape around current problems and possible general solutions to the knowability of organisms has been laid out, the advancement of fields that comprise systems biology as the currently most effective theoretical framework for understanding life are worth reviewing. In the following discussion, three components are of interest: phylogenomics (history of the information contained in the organism), structural bioinformatics (conditions of the expression of organismal information) and networks (relationships across those instances that feed back into existing information). While the present review is by far not exhaustive and restricted to genetic or structural information, it is representative of the classes of problems that reasoning about organisms brings to the surface for which progress can be sustained (hence the qualification of *promising*).

3.1. *The accuracy of phylogenomics in the context of living systems*

Inferring paths of molecular evolution for arbitrary genomes across long periods of time is hard due to the difficulties involved in tracing the impact of particular events and environments on populations or species in the distant past (Breen *et al.*, 2012). Despite such difficulties, a safe common ground exists on the concept of homology: conserved sequences (or features) are evidence of common ancestry (Nixon and Carpenter, 2012). Homology thus allows the construction of phylogenies by proposing candidate ancestries for a given set of organisms based on premises such as, for instance, the equivalence between observed frequencies of nucleotides or amino acid substitutions and biochemically plausible mutations.

Viewing biological information units at various levels across the organisms drastically increases the number of possible landscapes available for searching putative histories at the genome level. This is especially so thanks to the reticulation of collective histories caused by lateral exchange mechanisms permeating all evolutionary relationships. Speciation occurs when significant divergence takes place between members of a species sharing common ancestry and traits (Safran and Nosil, 2012), yet reticulation can make this limit diffuse; e.g. Hugall *et al.* (2005). Some events might impact the whole genome, part of it or only individual genes or domains. Determining each of which was the most probable cause can be determined only up to a certain degree of accuracy through *careful* phylogenomic analysis. Care, in this case, must be interpreted as the systematic elimination of unnecessary *ad hoc* hypotheses while remaining open to alternative explanations, later to assess which of them best fits the evidence and which of them to discard. A particularly illuminating example is exemplified in the use of synteny as a tool for understanding claims about whole-genome duplication events of *Saccharomyces cerevisiae* in two different organisms, *Ashbya gossypii* and *Kluyveromyces waltii* (Kellis *et al.*, 2004; Dietrich *et al.*, 2004). A closer look to prior analyses claiming whole-duplication events occurred in *S. cerevisiae* revealed discrepancies (such as the low proportion of synteny across all the genome, not just particular chromosomes) indicated that the claim is not plausible and that partial-genome duplication is more parsimonious with respect to the number of steps required to transform one genome into another (Martin *et al.*, 2007).

During the reconstruction of ancestries, hypothesizing about events with larger conserved regions simplifies the overall analysis process at the expense of biologically meaningless or wrong answers when multiple unknown factors intervened to preserve them. Multiscale interactions often explain these classes of issues, and therefore should not be ignored. How granular should the analysis be in order to obtain realistic results? A potential answer is proposed in DRIMM-Synteny (Pham and Pevzner, 2010) where a computational device, namely the A-Bruijn graph is used to detect synteny patterns at finer scales. While its results have shown effectiveness in detecting more evolutionary conserved segments across species than other tools, a mapping between biological events and the the semantics of the computational process is missing. For instance, there is no attempt to include structural domains in protein-coding genes in the analysis, a fundamental step according to existing knowledge about the conservation of function by virtue of increasing the fitness of the organism when an innovation is not deleterious.

Growing amount of data in molecular biology also brings a concern of practicality: can valid and accurate reconstructions be achieved by increasingly automatic means? The exponential-over-exponential trend of genomic information growth points to an apparently hopeless task for humans alone (Ciccarelli *et al.*, 2006), many decision steps seem tied to human expertise. One of the automatic reconstruction algorithms that attempts to bridge the gap between data excess and automation of phylogenomics is SYNERGY, which has proven successful in explaining gene duplication and loss through various mechanisms in fungi through various mechanisms (Wapinski *et al.*, 2007a,b). The introduction of two key elements, namely a novel method for computing gene similarity and the measurement of putative orthogroups, explains the success rate of the algorithm when compared against annotated data.

3.2. *Structural bioinformatics: function robustness in a sea of opportunity*

Plausible operational hypotheses must not only explain how existing genes have moved across known or potential ancestors through known mechanisms, but include suitable mechanisms for accounting for new genes that gradually appear in the population; bacteria have been central for studying molecular evolution both theoretically and experimentally due to their wide diversity (Woese, 1987). An interesting model of *de novo* gene birth from proto-genes suggests, for example, that the distinction between genes and proto-genes is not clear cut, rather being closer to a continuum (Carvunis *et al.*, 2012). These types of models not only are critical for explaining evolutionary innovation triggered by *de novo* gene generation, amplification and divergence along the history of particular species (Näsvall *et al.*, 2012); they allow distinguishing between tensions arising between innovation by caused divergence and those due to duplication (Bergthorsson *et al.*, 2007).

Sometimes, the history of evolutionary innovation can be harnessed by organisms to fullfil certain functions, such as driving their early development stages. The parallel between phylogeny and ontology found in zebra fish is a striking example: in an hourglass manner, each developmental step exhibits a positive correlation between the transcriptome age index (i.e. a measure of how ancient the expressed genes are present at a certain point) and the abundance of the relatively recent genes in a particular species (Domazet-Loso and Tautz, 2010). The fact that those genes consistent with the hourglass model are also central to the organism developmental processes is striking in several ways (Kalinka *et al.*, 2010). First, genomes do show flexible expression while remaining relatively stable across generations and populations. Second, the more conserved repertoire of genes appears to be modulated by *de novo* genes, providing flexibility; this appears as if old encodings can be repurposed by introducing new tokens that rebalance the probabilistic information landscape across the organism. Third, flexibility allows changes that might be beneficial while providing enough robustness to cope with potentially negative consequences at the species level.

Reconciling this apparent tension between change and stability in molecular evolution requires extending views around the building blocks of life -such as proteins- in terms of how structure and function are intertwined. Drawing from research in organization science (Feldman and Pentland, 2003), both structure and function are analogous to routines established across organizations: they emerge as sustained patterns of repeated interactions, they provide a scaffolding for safe platform to undertake larger explorations capable of bringing useful novelty, and they admit both ostensive (i.e. abstract specification) and a performative (i.e. particular instances) descriptions that separate timeless and time-full reasonings. Both structure and function in organisms are responsible for ensuring stability and endogenous change and can be encoded and exported to other organisms and species. In that case, the interpretation of the ostensive description within an ancestor organism may radically differ from the interpretation of that to which structure and function have been exported.

In organisms, structure is determined to a large extent by a degenerate coding scheme (multiple codons per residue) and a hierarchy of structural arrangements that can assemble into similar shapes despite having very different constituents. As in the case of phylogenomics, cladistics can be applied to the secondary structure of proteins in order to determine whether new classifications match annotated data, and to which degree of accuracy (Przytycka *et al.*, 1999). The results of such exercise hint at another way in which nature provides a fertile ground for variation: even with a limited number of motifs and combinatorial rules that operate in a biochemical context, many structures exist in a sea of possibility. The sea seems not to be, however, unbounded or continuous. Simplified backbone-based models of protein folding indicate that self-organization does occur within a limited (and possibly dense) set of structural constraints (Rose *et al.*, 2006). While certain biophysical facts are oversimplied (for example, exaggerating the role of hydrogen bonds beyond experimental observations), they appear to have some bearing in this issue when contrasted with the identification of families, superfamilies and hyperfamilies of structures in proteins. A limited but well-connected repertoire of structures explains a large share of protein structures and functions, in particular when biophysics is explicitly included in the quantification of structural similarity (Harrison *et al.*, 2002).

However, current understanding of the relation between structure and function remains limited by limited amount of data for protein structure as well as the computational complexity of performing structural comparisons. Moreover, since not all proteins can be (yet) brought to a stable crystal structure suitable for experimental analysis, the universe of observable folds remains restricted. Despite these limitations, the growth rates of databases such a SCOP and CATH (Schaeffer and Daggett, 2011) suggests at least that a significant portion of the fold universe has been already covered. The second challenge appears to depend only on two factors: cost and availability of computing resources and the quality of the algorithms used for structural comparison. Since cost constantly decreases thanks to Moore's Law, most of the advances are expected to come from the development of new and efficient algorithms. A representative example is the application of advanced techniques such as using Gauss invariant integration kernels for measuring distances in atomic coordinate space which has gained traction in biology (Røgen and Fain, 2003), and outperforms RMSD in efficiency. Finally, several new crystallographic methods (Fromme *et al.*, 2015; Wu *et al.*, 2016) have shown that more types proteins can be resolved under the right conditions.

Another way of attacking the folding problem is via *structural grammars*, an avenue suggested in the past and recently revisited (Jiménez-Montaño, 1984). A basic set of symbols and rules (i.e. folds and self-assembly affinities) is assumed to be ancestral and structural complexity is derived by applying the rules to the symbols. For example, hypothesis of a prebiotic RNA-peptide world appears to match an scenario in which RNA combined with freely available peptides in order to gain structural stability at first, and later a large portion of the universe of known proteins emerged through repetition and variation of domains taken from that primordial vocabulary (Alva *et al.*, 2016). When folds are clustered into galaxies according to similarity either in sequence space or structure space, hypothesis of a limited set of initial origins for all proteins arises naturally (Alva *et al.* 2010). An even more idealized model is that of a periodic table of folds where self-organization drives protein assembly, yet the rules operate locally on folds and depend on the particular shape and structure of the fold elements (Taylor, 2002; Ahnert *et al.*, 2015).

Structure is the substrate for the manifestation of function, hence function evolution depends on biophysically available structures. Elementary loops and elementary functional domains in Archaea support the prebiotic peptides hypothesis (Goncearenco and Berezovsky, 2012), the hypothesis that elementary functional loops were among the first elementary reactions available for biochemical transformation, constituting the first building blocks of existing protein functions. The study of enzymes becomes thus central, since catalysis is fundamental to the biochemistry of life. For an elementary functional loop, being highly designable (for which a plethora of sequences might exist) tends to correlate with being more fundamental (Goncearenco and Berezovsky, 2015). However, how can innovation happen where robustness and degeneracy are prevalent? The analysis of the relations between genotype space and structure space reveals that diversity of unique functions occurs even within close neighbourhoods in genotype space (i.e. similar sequences), and diversity increases dramatically only when their corresponding sequences differ at least by 80–90% (Ferrada and Wagner, 2010). Even simple *in silico* experiments where short binary sequences of idealized amino acids are subject to the action of simple canonical potentials indicate that innovation can happen, but certainly not at the expense of survivability (Ferrada, 2014).

An explosion in variety and possible innovation seems to be counterbalanced by a massive systemic coding redundancy across all levels of the organism. Back to information theory and probability, this an error-correction mechanism of sorts that treats structure and function as signals and ensures the encoding is preserved within some acceptable noise level. However, natural selection constantly replaces existing encodings by more efficient ones when these are found across many levels, all at once. Understanding structure and function at the genomic level demands awareness of their couplings across organismic scales.

3.3. *Network theory: the path beyond trees*

A surprising feature of living beings is their ability to survive in a wide range of extremes (Kitano, 2007). They exhibit inherent robustness to mishaps, unexpected changes and any other threats. If an organism may be abstracted as a complex system for which a network is a good representation, survivability may be translated to the capacity to withstand failures and attacks (Albert *et al.*, 2000). In this setting, failures are random removals of edges, and attacks are random removal of vertexes, mostly improbable but catastrophic events. Consider the effect of inhibiting the expression of an arbitrary protein in a cell at the level of DNA transcription. Recalling that proteins and functions are tightly related, the latter is equal to losing one or more functions. If the protein belongs to the essential part of genome, blocking its expression most probably affect organismal survivability except in two cases: when other proteins perform the same function, or when other sequences encode the same protein. Essentially, surviving loss of functions would be reducible to preserving reachability in the network of molecular products in the essential genome. The same argument can be extended to higher levels (e.g. cells, tissues) in terms either of adaptation of existing components or alternative encodings for them. From many possible topologies (arrangements of vertexes and edges), scale-free networks are robust with respect to large quantities of failures, but are vulnerable to attacks. An example of this are metabolic networks (Jeong *et al.*, 2000). An evolutionary rationale can be constructed in the following way: random challenging events have been mostly already selected by evolution (failures) at great cost, which pushes for good the preservation general solutions; on the contrary, disruptive events that force either innovation or extinction come at the expense of navigating again through the vast ocean of genotypical and phenotypical alternatives.

The tension between robustness and innovation is best captured by the evolution of a general architectural pattern known as hierarchical modularity. Sub-networks of interacting entities specialize on certain types of problems, and these compose at higher scales to produce more and more indirect responses that remain adequate across many scales. It must be stressed that composition does not mean analytic reducibility, but rather the presence of emergent generative effects as it has been discussed previously. Turning to biological examples, metabolic networks not only exhibit this pattern, but possibly exploit it for achieving new functions (Ravasz *et al*., 2002). One way to capture how functions may arise concretely is by observing how entities organize internally to solve problems. There may not always be an explicit or accessible representation of the fitness landscape (such as in the case of natural selection), but its outcomes are. Robustness and innovation must therefore be studied under the lens of collectives as well as of individual organisms, since each contains partial matching information about the other. Hierarchically modular communities appear to play an important role in the dynamics of composition and, when passed through the filter of adequacy, they perform multiple functions across a vast range of scales from biochemistry to social systems (Girvan and Newman, 2002). Recent evidence (Aziz *et al*., 2016) indicates that the emergence of functions triggers hierarchical modularity when protein networks reach complexity and size thresholds. Finding new structures that encode functions reduces the number of degrees of freedom by constraining which regions of the space of potential energy surfaces is *locally* available to that structure in the future, but hierarchical modularity allows *globally* regaining degrees of freedom for the organisms as a whole.

In the complex web of information provided by the network perspective, interpreting data is far from trivial. Under the premise of general systems theory, the same network representation also brings a plethora of useful constructs that allow reduction of interrelations into adimensional quantities with both abstract and biological significance. Centrality, betweenness and other observables that are computable from the network allow the construction of *organism maps* needed to navigate datasets and obtain new knowledge, for instance, about metabolic networks (Guimera and Amaral, 2005). Knowledge is, again, interpreted in the sense of complete descriptions of entities for which all possible and relevant instances can be computed (Fagin *et al.*, 2003). The map is dynamic, crucial for understanding evolution beyond the tree metaphor and must be consistent. Different organismal scales in the map can follow differing power laws due to different types of events (e.g. genome editions such as deletion, duplication, reversion), yet the map must remain consistent under different experimental interrogation approaches; the latter has been found to describe (within various simplifications) the evolution of protein networks (Wagner, 2003).

Considering the role of dynamical environment in which the network unfolds adds another significant perspective. Evolutionary operations take place amidst noisy environments in which signals must survive and reach their destination. The evolution of complex modular networks is a probable response to information-theoretic efficiency pressures: in order for messages (signals) to have causal effects across the organism, both modularity and redundancy are required (Hintze and Adami, 2008). This transcends living system: information-based self-organization pertains to any system that bears internal communication as an adaptive response (Pirolli, 2007). The effect of hierarchical modularity also bears resemblance to how self-organized criticality operates, in the sense of reaching higher complexity states that allows entropy to propagate at longer distances (Bak *at al.*, 1987); organisms minimize internal entropy in exchange for the ability to exert larger effects on their environment. An additional problem organisms appear to have found good solutions for is how to optimize the message passing process when tokens are susceptible to degradation, loss or interception. The longer the path, the higher the probability of mishaps. Longer sustained paths are also expensive in terms of energy use. Organisms leverage network connectivity by reaching beneficial equilibria in trade-offs between risk and energy expenditure. Instantiations of the contents of their genetic memory (Pfeiffer *et al.*, 2005) are variable in time and dependent on external conditions, such as availability of nourishment sources and the context of fluctuations across all gradients relevant for survival. Essentially, not only organisms are stochastic networks, but such networks are self-contained blueprints for communications infrastructure, deployed when needed and retracted when energy the expenditure becomes excessive.

Variations in the blueprint or in how it is executed also require better understanding. Simplicity may appeal to human aesthetics when describing biological phenomena e.g. parsimony operates upon such premise, but it also seems to correctly characterize certain aspects of biological networks. Energy efficiency, resilience and reproductive capacity are examples of goals that organisms seek to satisfy. While the goals are not explicitly stated, they are enforced by the hierarchy of natural laws that govern the lives of organisms and may be thought of as "design goals," where "design" refers to the architecture provided by the network. The evolution of modularity relates strongly to the ability of switching between "design goals" and satisfying them, even in electronic circuits (Kashtan and Alon, 2005). In organisms, modularity implies that the cost of switching between "design goals" decreases if the coupling can be flexibly established, which implies a decomposition into functions that can be re-utilized in different arrangements; to that extent, organisms are nearly decomposable systems (Simon, 1962).

Let us define *architectonic simplicity* as the property of requiring a certain number of composable elements to specify a robust system of constraints and opportunities for efficient sustained information, energy and matter transduction,[3] such that the number of resulting macroscale system states is much larger than the number of composable elements used to build it and the architecture is invariant under random fluctuations during any transduction processes. Architectonic simplicity relates to Kolmogorov complexity, or the complexity of the smallest "program" (i.e. set of instructions) that produces a given outcome (Li and Vitányi, 2013). Proper evidence of architectonical simplicity would include, for instance, the existence of quantities that capture the amount of information contained in the relations between composable elements, or the ratio between the compressibility of the information contained in the system and the information contained in the composable elements. Barabási and Albert (1999) showed that such ratios produce systems with power-law scaling relations; power law functions happen to be convex and characterize various state variables across thermodynamics (Schlögl, 1980). The convexity of the entropy in a distributed system, in particular, determines the difficulty of stabilizing steady states: distributed control mechanisms are non-linear and system size implies longer stabilization times and more complexity of the control mechanism (Alonso and Ydstie, 2001). However, organisms appear to have found in modularity and composition a way to harness composition and modularity such that the resulting scaling laws permit reaching stable states much faster and more robustly (Slotine and Lohmiller, 2001). Rentian scaling, the ratio between the number of edges between nodes within a given topological partition of the system and the number of edges connecting nodes

[3]For an extensive and rigorous description of energy transduction, see Hill (1977).

between partitions, is such a measure describing how efficient the particular scaling of a system is (Bassett *et al.*, 2010). The brain, one of the most complex pieces of biological machinery, is a hallmark of hierarchical modularity and Rentian scaling from the purely biophysical up to the neurological (Meunier *et al.*, 2010).

Despite the need for and effectiveness of convenient abstractions as well as evidence of any type of architectonic simplicity, the need for proper grounding on the physical and chemical substrates of life still holds. All decisions which organisms must make appear to depend on careful–yet stochastic–energy budgeting, or rather, energy control and flow which have been traditionally studied by cybernetics (O'Malley and Dupré, 2005; Gershenson *et al.*, 2013). When networks are placed in the context of energy fluxes in living systems, for instance, *E. coli* (Almaas *et al.*, 2004), an interpretation of modularity becomes apparent: highly connected hubs of activity allow for lower energy expenditure by moving hard, global decisions up into highly interconnected regulatory structures (e.g. metabolic pathways, signal transduction) and allowing increasingly specific bidirectional communication systems aggregate and segregate responses, using molecular tokens to mediate interactions between modules (Papin *et al.*, 2003). To a greater extent, networks have revealed that evolution has found clever ways to vary structure in order to minimize the consequences of the law of requisite variety (Ashby and Goldstein, 2011), and still comply with it. Rentian scaling may serve as a general mechanism that reduces energy consumption by grouping certain elements as operational units, and using other elements as information tokens that carry messages; only those tokens that are structurally stable in noisy environments, and those units that discriminate between types of tokens and adequately convey the message are suitable evolvable components of living systems.

Even with the advantageous view brought by networks, there are significant limitations that motivate the exploration of new perspectives. For example, networks only provide either a static view of processes or of their summarization. To construct them, data analysis reduces spatial and time dimensions to make them tractable or closer to direct interpretation. Details are lost, and depending on the assumptions behind the models, irreversibly so. Networks are incomplete also because other types of unknown, such as the existence of biology's *dark matter* and lack of information about how it alters the course of existence for organisms (Williams and Embley, 2014; da Silva *el al.*, 2015; Ross, 2016). Furthermore, network descriptions connect composable entities, yet the specifics of their interactions remain shrouded in mystery.

4. Biological entities as complex systems: some epistemological conclusions

The task of systems biology and connected research areas — phylogenomics, structural bioinformatics and network theory — is finding integrative explanations of biological phenomena by establishing novel data associations that, under proper assumptions and a falsification mechanism, help explain how organisms function, adapt and diversify in the environment (Reshef *et al.*, 2011). Beyond the immediate interest in the practical applications for solving pressing and relevant problems such as improving drug design (Yıldırım *et al.*, 2007), a more ambitious goal lies beyond yet unsurpassed mountains: the development of a predictive theory of biology, or rather, the mechanics of it. Similar to other historical points in the development of science, new principles, concepts, methods and tools are required. Tools and methods tend to emerge more rapidly by virtue of more immediate necessity. Principles and concepts, on the other hand, require more effort and appear sparsely, and yet are essential for any sort of fundamental progress.

As with Hennig's Auxiliary Principle for phylogenetics, new principles will be required as platforms to support and accelerate inferences in systems biology. Gathering the ideas from the discussion above, here is a crude attempt to sketch three of them:

Auxiliary Principle 1:

> *Evidence of architectonic simplicity, information exchange, and the ability to self-replicate across multiple scales suffices to characterize an entity as an organism.*

Auxiliary Principle 2:

> *Probability distributions of events, obtained by computing consequences of systems using hierarchically coupled stochastic models derived from approximate relations between composable elements across relevant scales over statistically significant ensembles, are adequate descriptions of organisms.*

Auxiliary Principle 3:

> *Predictions about isolated composable elements and phenomena, although statistically significant themselves, are not composable into predictions about the larger phenomenology of the organism.*

These (sketches of) principles mandate openly embracing probability and uncertainty not as inconveniences, but as the inherent and most natural substrate of meaningful analysis for biological systems. As corollary of this, scientific inferences should focus less on *particular events* and more on *ensembles of events* and their probability density function. Clearly, an adjustment of the pragmatics and the expectations during the research process will be required. Furthermore, adequate auxiliary principles may improve the design of experiments and instruments towards shifting the experimental focus from measuring entities (DNA, proteins, cells) to stronger efforts in directly measuring both their interactions and the biophysical context simultaneously. Having direct access to the interactions is critical, and cannot be replaced by indirect measurements or inferences. Third and most significant, partitioning the space of entities in processing units and signalling tokens opens up a new perspective where information content may help better elucidate several questions through appropriate metaphors, for instance attacking the evolution of molecular signalling as an instance of the code matching problem in information theory. As suggested by the history of science, the complete form of adequate auxiliary principles for systems biology will likely be concise and elegant, yet comprehensive and predictive.

Conceptually, across biology and other sciences, there is growing need to better understand what is meant by *interaction*. In biology, interactions are advantageously accessible directly in many cases (e.g. protein–protein interactions) and their interrogation has become routine. This is not the norm, however, and quickly experiments reduce to idealized laboratory conditions that may not be representative of interactions inside organisms. In addition, the formal machinery necessary to describe them efficiently has not yet appeared. In general, *inferring* interactions has usually being performed by discovering or fitting against non-linearities. Clearly, both processes describe temporal or spatial *consequences* of interactions, but not interactions themselves. A hint of this need is the fact that mathematical theories of biological events across scales, contrary to the composable elements they attempt to describe, are not straightforwardly scale-wise composable for most situations. Some new ideas are needed to unravel biological interactions.

Systems biology is an evolving field, already moving from infancy to adolescence. Unprecedented data volumes of unprecedented detail await integrated methods with potential to produce data driven discovery about the organism (Haas, 2015). Doing so requires principles or axioms to prevent infinite regress while providing some initial grounds to construct more elaborate reasonings. Biology is in need of a new type of pursuit of soundness and completeness (even if these are approximate), one in which discreteness, stochasticity and dynamics are the norm. A sound inference system (within honest accounts of experimental error and uncertainty) would ensure that the number of derivable statement about the ontology and phylogeny of an organism which are false in reality is minimized. A system would be complete if most true facts are accessible through inferences drawn from it. What systems biology aspires to is to become an architecture for systems problem-solving where routine operations of today become natural inferences in the science of tomorrow (Klir, 2013). In the words of the late philosopher and logician Alfred North Whitehead (Whitehead, 1958),

> *"It is a profoundly erroneous truism, repeated by all copybooks and by eminent people when they are making speeches, that we should cultivate the habit of thinking of what we are doing. The precise opposite is the case. Civilization advances by extending the number of important operations which we can perform without thinking about them. Operations of thought are like cavalry charges in a battle: they are strictly limited in number, they require fresh horses, and must only be made at decisive moments."*

Acknowledgments

The author wishes to thank Gustavo Caetano-Anollés (Department of Crop Sciences and C.R. Woese Institute for Genomic Biology, University of Illinois) for his encouragement and continuous support in the genesis and subsequent writing of the ideas contained in this manuscript, as well as to Eric Jakobsson (Department of Molecular and Cellular Biology, NCSA and Beckman Institute,

University of Illinois) for his guidance and Bertram Ludäscher (School of Information Science and NCSA, University of Illinois) for an introduction to HoH research. This work is dedicated to Les Gasser† (School of Information Science, NCSA and Beckman Institute, University of Illinois) *in memoriam*, without whom the development of many of the ideas would not have been possible. Funding sources: Illinois Informatics at the University of Illinois at Urbana-Champaign, SIGHPC ACM/Intel Fellowship for Computational and Data Science (2017) and the RDA-US/Whole Tale Early Career Fellowship 2018.

References

Adami C, Ofria C, Collier TC (2000) Evolution of biological complexity. *Proc Natl Acad Sci USA* **97(9)**: 4463–4468.

Adams K (2013) Genomic clues to the ancestral flowering plant. *Science* **342(6165)**: 1456–1457.

Aderem A (2005) Systems biology: its practice and challenges. *Cell* **121(4)**: 511–513.

Ahnert SE, Marsh JA, Hernández H, Robinson CV, Teichmann SA (2015) Principles of assembly reveal a periodic table of protein complexes. *Science* **350(6266)**: aaa2245.

Albert VA, Barbazuk WB, Der JP, Leebens-Mack J, Ma H, Palmer JD, Rounsley S, Sankoff D, Schuster SC, Soltis DE, *et al.* (2013) The *Amborella* genome and the evolution of flowering plants. *Science* **342(6165)**: 1241089.

Albert R, Jeong H, Barabási A-L (2000) Error and attack tolerance of complex networks. *Nature* **406(6794)**: 378–382.

Almaas E, Kovacs B, Vicsek T, Oltvai Z, Barabási A-L (2004) Global organization of metabolic fluxes in the bacterium *Escherichia coli*. *Nature* **427(6977)**: 839{843.

Alonso AA, Ydstie BE (2001) Stabilization of distributed systems using irreversible thermodynamics. *Automatica* **37(11)**: 1739–1755.

Alva V, Remmert M, Biegert A, Lupas AN, Söding J (2010) A galaxy of folds. *Protein Science* **19(1)**: 124–130.

Alva V, Söding J, Lupas AN (2016) A vocabulary of ancient peptides at the origin of folded proteins. *eLife* **4**: e09410.

Arthur Woods H, Martin LB, Ghalambor CK (2015) Conclusions: the central role of the organism in biology. In: *Integrative Organismal Biology*, Martin LB, Ghalambor CK, Woods HA (eds). Wiley Blackwell, New York, pp. 309–317.

Ashby WR, Goldstein J (2011) Variety, constraint, and the law of requisite variety. *Emergence: Complexity and Organization* **13(1/2)**: 190.

Aziz MF, Caetano-Anollés K, Caetano-Anollés G (2016) The early history and emergence of molecular functions and modular scale-free network behavior. *Sci Rep* **6**: 25058.

Bak P, Tang C, Wiesenfeld K (1987). Self-organized criticality: An explanation of the 1/f noise. *Phys Rev Lett* **59(4)**: 381.

Barabási A-L, Albert R (1999) Emergence of scaling in random networks. *Science* **286**: 509– 512.

Bassett DS, Greenfield DL, Meyer-Lindenberg A, Weinberger DR, Moore SW, Bullmore ET (2010) Efficient physical embedding of topologically complex information processing networks in brains and computer circuits. *PLoS Comput Biol* **6(4)**: e1000748.

Bassler BL (2002) Small talk: cell-to-cell communication in bacteria. *Cell* **109(4)**: 421–424.

Bates P, Chen Z, Sun Y, Wei G.-W, Zhao S (2009) Geometric and potential driving formation and evolution of biomolecular surfaces. *J Math Biol* **59(2)**: 193–231.

Bergthorsson U, Andersson DI, Roth JR (2007) Ohno's dilemma: evolution of new genes under continuous selection. *Proc Natl Acad Sci USA* **104(43)**: 17004–17009.

Bialas P, Burda Z, Jurkiewicz J, Krzywicki A (2003) Tree networks with causal structure. *Phys Rev E* **67(6)**: 066106.

Boulding KE (1956) General systems theory — the skeleton of science. *Management Sci* **2(3)**: 197–208.

Breen MS, Kemena C, Vlasov PK, Notredame C, Kondrashov FA (2012) Epistasis as the primary factor in molecular evolution. *Nature* **490(7421)**: 535–538.

Carvunis A-R, Rolland T, Wapinski I, Calderwood MA, Yildirim MA, Simonis N, Charloteaux B, Hidalgo CA, Barbette J, Santhanam B, Brar GA, Weissman JS, Regev A, Thierry-Mieg N, Cusick ME, Vidal M (2012) Proto-genes and de novo gene birth. *Nature* **487(7407)**: 370–374.

Ciccarelli FD, Doerks T, Von Mering C, Creevey CJ, Snel B, Bork P (2006) Toward automatic reconstruction of a highly resolved tree of life. *Science* **311(5765)**: 1283–1287.

da Silva RR, Dorrestein PC, Quinn RA (2015) Illuminating the dark matter in metabolomics. *Proc Natl Acad Sci USA* **112(41)**: 12549–12550.

Dave K, Davtyan A, Papoian GA, Gruebele M, Platkov M (2016) Environmental fluctuations and stochastic resonance in protein folding. *Chem Phys Chem* **17**: 1341–1348.

Dietrich FS, Voegeli S, Brachat S, Lerch A, Gates K, Steiner S, Mohr C, Pöhlmann R, Luedi P, Choi S, Wing RA, Flavier A, Gaffney TD, Philippsen P (2004) The *Ashbya gossypii* genome as a tool for mapping the ancient saccharomyces cerevisiae genome. *Science* **304(5668)**: 304–307.

Dobzhansky T (2013) Nothing in biology makes sense except in the light of evolution. *Am Biol Teach* **75(2)**: 87–91.

Domazet-Loso T, Tautz D (2010) A phylogenetically based transcriptome age index mirrors ontogenetic divergence patterns. *Nature* **468(7325)**: 815–818.

Edmonds B (1995) What is complexity? — the philosophy of complexity per se with application to some examples in evolution. The evolution of complexity. Available at: http: //cogprints.org/357/

Efron B (2005) Bayesians, frequentists, and scientists. *J Amer Stat Assoc* **100(469)**: 1–5.

Ehresmann AC, Vanbremeersch JP (2007) *Memory Evolutive Systems; Hierarchy, Emergence, Cognition* (Vol 4). Elsevier Science.

Fagin R, Moses Y, Vardi MY, Halpern JY (2003) *Reasoning About Knowledge*. MIT Press.

Farji-Brener AG, Amador-Vargas S. (2014) Hierarchy of hypotheses or cascade of predictions? A comment on Heger *et al.* (2013). *Ambio* **43(8)**: 1112–1114.

Feldman MS, Pentland BT (2003) Reconceptualizing organizational routines as a source of flexibility and change. *Administrative Science Quarterly* **48(1)**: 94–118.

Ferrada E (2014) The amino acid alphabet and the architecture of the protein sequence-structure map. I. Binary Alphabets. *PLoS Comput Biol* **10(12)**: e1003946.

Ferrada E, Wagner A (2010) Evolutionary innovations and the organization of protein functions in genotype space. *PLoS One* **5(11)**: e14172.

Fong B, Spivak DI (2018) Seven sketches in compositionality: An invitation to applied category theory. arXiv preprint arXiv: 1803.05316.

Friedman G, McCarthy S, Rachinskii D (2014) Hysteresis can grant fitness in stochastically varying environment. *PloS One* **9(7)**: e103241.

Fromme R, Ishchenko A, Metz M, Chowdhury SR, Basu S, Boutet S, Fromme P, White TA, Barty A, Spece J, Weierstall U, Liu W, Cherezov V (2015) Serial femtosecond crystallography of soluble proteins in lipidic cubic phase. *IUCrJ* **2(5)**: 545–551.

Fronczak A, Fronczak P, Hołyst JA (2007) Thermodynamic forces, flows, and Onsager coefficients in complex networks. *Phys Rev E* **76(6)**: 061106.

Gershenson C, Csermely P, Érdi P, Knyazeva H, Laszlo A (2013) The past, present and future of cybernetics and systems research. arXiv preprint arXiv: 1308.6317.

Girvan M, Newman ME (2002) Community structure in social and biological networks. *Proc Natl Acad Sci USA* **99(12)**: 7821–7826.

Goncearenco A, Berezovsky IN (2012) Exploring the evolution of protein function in archaea. *BMC Evol Biol* **12(1)**: 1.

Goncearenco A, Berezovsky IN (2015) Protein function from its emergence to diversity in contemporary proteins. *Phys Biol* **12(4)**: 045002.

Grazioli G, Andricioaei I (2016) Calculating Watson-Crick to Hoogsteen transition kinetics in DNA with Langevin dynamics and Fokker-Planck diffusion in reduced configuration space. *Biophys J* **110(3)**: 404a-405a.

Guimera R, Amaral LAN (2005) Functional cartography of complex metabolic networks. *Nature* **433(7028)**: 895–900.

Haas LM (2015) The power behind the throne: Information integration in the age of data-driven discovery. In: Proc 2015 ACM SIGMOD Intl Conf Management Data. ACM, pp. 661–661.

Harrison A, Pearl F, Mott R, Thornton J, Orengo C (2002) Quantifying the similarities within fold space. *J Mol Biol* **323(5)**: 909–926.

Hill T (1977) *Free Energy Transduction in Biology: The Steady-state Kinetic and Thermodynamic Formalism*. Elsevier.

Hinrichsen D, Pritchard AJ (2011) *Mathematical Systems Theory I: Modelling, State Space Analysis, Stability and Robustness* (Vol 48). Springer Science & Business Media.

Hintze A, Adami C (2008). Evolution of complex modular biological networks. *PLoS Comput Biol* **4(2)**: e23.

Hoover WG, Hoover CG (2012) *Time Reversibility, Computer Simulation, Algorithms, Chaos* (Vol 13). World Scientific Publishing, Singapore

Hugall A, Stanton J, Moritz C (1999) Reticulate evolution and the origins of ribosomal internal transcribed spacer diversity in apomictic *Meloidogyne*. *Mol Biol Evol* **16(2)**: 157–164.

Huopaniemi I, Luo K, Ala-Nissila T, Ying SC (2006) Langevin dynamics simulations of polymer translocation through nanopores. *J Chem Phys* **125(12)**: 124901.

Jeong H, Tombor B, Albert R, Oltvai ZN, Barabási A-L (2000) The large-scale organization of metabolic networks. *Nature* **407(6804)**: 651–654.

Jeschke JM, Heger T (2018) *Invasion Biology: Hypotheses and Evidence* (Vol 9). CABI.

Jiménez-Montaño MA (1984) On the syntactic structure of protein sequences and the concept of grammar complexity. *Bull Math Biol* **46(4)**: 641–659.

Kalinka AT, Varga KM, Gerrard DT, Preibisch S, Corcoran DL, Jarrells J, Ohler U, Bergman CM, Tomancak P (2010) Gene expression divergence recapitulates the developmental hourglass model. *Nature* **468(7325)**: 811–814.

Kashtan N, Alon U (2005) Spontaneous evolution of modularity and network motifs. *Proc Natl Acad Sci USA* **102(39)**: 13773–13778.

Kellis M, Birren BW, Lander ES (2004) Proof and evolutionary analysis of ancient genome duplication in the yeast *Saccharomyces cerevisiae. Nature* **428(6983)**: 617–624.

Kitano H (2002) Computational systems biology. *Nature* **420(6912)**: 206–210.

Kitano H (2007) Towards a theory of biological robustness. *Mol Syst Biol* **3(1)**: 137.

Klir G (2013) *Architecture of Systems Problem Solving*. Springer Science & Business Media.

Koonin EV (2016) The meaning of biological information. *Phil Trans R Soc A* **374(2063)**: 20150065.

Li M, Vitányi P (2013) *An Introduction to Kolmogorov Complexity and its Applications*. Springer Science & Business Media.

Liepe J, Filippi S, Komorowski M, Stumpf MP (2013) Maximizing the information content of experiments in systems biology. *PLoS Comput Biol* **9(1)**: e1002888.

Losos JB, Arnold SJ, Bejerano G, Brodie III E, Hibbett D, Hoekstra HE, Mindell DP, Monteiro A, Moritz C, Orr HA, Petrov DA, Renner SS, Ricklefs RE, Soltis PS, Turner TL (2013) Evolutionary biology for the 21st century. *PLoS Biol* **11(1)**: e1001466.

Magnani L, Nersessian N, Thagard P (2012) *Model-based Reasoning in Scientific Discovery*. Springer Science & Business Media.

Martin N, Ruedi EA, LeDuc R, Sun FJ, Caetano-Anollés G (2007) Gene-interleaving patterns of synteny in the Saccharomyces cerevisiae genome: are they proof of an ancient genome duplication event? *Biol Direct* **2(1)**: 1.

Martyushev LM, Seleznev VD (2006) Maximum entropy production principle in physics, chemistry and biology. *Phys Rep* **426(1)**: 1–45.

McAuliffe JD, Pachter L, Jordan MI (2004) Multiple-sequence functional annotation and the generalized hidden Markov phylogeny. *Bioinformatics* **20(12)**: 1850–1860.

Meunier D, Lambiotte R, Bullmore ET (2010) Modular and hierarchically modular organization of brain networks. *Front Neurosci* **4**: 200.

Michod RE (1997) Cooperation and conflict in the evolution of individuality. I. Multilevel selection of the organism. *Am Naturalist* **149(4)**: 607–645.

Michod RE (2007) Evolution of individuality during the transition from unicellular to multicellular life. *Proc Natl Acad Sci USA* **104(suppl 1)**: 8613–8618.

Moodie S, Le Novere N, Demir E, Mi H, Villeger A (2015) Systems biology graphical notation: process description language level 1 version 1.3. *J Integ Bioinfo* **12(263.10)**: 1038.

Näsvall J, Sun L, Roth JR, Andersson DI (2012) Real-time evolution of new genes by innovation, amplification, and divergence. *Science* **338(6105)**: 384–387.

Nedelec F, Foethke D (2007) Collective Langevin dynamics of flexible cytoskeletal fibers. *New J Phys* **9(11)**: 427.

Nicolis G, Prigogine I (1977) Self-organization in nonequilibrium systems. In: *Bifurcation Analysis*, Hazewinkel M, Jurkovich R, Paelinck JHP (eds). Springer, Dordretch.

Nixon KC, Carpenter JM (2012) On homology. *Cladistics* **28(2)**: 160–169.

Norman TM, Lord ND, Paulsson J, Losick R (2015) Stochastic switching of cell fate in microbes. *Annu Rev Microbiol* **69**: 381–403.

Núnez-Corrales S, Jakobsson E (2011) Hierarchical modularity: The description of multi-level complex systems as nested coupled Fokker-Planck equations. Proc 8th Intl Conf Complex Systems, Quincy, MA, pp. 967–981.

O'Malley MA, Dupré J (2005). Fundamental issues in systems biology. *BioEssays* **27(12)**: 1270–1276.

Papin JA, Price ND, Wiback SJ, Fell DA, Palsson BO (2003). Metabolic pathways in the post-genome era. *Trends Biochem Sci* **28(5)**: 250.258.

Paquet E, Viktor HL (2015) Molecular dynamics, Monte Carlo simulations, and Langevin dynamics: a computational review. *BioMed Res Intl* 2015: 183918.

Pfeiffer T, Soyer OS, Bonhoeffer S (2005) The evolution of connectivity in metabolic networks. *PLoS Biol* **3(7)**: e228.

Pham SK, Pevzner PA (2010) DRIMM-Synteny: decomposing genomes into evolutionary conserved segments. *Bioinformatics* **26(20)**: 2509–2516.

Pigliucci M (2012) Philosophical reflections on Darwin and evolutionary theory. *Trends Ecol Evol* **5(27)**: 258.

Pouvreau D (2013) The project of "general systemology" instigated by Ludwig von Bertalanffy: Genealogy, genesis, reception and advancement. *Kybernetes* **42(6)**: 851–868.

Przytycka T, Aurora R, Rose GD (1999) A protein taxonomy based on secondary structure. *Nature Struct Mol Biol* **6(7)**: 672–682.

Ravasz E, Somera AL, Mongru DA, Oltvai ZN, Barabási A-L (2002) Hierarchical organization of modularity in metabolic networks. *Science* **297(5586)**: 1551–1555.

Reshef DN, Reshef YA, Finucane HK, Grossman SR, McVean G, Turnbaugh PJ, Lander ES, Mitzenmacher M, Sabeti PC (2011) Detecting novel associations in large data sets. *Science* **334(6062)**: 1518–1524.

Røgen P, Fain B (2003) Automatic classification of protein structure by using Gauss integrals. *Proc Natl Acad Sci USA* **100(1)**: 119–124.

Rose GD, Fleming PJ, Banavar JR, Maritan A (2006) A backbone-based theory of protein folding. *Proc Natl Acad Sci USA* **103(45)**: 16623–16633.

Ross JL (2016) The dark matter of biology. *Biophys J* **111(5)**: 909–916.

Safran R, Nosil P (2012) Speciation: The origin of new species. *Nature Education Knowledge* **3(10)**: 17.

Salmon WC (1971) *Statistical Explanation and Statistical Relevance* (Vol 69). University of Pittsburgh Press.

Salmon WC (1990) Rationality and objectivity in science or Tom Kuhn meets Tom Bayes. *Sci Theor* **14**: 175–204.

Samadi S, Barberousse A (2006) The tree, the network, and the species. *Biol J Linnean Soc* **89(3)**: 509–521.

Schaeffer RD, Daggett V (2011) Protein folds and protein folding. *Protein Eng Design Selection* **24(1–2)**: 11–19.

Shannon CE (1948) A mathematical theory of communication. *Bell System Tech J* **27(3)**: 379–423.

Shlesinger MF (1987) Fractal time and 1/f noise in complex systems. *Ann New York Acad Sci* **504(1)**: 214–228.

Schlögl F (1980) Stochastic measures in nonequilibrium thermodynamics. *Phys Rep* **62(4)**: 267–380.

Shou W, Bergstrom CT, Chakraborty AK, Skinner FK (2015) Theory, models and biology. *eLife* **4**: e07158.

Simon HA (1962) The Architecture of Complexity. *Proc Amer Phil Soc* **106(6)**: 467–482.

Slotine JJ, Lohmiller W (2001) Modularity, evolution, and the binding problem: a view from stability theory. *Neural Networks* **14(2)**: 137–145.

Tang CY, Hung C-L, Hsu C-H, Zheng H, Lin C-Y (2014) Novel computing technologies for bioinformatics and cheminformatics. *BioMed Res Intl* **2014**: 392150.

Thagard P (1989) Explanatory coherence. *Behav Brain Sci* **12(03)**: 435–467.

Thagard P (1993) *Computational Philosophy of Science*. MIT Press.

Transtrum MK, Machta BB, Brown KS, Daniels BC, Myers CR, Sethna JP (2015). Perspective: Sloppiness and emergent theories in physics, biology, and beyond. *J Chem Phys* **143(1)**: 010901.

Taylor WR (2002) A 'periodic table' for protein structures. *Nature* **416(6881)**: 657–660.

Tkačik G, Callan CG, Bialek W (2008) Information flow and optimization in transcriptional regulation. *Proc Natl Acad Sci USA* **105(34)**: 12265–12270.

Wagenmakers EJ (2007) A practical solution to the pervasive problems of p values. *Psychon Bull Rev* **14(5)**: 779–804.

Wagner A (2003) How the global structure of protein interaction networks evolves. *Proc Royal Soc London B: Biol Sci* **270(1514)**: 457–466.

Wapinski I, Pfeiffer A, Friedman N, Regev A (2007a) Automatic genome-wide reconstruction of phylogenetic gene trees. *Bioinformatics* **23(13)**: i549–i558.

Wapinski I, Pfeiffer A, Friedman N, Regev A (2007b) Natural history and evolutionary principles of gene duplication in fungi. *Nature* **449(7158)**: 54–61.

Weinberger LS (2015) A minimal fate-selection switch. *Curr Op Cell Biol* **37**: 111–118.

Wendel JF, Doyle JJ (1998) Phylogenetic incongruence: window into genome history and molecular evolution. In: *Molecular Systematics of Plants II*. Springer, pp. 265–296.

Williams TA, Embley TM (2014) Archaeal "dark matter" and the origin of eukaryotes. *Genome Biol Evol* **6(3)**: 474–481.

Whitehead AN (1958) *An Introduction to Mathematics* (Vol 18). Oxford University Press, New York.

Woese CR (1987) Bacterial evolution. *Microbiol Rev* **51(2)**: 221.

Woese CR (2004) A new biology for a new century. *Microbiol Mol Biol Rev* **68(2)**: 173–186.

Wu P, Noland C, Ultsch M, Edwards B, Harris D, Mayer R, Harris SF (2016) Developments in the implementation of acoustic droplet ejection for protein crystallography. *J Lab Automation* **21(1)**: 97–106.

Yıldırım, MA, Goh KI, Cusick ME, Barabási AL, Vidal M (2007) Drug-target network. *Nature Biotechnol* **25(10)**: 1119–1126.

Zenko Z, Rosi B, Mulej M, Mlakar T, Mulej N (2013) General systems theory completed up by dialectical systems theory. *Syst Res Behavioral Sci* **30(6)**: 637–645.

Zotin AI, Zotina RS (1967) Thermodynamic aspects of developmental biology. *J Theor Biol* **17(1)**: 57–5.

Power

Life as a tendency to go beyond itself — The struggle for power

J. Gómez[1,2]
I. Cacho[1,2]
A. Ariza-Mateos[1,2]

Summary. Nietzsche found in Darwin's theory that the function of words and logic, the bases of human thought, was not to serve as a means of engaging with the truth, but to favor the biological success of the species. But even as he made use of Darwin's theory, he also developed his own critique of it. He identified the metaphysical root of natural selection in Spinoza's notion of self-preservation — to which all things tend — and proposed a different impulse more suited to his own view of the world, that of "becoming more." This impulse upheld the process of evolution by natural selection in the case of organisms which were only capable of playing a passive role, but also provided a framework for Nietzsche's understanding of the fundamental orientation of life as an active striving for growth — in number, in frequency, in space, in access to resources, in

[1] *Laboratory of RNA Archaeology, Instituto de Parasitología y Biomedicina 'López-Neyra', Consejo Superior de Investigaciones Científicas*
[2] *Centro de Investigación Biomedicina En Red de enfermedades hepáticasy digestivas (CIBERehd), Armilla, Granada, Spain*

the harnessing of the work of others, in overcoming all and any resistance, for all of which organisms frequently sacrifice part of their own individuality, and even their life. Thus, according to Nietzsche, the significant element in evolution is not really the individual. The important thing is victory in conflict, and not simply external conflict but also any conflict taking place within the individual, which follow Roux's reading. These are the key elements of Nietzsche's philosophy of evolution, forming part of the vast conceptual edifice of "the will to power," which asserts that, in philosophical terms, the struggle is not for life, but for power. Projecting his philosophical intuitions into the realms of science, Nietzsche put forward a testable — if politically incorrect — hypothesis which contended, contrary to what Darwin had proposed, that the winners of conflict were the weak. He gathered preliminary data by observing the great men and the history of civilizations as if they were biological individuals and evolutionary processes respectively. In order to do so, he revived the pre-Christian cultural notions of "good" and "evil" in a genealogical analysis, which he then associated with the biological terms "strong" and "weak." Later on, intuition led him to extend his hypothesis across the whole of biological evolution. However, while the interpretation of Nietzschean philosophy has proved a fruitful pursuit in many areas of knowledge, his contribution to biological evolution, one of the fields of greatest concern to him, has remained unexplored. In this chapter, we will situate Nietzsche's work on biology in the current context, highlighting the widespread adoption of the terminology of power in molecular biology, and will endeavour to map the relationship between Nietzsche's propositions and results drawn from work in evolutionary biology. Evolutionary processes involving symbiosis, parasitism, mimesis, and competition based on collaboration all align comfortably with Nietzsche's conception.

1. Introduction

1.1. *Why we need to talk about Nietzsche*

Friedrich Nietzsche (1844-1900) was the first great philosopher to take Darwinist theory completely seriously (Gayon, 1999b), eventually producing a profound critique of its metaphysical roots. His aim was to surpass or rather to subsume it as a special case within his own argument founded on a vision of life as the expression of the "*will to power*," the central principle of his philosophy (Nietzsche, 1968g; Sánchez-Meca, 2004). Although it might appear contradictory, he took culturally-situated Man as a biological model (Gayon, 1999b) and used a genealogical approach to turn the spotlight — or as he himself put it, a magnifying glass — onto the great thinkers, the great men and civilizations. He did this not to create a history of philosophy or to construct a complex conceptual structure. Instead, his focus was to understand the properties that are fundamental to and inherent in life: the absolute perspectivism of the living being (Aydin, 2007; Ferraris, 2000i; Tauber, 1994e), the changing nature of every individual (Tauber, 1994c), the characteristics of the hierarchy of living entities (Stiegler, 2001c), the tendency of each living being to create and exploit difference, to struggle to overcome challenges, even the struggle to overcome itself (Ferraris, 2000g; Stiegler, 2001c).

Trained in classical philology, and a follower of Schopenhauer in his early years, Nietzsche found in Darwin's natural selection nothing more and nothing less than the platform on which to build a refutation of Kant's categories of human understanding — causality, identity, and so forth. These features of thought, that Kant deemed not to be derived from experience but instead from universal rules, and which he described as *a priori* of experience, were transformed by Nietzsche into *a posteriori* features of evolution (Stack, 2012a; Stiegler, 2001d).

In this, and in many other examples of the interweaving of philosophy with science, Nietzsche locates exclusively within the biological domain of the physical body that which Descartes had separated into thought and matter (Stiegler, 2001f). For Nietzsche, it was the living organism that provided the model for both an ontology and a vision of the world (Ferraris, 2000e; Stiegler, 2001f; Tauber, 1994g).

However, Nietzsche's thought regarding evolution and biology remains entirely overlooked in current scientific debate (Stack, 2012b).

There are no references to his work among the original scientific articles in PubMed. All we find are a few citations in essays (Gómez *et al.*, 2015), or occasional reviews (Goodrich, 2010; Megill, 2016; Pence, 2011). This fact has attracted the attention of a small number of authors, who have been struck by the seriousness with which Nietzsche engaged with the natural sciences, how he immersed himself particularly in the physics, biology and the medicine of his day (Ferraris, 2000b; Gayon, 1999f; Pence, 2011; Stiegler, 2001a). The reasons for this silence regarding Nietzsche's critique and exegesis of natural selection have been blamed on his writing style, on the possible motives of those interpreting him, and on the historical and cultural context.

Nietzsche is, in the first instance, politically incorrect, and sometimes even insulting (Dennett, 1995). The way he expresses his thought is not the way a scientist expresses him- or herself. He uses short aphorisms packed with enormous amounts of information from a range of sources and writers, for which he almost never supplies references and which are often from widely disparate disciplines, giving the impression to a scientific reader unfamiliar with Nietzsche's style of being completely thematically disorganized. When he writes on a scientific subject, he makes no effort to present a systematic sequence of propositions regarding a defined set of testable properties (Gómez, 2001). He does not use a propositional development or follow any kind of system. Instead, he actively avoids such things, believing them to be a permanent source of falsehood (Aydin, 2006). In contrast to such classical styles, his texts offer the reader a broad cosmological sweep, through which one progresses via a series of metaphors and brilliant ideas. Furthermore, he opposes the mechanistic view that has undeniably been the dominant approach taken in biology because it sustains the dualistic Cartesian separation of thought and extension which his own philosophy ultimately aimed to supersede (Ferraris, 2000h; Tauber, 1994g). But the lack of awareness of Nietzsche's biological thinking can also be attributed to those who interpreted his work after the Second World War but before the arrival of the new generation of philosophers that included Foucault and Deleuze. Nietzsche's work was clearly rejected in the philosophy of science, particularly by Mario Bunge in his publication *Revista Continental de Filosofia* 2, nº4, 1944 [commented in Nature (Editorial 1945)]. Similarly, his pro-scientific work was taken as metaphysical, where Heidegger, the grand interpreter of Nietzsche (Heidegger, 2000d), sought to portray his interest in science as being of secondary importance and to disparage his key concepts in biologistic terms in order to "de-biologize" his thought and to move it onto the terrain of metaphysics (Ferraris, 2000c), something that Nietzsche would have never accepted. Independently, the cultural climate for biology probably did not help his cause either (Gayon, 1999c). Biology has suffered from an intense inferiority complex vis à vis physics and chemistry due to its more recent inception and because, until recently, successful prediction and intervention in the field have been very limited. This complex has made the discipline wary of addressing any aspect of biological processes that might be considered as not purely physical or chemical (Mayr, 1982). This includes matters relating to biological information, where a quantitative approach has always been preferred over any interpretative treatment, at least up until the advent of biosemiotics (Barbieri, 2007; Emmeche and Kull, 2011; Witzany, 2002).

All these factors have meant that Nietzsche's ideas on biology have been excluded from the principal currents of biological thinking, reduced merely to influences on the great thinkers in the discipline, such as Jacob Von Uexküll, Canguilhem and Simondon, and also left out of any discussion within molecular biology.

1.2. *What nobody will say about molecular biology today*

One of the central themes of research in classical biochemistry has been the control of metabolism. The emergence of biological information as part of the discipline, leading to molecular biology development, has increased interest in the control theme still further. As more has become known about the mechanisms through which the information expressed in the cell is encoded, edited, silenced and so forth, we have seen the opening up of a huge array of possible forms of control, not only of metabolism, but also of communication. In effect, what we can now see in the cell are areas of control analogous to those of a modern society: metabolism (production and consumption), communication and control itself (government). The question of control permeates the whole of molecular biology, as can be observed in the terminology it uses (Gómez and Cacho, 2004): Regulate, induce, suppress, silence, compartmentalize. These are not terms unique to this discipline, as might be the case, for example, with virology: vaccine, infection, capsid, and so on. This terminology, a wider sample of which appears in Table 1, is in fact the vocabulary of power relations in any situation that involves domination, confrontation and politics. Furthermore, these terms and concepts are not only used very frequently to describe and explain molecular phenomena, but some of them are clearly very accurate. We could not explain molecular biology without them (Cacho and Gómez, 2010; Gómez and Cacho, 2004).

It might appear that the rapid development of molecular biology has outstripped the discipline's ability to produce enough of its own necessary terms and concepts, and that it has raided other disciplines such as sociology, political geography and the law. But if this was the case, the assumption would have to follow that the use of the terms in the context of molecular biology was only superficially suitable, that the meanings would be inadequate, requiring refinement and eventual replacement with more appropriate "home-grown" expressions. While it may be true that these terms have not always been used with the precision one might wish, or in the full knowledge of their origins, the idea to which all of them directly point is that of power. The question of power exists in a number of

Table 1. Metaphors from the fields of power relations employed in molecular biology.

Mechanical power	Spatial metaphors	Political metaphors
Induce or repress activities	Domains	Hierarchical structure in the organization
Control the expression, the growth		Control strategies, avoidance
Restricting entry and output	Regions	Mechanism that governs
	Signaling pathways	
Make more or less likely an action	Compartmentalize activities	
Positive/negative regulation	Barriers, gates, forcing	
Imitate, simulate, interfere, modulate, resist, assimilate, acquire, incorporate, attach, select, reduce, force, coordinate, release, arrange, immobilize, destabilize, struggle, protect, promote, displace, suppress, reprogram, defend, subvert, attack, destroy, trigger, restrict, limit, neutralize, mediate, stimulate, hijack, conflict, damage, manipulation, invasion, risk factors, activation, rescue, exploit, reorganize, confine, explore, conquer. Latent, dependency, tolerance, agonist/antagonist		

the so-called "human" or "social" sciences, and forms the central axis of the philosophical work of certain recent authors. At the head of these is Michel Foucault, widely known to have taken inspiration from the philosophy of Nietzsche (Cacho and Gómez, 2010).

It is hard to disagree that the question of power is already implicitly a part of molecular biology. What is more, it may actually be the scaffolding that gives the discipline shape. And while this might be a slightly audacious hypothesis, it has to be pointed out that Michel Foucault himself in the 1970s was already putting forward the need to address the question of power in biochemistry in the same way as he and others had done for psychiatry, sociology and geography (Foucault, 1979c).

1.3. *Objective*

It is not customary academic practice to rescue paradigms rejected more than a hundred years ago. However, we believe that some of Nietzsche's criticisms of Darwin remain great unknowns, and a number of them are still relevant. We will take up those discussions which relate to nature, focusing on the origins and foundations of the divergences in thinking between the two writers. This is followed by a review and discussion of some of Nietzsche's concepts, along with those of other writers who either diverge from or converge with him. Viruses, about which no consensus yet exists among the specialists as to whether they are living beings or not (Koonin *et al.*, 2009; Moreira and Lopez-Garcia, 2009; Nasir and Caetano-Anolles, 2015), assume new meaning in a biology founded not so much on the preservation of the self as on the furtherance of the self.

We believe that revisiting the discussions on biological matters found in Nietzsche's philosophical writings, and in those of others who either preceded or followed on from him, may help us to draw together the increasing number of experimental observations and theoretical reflections that are now breaking through the rigid bounds of biology in its strictly mechanistic and neo-Darwinian terms. In particular, we cannot continue to see the organism as a passive object of actions in the environment, nor can we continue to separate biological information from its interpretation. The organism is both content-dependent and context-dependent. The determining factors of the context are the social/molecular group in which communication takes place (Villarreal and Witzany, 2013), their evolutive history (Markoš *et al.*, 2008) and the power relations that pertain within that group, which involve each and every organism and molecular agent. In this chapter, we propose a dynamic, biosemiotic approach that heals the Cartesian dualist split and reasserts the existence of a strong link between biology and the human and social sciences. To complete the discussion, we rescue a point of intersection between the perspectivist theories of Pierce and Nietzsche, a nexus we can use to further our understanding of evolution.

2. Biology and the will to power

Although we are attempting to draw together those aspects of Nietzsche's philosophy most closely related to biology, we cannot but reiterate how the enormous conceptual edifice of the will to power subverted the concepts and value systems of modern bourgeois society and of western civilization. This edifice served as an inspiration to some of the great thinkers of the 20th century and provided the fundamental impulse behind the so-called postmodernism.

Nietzsche proposed the idea of will to power as a tendency that was universal, with a particularly acute presence in living beings, which he presented as a hypothesis (Nietzsche, 1968b; Sanchez-Meca, 2004, 2005). Put very briefly, the will to power is at one and the same time the substrate of reality and its source of dynamism (Sánchez-Meca, 2004). Three of its characteristic properties are:

(1) Its modality is the enactment of the desire of a living being, which means that it manifests the ability to act or to be acted upon. It seeks to maximize its own possibilities against that which resists it, or that which impinges upon it. Where there is power, there is resistance (Foucault, 1979b). Opposite of what happens in classical mechanics, where differences in potential tend towards equilibrium, the will to power expresses an urge to create difference,

to always seek to increase — to be more numerous, more frequent, more diverse (Nietzsche, 1968j) — and to have more space available, even if this leads to conflict and eventual self-annihilation (Nietzsche, 2008). Nietzsche based his will to power idea on the struggle between individuals of one species, in line with the theory of Darwin, but also in the internal struggle of each living being, as described by Roux (Gayon, 1999e; Heams, 2012; Roux, 2012; Stiegler, 2001c; Tauber, 1994b).

(2) The will to power is not one single force. It is always a multiplicity of differing forces in a constant process of interaction and confrontation, of self-organization and self-destruction, with no other aim than to increase its own ability to act and to appropriate other forces, or to resist other forces that impinge upon it. In this way Nietzsche overcomes the Cartesian dualism of mind and body. For him, thought and nutrition are equivalent notions, two forms of incorporation, two different manifestations of the will to power. Along this line, Foucault begins his interpretation by denying that any such thing as power exists. For him there are only power relations, the actions of one force upon another. There is power only where there are power-relationships between two or more forces (Schacht, 1983b).

(3) Foucault insists that the will to power is not a metaphysical concept: nothing is more material, physical, corporal than the exercise of power (Foucault, 1979a). He distinguishes between power relations [be it reproductive, intellectual, etc. (Nietzsche, 1968a)] and the action of a force on an object (Foucault, 2001), on which point he coincides with Canetti (Canetti, 1977). This latter kind of action should be treated as a capability. The will to power may involve creation and destruction, but it should not be understood as an impetus toward either creativity or its opposite, as with Bergson (Bergson, 1987), but rather in terms of the dynamic interplay of forces and chance (Schacht, 1983a) and with no final aim (Schach, 1983). Deleuze's interpretation distinguishes between active and reactive forces (Deleuze, 1986b), and while it is outside the scope of this manuscript, it may help

to fill the gap between the philosophical interpretation of force and the scientific one.

On occasions, both in Nietzsche's own writings and in those of his interpreters, the will to power is equated with the will to live (Nietzsche, 2008; Schacht, 1983c). In the course of the struggle, the weaker forces combine. This point is key, because they are able to sacrifice their own identity in order to assume specialized roles and subordinated functions, thereby allowing forces that were originally weaker to subsume and triumph over the force that was originally stronger. This dynamic is the key element in any understanding of Nietzsche's concept of the living, and how it goes beyond the theories of Darwin.

In any case, his remarks owed a large debt to Spinoza. As a summary to this, we might look to two passages from Deleuze's Nietzsche in which he notes and endorses the way in which Nietzsche draws on Spinoza's writings (Deleuze, 1986a):

> *"Spinoza suggested a new direction for the sciences and philosophy. He said that we do not even know what a body can do, we talk about consciousness and spirit and chatter on about it all, but we do not know what a body is capable of, what forces belong to it or what they are preparing for..."*
>
> *"What is the body? We do not define it by saying that it is a field of forces, a nutrient medium fought over by a plurality of forces. For in fact there is no "medium", no field of forces or battle. There is no quantity of reality, all reality is already quantity of force. There are nothing but quantities of force in mutual "relations of tension" (Nietzsche 1968d). Every force is related to others and it either obeys or commands. What defines a body is this relation between dominant and dominated forces. Every relationship of forces constitutes a body — whether it is chemical, biological, social or political.*

> *Any two forces, being unequal, constitute a body as soon as they enter into a relationship. This is why the body is always the fruit of chance, in the Nietzschean sense, and appears as the most "astonishing" thing, much more astonishing, in fact, than consciousness and spirit. But chance, the relation of force with force, is also the essence of force."*

It is also fair to say that this book chapter takes the exact opposite position to the most influential Western interpretation of Nietzsche — that of Martin Heidegger (Ferraris, 2000c; Heidegger, 2000d). Heidegger understood the will to power as a purely metaphysical concept. He interpreted it in terms familiar from classical metaphysics as the "equivalent" of essence, with Nietzsche's other great concept, the eternal return similarly seen as the "equivalent" of existence (Heidegger, 2000a).

3. Nietzsche' criticism of Darwin

One element of Nietzsche's criticisms of Darwin is incorrect. This is because he acquired his knowledge of Darwin's theories indirectly through the German biologists of the time, and also because the only book of Darwin that he may have read for himself was "*The variation of animals and plants under domestication*" (1886), which is the most Lamarckian of them all (Ferraris, 2000d). The criticisms in question refer to a supposed teleologism in Darwin's work to which Nietzsche was totally opposed. There is wide agreement that this criticism must be disregarded, given that natural selection is not a teleological process. One specific criticism he makes of Darwin refers to the unsubstantiated optimism of assuming that any creature that manages to get through the process of natural selection is necessarily the best, to which Ferraris responds that it cannot be otherwise, as no other principle appears to be applicable (Ferraris, 2000f). Consequently, rather than being directed at specific aspects of the theory, this criticism is bound up, we believe, with the Popperian impossibility of refuting the theory of natural selection. According to Popper, such impossibility casts doubt on the scientific nature of Darwin's theory, which should more properly be considered a program of metaphysical investigation (Popper, 1988).

The second series of criticisms of Darwin derives from Nietzsche's belief that victory is more likely secured by the average members of the population than by the exceptional individual (Ferraris, 2000a; Gayon, 1999d; Moore, 2002; Nietzsche, 1968e, 2005, N2008; Tauber, 1994a). Examples are for example "The Christians (weaker) defeated Caesar (stronger)" or in experimental virology "RNA virus quasispecies populations can suppress vastly superior mutant progeny" (de la Torre and Holland, 1990; Gómez, 2001). This criticism is also incorrect as a criticism of Darwin's theory because it confuses ontological levels, the adjective used refers to the biological properties of the particular individuals that make up the unit of selection, not the overall properties of the unit of selection that has gone on to become, respectively, a sect or a quasispecies.

What this criticism makes clear is the fundamental importance for Nietzsche of all kinds of internal relations of interaction between individuals who find themselves in competition with each other, including cooperation, mimesis, intelligence (Nietzsche, 2005). Darwinian theory, and particularly neo-Darwinism, have emphasized competition and understated the importance of cooperation. This has begun to change over the last twenty years. This criticism that Nietzsche makes also raises two further issues: first, that mediocre individuals have a greater number of relations of interaction because they are more similar to each other (Nietzsche 1968k); and second, that the collaborative trait is in itself evidence of weakness (Tauber, 1994a).

> *"My idea is that every specific body strives to become master over all space and to extend its force (its will to power) and to thrust back all that resists its extension. But it continually encounters similar efforts on the part of other bodies and ends by coming to an arrangement ("union") with those of them that are sufficiently related to it: thus they conspire together for power. And the process goes on."* (Nietzsche, 1968k)

In Nietzsche's view, an advantageously-constituted individual may for a time lead the process of evolutionary transition, but will subsequently be reabsorbed and removed by its community.

> *"A people is nature's roundabout way of getting six or seven great men. — Yes: and then of getting around them."* (Nietzsche, 2002).

The conclusion we draw from this criticism is that the unit of selection is shaped by power relations, which are internal relations of interaction, while the more fit unit of selection will be the winner in the process of natural selection, which is an external relation of comparison (Gómez, 2001).

The third type of criticism Nietzsche makes of Darwin arises out of his reading of Roux, whom he followed very closely. Roux (1850-1924) had proposed in his book, *La lutte des parties dans l'organisme* (Roux, 2012) that struggle takes place on every level: at the corpuscular (molecular) and cellular levels, and at the level of the organism, the organismic level, and so on.

The great evolutionary discontinuities found in the fossil record continue to be another factor leading to his discrepancies with Darwin (Erwin, 2000). In this case, Nietzsche's points of reference are historical, relating to the history of morality, but given that his model is Man, it is not inappropriate to compare evolution with history. Nietzsche strongly opposes a linear approach to history, and sequential actions. In his book *Nietzsche, Genealogy, History*, Foucault clarifies the specific meanings of the terms used by Nietzsche to discuss historiography, descendance, ascendance and emergence (Foucault, 2008). In Nietzsche's writing, emergence forms part of a new approach to history, an approach that Foucault shared and extended; it attributes a degree of importance to singular events, to randomness that produce a non-lineal and a non-gradual history.

Finally, where the two thinkers perhaps differ most is over Darwin's assertion of the entirely passive role of the individual. In response to changing external circumstances, the survival of the species can only be entrusted to successive randomly-generated variations. Adaptation in this case is a completely passive function, whereas Nietzsche proposes an evolving subject that is active, that strives to transform itself or to transform its surroundings, or to incorporate the external circumstances into itself (Stiegler, 2001b; Stiegler, 2001c; Tauber, 1994d): *"Life is not the adaptation of inner circumstances to outer ones, but will to power, which working from within, incorporates and subdues more and more of that which is outside"* (Nietzsche, 1968i). Commentators agree that it is here where Nietzsche felt a greater affinity with Lamarck than with Darwin. In recent years we have encountered more and more examples of how organisms are able to adapt to certain external challenges in real time and to transmit information to their offspring. These are the epigenetic modifications that are transmitted to subsequent offspring, and the horizontal transfer of genes, although this does not involve the transmission of phenotypic characteristics as Lamarck believed, but rather of its determinants, some of which are genomic and others are cytoplasmic microRNAs (Bapteste *et al.*, 2012; Noble, 2015; Sharov, 2014; Villarreal, 2009; Villarreal and Witzany, 2013). We must also add here that for Nietzsche not even randomness was a constant — rather, the organism seeks it out. It was the *"iron hands of necessity which shake the dice-box of chance"* (Nietzsche, 2006b). The facts referred to in this magnificent mixed metaphor of an active organism is somehow related to the concept of evolvability.

4. Nietzsche's subversion

From the foregoing and subsequent discussion, it is evident that the results of biological experimentation have generated concepts which have subsequently expanded, and have on occasion overcome the straitjacket of neo-Darwinian evolution, in which natural selection is seen as the only mechanism of evolution, producing an increasing number of parallels with Nietzsche's thoughts and proposals on evolution. The existence of so many parallels is not simple coincidence, nor is it part of any attempt to match the results of experiments on evolution to Nietzsche's propositions. There is an explanation: Nietzsche found the deepest roots of Darwinian evolutionary theory in the work of Spinoza and responded by producing his own counterproposal (Nietzsche, 2008).

Whilst there is considerable consensus on the influence of Malthus' work on the theory of Darwin's natural selection, Nietzsche traces its metaphysical roots to Spinoza's concept of conatus. This had been a nuclear concept in the debates in metaphysics and Western physics since the Middle Ages and relates to inertia. Nietzsche traces the Darwinian fight for existence that underpins the theory of natural selection to Spinoza's conatus: the effort by which each being enforces its own preservation. This appears to be similar to an inherent tendency. Take, for example, a dab of oil paint of a given color on a canvas. It will stay where it is put, may possibly even spread across the material, but it remains in place as new layers of paint are applied on top of it until it can no longer be seen. This models the basis of Darwin's theory: organisms do not wish to change, they resist change, they want to continue to be what they are, and change only comes about as a last resort. As he makes this genealogical analysis of Darwinism, Nietzsche turns it into a branch of metaphysics. It is this Spinozan principle that Nietzsche opposes, and against which he advances his own "conatus": self-affirmation, the desire to go beyond itself (Heidegger, 2000c, 2000d; Nietzsche, 1968f). This simple change will allow Nietzsche to include evolution by natural selection into a much wider framework, in which adaptation comes to be only one of the causes of change. Active change will come to be inherent to all living beings. It is a continuous tendency towards self-strengthening. This is nothing more and nothing less than a way of expressing his thinking on the will to power: The struggle is not for life, but for power, there is only a struggle for life when the latter has been put in danger (Nietzsche, 2008).

5. Nietzsche' update

Having looked, up to this point, at Nietzsche's criticisms of Darwin, and having traced the origin of his divergence from Darwin in the writings of Spinoza, the question we must now ask is: Which theory or concepts can Nietzsche help us with?

5.1. *Multiplicity*

In biology, organisms and cells are conceived of as coherent units which bring together and stabilize the processes occurring within them in a hierarchy and maintain a memory of past events. In this way biology turns the organism, or the cell, into both the functional unit and an active subject. However, this attributes nominal unity to a group of complex, dynamic and internally-diverse phenomena (Roux, 2012). Wilhelm Roux, the nineteenth-century German biologist, suggested that some of these phenomena might have sufficient entity to participate independently as units of selection. He thus posited that the Darwinian struggle must not only take place between individual organisms of the same species, but also on other, lower levels, such as the cellular or molecular levels.

Examples at the cellular level could include the outgrowth of a cellular clone expressing a specific antibody after an antigen presentation or the selection of subsets of neuronal webs during development (Heams, 2012). At the subcellular level these include the sequestered functional chloroplasts that can still be found in sea slugs, which use them as a source of stored food reserves for sustenance during starvation periods (Christa *et al.*, 2014a, 2014b) or in a reverse way by losing the mitochondria (Karnkowska *et al.*, 2016). The molecular struggle has also been related to several different forms of intragenomic conflicts. Examples include genes that do not follow the same laws of transmission (Crespi and Nosil, 2013), the hypothetical genome reduction occurring during organelle adaptation (Bapteste and Gribaldo, 2003) or other endosymbiosis (Husnik *et al.*, 2013), and potential antagonist interaction between biomolecules within the same cell, for example in addiction modules of bacteria (Villarreal, 2009; Villarreal and Witzany, 2013).

Roux's idea was taken up by Nietzsche, who incorporated it into his metaphysical reflections: *"Although the units are required to tell the tale, this doesn't mean that such units exist. The concept of the unity is derived from our concept of I, which is our oldest article of faith. If we never considered ourselves as units, we would never have reached the concept of thing. Maybe we have been warned too late that our conception of I does not guarantee anything that can be referred to as a real unit"* (Nietzsche, 1968g).

This instance of criticism of the unity of the subject that began with Nietzsche has had a huge impact on twentieth-century thought, including on Freud (divergent opinions), Lacan, Althusser, Lévi-Strauss, Guattari and others. The individual subject is not capable of controlling the life for which it is held to be responsible. From Nietzsche's perspective, that instance of life, expressed biologically in the form of multiple internal conflicts, bursts through the limits of such unity, introduces chaos, and even endangers the life of the cell or organism. But it is perhaps this continual destabilization that allows creativity and change. *"One must still have chaos in oneself to give birth to a dancing star"* (Nietzsche, 2010).

The way in which these conflicting internal processes can be organized is through hierarchy (Stiegler, 2001c), but this is a conception of hierarchy that differs from that used in formal logic. It is a Nietzschean interpretation of hierarchy in which hierarchical organization always appears as the result of a prior struggle in which the contenders have reached a state of equilibrium with the winner controlling, containing, or to some degree repressing the loser, but where victory has not totally destroyed or assimilated the loser. Furthermore, in Nietzsche's version of hierarchy, those agents at the base of the pyramid are able to keep the top in check, and may overturn the hierarchical order using new interactions between elements within the base to do so. The hierarchical structure is thus in a state of constant renewal (Deleuze, 1968; Stiegler, 2001c), meaning change would be guaranteed, either by the partial or total renewal of the hierarchical structure, or by its collapse. In summary, Nietzschean hierarchy is a dynamic form of organization based on perpetual conflict leading to continual change.

5.2. *The fluid self*

In some biological models, it might even be said to be impossible to find any definite biological entity. This is the case when the model incorporates evolutionary interactions and ecological interactions between organisms or entities of different levels of complexity and with different rates of evolutionary change. Such cases require a change of ontological model, a processual model (Bapteste and Dupre, 2013; O'Malley and Dupre, 2007).

Nietzsche suggested an ontology for interpreting organisms as a temporary result of a never-ending conflict, either inner or with the external world (Stiegler, 2001c; Tauber, 1994f). *"Continuous becoming does not allow us to speak of individuals, etc.: the number of beings varies constantly..."* (Nietzsche, 1968c). In fact, the affirmation that all life strives for self-preservation presupposes that there is a substantial self or organism that wants to preserve itself, to which Nietzsche opposes (Aydin, 2007).

Three examples are of particular biological relevance: RNA viruses, bacterial communities and cancer.

The genome of RNA viruses cannot be defined by a single sequence, but rather by a population of variant sequences closely related to one another which are simultaneously present in an infected organism. This way of organizing the genetic information is referred to as quasispecies (Domingo, 2007; Domingo *et al.*, 1978; Eigen, 1971, 1988). Throughout infection, the number and composition of the variants in the viral population keep changing owing to environmental influences, including genomic variations of the host cell, and internal perturbations: the presence of interferent genomes, recombination, etc., resulting in a virus that is constantly redefining itself both genetically and phenotypically. Whenever natural selection faces a quasispecies, it is confronted with something that is similar, but different and changing, requiring selection of a dynamic rather than a static individual phenotype (Gómez *et al.*, 1999). We have defined viral quasispecies as a process (Mas *et al.*, 2010).

Peter Nowell proposed in 1976 that cancer was a complex evolutionary system that showed high heterogeneity and clonal evolution (Nowell, 1976). In fact, clonal diversity in cancer cells could be used as a factor for predicting progression in an oesophageal adenocarcinoma cancer model (Maley *et al*., 2006). Cancer studies can be based on an evolutionary and ecological context (Maley and Forrest, 2000; Merlo *et al*., 2006) and theoretical studies have correlated cancer and genetic instability (González-García *et al*., 2002; Maley and Forrest, 2000) with quasispecies models of minimal replicators (Mas *et al*., 2010; Solé and Deisboeck, 2004). These studies show the high genetic heterogeneity of tumor cells as the source of adaptation for cancer to fight against the immune system, become resistant to different treatments, invade adjacent tissues, and sometimes metastasize and invade other organs (Koren and Bentires-Alj, 2015; Mas *et al*., 2010).

The microbiological biofilms are communities of microorganisms encased within an extracellular polymeric slime matrix living on surfaces (Branda *et al*., 2005), which are endowed with structural heterogeneity, genetic diversity, and complex community interactions. In such communities, also known as superorganisms, a subpopulation of microorganisms within the biofilm may form a unique and highly protected, phenotypic state, may acquire division of labor, or resistance to phages and antibiotics (Branda *et al*., 2005; Mas *et al*., 2010).

The basic idea in all of these examples is that either cooperative or hostile interactions within the community, which might include interactions between entities from distinct levels of biological organization, determine a Nietzschean concept of *being in flux* (Nietzsche, 1968c).

According to Nietzsche, what is important in competition is no longer the individual or the member of the species that takes part, but the outcome of the competition, for which the individual will even sacrifice part of its individuality. Individuality is sacrificed in the process of collaboration.

5.3. *The pre-individual being and viruses*

Simondon proposed that the individual is not merely the end result of a process, but that within the individual there exists a *"pre-individual"* element. This vestigial imprint is what endows the individual with the capacity to create new relationships, perform transformations and produce other new individuations (Penas-López, 2014).

From the point of view of power relations, and in response to the question of what viruses are, we have proposed a cell model involving internal tension very much in accordance with Simondon's theory. The suggestion is that the cell contains numerous activities, elements that must be kept repressed, silenced, compartmentalized, slowed down, hierarchized, and so on, in order for the cell to continue to function correctly. Any miscontrol would lead to the malfunction of its mechanisms. Thus, we have defined the habitual state of the cell as one of latent pathology. This hypothesis can be explained by looking to what must have been the "trauma" undergone by the different molecules and organizations present at the origin of life when they were enclosed within the cell, or when RNA cells became DNA. In the thermodynamic sense, this "trauma" results from the loss of a vast range of freedoms for these early biomolecules and organizations, given that only cooperative activities between the various agents could be expressed, and all those activities that might turn out to be toxic would have to be repressed. It is very possible that many of these inactive functions have now irreversibly disappeared. For example, RNA is no longer responsible for hereditary transmission, but other functions may remain within the cell, silenced, and in our hypothesis these represent a part of those which are activated, expressed, and so forth, when viruses come into play. Therefore, our hypothesis is that molecular structures, their activities, and the majority of the information that makes up a viral induced pathological state does not follow the entry of the virus and is not a consequence of viral activity, but exactly the opposite. We posit that latent pathology within the cell must be placed ahead of it, as the basis of the infection. In other words, it is what makes infection possible. The basis of the infection is already there because it is also the basis of the cell itself. Basis in the sense of origin: this latent pathological cellular state would have been there all along since the beginning of the cell, and in its very development (Gómez *et al.*, 2015). Thus, in this new scenario, the battle is considered intracellular rather than between the virus and the cell. And so the virus is treated as the signal (a folder of signals) that precipitates the cell's change from a latent to an active pathological state (Figure 1).

In line with Simondon, each virally-dictated pathological state partially restores an early critical life situation, an ancient pre-individual reality. Viruses are signals derived from the molecular organizations, activities and structures of life that existed in the remote past, but which are still present in our cells, although in a fragmented, disjointed and inhibited form. In our hypothesis, these pre-pathological states would be sustained by hidden potentialities, awaiting the arrival of an appropriate virus signal that tells them how to self-organize, express themselves, and so on.

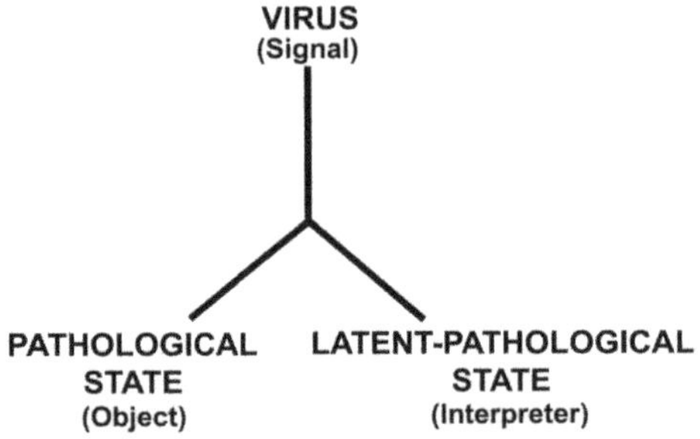

Figure 1. A triadic model for a viral infection in biosemiotic terms: virus ↔ pathological state↔ transformation of the latent pathological state into a pathological state. The model is related to Peirce's triadic sign (signal ↔ object ↔ interpretant). The interpretant is that what changes of the interpreter.

5.4. *Perspectivism and pragmatism*

> *"There are no facts, only interpretations"* (Nietzsche, 1998).

For Nietzsche, there is no *"thing-in-itself."* Rather, for whatever might exist, and by whom or whatever it may be perceived, what we have are differing interpretations. The biological organism and the philosophical self are also products of interpretation, or better still, they are the result of multiple interpretations. There is no self, no interpreting organism. In fact, the opposite is the case: these consist entirely of interpretations (Heidegger, 2000b; Nietzsche, 1998). Since there is nothing requiring the interpretations to coincide, they come into conflict, where *"the strongest imposes its interpretation"* (Tauber, 1994e). The thing that interprets is always a force, but it is a force that acts upon other forces, performing actions that are interactive and which develop a hierarchical structure, as described earlier (Deleuze, 1986b).

> *"In fact, interpretation is itself a means of becoming master of something. (The organic process constantly presupposes interpretations.)"* (Nietzsche, 1968h).

It should also be noted that according to Nietzsche, the result of an interpretation is not objective and even a misguided interpretation of what is happening can in many cases be a means of adjusting the circumstances to suit oneself (Heidegger, 2000b; Stiegler, 2001c).

> *"But what identifies something as a living thing is the faculty of memory, which allows it to see the same thing as before in something completely other..."* (Stiegler 2001e).

The "organism" interprets according to its biological interest in surviving and in advancing itself. For this reason, Nietzsche's perspectivism is not entirely relativist, but pragmatic (Stack 2012c) and is therefore one of the founders of pragmatism (Stack, 2012d). A theoretical and practical example can be found in the quasispecies of the theoretical replicons at the origins of life and in RNA viruses. Each quasispecies consists of a multitude of replicative variants, all of which compete for the resources necessary for replication, while simultaneously cooperating in multiple ways between them (Domingo 2007). The properties of the quasispecies are determined by the combined variants in their varying proportions and according to their particular qualities, but not by a wild type sequence. The composition of the variant population also undergoes a constant process of self-redefinition due to its ability to move into gaps in the sequences and to follow paths leading to positions of greater fitness (gradient ascent) (Eigen, 1988). Here we find an example of how the multiplicity of variants (perspectives) is what makes it possible to "look ahead" and seek out the paths that confer the greatest biological efficacy (pragmatism). Interpretations thus arise from social interplay (Villarreal, 2015; Villarreal and Witzany, 2013; Witzany). *"It is the context (pragmatics) in which the living agent is concretely interwoven that determines the meaning (function) of a given sequence"* (Witzany, 2012).

Around the same time, the mathematician Charles S. Pierce developed pragmatism as a logical system (de Waal, 2013a), and in the same way as Nietzsche, departed from the Kantian understanding of what is. In relation to evolution Pierce put forward the idea of habit. Instead of random variation as the primordial element — as in the Darwinian model — Pierce placed the formation of a habit and its transmission at the center of the process, a view previously advanced by Lamarck (de Waal, 2013b).

Nietzsche and Pierce both move away from the Aristotelian substantialist view that posits the existence of a "self" on the one hand, and the world with which it interacts on the other. Neither of them sees any end point — the chain of interpretations is infinite. For both philosophers interactions are not additional and external aspects of things, but are essential and constitutive (Aydin, 2006).

The biological expression of the preceding theoretical concepts is to be found in the notion of *Umwelt* proposed by the early-twentieth-century German biologist, Jacob von Uexküll, which approximates a "subjective universe" (Kull, 2001; Sharov, 2001; Tønnessen, 2016; Witzany, 2006). Organisms living in the same place do not necessarily live in the same ecosystem, since each has its own *Umwelt* and interacts with different elements in different ways. There is agreement among specialists in the field that this concept is complementary to Pierce's position (Sharov, 2001), although Uexküll extends it to the animal world and even the unicellular level. However, while references to the thought of Nietzsche are not as frequent, there does seem to be a notable precedent for Uexkull's Umwelt concept: according to Nietzsche, *"Each centre of energy — not only man — constructs the world in terms of itself"* (Nietzsche, 2006a), whereas Uexküll asserts that *"Nobody is a product of their environment — everybody is the master of [their] Umwelt"* (Kull, 2001).

In fact, in his book on Nietzsche, Heidegger attempts to explain the latter's perspectivism using an illustration in which we can recognize an example of Umwelt:

> *"For example, a lizard hears the slightest rustling in the grass but it does not hear a pistol shot fired quite close by. Accordingly, the creature develops a kind of interpretation of its surroundings and thereby of all occurrence, not incidentally, but as the fundamental process of life itself: The perspectival [is] the basic condition of life"* (Heidegger, 2000b).

While this concept has been widely developed by some schools of ethology, examples from molecular biology are rare, although it may appear implicitly in discussion of the immune system as an agent that distinguishes between foreign matter and self with reference to autoimmune diseases. Another case might be that of molecular mimicry, especially RNA mimicry, which is common in RNA viruses and in the mimetic structures of its tRNA (Díaz-Toledano and Gómez, 2015; Nadal *et al.*, 2002; Yot *et al.*, 1970). A molecule is considered to mimic another if it is capable of competing for the biological element that recognizes it. This means that the mimetic function is not defined by the degree of similarity that two molecules might have in the eyes of the scientist who observes them.

Table 2. Summary of the concepts reviewed.

Classic view	Nietzsche view
Self-preservation	Enhancement beyond itself.
What is essential is what is conserved within the individual	Conservation is a consequence of having placed the individual at risk.
The metaphysical view of the fixed individual with its attributes (substantialism)	Entities which are rendered indeterminate by their relations with their environment and by chance, always in flux: agent, individual, superorganism, ecosystem. Unity depends on perspective, and is always transitory.
Individual passive vis à vis environment	"Individual" as active interpreter (to interpret is to apply value criteria).
Variation through chance	Variation resulting from chance and power relations.
Variation and selection	Variation, mimicking, learning, discipline, intelligence, collaboration, participate in modulation of the unit of selection.
Cartesian dualism of mind and body	Man and his psyche are one; Man and his culture can be subject of research, along with other forms of life and nature, from the perspective of power relations.
Human culture and Nature are divorced	Power relations intensely studied in humanities, as new paradigm for the study of the natural world. Ways of analyzing human societies, taking account of Man's biological aspect: biopower.

Here, it is the biological feature in question — the "interpreter" in Pierce's tripartite model — that decides.

Table 2 summarizes some of the concepts reviewed.

Acknowledgments

We thank critical reading of Drs. Esteban Domingo and Günther Witzany. The work was funded by the National Ministry of Competitiveness (which includes science and development) BFU2012–35898, Proyecto Excelencia Junta de Andalucía CVI-03050 and CIBERehd (Centro de Investigación en Red de Enfermedades Hepáticas y Digestivas) to A-AM.

References

Andrade E (2003a) Los demonios de Darwin. Semiótica y termodinámica de la evolución biológica. Universidad Nacional de Colombia. Unibiblos, Bogotá D.C.

Andrade E (2003b) Semiótica de la evolución biológica. In: *Los demonios de Darwin. Semiótica y termodinámica de la evolución biológica*, Rubiano G (ed). Universidad Nacional de Colombia. Unibiblos, Bogotá D.C., pp. 111–212.

Aydin C (2006) Beyond Essentialism and Relativism: Nietzsche and Peirce on Reality. Cognitio 7: 25–47.

Aydin C (2007) Nietzsche on reality as will to power: Toward an "Organization–Struggle" model. *J Nietzsche Stud* **33**: 25–48.

Bapteste E, Dupre J (2013) Towards a processual microbial ontology. *Biol Philos* **28**: 379–404.

Bapteste E, Gribaldo S (2003) The genome reduction hypothesis and the phylogeny of eukaryotes. *Trends Genet* **19**: 696–700.

Bapteste E, Lopez P, Bouchard F, Baquero F, McInerney JO, Burian RM (2012) Evolutionary analyses of non-genealogical bonds produced by introgressive descent. *Proc Natl Acad Sci USA* **109**: 18266–72.

Barbieri M (2007) *Introduction to Biosemiotics. The New Biological Synthesis.* Springer, Dorderecht.

Bergson H (1987) La vida o la diferenciación de la duración. Memoria y vida. Textos escogidos por Gilles Deleuze. *Alianza Editorial*, Madrid, pp. 88–123.

Branda SS, Vik S, Friedman L, Kolter R (2005) Biofilms : the matrix revisited. *Trends Microbiol* **13**: 20–26.

Cacho I, Gómez J (2010) Biopuvoir et virus à ARN. De l' usage des métaphores en biologie moleculaire, à contre-courant des dogmes. *Cahiers internationaux de symbolisme* **125–127**: 89–100.

Canetti E (1977) *Elementos del Poder. Masa y Poder*. Muchnik Editores, Barcelona, pp. 277–295.

Crespi B, Nosil P (2013) Conflictual speciation: species formation via genomic conflict. *Trends Ecol Evol* **28**: 48–57.

Christa G, de Vries J, Jahns P, Gould SB (2014a) Switching off photosynthesis: The dark side of sacoglossan slugs. *Commun Integr Biol* **7**: e28029.

Christa G, Handeler K, Schaberle TF, Konig GM, Wagele H (2014b) Identification of sequestered chloroplasts in photosynthetic and non-photosynthetic sacoglossan sea slugs (Mollusca, Gastropoda). *Front Zool* **11**: 15.

de la Torre JC, Holland JJ (1990) RNA virus quasispecies populations can suppress vastly superior mutant progeny. *J Virol* **64**: 6278–81.

de Waal C (2013a) Peirce Bloomsbury Academy, London.

de Waal C (2013b) *Three Modes of Evolution Pierce*. Bloomsbury Academy, London, pp. 144–148.

Deleuze G (1968) *Activo y reactivo. La jerarquía. Nietzsche y la Filosofía*. Anagrama SA, Barcelona, pp. 87–89.

Deleuze G (1986a) *Activo y reactivo. El cuerpo. Nietzsche y la Filosofía*. Anagrama SA, Barcelona, pp. 59–61.

Deleuze G (1986b) *Activo y reactivo. La distincion de las fuerzas. Nietzsche y la filosofía*. Anagrama SA, Barcelona, pp. 61–67.

Dennett DD (1995) Friedrich *Nietzsche's Just So Stories. Darwin's Dangerous Idea. Evolution and the Meanings of Life*. Penguin group, London, p. 464.

Díaz-Toledano R, Gómez J (2015) Messenger RNAs bearing tRNA-like features exemplified by interferon alfa 5 mRNA. *Cell Mol Life Sci* **37**: 3747–68.

Domingo E (2007) Virus Evolution. In: *Virology*, Knipe DH, PM Fields (eds). Walters Kluwer/Lippincott Williams & Wilkins, New York, pp 389–421.

Domingo E, Sabo D, Taniguchi T, Weissmann C (1978) Nucleotide sequence heterogeneity of an RNA phage population. Cell **13**: 735–44.

Editorial (1945) Nietzsche and Science. *Nature* **156**: 263.

Eigen M (1971) Self-organization of matter and the evolution of biological macromolecules. *Naturwissenschaften* **58**: 465–523.

Eigen M, Biebribricher, CK (1988) Sequence space and quasipecies distribution. In: *RNA Genetics. Variability of RNA Genomes*, Domingo E, Holland JJ, Ahlquist P (eds). CRC press, Boca Raton.
Emmeche C, Kull K (2011) *Towards a Semiotic Biology. Life is the Action of Signs*. Imperial College Press CP, London.
Erwin DH (2000) Macroevolution is more than repeated rounds of microevolution. *Evol Dev* **2**: 78–84.
Ferraris M (2000a) El paradigma Biológico. In: *Nietzsche y el Nihilismo*, Duque F (ed). Ediciones Akal, Madrid, p. 60.
Ferraris M (2000b) El paradigma biológico. In: *Nietzsche y el Nihilismo*, Duque F (ed) Ediciones Akal, Madrid, p. 56.
Ferraris M (2000c) El paradigma biológico. In: *Nietzsche y el Nihilismo*, Duque F (ed). Ediciones Akal, Madrid, p. 58.
Ferraris M (2000d) El paradigma biológico. In: *Nietzsche y el Nihilismo*, Duque F (ed). Ediciones Akal, Madrid, p. 63.
Ferraris M (2000e) El paradigma biológico. In: *Nietzsche y el Nihilismo*, Duque F (ed). Ediciones Akal, Madrid, p. 55.
Ferraris M (2000f) El paradigma biológico. In: *Nietzsche y el Nihilismo*, Duque F (ed). Ediciones Akal, Madrid, p. 64.
Ferraris M (2000g) El paradigma biológico. In: *Nietzsche y el Nihilismo*, Duque F (ed). Ediciones Akal, Madrid, pp. 64–65.
Ferraris M (2000h) El paradigma biológico: la crítica al mecanicismo. In: *Nietzsche y el Nihilismo*, Duque F (ed). Ediciones Akal, Madrid.
Ferraris M (2000i) Perspectivismo y monadología. In: *Nietzsche y el Nihilismo*, Duque F (ed). Ediciones Akal, Madrid, pp. 55–65.
Foucault M (1979a) *Poder-Cuerpo Microfísica del poder*. Las Ediciones de La Piqueta, Madrid, p. 105.
Foucault M (1979b) *Poderes y estrategias Microfisica del poder*. Las Ediciones de La Piqueta, Madrid, p. 171.
Foucault M (1979c) *Preguntas a Michael Foucault sobre la geografía. Microfísica del poder*. Las Ediciones de La Piqueta, Madrid, p. 112.
Foucault M (2001) El sujeto y el poder ¿en que consiste la naturaleza específica del poder? In: *Arte después de la modernidad*, Wallis B (ed). Ediciones Akal, Colmenar, Madrid, p. 431.
Foucault M (2008) *Nietzsche, la genalogía, la historia*. Pre textos, Valencia.
Gayon J (1999a) Nietzsche and Darwin. In: *Biology and the Foundation of Ethics*, Maienschein J, Ruse M (eds), Cambridge University Press, Cambridge. UK, p. 162.

Gayon J (1999b) Nietzsche and Darwin. In: *Biology and the Foundation of Ethics*. Maienschein J, Ruse M (eds). Cambridge University Press, Cambridge.UK, pp. 154–197.

Gayon J (1999c) Nietzsche and Darwin. In *Biology and the Foundation of Ethics.* Maienschein J, Ruse M (eds). Cambridge University Press, Cambridge. UK, p. 156.

Gayon J (1999d) Nietzsche and Darwin. In: *Biology and the Foundation of Ethics*, Maienschein J, Ruse M (eds). Cambridge University Press, Cambridge. UK, pp. 164–165.

Gayon J (1999e) Nietzsche and Darwin. In: *Biology and the Foundation of Ethics*, Maienschein J, Ruse M (eds). Cambridge University Press, Cambridge. UK, p. 170.

Gayon J (1999f) Nietzsche and Darwin: Some biographical landmarcks In: Maienschein J, Ruse M (eds) *Biology and the Foundation of ethics*, Maienschein J, Ruse M (eds). Cambridge University Press, Cambridge. UK, pp. 158–161.

Gómez J, Ariza-Mateos A, Cacho I (2015) Virus is a signal for the host cell. *Biosemiotics* **8**: 483–491.

Gómez J, Cacho I (2004) La pregunta pel poder i els virus de RNA, coincidencies de vocabulari, conceptes i resultats experimentals. In: *Virologia i Filosofia del Poder*, Gomez J, Cacho I, Matínez MA, Montserrat M (eds). Societat Catalan de Biologia, Institut d'Estudis Catalans, Barcelona.

Gómez J, Martell M, Quer J, Cabot B, Esteban JI (1999) Hepatitis C viral quasispecies. *J Viral Hepat* **6**: 3–16.

Gómez J, Cacho I (2001) Can Nietzsche power relationships be experimentally approached with theorethical and viral quasispeices? *Contributions to Science* **2**: 103–108.

González-García I, Solé RV, Costa J (2002) Metapopulation dynamics and spatial heterogeneity in cancer. *Proc Natl Acad Sci USA* **99**: 13085–13089.

Goodrich BG (2010) We do, therefore we think: time, motility, and consciousness. *Rev Neurosci* **21**: 331–361.

Heams T (2012) Selection within organisms in the nineteenth century: Wilhelm Roux's complex legacy. *Prog Biophys Mol Biol* **110**: 24–33.

Heidegger M (2000a) *Existencia. Nietzsche*. Ediciones Destino, p. 388.

Heidegger M (2000b) *La nueva interpretación de la sensibilidad y la provocativa discrepancia entre arte y verdad. Nietzsche*. Ediciones Destino, Barcelona, pp. 198–207.

Heidegger M (2000c) *La voluntad de poder como principio de una nueva posición de valores. Nietzsche*. Editorial Destino, Barcelona, p. 394.

Heidegger M (2000d) *La voluntad de poder como principio de una nueva posición de valores. Nietzsche*. Ediciones Destino, Barcelona.

Husnik F, Nikoh N, Koga R, Ross L, Duncan RP, Fujie M, Tanaka M, Satoh N, Bachtrog D, Wilson AC, von Dohlen CD, Fukatsu T, McCutcheon JP (2013) Horizontal gene transfer from diverse bacteria to an insect genome enables a tripartite nested mealybug symbiosis. *Cell* **153**: 1567–78.

Karnkowska A, Vacek V, Zubacova Z, Treitli SC, Petrzelkova R, Eme L, Novak L, Zarsky V, Barlow LD, Herman EK, Soukal P, Hroudova M, Dolezal P, Stairs CW, Roger AJ, Elias M, Dacks JB, Vlcek C, Hampl V (2016) A eukaryote without a mitochondrial organelle. *Curr Biol* **26**: 1274–1284.

Koonin EV, Senkevich TG, Dolja VV (2009) Compelling reasons why viruses are relevant for the origin of cells. *Nat Rev Microbiol* **7**: 615; author reply 615.

Koren S, Bentires-Alj M (2015) Breast tumor heterogeneity: Source of fitness, hurdle for therapy. *Mol Cell* **60**: 537–46.

Kull K (2001) Jakob von Uexküll: An introduction. *Semiotica* **134**: 1–59.

Maley CC, Forrest S (2000) Exploring the relationship between neutral and selective mutations in cancer. *Artificial Life* **6**: 325–345.

Maley CC, Galipeau PC, Finley JC, Wongsurawat VJ, Li X, Sanchez CA, Paulson TG, Blount PL, Risques RA, Rabinovitch PS, Reid BJ (2006) Genetic clonal diversity predicts progression to esophageal adenocarcinoma. *Nat Genet* **38**: 468–473.

Markoš A, Grygar F, Kleisner K, Neubauer Z (2008) Towards a Darwinian Biosemiotics. Life as Mutual Understanding. In: *Introduction to Biosemiotics*. Barbieri M (ed). Springer, Dordrecht, pp. 235–255.

Mas A, López-Galíndez C, Cacho I, Gómez J, Martínez MA (2010) Unfinished stories on viral quasispecies and Darwinian views of evolution. *J Mol Biol* **397**: 865–877.

Mayr E (1982) *The Position of Biology within the Sciences. The Growth of Biological Thought*. The Belknap Press, Harvard University Press, Cambridge, Massachusetts, pp. 32–36.

Megill A (2016) Theological presuppositions of the evolutionary epic: From Robert Chambers to E. O. Wilson. *Stud Hist Philos Biol Biomed Sci* **58**: 24–32.

Merlo LM, Pepper JW, Reid BJ, Maley CC (2006) Cancer as an evolutionary and ecological process. *Nat Rev Cancer* **6**: 924–935.

Moore H (2002) *Evolution and the Increase of Life. Nietzsche, Biology and Metaphor*. Cambridge Universtiy Press, Cambridge, UK, pp. 46–55.

Moreira D, Lopez-Garcia P (2009) Ten reasons to exclude viruses from the tree of life. *Nat Rev Microbiol* **7**: 306–11.

Nadal A, Martell M, Lytle JR, Lyons AJ, Robertson HD, Cabot B, Esteban JI, Esteban R, Guardia J, Gómez J (2002) Specific cleavage of hepatitis C virus RNA genome by human RNase P. *J Biol Chem* **277**: 30606–30613.

Nasir A, Caetano-Anolles G (2015) A phylogenomic data-driven exploration of viral origins and evolution. *Sci Adv* **1**: e1500527.

Nietzsche F (1968a) Aphorism 656 In: *The Will to Power*, Kauffmann W (ed). Random House, New York, p 364.

Nietzsche F (1968b) Critique of the concept: cause. Aphorism 689. In: *The Will to Power*, Kauffmann W (ed). Random House, New York, pp. 367–368.

Nietzsche F (1968c) Principles of a new evaluation. Aphorism 620. In: *The Will to Power*, Kauffmann W (ed). Random House New York, p. 281.

Nietzsche F (1968d) Principles of a new evaluation. Aphorism 635. In: *The Will to Power*, Kauffmann W (ed). Random House, New York, p. 338.

Nietzsche F (1968e) Principles of a new evaluation. Aphorism 685. In: *The Will to Power*, Kauffmann W (ed). Random House, New York, pp. 364–365.

Nietzsche F (1968f) Theory of the Will to Power and of Values. Aphorism 688. In: *The Will to Power*, Kauffmann W (ed). Random House, New York, pp. 366–367.

Nietzsche F (1968g) The will to power. Aphorism 520. In: *The Will to Power*, Kauffmann W (ed).Random House, New York.

Nietzsche F (1968h) The will to power as life. Aphorism 643 In: *The Will to Power*, Kauffmann W (ed). Random House, New York, p. 342.

Nietzsche F (1968i) The Will to Power in Nature. Aphorism 681. In: *The Will to Power*, Kauffmann W (ed). Random House, New York, pp. 360–361.

Nietzsche F (1968j) The will to power in nature. Aphorism 685. In: *The Will to Power*, Kauffmann W (ed). Random House, New York, p. 364.

Nietzsche F (1968k) The world and life. Aphorism 636. In: *The Will to Power*, Kauffmann W (ed). Random House, New York, pp. 339–340.

Nietzsche F (1998) Aphorism 7. In: *El Nihilismo: escritos postumos*, Cornudella J (ed). Península, Barcelona, pp. 60–61.

Nietzsche F (2002) Epigrams and entr'actes: Aphorism 126. In: *Beyond Good and Evil*, Horstmann & Norman (ed). Cambridge University Press., New York, p. 66.

Nietzsche F (2005) Skirmishes of an ultimately man. Aphorism 14. In: *The Anti-Christ, Ecce Homo, Twilight of the Idols*, Ridely; Norman (ed). Cambridge University Press., Cambridge, p. 199.

Nietzsche F (2006a) Aphorism 14. In: *Fragmentos Postumos*, Meca DS (ed). Tecnos, Madrid.

Nietzsche F (2006b) Purposes? Will? Aphorism 130. In: *Daybreak*, Clark M, Leiter B (ed). Cambridge University Press, Cambridge. UK, p. 80.

Nietzsche F (2008) We fearless ones. Aphorism 349. In: *The Gay Science*, Williams B (ed). Cambridge University Press, New York., pp. 207–208.

Nietzsche F (2010) Prologue 5 In: *Thus Spoke Zarathustra*, Chapko B (ed). Feedbooks, pp. 16–18.

Noble D (2015) Evolution beyond neo-Darwinism: a new conceptual framework. *J Exp Biol* **218**: 7–13.

Nowell PC (1976) The clonal evolution of tumor cell populations. *Science* **194**: 23–28.

O'Malley MA, Dupre J (2007) Towards a philosophy of microbiology. *Stud Hist Philos Biol Biomed Sci* **38**: 775–779.

Penas-López M (2014) Thesis. Individuación, individuo y relación en el pensamiento de Simondon Filosofía. *Autónoma de Barcelona y Université de Tolouse, Bellaterra*. Barcelona, p. 325.

Pence CH (2011) Nietzsche's aesthetic critique of Darwin. *Hist Philos Life Sci* **33**: 165–190.

Popper SK (1988) Darwinism as a metaphysical research program. In: *But it is Science*, Ruse M (ed). Prometheus Books, Buffalo, New York, pp. 144–155.

Roux W (2012) *La lutte des parties dans l'organisme* (original 1881), Paris.

Sanchez-Meca D (2005) *Capitulo 3. El cuerpo y la cultura. La hipótesis de la voluntad de poder*. Editorial Tecnos, Madrid.

Sánchez-Meca D (2004) Nietzsche: la vida como voluntad de poder. In: *Virología y filosofía del poder*, Gómez C, Cacho I, Martinez MA, Montserrat M(eds). Societat Catalana de Biología, Barcelona, pp. 27–35.

Schach R (1983) *The World and Life*. In: Honderich T (ed) Nietzsche. Routledge & Kegan, Illinois, p. 211.

Schacht R (1983a) *The World and Life*. In: Honderich T (ed) Nietzsche. Routledge & Kegan, Illinois, p. 220.

Schacht R (1983b) *The World and Life*. In: Honderich T (ed) Nietzsche. Routledge & Kegan, Illinois, p. 217.

Schacht R (1983c) *The World and Life*. In: Honderich T (ed) Nietzsche. Routledge & Kegan, Illinois, p. 247.

Sharov A (2001) Umwelt-theory and pragmatism. In: *Jakob von Uexküll: A Paradigm for Biology and Semiotics*, Kull K (ed). Semiotica (Special Issue) **134(1)**: 211–228.

Sharov A (2014) Evolutionary constraints or opportunities? *Biosystems* **120C**: 21–30.

Solé RV, Deisboeck TS (2004) An error catastrophe in cancer? *J Theor Biol* **228**: 47–54.

Stack G (2012a) An evolutionary epistemology. In: *Nietzsche and Modern German Thought*, Ansell-Pearson K (ed). Routledge, London, pp. 36–46.

Stack G (2012b) An evolutionary epistemology. In: *Nietzsche and Modern German Thought*, Ansell-Pearson K (ed). Routledge, London, p. 36.

Stack G (2012c) Kant, Lange and Nietzsche: critique of knowledge. Kant and Lange: intial reactions. In: *Nietzsche and Modern German Thought*, Ansell-Pearson K (ed). Routledge, London pp. 32–36.

Stack G (2012d) Kant, Lange and Nietzsche: critique of knowledge. In: *Nietzsche and Modern German Thought*, Ansell-Pearson K (ed). Routledge, London, p. 36.

Stiegler B (2001a) *Introduction La vie comme "lutte interne des parties de l'organisme"*. Presses Universitaires de France, Paris, p. 7.

Stiegler B (2001b) *La vie comme "comme pouvoir interne de creér des formes" Nietzsche et la biologie*. Presses Universitaires de France, Paris, pp. 45–50.

Stiegler B (2001c) *La vie comme "lutte interne des parties de l'organisme" Nietzsche et la biologie*. Presses Universitaires de France, Paris, pp. 50–57.

Stiegler B (2001d) *La vie comme "lutte interne des parties de l'organisme" Nietzsche et la Biologie*. Presses Universitaires de France, Paris, pp. 11–12.

Stiegler B (2001e) *La volonté de puissance et la nécessité de la mort Nietzsche et la Biologie*. Presses Universitaires de France, Paris, p. 79.

Stiegler B (2001f) *Partir du corps et de la physiology Nietzsche et la biologie*. Presses Universitaires de France, Paris, pp. 17–43.

Tauber A (1994a) The immune self: theory or metaphor? In: *The Self as an Organism: A Philosophical Consideration*, Ruse M (ed). Cambridge University Press, New York, pp. 254–255.

Tauber A (1994b) The self as an organism: a philosophical consideration. In: *The Immune Self: Theory or Metaphor*? Ruse M (ed). Cambridge University Press, New York, p. 253.

Tauber A (1994c) The self as an organism: a philosophical consideration. In: *Organism and the Origins of the Self*, Ruse M (ed). Cambridge University Press, New York, p. 266.

Tauber A (1994d) The self as an organism: a philosophical consideration. In: *The Immune Self. Theory or Metaphor*? Ruse M (ed). Cambridge Universtiy Press, New York, p. 232.

Tauber A (1994e) The self as organism: A philosophical consideration. In: *The Immune Self: Theory or Metaphor*? Ruse M (ed). Cambridge University Press, New York, p. 234.

Tauber A (1994f) The self as organism: A philosophical consideration. Nietztsche's biologicism. In: *The Immune Self: Theory or Metaphor*? Ruse M (ed). Cambridge University Press, New York, pp. 249–261.

Tauber A (1994g) The self as organism: A philosophical consideration. Nietztsche's biologicism. In: *The Immune Self: Theory or Metaphor*? Ruse M (ed). Cambridge University Press, New York, p. 252.

Tønnessen M MR, Brentari C (2016) The Biosemiotic Glossary Project: Umwelt. *Biosemiotics* **9**: 129–149.

Villarreal LP (2009) *Origin of Group Identity. Viruses, Addiction and Cooperation.* Springer, Irvine.

Villarreal LP (2015) Force for ancient and recent life: viral and stem-loop RNA consortia promote life. *Ann NY Acad Sci* **1341**: 25–34.

Villarreal P, Witzany G (2013) The DNA habitat and its RNA inhabitants: At the Dawn of RNA sociology. *Genomics Insights* **6**: 1–12.

Witzany G (2014) RNA sociology: group behavioral motifs of RNA consortia. *Life (Basel)* **4**: 800–818.

Witzany G (2002) Reduction of biological phenomena? Deficits of systems theory and the alternatives. In: *Sign Processes in Complex Systems*, Schmitz W (ed). Mouton de Gruyter, Berlin, New York, pp. 303–307.

Witzany G (2006) From Umwelt to Mitwelt: Natural laws versus rule-governed sign-mediated interactions (rsi's). *Semiotica* **158**: 425–438.

Witzany G (2012) From molecular entities to competent agents: Viral infection-derived consortia act as natural genetic engineers. In: *Viruses: Essential Agents of Life*, Witzany G (ed). Springer, Dordrecht.

Yot P, Pinck M, Haenni AL, Duranton HM, Chapeville F (1970) Valine-specific tRNA-like structure in turnip yellow mosaic virus RNA. *Proc Natl Acad Sci USA* **67**: 1345–1352.

Temporal parts and biological change

G. Caetano-Anollés[1]
F. Mughal[1]
K. Caetano-Anollés[2]
D. Caetano-Anollés[3]

Summary. Persistence is an important property of biological systems. Systems persist when their identity is maintained through time. The problem of identity and change unfolding temporally and spatially has been addressed by the endurantist and perdurantist schools of thought. For endurantists, things are wholly present whenever they exist; they are 3-dimensional. For perdurantists, things have temporal parts and persist through both time and space; they are 4-dimensional and atemporal. Inspired by Verlinde's conjecture on the entropic origins of gravity, we here propose a theory of entanglement that explains causal relationships responsible for the increasingly extended and complex temporal parts of biological systems.

[1]*Department of Crop Sciences, University of Illinois, Urbana, Illinois, USA*
[2]*Division of Biomedical Informatics, College of Medicine, Seoul National University, Republic of Korea*
[3]*Broad Institute of MIT and Harvard, Cambridge, Massachusetts, USA*

1. Introduction

"The ship wherein Theseus and the youth of Athens returned had thirty oars, and was preserved by the Athenians down even to the time of Demetrius Phalereus, for they took away

> *the old planks as they decayed, putting in new and stronger timber in their place, insomuch that this ship became a standing example among the philosophers, for the logical question of things that grow; one side holding that the ship remained the same, and the other contending that it was not the same."*
>
> — *Theseus*, Plutarch

Space, time, identity and change are interlinked. While these concepts have been collectively addressed in some way or another by philosophy (perhaps beginning with Presocratic scholars such as Parmenides, Heraclitus and Empedocles) the difficulties that they bring are of current and central interest, particularly for evolutionary genomics. Here we discuss the *perdurantist* theory that proposes that systems extend through space by having "spatial parts" in different places, and through time by having "temporal parts" manifesting at different time frames (Lewis, 1986; Haslanger, 2003; Balashov, 2011; Effingham, 2012; Hawley, 2018). This philosophical doctrine of temporal parts "perduring" through time has also been called *4-dimensionalism*. Opponents of the theory, *endurantists* or *3-dimensionalists*, explain systems as wholly present at every moment of their existence. While perdurantists think atemporally about parts, endurantists think of parts that a system has at different times. As we will discuss, both philosophical views present an ontological disagreement of "existence" and bring difficulties of their own. However, the perdurantist view appears in line with modern physics (Balashov, 2011) and the *B-theory* of time (McTaggart, 1908), which is also compatible with theoretical physics arguments and *eternalism*. A glossary helps summarize philosophical doctrines discussed (Box 1). Figure 1 illustrates the main distinction of the perdurantist and endurantist philosophical views. Here we make the point that the molecular makeup of the cell is a patchwork that complicates definition of biological systems and can be better explained by a

4-dimensional space-time "worm" theory of temporal parts than by a *presentist* view of reality.

BOX 1: Glossary of major philosophical doctrines of existence and time

Perdurantism: A metaphysical doctrine that considers things have temporal parts and persist (survive change) through both time and space by "perduring." While a thing that exists in a region of space has at least one physical part in that region, that same thing will exist in a region of time and will have distinct temporal parts throughout its existence, past, present and future. Perdurantism is also known as "4-dimensionalism" or "worm theory" and is compatible with "eternalism."

Endurantism: A doctrine that considers things are wholly present whenever they exist. These things, which persist by "enduring," have spatial parts at different times, past, present or future. Endurantism reject an atemporal view of parthood. Consequently, the concept of temporal parts is meaningless. Endurantism is also known as "3-dimensionalism" and is compatible with "presentism."

Exdurantism: A 4-dimensionalism-related doctrine known as "stage theory" that views persisting things with temporal parts stacking up in succession through time (especially processes and events). However, the identity of some persisting things is not regarded as the sum of all temporal parts. Instead, the identity manifests as a single "stage" or "time-slice," especially for ontologically self-sufficient or intrinsically unchanging things, which persists when it finds "counterparts" in other stages and times.

Presentism: A doctrine that considers that only the present is real. Things or events that occurred in the wholly past or wholly future do not exist. The doctrine follows the "A-theory" of time, a temporal ordering relation of "tensed" events in which events continuously transform from the future to the present and from the present to the past as time proceeds. Time is therefore a series of continuous transformations.

Eternalism: A doctrine that considers all points in time are equally real, just as spatially distant things are equally real. The doctrine follows the "B-theory" of time, a philosophy of temporal ordering relation of events. The B-theory considers time is both an illusion and "tenseless" (temporal becoming is not an objective characteristic of reality). Events are ordered as a series of temporal positions by two-term relationships ("earlier than" and "later than") that are anti-reflexive (no element relates to itself; e.g. the "greater than" relationship), transitive (translation of element relationships) and asymmetric (element relationships are not reciprocated). In contrast with A-series events, B-series relationships do not change with time.

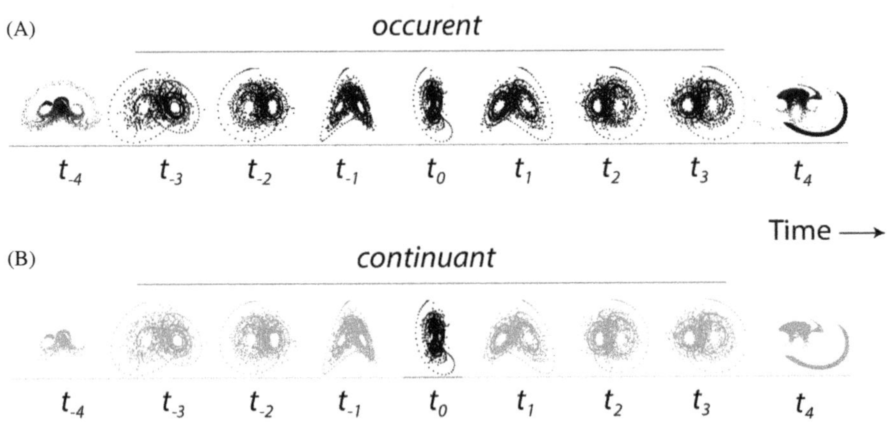

Figure 1. Two views of a "persistent," an entity that exists in time and space. (A) A 4-dimensionalist (perdurantist) view sees the persistent as an occurrent, a collection of temporal parts that extend between times t_{-3} and t_3. (B) A 3-dimensionalist (endurantist) view sees the persistent as a continuant, an entity that is wholly present at any instant at which it exists. In the diagram, the endurantist sees the persistent only at the present time, t_0. All other entities that are wholly present in the past or in the future are shaded. The persistent entity used in this example is a Lorenz chaotic attractor that produces a dynamic 3-dimensional "butterfly effect" between times t_{-3} and t_3.

2. Temporal parts

Things, which we will refer to as "systems" within the context of biology, are internally organized wholes (Simon, 1962). These wholes are made of parts, elemental entities of the system at some level of organization, which are somehow cohesively structured to collectively endow the whole with individual properties (Collier, 2003). This is the subject matter of mereology [from the Greek μέρος meros (root: μερε- mere-, "part") and the suffix-logy "study, discussion, science"] and the foundational part-whole theory formalized by Leśniewski (1916) and extended by Leonard and Goodman (1940). Regardless of ontological definitions of "part," mereology begins by addressing parthood and its three partial order relationships: (i) *Reflexivity*: everything is part of itself; (ii) *Transitivity*: any part of any part of a thing is itself part of that thing; and (iii) *Antisymmetry*: two distinct things cannot be part of each other (Varzi, 2019). Five additional mereological predicates arise from these partial ordering relations: equality, proper parthood, proper extension, overlap, and underlap, which we will not discuss. Though mereology appears straightforward, some concepts become paradoxical and philosophically problematic, sometimes necessitating axiomatic formulations. For example, when referring to biologists using "parts" to describe "functional subunits" of organisms, Reschner (1955) famously objected to the reflexive property contending that no organism is a functional subunit of itself. The argument however simply regards identity as limit case of parthood and can be contested by taking proper parthood (combination of parts accounting for the whole) as a primitive. Similarly, parthood often necessitates a "function" or some predicate modifier to fulfill the transitive property, and in the case of antisymmetry, the use of constitution as a genuine case of proper parthood may not be symmetric: the functional subunit is part of the organism but not vice versa. While much of this complexity concerns ontological difficulties that we will discuss elsewhere, the notion that parthood is a perfectly determinate relation often brings objections of indeterminacy and fuzziness. After all, parthood appears to be a matter of convention.

Mereological theories of parts and wholes are further challenged by the impact of time, especially when time interfaces with space, identity and change. One first problem is that things can be in more than one place at a time. The atoms of a protein molecule for example occupy different locations of space at any particular time, with spatial positions appropriately defined by for example the 3D coordinates of crystallographic atomic models of the Protein Data Bank (PDB). Similarly, the different domains that make up a multidomain protein occupy non-overlapping volumes of the molecule. They also exhibit different structural identities and functional properties. Nothing appears special about this "spatial" property of parts of a physical system. The problem however is that things (including people) take up both space and time (Hawley, 2018). A protein exhibits a multitude of 3-dimensional atomic conformations as the molecule moves in space. These conformations can be considered "temporal parts" of the molecule. For example, the conformation that the molecule takes before binding a ligand changes in time when the ligand docks into one of its molecular pockets and then changes again when the ligand is released. These different conformations occurring at different times are considered temporal parts of a same molecule: before ligand docking, with the docked ligand, and after the docking takes place. The identity of the thing, the molecule, is the same but expresses in different ways in space and time. Protein conformations are different but do not detract from the identity of the protein. Moreover, as the abundance of a protein increases, the identity of particular conformations that are more frequently visited become less fuzzy and may attain identities of their own. Recalling Plutarch's ship of Theseus paradox, the ship's identity is maintained by replacing old planks with new ones upon decay. The ship at different times is simply a collection of different temporal parts with collections of different spatial parts. This perdurantist view of persistence is therefore atemporal. Four-dimensional entities (*occurrents*) persist by existing along time and having consecutive temporal parts with different properties and spatial parts (Figure 1A).

In contrast, endurantists consider parts are present in a thing at different times, during the past, the present and the future, and do so as the window of the present moves along time. At any time, a thing (*continuant*) exists as a whole, i.e. it is "wholly present" (Figure 1B). All the parts that make up the thing also exist at that time. Thus, things and their parts persist by existing at consecutive times and change when things have different parts or properties at different times. Thus, endurantists consider time as an ordered series of transformation events. While endurantists agree with perdurantists that processes or events embody temporal parts, they quarrel with the idea of atemporality. In their view, the ship of Theseus changes with plank replacement because the ship becomes an increasing patchwork of old and newer planks. Some argue that the perdurantism-endurantism debate is moot ("*a storm in a teacup*"), that metaphysical enquiry is meaningless, or that there is uncertainty about the existence of temporal parts (McCall and Lowe, 2006). Reydon (2008) disagrees when he discusses the impact that the biological species problem has on the temporal parts debate. He claims that the perdurantism-endurantism descriptions of reality are inter-translatable. Mixed views also exist, including exdurantists and their doctrine of "stages." This doctrine resembles perdurantism in that persistence is a spatial extension through time that defines entities as aggregates of temporal and spatial parts (Lewis, 1986). However, persistence occurs without subdivision into temporal parts or with very brief temporal stages. Instead, stages persist by having distinct stage counterparts at other times. Exdurantism mimics David Lewis' *modal counterpart theory*, which posits that things exist in one world but that their properties can exist in possible worlds through "counterpart" similarity relationships that do not need to be transitive or symmetric. For example, the sequences of extant proteins with their corresponding folds represent a minuscule fraction of possible proteins that arise from a vast sequence space that was never visited in evolution and may also contain those same fold properties. Finally, there are those that believe continuants and occurrents are equally basic (e.g. Wiggins, 2001), though one or the other can be given metaphysical priority.

Both perdurantists and endurantists agree that there are perduring things (*persistents*), things that persist. The debate is about how things persist. Perdurantists see the world (reality) in 4-dimensions, always thinking historically or predictively. In contrast, endurantists think in 3-dimensions and consider time as a transformative force. They are more interested in the mechanics of the present than in the history of things. In their view, only the present exists. Remarkably, this debate follows the two different scientific approaches proposed by Windelband (1894), the *ideographic* science of studying process and history with retrodictive statements, and the *nomothetic* science of explaining present events with high predictive power using universal and predictive statements.

3. Temporal parts and growth

While a temporal part can be considered simply a part of the history of a thing or system (Thomson 1983), temporal parts are proper parts of a whole spatial (physical) entity, a spatial thing with a temporal dimension. Proper parts in this regard are parts that when combined together make up the whole, i.e. they account for the whole entity. In contrast spatial parts are also proper parts, but they are proper parts of a given temporal part. Spatial parts are therefore indirectly proper parts of the whole 4-dimensional object over its lifetime, i.e. over its history. An example can be helpful.

If three protein domains (D_1, D_2 and D_3) are added one by one to make a 3-domain protein (P_3) in the course of evolution, the first domain constituents (D_1 and D_2) that were created prior to the formation of P_3 (by fusion with D_3) must be considered temporal parts of P_3, since they are spatial and temporal parts of it. While it seems unproblematic to claim that parts exist prior to the existence of a whole and to reject that they are parts of the whole at that earlier time before the whole materialized, under a perdurantist view the spatial things that make up the whole gain "parthood" when the whole thing is created. The issue is ontological and related to "*things that grow*" (*sensu* Plutarch) since we could make a similar argument of the intermediate 2-domain protein (P_2), a whole composed of D_1 and D_2. In our example, domains exist prior to the P_3 multidomain protein but they are not yet part of it. They can gain or lose parts without their identities being affected, but they can gain or lose parthood by being a spatial part of a temporal part. Thus, a temporal part of D_1 or D_2 is a spatial part of a temporal part of P_3, the whole 4-dimensional thing. This lifts most endurantist objections (Thomson, 1983).

Physical things are composed of temporal parts, which are in turn composed of physical parts. They are perduring things that can gain or lose properties. In one extreme, these things can be devoid of its physical standing. If we remove all timber of the Theseus ship that becomes rotten replacing planks with an abstraction, say a node of a network describing how planks relate to each other, then the ship as a system with all its physical constituents vanishes but the relational abstract entity of the ship remains as a collective of temporal parts with physical parts that do not exist. One can therefore argue that it is the relational abstraction of parts what actually counts as defining the Theseus ship. Similarly, it is the relational abstraction of how domains D_1, D_2 and D_3 interact what defines the temporal parts of P_2 and P_3 multidomain proteins and how these parts accumulate.

4. Temporal parts and change

Because things change, their existence changes accordingly. Things change through space and time by for example changing parts, properties, relations to other things, and by changing dynamics. The passage of time also involves change. A protein molecule's conformation at one time will be different some nanoseconds later. Conformations barely persist through time but the persistence of the protein as an entity is longer. Temporally, the protein persists because it is older when it is about to be degraded by enzymatic or random breakage than when it is released through the exit pore of the ribosome. Ontologically, the protein persists through change because its identity is unchanged despite the multiple atomic conformations it adopts throughout its existence. The relative location of parts in space are changed. Similarly, the protein also persists when gradually built by the ribosome, amino acid by amino acid, or when gradually degraded by the proteasome, peptide by peptide. In other words, the protein persists as it grows and withers. However. Gottfried W. Leibniz argues: "*We cannot say, speaking according to the exact truth of things, that the same whole is preserved when a part is lost... objects of this kind only appear to persist*" (Leibniz, 1937). To complicate matters, new identical versions of the protein are likely to be transcribed and translated throughout the lifespan of an organism. While the individual protein molecules may have different atomic conformations or constitutions, the identity of the protein persists much longer as a collective, during periods of perhaps years or decades. If the protein structure is what defines protein identity, then the protein persists even longer, perhaps millions of years, by resisting mutational changes at DNA level.

The "problem of change" demands an explanation of change (Lewis, 1986; Wasserman, 2006). How is it possible that time causes a thing to have different properties? Perdurantists explain change through temporal parts. A protein with conformation *A* has temporal part *A'*, which changes nanoseconds later to conformation *B* with temporal part *B'*. It changes by adopting different properties at different times. In the example, change is both spatial and temporal. One objection, however, is that perdurantists look at things atemporally as a limited or vast collective of temporal parts and their spatial extensions. The 4-dimensional view of the protein includes temporal parts in the future, present and past, i.e. all possible conformations, parthoods and properties that the protein attains along its existence. This 4-dimensional view is extensive and does not embody change because it holds all possible spatial and temporal variants. Perdurantists however lift the objection by also holding a temporal view that shows that temporal parts have properties that change with time. In contrast, endurantists reject temporal parts but have more problems explaining change because their view is that only present states or events exist. Change for them is irrelevant since time is simply an ordered series of transformation events in which only the present exists. Since irrelevance is objectionable, endurantists also explain change by considering things exhibit different relations to other things at different times. A protein conformation is simply different from another while both only exist at specific times or at a same time if different conformations are of identical molecules. They carry different "extrinsic" or "intrinsic" properties if they interact or not with the outside molecular world. Other endurantists consider *adverbialism* as an explanation of why things can have different properties. Instead of explaining change with temporal parts or by subdividing properties of things, adverbialists subdivide the way how things have properties. However, this classification-like effort of properties that are generically endowed to things raises problems of instantiation of things.

The ship of Theseus paradox can be used to summarize the problem of change, growth and identity and the difficulties faced by the *endurantist* school (Figure 2). Each event of replacement of

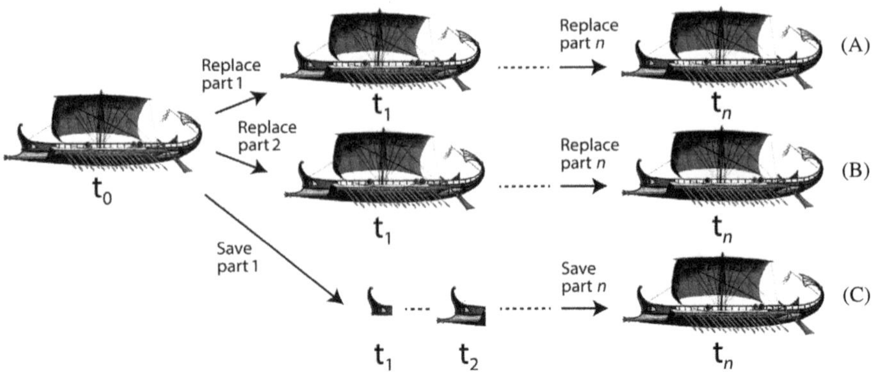

Figure 2. The ship of Theseus and the multiple paradoxes of things that grow. The ship of Theseus was kept in the harbor of Athens as a memorial since its arrival at time t_0 and until the time of Demetrius Phalereus t_n. During that time period Athenians replaced *n* planks that were decaying with new ones, beginning with part 1, then with part 2, and so on until part *n*. Philosophers then questioned if **Ship A** was the same as the ship of Theseus at t_0. The two ships have the same makeup in terms of parts. If the age of parts is considered irrelevant, then the two ships should be considered identical and **Ship A** is still the ship of Theseus. However, one had all its parts gradually replaced by new ones (replacement is what grows). The ships should be considered different if old and new parts are considered different or if what matters is the arrangements of parts that make the Theseus ship. In an alternative world the *n* planks could have been replaced following a different order, beginning with part 2 instead of part 1 and leading to **Ship B**. Philosophers could have also questioned if **Ship A** and **Ship B** were the same. The two ships have the same makeup in terms of parts but the history of part replacement is different. Finally, Athenians could have also saved the old decaying planks and using a special technique (similar to that used to recover the planks of the Vasa Swedish warship that sank in 1628) preserve each original part that was being replaced to build **Ship C** (the Hobbes' scenario). Philosophers could have questioned if **Ship C** was the same as the ship of Theseus at t_0. **Ship C** is made of the same parts as the original ship of Theseus and preserves the same arrangement of planks. Consequently, the two ships should be identical in terms of makeup. However, **Ship C** was built after all parts of the original ship were replaced by new ones. All ships are historically different.

decaying planks changes the original properties of the ship but appear not to change the identity of the ship. Replacements embody gradual change accumulating in time. One interesting twist of the paradox is the idea of saving planks that are being replaced and using them to build a new ship. This is Thomas Hobbes recount: "*Some man had kept the old planks as they were taken out, and by putting them afterwards together in the same order, had again made a ship of them*" (Hobbes, 1839). Under this scenario, only parts are needed to preserve the identity of things. Let us now make explicit three main assumptions:

(i) *Gradual change principle*: Ordinary things such as the ship of Theseus survive gradual change. Even if things gradually change, their identity does not need to change. The ship of Theseus can survive plank replacements as long as the identity (and inter-relationship) of the parts that are being replaced is preserved. This assumption motivates the conclusion that the ship with all planks replaced (**Ship A**) is identical and as real as the original ship of Theseus (Figure 2). A solution to this impossibility is for endurantists to deny the change principle. This can be done strongly by claiming that as soon as a single plank is replaced the identity of the ship is destroyed (Leibniz proposal). Consequently, things can only persist in a loose ontological sense (Chisholm, 1970). Alternatively, each ship with a newly replaced part can be considered a newly created ship. Both of these two scenarios appear implausible. Denying the change principle can also be done weakly by claiming that things survive some gradual change but not all, i.e. that as planks are being replaced at some specific point the ship of Theseus ceases to exist. This explanation seems arbitrary.

(ii) *Parts principle*: While things persist through changes in parts, things with the same parts and part relationships should be considered identical. Things go where their parts go. Following Thomas Hobbes recount: saving planks that are being replaced and using them to build the ship (**Ship C**) results in a real ship of Theseus (Figure 2). However, the ship with all planks replaced (**Ship A**) being identical to the ship built from old planks (**Ship C**) is an impossibility because both ships cannot be in existence at one time and be real ships of Theseus.

(iii) *Transitivity principle*: If two things *A* and *B* are identical and *A* is identical to another thing *C*, then *B* must be identical to *C*. This is similar to the transitivity property of mereology. Endurantists could negate the transitivity principle to salvage both the change and parts principle: the ship of Theseus is identical to **Ship A** and to **Ship C**, but **Ship A** and **Ship C** are not identical to each other. However, rejecting a principle that seems uncontroversial can have dire consequences.

While the three principles are plausible, endurantists face the dilemma of having to reject (or loosen) any one of the three principles in order to explain the Theseus paradox. In doing so they seriously weaken explanatory power. In contrast, perdurantists offer a solution with their "worm" theory. For perdurantists, each event of plank replacement defines a temporal part and explains changes in the ship's temporal and spatial properties. The ship's identity is atemporal. However, plank replacements are seen as atemporal ship abstractions instantiated by time in the form of a "process." In contrast, endurantists cannot explain identity and change in such a natural way. Change is considered irrelevant, or alternatively, a relation or a class of properties.

5. Evolution of parts

The Theseus paradox is complicated by ontological convention and history, none of which were properly discussed above. Identity can be addressed at a particular time (synchronic identity) or at multiple times (diachronic identity) (Gallois, 2016). While synchronic identity has been claimed to be unproblematic, diachronic identity brings difficulties that appear intractable, mostly because persisting things have the potential to change their intrinsic properties (Lewis, 1986). Let us consider two genomes, one of them truncated by a chromosomal fission. Are these genomes identical? The genome before the time of truncation contains the untruncated chromosomal segments, including the segment that will be truncated. The genome after truncation also contains the untruncated chromosomal segments plus the segment that was truncated, which is a part that is now separated. Comparing things that change according to their parts (mereologically) suggest both genomes are identical despite the episode of change. Broadly speaking, their atomic composition before and after truncation appears the same given the intrinsic properties of the things that are being compared. Endurantists reject this "doctrine of arbitrary undetached parts" (van Inwagen, 1981). They reject the proposition that the excised genome existed as part of the complete genome prior to truncation, claiming it is illegitimate to excise a part from any object and claim it is an undetached part. Instead, the complete genome goes out of wholly existence after truncation. As noted above, however, all of this brings back multiple difficulties of existence that the endurantists have problems to explain. Returning to a perdurantist view, let us pose the question: Is the untruncated genome equal to the segmented genome following truncation? A careful study of the chemistry of the truncation process shows that the transesterification reaction catalyzed by the restriction endonuclease responsible from truncation involves either hydrolysis or alcoholysis depending on environmental context (Sasnauskas *et al.*, 2007). Thus, extrinsic properties that are alien to the changing thing and its intrinsic properties are involved and challenge identity.

To address these kinds of difficulties, Geach (1967) proposed denying any kind of absolute relation of identity. Instead, identity becomes a relative concept of relationship, i.e. one thing is identical to another given some "sortal" or "modal" concept of the thing.

Thing *X* cannot be the same as thing *Y*. Instead, there must be a concept of "kind of thing" related to *X*, such that we can propose: *X* is the same "what" as *Y*. This "relativization" approach eliminates the problem of explaining if synchronic or diachronic identities are the same or different kinds of identity. More specifically, the fact that a thing *x* is not *A* before an event but is *A* after the event does not mean that the event constitutes a change in *x*. In other words, there are "real" changes occurring according to the following definition:

> *"A thing undergoes real change following an event if, and only if (iff), there is some intrinsic property it satisfied before the event but not afterwards."* (1)

However, as we made clear with the genomic truncation example there are also "mere" indirect changes. Lewis (1986) took advantage of Geach's proposals to deny identity between things wholly existing at different times, strengthening the perdurantist view. His argumentation against endurantism is that denying strict identity is inconsistent with things undergoing real change. If the thing can be *A* at one time and not *A* at another, then the property of *A* must be related to time and cannot be an intrinsic property. Recall that intrinsic properties are those related to the whole and its proper parts, i.e. to changes that do not result from interactions with the outside world. One often touted example is the shape of an object. But is the shape of an object truly intrinsic? While shape appears intrinsic, Nerlich (1979) suggests that shape depends on the curvature of the space that embeds the object, making shape an extrinsic property. In another example, an amino acid residue in a polypeptide chain (thing *x*) contributes to the hydrophobicity of the molecule (property *A*). The amino acid residue can be considered an intrinsic molecular and physicochemical property of the polypeptide chain. However, hydrophobicity is not. When the molecule changes by adopting a different conformation, the hydrophobicity property of the molecule changes because it distributes vectors of hydrophobicity (collectively known as "hydrophobic moment") differently depending on molecular context. Thus, hydrophobicity is an extrinsic property and molecular change induces an extrinsic property in the molecule. By the same token, segmentation is an extrinsic property of genomes that depends on molecular context. Lewis (1986) calls the difficulty of identity through time "temporary intrinsics," which he puts to test with presentist and eternalist viewpoints.

The problem of change was previously discussed by Chisholm (1970) when treating identity in a "loose and popular" sense and introducing his mereological theory of persistence. The theory addresses logical judgements concerning identity as being conventional. Things such as the ship of Theseus exist when a certain collection of parts exist. At any moment, this composite of parts is an "*ens per se*," it exists in its own right (in virtue of its own inherent nature). When a composite of parts unfolds in time, it becomes an "*ens successivum*," a successive composite of temporal parts. A thing however is also an "*ens per alio*," it exists "through another" thing or "by convention." The thing exists in virtue of other things existing, or has properties at a time in virtue of things happening at other times. A specific enzyme, for example, exists when it is named according to some molecular function typical of enzymes [e.g. EC 1.1.1.37 describing Enzyme Commission classification: oxidoreductase (1) — acting on the CH–CH group of donors (1) — with NAD+ or NADP+ as acceptor (1) — malate dehydrogenase (37)]. The enzyme has typical properties of other enzymes embodied in Enzyme Commission classification. This enzyme also embodies a composite of atomic parts that exists in its own right (PDB entry 5MDH, with the LDH-C C-terminal domain-like fold under SCOP classification d.162.1.1) and manifests in time as a composite of molecular conformations (temporal parts uncovered for example with molecular dynamics simulations) in virtue of other conformations that existed beforehand. Similarly, an organism exists when it is named according to some definition of species (e.g. the fruit fly *Drosophila melanogaster* 7227) in virtue of some species classification scheme that describes a successive composite of temporal parts. Mereologically, species 7227 is defined at time of analysis by 13,793 genes located in a 180 Mb reference genome, which encodes 21,982 proteins and is part of a collective of temporal parts that define its lineage (NCBI taxonomy: Cellular organism — Opisthokonta — Metazoa — Eumetazoa —.... Drosophila — melanogaster subgroup). This composite of parts exists in virtue of its own right.

Composites of temporal parts require defining change. Chisholm (1970) introduced the concept of "*direct evolution*" within the context of "*one compositum evolving from another*," where a "*compositum*" refers to "*anything that has a part*":

> "*X evolves directly from Y if, iff, either X is identical to Y or there is no time at which X and Y both exist but there is a z such that z is a part of Y at one time and z is a part of X at a later time*" (2)

Chisholm's definition of evolution follows Frege's logical conceptualization of ancestral relation (Frege, 1879). In order to explain natural numbers, Frege in his "Proposition 76" defined for every 2-place relation *R*, the general concept: "*x is ancestor of y in the R-series*." This relation is called "*the ancestral of the relation R*," which can be considered a "*fatherhood*" relation of the kind *a* is the father of *b*, *b* is the father of *c*, and so on, for a "*fatherhood series*," where "*is a father of*" refers to "*preceding*" in a predecessor series. Here, Chisholm proposes that parts are important elements not only of identity but of evolutionary change. He uses evolution to explain persistence (identity through time), which he only promotes in a "a loose and popular sense." However, he concedes that the conceptualization explains persistence imperfectly. Several problems of this line of thinking have been raised by Michaels (1980), sometimes within the context of the ship of Theseus paradox. One problem is that explanation (2) or other compositum-based alternatives that Chisholm proposed do not address the temporal order of direct evolution. Adding or subtracting parts does not provide a directionality for evolution that would establish unequivocally the flow of time. Another problem is that of "stray" evolvers, composita in a predecessor series with missing parts that cannot be accounted for by direct evolution statements. A final problem is the problem of "fission" or, more appropriately, the problem of lineage splitting. In Figure 2, **Ship A** and **Ship B** represent the products of alternative histories. In one case the order of plank replacement is different from the other. In contrast, **Ship A** and **Ship C** represent the products of parallel histories. While planks are being replaced in the history of **Ship A**, the replaced planks are being used to build a new ship in the history of **Ship C**. The evolutionary "split" that occurs when planks are being replaced and used for ship construction follows very much the thinking used to build phylogenies in for example evolutionary genomics. In biology, for example, the history of **Ship A** embodies replacement of the kind you would see when studying mutational change in protein sequences under saturation (Caetano-Anollés *et al.*, 2018). The history of **Ship C** embodies evolutionary growth of the kind you observe when studying the growth of protein domains in proteomes. When Chisholm considers a split that generates two parallel evolving systems of composites, the questions of identity can be settled by agreeing on a set of conventions. This has been exactly the kind of approach used in evolutionary bioinformatics.

Previously, Hennig (1966), the father of systematic biology, elaborated solutions that could have addressed difficulties of Chisholm's logical composite constructions. He proposed the use of "shared and derived" properties of evolving things (phylogenetic characters), which he named "synapomorphies," to test evolutionary hypotheses in the form of homologies (hypotheses of ancient origin of identities), transformations between homologs, and phylogenies (evolutionary ground plans). Characters are given different properties-values according to a set of character states. His methodological approach of retrodiction unfolded a trilogy of data-observations, models of change, and phylogeny in the form of a network representation of change (Caetano-Anollés *et al.*, 2018). This trilogy defined the fundamental operations of retrodiction. We can modify Chrisholm definition of direct evolution (2) to match Hennig's shared and derived concept:

> *"X and W evolve directly from Y iff either X is identical to Y and W or there is no time at which X and Y or W and Y both exist but there is a z such that z is a part of Y at one time and z is a part of X and W at a later time."* (3)

Here, *z* represents a historical event (a character state) in the direct evolution of the observable feature *z* along splitting branches of a network hypothesis that describes the evolution of *X* and *W* from *Y*. Testing homology relationships defined by *z* provides a time directionality to the minimal tree structure of an unfolding network. Another way to describe this relationship of existence is Hennig's definition of "monophyletic" group (Hennig, 1966): "*A taxon X is more closely related to another taxon Y than it is to a third taxon Z iff it has at least one stem species in common with Y that is not also a stem species of Z.*" Hennig's taxa are Chisholm's composita, the evolving things. Here, monophyly is expressed as a function of a "stem species" or ancestral relation that holds all of its ancestral properties. All descendants of this common ancestor (including descendants of the stem species) make up the monophyletic group.

From a Hennigian perspective, what is missing however in this definition is a model that establishes a relationship between the character and its states and provides a mathematical function or "transformation series" that can be used to optimize evolutionary change, both of which provide boundary conditions to the possible historical transformations of character states (Grant, 2002). These aspects have been fully elaborated by the application of maximum parsimony, maximum likelihood and Bayesian approaches to phylogeny reconstruction.

One difficulty that relates to the issue of "stray" evolvers and the problems of fissions and fusions is that the ground plan conceived in explanation (3) is a tree-like statement of splitting (fission) that involves a single ancestral origin. However, if phylogenies have a ground plan of networks that form reticulations and statements of multiple ancestral origins, then we should also account for both fusions and evolutionary loss:

> *"Y evolves directly from X and W iff either Y is identical to X and W is identical to Y or there is no time at which Y and X or Y and W both exist but there is a z such that z is a part of X and W at one time and z is a part of Y at a later time."* (4)

Note that in explanations (3) and (4), characters transmit change through changes in character states as modified or unmodified characters in a phylogeny. Characters also show at least two states, known as transformational homologs, one ancestral (a plesiomorphic state) and the other derived (an apomorphic state). As anticipated by Hennig, this provides a temporal directionality to evolutionary statements.

In explanations (2), (3) and (4), identity can be strict or loose according to different criteria. For example, Hull (1978) considered that the individuality of material bodies was constrained by (i) retention of mereological substance, (ii) retention of structure through patterns of interaction, and (iii) continuous existence through time. He conceived diachronic identity, which he referred to as "genidentity," as a succession of states through time that followed a continuity of internal organization. This continuity could be disrupted by more or less temporally stable changes and by living entities splitting and merging at different temporal scales. Causes of splitting involved ontogenetic and phylogenetic changes leading for example to differentiation or speciation. Total or partial merging events involved fusions and integration or parts and wholes in biology, ranging from episodes of horizontal gene transfer or symbiosis to cellular or lineage fusions. As we will later discuss, these patterns of diversification and unification often result in atypical parthood relationships that complicate diachronic identity relationships.

6. Causation

Life requires continuous change and a multitude of overlapping processes for causal explanations of its multilayered complexity. We could call these evolutionary processes "activities" and the temporal ordering of dependencies between complexity layers "causation." How does causation manifest in a world? Three criteria must be met to validate a causal relationship in scientific exploration: empirical association, temporal priority, and non-spuriousness. In addition, establishing a mechanism for causation and the context of this mechanism strengthen the causal relationship but are not necessary. The first criterion of *empirical association* typically requires finding a correlation between dependent and independent variables establishing how they vary together. Without association there is no causal effect. For example, the replacement of planks of the Theseus ship is correlated with its persistence; failure to take away "old planks as they decayed" will fail to preserve the ship in time. The second criterion of *temporal priority* requires that variation of the independent variable comes before variation in the dependent variable. Cause must come before its effect. Replacement of planks (cause) extend the ship's persistence (effect) by strengthening the ship's timber with time. The third criterion of *non-spuriousness* attempts to eliminate falsehood in apparent cause-effect inferences that would explain empirical associations. The criterion ensures the falsification of other possible variables that would explain a cause-effect association. For example, the contention that timber strength is actually responsible for the association of plank replacement and ship's persistence must be falsified, including instances of timber strength correlating with plank replacement. As we already mentioned, two additional but non-necessary criteria strengthen causality. A fourth criterion of *mechanism* attempts to unravel the mechanistic process by which the independent and dependent variables are connected. For example, the association of plank replacement and persistence can be explained by a mechanism by which "stronger timber" lengthens the preservation of the ship. The final and fifth criterion of *context* analyzes the contextual environment within which a cause-effect relationship unfolds. A cause is embedded within a network of cause-effect relationships. This context is external to the causal relationship that is being examined but strengthens our understanding of causation. For example, the Athenians decided to preserve the ship of Theseus "down even to the time of Demetrius Phalereus." This causal decision is external to the cause-effect relationship of gradual plank replacement that "became a standing example among the philosophers."

The standard philosophical view of causation establishes "causal relata" and a "causal relationship" (Schaffer, 2016). We will illustrate these concepts and show how they are also related to the problem of change with a "domino show" that stands up dominoes and topples the first domino in a sequence to create a chain reaction.

(i) *Causal relata.* Relata represent one or more concrete events occurring in space-time (immanent entities) or one or more non-concrete facts that are abstract (transcendent entities). Usually, events or facts are activities that come in sets of two in the form of a *cause* and an *effect*. In a "domino show," an actual event e_1, the actual knocking action that a toppled domino has on another, causes a distinct actual event e_2, the knocking over of a domino. e_1 is cause (*C*) and e_2 is effect (*E*). Note that in the "domino show" example there is a domino effect, a mechanical chain of causes and effects. Relata can be categorized according to some level of granularity (individuation) that ranges from coarse-grained (events of certain stripe) to fine-grained (facts or events of other stripes). The need for categorization comes from a need to individuate the relata, i.e. find out how distinct they are. Discussions about granularity of immanent and transcendent relata likely span a continuum of the individuation of fine-grained properties. Relata can also vary in number and roles, a philosophical issue known as "adicity." Two kinds of arguments have been raised in this regard, one that posits contrasts and the other that posits "secondary relata" (descriptions, models, or default states): (a) Arguments of contrasts address the issue of pairs of events or facts being insufficient, suggesting instead that causal relata must be four to resolve problems of absences and causal differences. In causal chains for example the effect of the first cause-and-effect relationship serves as the cause of the second, and so on. For this to be possible the cause-and-effect must be exchangeable. The first domino knocks over the second, the second knocks over the third, and so on until the chain is complete. Invoking a binary relata scenario implies that the toppling of the first domino causes the toppling of the second. Philosophers complain that causal relations of this type are ill defined. Instead, invoking a quaternary relata scenario implies that the toppling of the first domino rather than the domino remaining upright causes the toppling of the second rather than the second remaining upright. The presence of contrasts of the form *C rather than C* causes E rather than E** enable causal relations that are exchangeable. Contrasts have determinate truth value, reconcile immanence with absence causation, and address the "causal differences" argument demanding that relata be fine grained to mark differences in causal relatedness. (b) Arguments of secondary relata stress the relative nature of causal relations. These arguments have been disputed for not being compatible with the objectivity of causation. In other words, any relative statement that is made deviates from the simple determinative fact that *C* causes *E* and should not be affected by how it is framed. However, while causation can be framed as a deficient model of real causal facts, causation must be explored in some way, either with descriptions, models, or default states. In fact, causal modeling techniques have been very successful, especially when incorporating probabilistic or Bayesian network approaches and structural views of causal and counterfactual inference (Pearl, 2000). These kinds of approaches are successfully impacting the emerging field of Artificial Intelligence (AI). In addition, one could rightly argue that causation must be relative to a model so that modeling can be enabled. Indeed, modeling provides accounts of causal epistemology and can help find canonical models that explain particular facts.

(ii) *Causal relations*. Relations describe cause-effect sequences in cause-effect networks of the kind illustrated in Figure 3A. Pairs of nodes in these causal graphs establish relata and cause-effect relations. An example would be the toppling of the second domino by the first toppled domino in the toppling sequence of the domino show. *Connections* make explicit how relata are related to each other through links of the network. They depend on accounts of causation. Nomological, statistical, agential and counterfactual views can be described as indeterministic accounts that consider links of a network in terms of "probability." Views that stress transference, energy and change can be

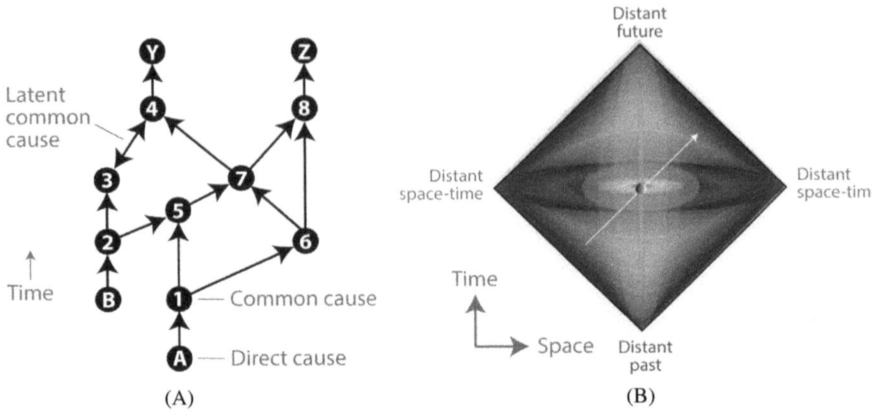

Figure 3. Causation in space-time. (A) Graphical causal model. To illustrate a causal model, we define a directed graph by a set of variables V = {A,B,1,2,3,4,5,6,7,8,Y,Z} and causal-effect connections (arcs). In the graph, time unfolds from A and B toward Y and Z. In addition, A is *direct cause* of 1; A is parent and ancestor of 1 and 1 is descendant of A. For the set {6,7,8}, *common cause* 6 establishes causal relationships with 7 and 8 and influences 8 along two different directed paths. If a Faithfulness Condition is applied to these causal relations, the two paths cancel their respective influences when they make 6 and 8 probabilistically independent. The double-headed arrow implies a *latent common cause* of 3 and 4, i.e. a common cause that is not included in set V and does not give rise to a "genealogical" relationship; 3 is not parent, ancestor or descendant of 4. (B) Penrose 2-dimensional diagram of de Sitter space-time capturing causal relations in an infinite 4-dimensional Minkowski universe. Lines at 45° angles correspond to light rays. Every point in the diagram of a spherically symmetric space-time is a 2-dimensional sphere. According to Verlinde's theory, mass M at the center displaces entropy around it, enriching the horizon areas at the edges of the diagram.

described as deterministic accounts that treat links as "processes." *Direction* is established with arrowed links (arcs), often unfolding time. The existence of joint effects of common causes makes direction explicit. For example, nodes 6 and 7 elicit *joint effects* in node 8, while 6 is *common cause* of effects 7 and 8. An example of a common cause is gravity; it affects all immanent entities of our planet at different degrees and at different times. *Selection* are background conditions that constrain causal relations in the cause-effect network. They are sometimes considered paths of the network, such as the path defined by the sequence {A,1,6,8,Z} in Figure 3A. Causal relations can be implemented with network models in different ways. Sewall Wright implemented them as "path diagrams" (Wright, 1921). Lewis' neuron diagrams described events as firing neurons and arcs as causation synapses (Lewis, 1973, 2000). Causal graphs can be causal Bayesian networks or probabilistic graph models. In all cases variables in a model are represented by nodes and correlations between variables (when all other variables are held constant) or conditional dependencies as links. Associated with each node there is a causal mechanism that determines its value as function of its parents.

In biology, causation is often associated with levels of biological organization and with hierarchical classifications, in which higher "levels" influence or are influenced by lower levels (Love, 2012). This brings the issue of "top-down" or "bottom-up" relationships of causation occurring in biological systems (sometimes occurring concurrently) and better explanatory power of pluralistic dependencies that are based on local causal representations. Some of Lewis' philosophical tenets can help dissect causation as both process and probability (Lewis, 1973): (i) *Conventions*: Conventions are regularities in action that solve co-ordination problems. They have been nicely illustrated with language. Conventions do not require agreement but simply arise when stumbling on persisting regularities. (ii) *Counterfactuals*: Counterfactuals are conditionals that can play important roles in causation when establishing "possible worlds" ("*way things could have been*" or "*what-if questions*"). Counterfactual conditionals such as similarity are context dependent: "*Had A not occurred, B would not have occurred.*" (iii) *Modality*: It stresses the realism of a world being one of many like it: "*A proposition is true iff the proposition is true in one of these worlds.*" For example, the proposition "a protein sequence folds" is true for all counterpart sequences of proteins (a "*plenitudinous*" number of sequences that have never realized in the modern protein world) because realized extant protein sequences fold. (iv) *Properties*: Properties are sets of possibilities that must be taken at face value and sometimes lead to paradox. Under modal realism, properties can follow a principle of recombination that offer multiplicities in alternative worlds. Because an observer of those worlds is one of many, he could be deceived. This uncertainty fosters an inductive skepticism. (v) *Humean supervenience*: Supervenience is the *"doctrine that all there is to the world is a vast mosaic of local matters of particular facts, just one little thing and then another.*" Truth about a world supervenes properties and relationships of that world, which are intrinsic properties and spatiotemporal relationships.

These concepts help understand diversity and drivers of biological systems. When events are causally related to each other, causation can be defined in terms of one event being counterfactually dependent on another: event A causes event B, event B causes event C, and so on, such that within the chain, events succeeding A are counterfactually dependent on previous events. Causation is therefore the ancestral of counterfactual dependence. While this theoretical view has been highly successful over deductive-nomological alternatives, concerns about redundant causation and a proposal of quasi-dependence to address it forced Lewis (2000) to advance an alternative theory, *causation as influence*.

This new theory abandons the assumption of determinism and changes the definition of counterfactual dependence to state dependences of *whether*, *when* and *how* one event occurred on *whether*, *when* and *how* another event occurred. In addition, causation becomes a degree concept that can be defined as an ancestral relation of the type: *A* causes *E* iff there is a chain of stepwise influence from *A* to *E*. The new theory addresses cases of early, late and trumping preemption, and other difficulties. More recently, mathematical elaborations of causal graphs enabled algorithmic implementations of a hierarchy of queries of causal relations that identifies associative relations from joint distributions, cause-effect relationships from external interventions that are neither accidental nor contingent, and counterfactuals from distributions that span parallel worlds (Shpitser and Pearl, 2008).

Causally related events can be typically understood as occurrences more than changes (Lewis, 1986). Within the perdurantist metaphysical view, continual change within the occurrent is something that must be explained, perhaps through transitive and pluralistic causal chains of influence or through a processual model explained for example with a causal graph. Thus, the static view of a succession of discrete temporal parts must be complemented with a dynamic view that explains how temporal parts are related to each other. Lewis explained causation in terms of counterfactual conditionals, extrinsic properties that provide context to intrinsic properties of the whole and its proper parts.

7. Origin of causation

A network exists that accounts for all causal relations of the universe. It expands from the first causal relation, probably at the onset of the big bang, and narrows down to a vast multiplicity of more recent relations (such as those that end in X and Y of Figure 3A). While concurrent causal relations will be pervasive, this universal causal network unfolds space-time and is central to the perdurantist view of existence. Note that the causal network could be mapped onto a Penrose diagram (Figure 3B) depicting the curvy de Sitter space, the standard universal model in which 4D space-time of the universe is shaped like a sphere. In contrast, the anti-de Sitter (AdS) space is shaped like a horse saddle.

In physics, the standard view is that four apparently irreducible forces exist, the long-range *gravitational* and *electromagnetic* interactions, and the subatomic *strong* and *weak* interactions. Except for the gravitational force, the other interactions are mediated by elementary particles and explained with quantum mechanics with the standard model of particle physics. Recent years have seen growing support for quantum information being the ultimate cause of gravity and the process of accretion in the universe. Gravity appears an emergent phenomenon triggered by a powerful entropic diffusive force. This force arises from the quantum entanglement of small quantum bits (qubits) of space-time information (Van Raamsdonk, 2016) and can be explained with holographic and thermodynamic approaches (Carroll and Remmen, 2016). Einstein's traditional view considered gravity arising from the curvature of space-time that is induced by matter. This view has been challenged by a thermodynamic description, beginning with the formulation of the laws of black hole mechanics (Bardeen *et al.*, 1973), black hole thermodynamics (Bekenstein, 1973; Hawkin, 1976), and the demonstration that Einstein's field equations can be derived from thermodynamics and the equivalence principle (Jacobson, 1995). These initial important findings have been revived by recent work that suggest gravity is a byproduct of quantum interactions. Studies also addressed the problem that cosmic bodies appear to gravitate to something that is not there, "dark matter," which outweighs traditional matter by a ratio of 5:1. Recent work now suggests that gravity attributed to dark matter is actually an effect of "dark energy," a background energy in the spatiotemporal dimension of the universe (Verlinde, 2011, 2016). These radical ideas begin by treating space-time and the matter within it as a hologram projected by a network of qubits. In other words, space-time acts very much as a software encoded in the bits of a computer's silicon chip. Unlike classical bits, qubits coexist in states 0 and 1 with various degrees of probability. They interact with each other, becoming "entangled" when the state of a qubit determines the state of another in an interaction network (Van Raamsdonk, 2016). Since entanglement in quantum information is analogous to classic uncertainty about the state of a system, one natural measurements of entanglement is the system's entropy, which corresponds to the area of a particular surface in space-time geometry. Note that in this fascinating connection, entanglement resembles very much an immensely complicated and primordial causal graph. In addition, the "holographic" view of gravity implies that, given a region of space-time, the degrees of freedoms are proportional to the area of the region's boundary and not to its volume.

Entropic forces are effective macroscopic forces that originate in a system that holds numerous degrees of freedom when the system tends to increase entropy (disorder) by dissipation. These forces are generally described as entropic differences independent of microscopic dynamics. We previously discussed the connection between *dissipation*, the loss, dispersion or diffusion of energy, matter or information in natural systems, and the structuring of those systems (Caetano-Anollés *et al.*, 2010). Here we focus on the initial conditions where dissipation tendencies are minimal and associated with an entropic force. To illustrate entropic forces, we follow Verlinde (2011) and the elasticity of a polymer such as a polypeptide chain (Figure 4A). When the polymer is immersed in a heat bath with temperature T, the polymer folds into a randomly coiled configuration that is entropically favored. Pulling one of its ends (with force F) to stretch the polymer in an extended configuration along an x axis will develop a tendency of the molecule to return to a maximum entropic state. This tendency manifests in the form of a macroscopic elastic force that obeys Hooke's law, $F \sim -\mathrm{C}k_bT$, where C is a constant and k_b is Boltzman's constant. The elastic force is entropic and emergent and equals the pulling force. This polymer example can be used as a metaphor to illustrate that changes in

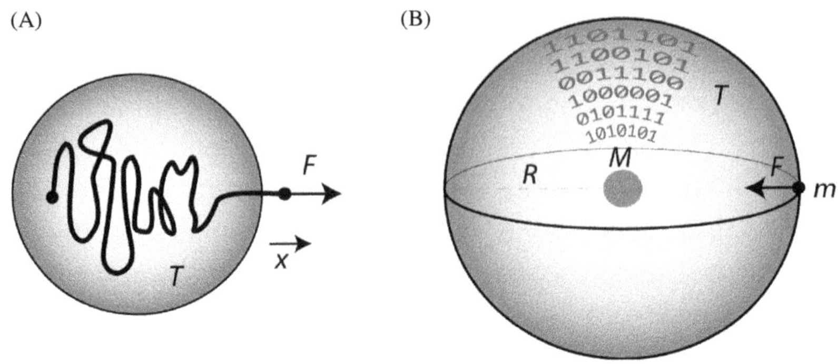

Figure 4. Entropic force. (A) Pulling out a polymer from its equilibrium in a heat bath with temperature T by an external force F generates an opposing entropic force (not shown). (B) A particle with mass m near a spherical holographic screen evenly distributes energy over occupied bits. This energy is equivalent to the mass M that emerges in the part of space that is surrounded by the spherical screen. Modified from Verlinde (2011).

information (measured as an entropy) can lead to an emergent entropic force.

The polymer metaphor helps Verlinde unfold gravity together with space-time in a holographic horizon. Space is the first storage device for information in the universe and holds positions and movements of particles and bodies. Following the most reliable string theory realization of the holographic principle, the AdS/CFT correspondence (Maldacena, 1998), information can be stored in surfaces (screens) as discrete bits without making the dynamics of microscopic storage explicit. These surfaces can be stretched horizons that separate microscopic data on one side from variables describing space-time on the other. This establishes a direction in which space is emergent. Verlinde's derivation of gravity was initially motivated by Bekenstein's entropy formula derived when a fictitious string holding a particle is lowered towards a black hole horizon (Bekenstein, 1973). When a particle with mass m approaches a holographic screen from the side in which space has already emerged, information on the other side is stored in microscopic degrees of freedom mapped onto the screen. The entropic changes (ΔS) induced by the particle travelling towards the screen equals $2\pi k_b$ at the boundary given a displacement perpendicular to the screen Δx, which equals $\hbar/mc$, with $\hbar$ being the reduced Planck constant and c the speed of light. ΔS can be re-written as $2\pi k_b\ mc\hbar^{-1}\Delta x$ when ΔS is linear with Δx near the screen. Since bits on the screen can be associated with a temperature, and according to the Unruh effect a temperature is experienced by an observer under acceleration, acceleration induced by ΔS can recover a force and the second law of Newton, $F = ma$. In other words, if the screen carries a temperature, the particle will experience an effective entropic force F towards the screen, mimicking the process of osmosis across a semipermeable membrane or the elastic force of the polymer metaphor previously introduced in Figure 4A. Thus, there needs to be a temperature to have a force.

To illustrate the derivation of entropic gravity, Figure 4B shows a screen that is not infinite but forms a closed surface, a spherical storage device for information, with emergent space in the outside. Assuming the holographic principle, bits of the screen induced by a particle with mass m will be proportional to the screen's area and total energy E will distribute evenly over bits according to the equipartition principle. Given the mass-energy equivalence $E = Mc^2$, the mass M will emerge in the space inside the spherical screen. While it will not be visible, its presence will be evident by the energy distributed on the screen. The speed of light is here only necessary to translate mass into an energy that provides a heat bath for an entropic force. Straightforward calculations then recover Newton's law of gravitation, $F = GMm/R^2$, a relativistic generalization of entropic gravity reproduces Einstein field equations, and the equivalence principle suggests the origin of Newton's law of inertia is also entropic (Verlinde, 2011). *To summarize, entropic gravity arises when space has one emergent holographic direction that holds entropic change, degrees of freedom are proportional to the area of the screen, and energy is evenly distributed over degrees of freedom following the equipartition principle.*

Verlinde (2016) generalized his initial proposal for AdS space, which relies on an area law for entanglement entropy, to the standard de Sitter space-time geometry that is typical of our universe. He claims that dark matter is an illusion that can be explained by an entropic effect of dark energy, the energy that permeates space-time and accelerates its expansion. He models dark energy as a thermal energy that heats the de Sitter universe to an excited state while AdS geometry describes its grounds state. Verlinde explains this dark energy as the long-range entanglement of qubits, which drives them apart. The competition of area-driven and volume-driven entanglements of the microscopic de Sitter states causes them to retain memory effects in the form of an entropy displacement caused by matter. This disruption by matter removes dark energy from the corresponding area of space (Figure 3B) causing an "elastic" response on the matter that is equivalent to a gravitational attraction. In other words, the law of emergent entropic gravity contains an additional "dark" gravitational force explaining the entropy displacement elastic response. The elastic response increases with space-time volume altering gravitational effects on celestial body movements such as galaxies and clusters currently attributed to dark matter.

The equations of "modified Newtonian dynamics" (MOND) explain phenomena of dark matter by modifying the inverse-square law of gravity. As predicted by Newton's and Einstein's theories, the gravitational pull of the massive center of a galaxy decreases in proportion to the square of the distance. That causes stars that are further away from the galactic bulge to orbit more slowly than stars closer to the center. However, star velocities do not drop steadily past a distance that poses a gravitational acceleration of 1.2×10^{-10} m/s^2.

To explain this mysterious anomaly, is has been postulated that giant "halos" of dark matter provide extra gravitational acceleration to the peripheric stars, flattening curves of rotational speed against distance from the galactic bulge. However, finding hypothetical dark matter particles such as WIMPs or axions that would explain the halos have been unsuccessful. MOND instead modifies the regime of the Newtonian dynamics and Verlinde's explanation of gravity as an emergent thermodynamic effect that includes the baryonic mass distribution and the Hubble parameter provide an explanation for the differential regime. Remarkably, an analysis of the rotational speed of visible matter at given distances from the center of 153 galaxies revealed a "radial acceleration" regularity that confirms predictions from the observed distributions of baryons (McGaugh *et al.*, 2016). Results supports visible matter as the exclusive culprit of gravity, falsifying the existence of dark matter and strengthening the MOND paradigm. More importantly, the gravitational distortion of light (lensing) from more than 33,000 galaxies correctly predicted Verlinde's theory (Brower *et al.*, 2016). While the couplings of entropic gravity and dark matter gravitational studies have been criticized for rendering in certain circumstances inconsistent results, a growing number of lines of evidence continue to support the idea that gravity is an entropic force.

8. Modular parts as entropically entangled occurrents

Verlinde's gravitational laws for de Sitter space are based on the following two premises: "(i) *There exists a microscopic bulk perspective in which the area law for the entanglement entropy is due to the short distance entanglement of neighboring degrees of freedom that build the emergent bulk spacetime.* (ii) *The de Sitter entropy is evenly divided over the same microscopic degrees of freedom that build the emergent spacetime through their entanglement, and is caused by the long-range entanglement of part of these degrees of freedom.*" (Verlinde 2016).

Remarkably, the interplay of "short distance" and "long distance" entanglements embodied in these premises for quantum particle physics can be considered generic for many explanations in biology. When biological systems are modeled with network abstractions, networks become structured by the emergence of subsets of highly connected nodes that are loosely connected to the rest of the network (Newman and Girvan, 2004). These structural communities of highly connected nodes form network "modules" that usually counterbalance tendencies towards scale-freeness that push the preferential attachment of new nodes to highly linked nodes of a network. The generation of modular structure could be considered a "short distance" entanglement of the networks (Caetano-Anollés *et al.*, 2019, 2020). Networks however are not static but rather dynamic entanglements in which matter/energy, information, language and time flow through the wiring diagrams. Connections are often directed (arcs). These connections define a directionality of travel along walks and paths of the network. Paths that begin and end in a same node form "cycles," which tend to organize an additional modular structure, the "bow-tie." In a bow-tie, strongly connected components with numerous cycle structures are stabilized by a balanced feedforward structure of inputs and outputs. In the metabolic networks of *Escherichia coli*, for example, only 93 input metabolites feed a central component of 273 metabolite precursors that reuse molecules and produce 161 output products (Ma and Zeng, 2003). Remarkably, networks become mathematically entangled when cycles in the directed graph are intertwined. Thus, the bow-tie separates input and outputs of a network from a highly entangled core.

The short distance entanglement of network modules and cycle structures is also organized by a "long distance" entanglement of the emergent modular structures. Crucially, this often takes the form of a hierarchical structure. In fact, a number of studies have shown that hierarchical modularity is an emergent property of biological networks manifesting in behavior, development and evolution at different scales of time and biological organization (Caetano-Anolles *et al.*, 2019). The "long distance" entanglement makes up a force of integration that can unify all subsystems of a system. Evidence for this long-range effect can be found in the Rentian scaling of networks (Bassett *et al.*, 2010; Ho and Nvlakha, 2018), cartographic depiction of biological networks (Guimerà and Amaral, 2005), or the hierarchical structure of many biological networks (e.g. metabolism; Kanehisa *et al.*, 2017) that describe different levels of organization.

The "short distance" and "long distance" entanglements are plainly evident in the folding dynamics that lead to the structure of biological molecules, especially the structure of RNA and proteins. RNA molecules are polyelectrolytes that are almost always soluble in water. The most significant contribution to folding into a stable nucleic acid structure is the formation of base pairs. An RNA molecule folds when its strand forms double helical duplexes separated by single stranded segments (Schuster, 2010). This process is highly frustrated (entangled). The use of fluorescence resonance energy transfer (FRET), atomic force microscopy, and optical tweezers to unfold an RNA molecule and study its capacity to refold help understand the dynamics of folding and conformational change (Li *et al.*, 2008). The application of a mechanical force to unfold an RNA molecule allows calculation of mechanical work and free energy changes, which can unravel folding pathways in large molecules or can be used to induce misfolding. Fluoresce techniques provide insights into tertiary structure and ribozyme activity. Typically, base pairs associate and disassociate at rates of less than 0.5 s^{-1} (Fang *et al.*, 1999), forming stems (also known as stacks) of variable length that often produce double-stranded helices of the A type. The rate of folding was found negatively correlated with the length of the RNA chain and its speed limited to ~1 ms, the folding speed of globular proteins (Hyeon and Thirumalai, 2012). The kinetics of RNA folding is initiated by local folding, which involves

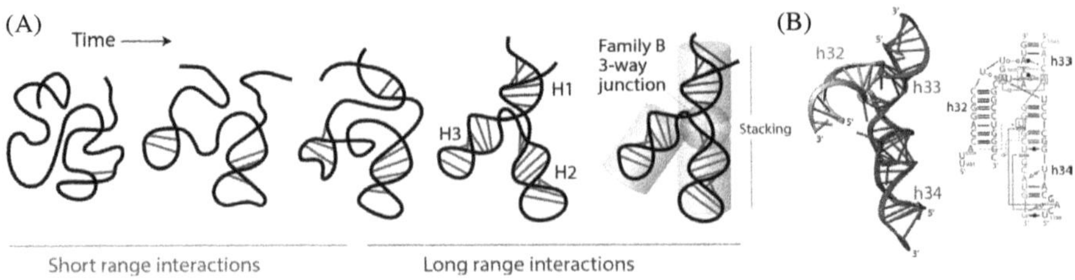

Figure 5. Short and long-range base pair interactions stabilize RNA structure. (A) A model of RNA dynamics. Short range base pair interactions (red lines) produce structural nuclei that lead to helical structures (H2 and H3) around which long range interactions (e.g. base pairing, stacking, formation of helix H1, coaxial helical stacking; in blue) are used to stabilize a branched RNA structure with a "family B" 3-way junction. (B) Atomic structural model of a typical "family B" 3-way junction that is present in the small subunit of the ribosome. The three rRNA helices that make up the junction are h32, h33, and h34. Note that h33-h34 helices (colored red) are coaxially stacked and make a trunk that holds functionally important pivot points of the subunit. The inset describes a network interaction diagram of the same junction showing the h33-h34 stacked helices (Lescoute and Westhof, 2006). The panel was redrawn from Caetano-Anollés and Caetano-Anollés (2017).

establishing base pairs between regions that are molecularly close (Figure 5A). Within the range of free energies accessible at a given temperature, an RNA molecule finds a diverse set of base pair configurations, each enabling a different fold conformation (Ancel and Fontana, 2000; Fontana, 2002). This "plastic repertoire" represents an ensemble of possible conformations carefully constrained by evolution. The repertoire defines a free energy landscape that characterizes the initial kinetics of folding along a morphogenetic trajectory of "canalized" sequences exhibiting low free energy barriers. Folding proceeds down a funnel to a single or small set of tertiary structures with low minimum free energy defined by the formation of a limited set of stems. Some of the initial stems then extend to form helical segments, which stabilize by forming base pair stacking interactions. These flexible helices then act as folding nuclei, bringing distant regions of the molecule together. This involves establishing additional long distance interactions through tertiary contacts and stabilization with counterions (e.g. Mg^{2+} binding). Large RNA molecules, such as tRNA, RNase P RNA or rRNA, organize helices by joining them to each other with multi-branch loops known as RNA "junctions" (Figure 5B). These regions form an ordered array of non-Watson–Crick base pairs that is structurally constrained (Leontis *et al*. 2001). These constraints enable coaxial stacking of pairs of helices and define different spatial alignments of helices and junction topologies that can be grouped into families (Lescoute and Westhof, 2006; Laing and Schlick, 2009a,b). Long range distance interactions enable the formation of helical communities that constrain RNA dynamics through structural flexibility (Hohng *et al*., 2004; Besseova *et al*., 2010) and interaction with proteins (Klein *et al*., 2004). We note that initial short distance base pair interactions are highly dynamic within the plastic repertoire, defining an entanglement of interacting states of base pairings. This entanglement becomes increasingly constrained with folding time and when the RNA structure becomes further constrained by long range interactions. This is illustrated in the model of Figure 5A.

As with RNA, the folding of proteins follows a well-established mechanism that optimizes the energy landscape of the molecule and favors the stability of the folded native state (Onuchic *et al*., 1997). The "funneled energy landscapes" paradigm, a concept that was proposed in the early 1990s, delimits a multiplicity of folding pathways of conformations that the protein take when emerging from the ribosomal pore or when proteins unfold or refold in the crowded environment of the cell (Leopold *et al*., 1992). This statistical "landscape" view, often illustrated with funnel representations, was soon experimentally supported by definition of kinetic folding intermediates and pathways (Baldwin, 1995). These explorations require the use of techniques other than crystallography and NMR spectroscopy, such as fluorescence, circular dichroism, and infrared spectroscopic methods. With time, this "new view" received strong experimental support and led to the current theory that proteins are multi-state objects that fold through the asynchronous formation of numerous folding intermediates and equilibrium analogs called "molten globules" over a range of time scales (Englander and Mayne, 2014). First, they cooperatively fold into a number of small-sized (~20 residues) "foldon" building blocks that fold and unfold repeatedly. This dynamic structuring resolves the Levinthal paradox, the impossibility of finding the native structure by random walks of folding. Foldons involve "short distance" hydrogen-bond interactions between the protein backbone chains responsible for elements of secondary structure, including helical elements (α, π, 3_{10} and polyproline helices), hydrogen-bonded turns, and bends. These emergent elements of secondary structure are then complemented by "long distance" interactions that for example form strands, bridges and sheet arrangements through foldon-foldon interactions. These interactions guide the free energy landscape through stepwise pathways towards the native state. Example of studies that follow this duality of short-range and long-range folding entanglements include identification of cooperative foldon interactions of cytochrome c with equilibrium native state hydrogen exchange pulse labeling experiments (Bai *et al.*, 1995) and analysis of folding trajectories of ribonuclease H that involve well-formed H-bonded foldons with advanced pulse labeling mass spectrometry experiments (Hu *et al.*, 2013).

The resemblance of quantum entanglement and optimization processes that are ongoing in biological macromolecules at different time scales is not a far-fetched concept. The mathematical framework of quantum mechanics describes properties of subatomic particles, atoms, molecules, molecular complexes and much more, including those of biological origin (Marais *et al.*, 2018). These properties operate at nanometer scales or below. Subatomic particles for example exhibit quantum entanglement when their states are interdependent regardless of distance separating them. These particles have also wave-like properties that follow a probabilistic wave function. Under certain conditions, particles such as electrons or photons can "tunnel" through energy barriers. Both entanglement and tunneling have been observed in living systems, which depend on transfer of energy and charge through electrons, protons and ions (Marais *et al.*, 2018). Quantum phenomena are associated with excitation energy transfer and charge transfer of photosynthesis and enzyme catalysis. Quantum coherence was observed in complexes of chlorophyll molecules as oscillatory signals (quantum beating) at low and ambient temperatures when complexes were extracted from photosynthesizing bacteria (Engel *et al.*, 2007; Panitchayangkoon *et al.*, 2019), marine algae (Collini *et al.*, 2010) and spinach (Callhoun *et al.*, 2009). While some aspects of the subject remain controversial, it appears photosynthetic machinery likely evolved mechanisms that exploit quantum phenomena. In metabolism, the oxidation of benzyl alcohol to benzaldehyde catalyzed by an alcohol dehydrogenase enzyme from yeast appears to involve quantum tunneling (Cha *et al.*, 1989). Hydrogen transfer of other reactions was later observed to take place with the assistance of tunneling (Klinman and Kohen, 2013). Even environmental sensing and animal behavior appears controlled by quantum biology processes. Magnetoreception, olfaction and cognition occur at quantum level. For example, Schulten *et al.* (1978) was the first to propose a radical-pair quantum mechanism for avian magnetoreception. Reactions in the cryptochrome protein generate a pair of radicals, each with lone electrons, which can be quantumly entangled with each other when responding to alignment of Earth's weak magnetic fields. Alignment changes trigger downstream chemical reactions, transferring information to the brain of for example migrating European robins (Ritz *et al.*, 2004). It is highly likely that many if not all phenomena in biology are impacted to some degree by quantum processes, including the rules that determine the genetic code.

Distance in short-range and long-range entanglements must be considered within a space-time dimension, i.e. spatial distances (such as interactions along a nucleic acid or polypeptide chain) must be linked with distances measuring time scales (nanosecond dynamics to billions of years of evolution). This implies that within the temporal parts paradigm, entanglements responsible for occurrents occur at both space and time levels and their consequences become evident in the makeup and dynamics of biological structure, with structure being broadly defined. For example, we explored the nanosecond dynamics of protein loops with molecular dynamics (MD) simulation tools (Mughal *et al*., 2017). Networks describing a dynamic cross correlation matrix of the motions of protein residues captured biases in atomic movements of the molecules. Remarkably, we found that the modularity of these highly dynamic network systems of protein folding increased with deep evolutionary time as networks decreased their randomness and heterogeneity (see Figure 6 in Caetano-Anollés *et al*., 2019). Thus, nanosecond dynamics appears constrained by an evolutionary history that spans billions of years of protein evolution. In other words, the short-long distance entanglement interplay responsible for the protein folding landscape of atomic conformations is constrained by temporal entanglements that manifest at time scales of billions of years. When considering deeper time frames, Guseo (2016) makes a surprising entropic connection between the static equilibria of the catenary function and the aggregate logistic model that explains the dynamics of innovation typical of biological and social systems. The catenary function explains the shape and static equilibrium of chains and cables and catenary arches that are common in bridges and architecture in general (e.g. the biomorphisms of the Catalonian architectures of Gaudí), when these contraptions are subjected to the effect of gravity. In turn, the aggregate logistic model is linked to the Bass model for diffusion of innovations that explains the dynamics of self-sustaining evolutionary growth (discussed by

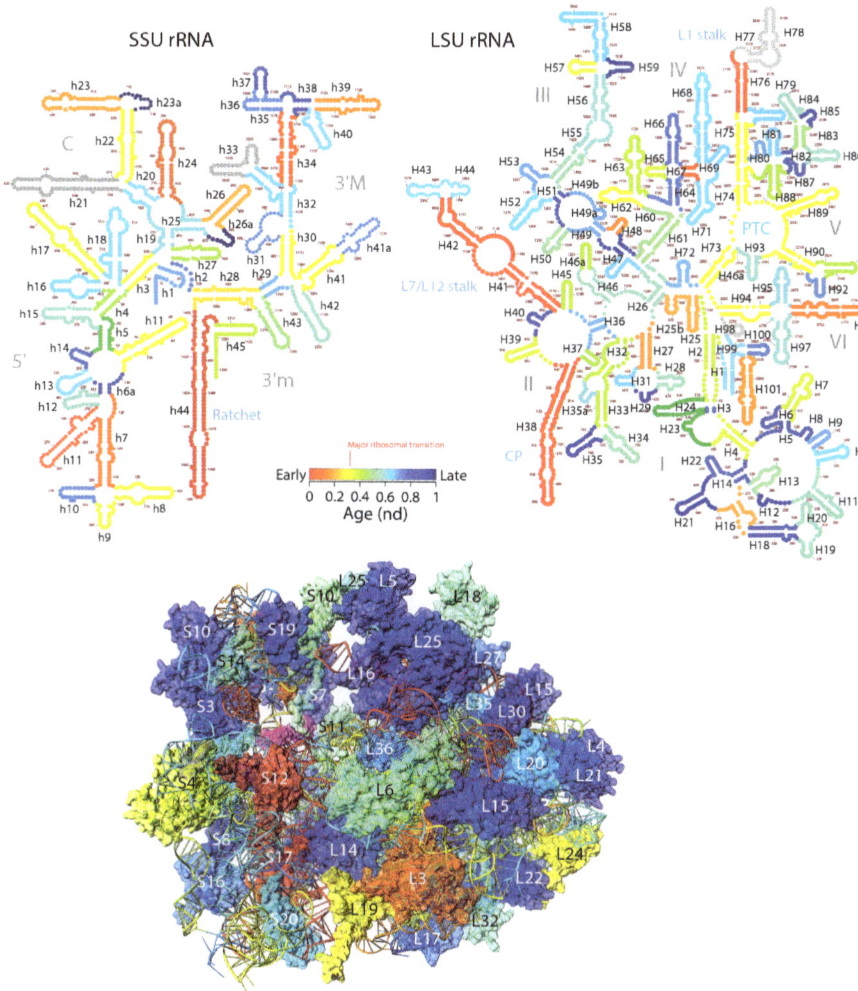

Figure 6. The ribosome is a historical patchwork of co-evolving ribosomal proteins and RNA helices. Secondary structure models of the small (SSU) and large (LSU) subunits of ribosomal rRNA from *Escherichia coli* are shown at the top with helical segments colored according to their relative evolutionary age. A 3-dimensional atomic view of the ribosomal complex with ages of ribosomal proteins and rRNA helices traced on an *Escherichia coli* structural model of the ribosomal core is shown at the bottom. Note that ribosomal proteins S12 and S17 and helix 44 of LSU are part of the most ancient ribosomal structural core. Data from Harish and Caetano-Anollés (2012).

Caetano-Anollés and Caetano-Anollés, 2020). Remarkably, the local entropy function related to the logistic distribution is proved to be a catenary, and vice versa. This surprising entropic correspondence between static and dynamic equilibria is congruent with Verlinde's conjecture of the entropic origin of gravity.

Assuming that entanglements manifest in the modular and hierarchical structure of biological systems, we now test the space-time entanglement hypothesis by examining examples of temporal part dynamics that could explain the complexity of macromolecular biosystems and how that dynamics is constrained by history.

9. Dynamic patchworks of temporal parts

Cuvier (1833), almost two centuries ago, was already cognizant of the central role that dynamics played in biological systems and how persistence was grounded in form. His recount of *"Des êtres vivans, et de l'organisation en général"* in his book, *"Le règne animal distribue d'après son organization,"* is prescient:

> *"Life then is a vortex, more or less rapid, more or less complicated, the direction of which is invariable, and which always carries along molecules of similar kinds, but into which individual molecules are continually entering, and from which they are continually departing; so that the form of a living body is more essential to it than its matter. As long as this motion subsists, the body in which it takes place is living — it lives. When it finally ceases, it dies."*

A focus on evolution of macromolecular structure and biological networks can clarify the frustrated dynamics of change and the constancy of "form" that is intimated by the early thinking of Cuvier. The most widespread and deepest evolutionary transformations involve recruitment, the reuse of gradually accumulating molecular innovations in different molecular and cellular contexts (Ycas, 1974; Jensen, 1976). As we will discuss, recruitment is a pervasive activity of persisting things, occurrences of 4-dimensional persistents. These persisting things are "structured" entities such as the 3-dimensional fold or the molecular function of a nucleic acid or a protein molecule. When persistents are useful, they are coopted for other tasks and become successful. Phylogenomic analysis of molecular structures and functions, both of which are highly conserved in evolution, provide decisive evidence of widespread recruitment of persistents across biological scales. Analyses of the ribosome (Caetano-Anollés, 2002; Harish and Caetano-Anollés, 2012), metabolic networks (Kim *et al.*, 2006; Caetano-Anollés *et al.*, 2007; Caetano-Anollés and Caetano-Anollés, 2013; Mughal and Caetano-Anollés, 2019), the metallome (Dupont *et al.*, 2010), the proteome (Wang *et al.*, 2007; Wang and Caetano-Anollés, 2009), or the functionome (Aziz *et al.*, 2017; Koç and Caetano-Anollés, 2017) reveal that biological systems are evolutionary patchworks. To illustrate this fact, we here present four examples of how recruitment is pervasive in biology:

(1) *Origin and evolution of the ribosome*. The ribosome is a ribonucleoprotein complex responsible for protein biosynthesis. The complex is assembled around a universal ribosomal core. The small (SSU) and large (LSU) subunits contain 30 – 40 and 30–45 ribosomal proteins (r-proteins) and about 50 and 100 helical rRNA core segments, respectively. Both r-protein and rRNA structures hold considerable historical information, which can uncover how the ribosome has grown in evolution. Phylogenomic analysis of r-proteins and rRNA allowed construction of models of ribosomal evolution from genomic data; reviewed in Caetano-Anollés and Caetano-Anollés (2015). Advanced hidden Markov models (HMMs) of structural recognition are powerful devices that assign fold structures to protein sequences with high accuracy and low error rates. These

bioinformatic annotations permit the generation of a structural census of structural domains in proteomes, with structural domains defined at various levels of protein structural abstraction in the hierarchies of the SCOP (Murzin *et al.*, 1995) and CATH (Orengo *et al.*, 1997) taxonomies, the gold standards of protein classification. Phylogenetic trees of domains and proteomes and their molecular functions were built from a genomic census of millions of encoded proteins and associated molecular functions. Direct calculation of ancestry values as "node distances" (nd) from the base of the trees allow to build explicit timelines of accretion of structural domains in proteomes, with nd = 0 representing the origin of proteins and nd = 1 the extant protein world. A molecular clock of folds coverts this relative timeline into geological timescales by correlating domain ancestry with domain structures linked to markers of the geological record (Wang *et al.*, 2011). RNA molecules also carry deep phylogenetic history. In fact, historical accounts can be directly inferred from the topology and thermodynamics of RNA structures. Here, the ages of helical substructures can be mapped onto the structures of RNA molecules, which can be visualized as color-painted 2-dimensional secondary structures or 3-dimensional atomic models. Figure 6 shows phylogenetic tracings of rRNA helical structures and r-proteins, with the substructures and r-proteins colored from red (ancient) to blue (derived). The color tracings clearly show a patchwork of ages suggestive of rRNA structural recruitments that gradually increased the size of the complex (Harish and Caetano-Anollés, 2012) through outward and inward episodes of ribosomal growth (possibly mediated by structural grafting; Caetano-Anollés and Caetano-Anollés, 2015). An analysis of helix insertion points in junctions revealed the existence of 19 independent origins of the rRNA molecules, reflecting the recruitment of at least 19 rRNA fragments to form the complex (Caetano-Anollés and Caetano-Anollés, 2017). Despite the apparently haphazard-like addition of molecular components to the growing ribosomal complex, the tracings reveal an early appearance of structures involved in mRNA decoding and tRNA translocation, the coevolution of ribosomal proteins and RNA, and a first evolutionary transition that brought ribosomal subunits together into a processive protein biosynthetic complex. These results prompt the study of molecular growth and information transfer of macromolecules and cellular complexes.

(2) *Emergence of metabolic networks:* Chronologies also help reveal patterns of enzymatic recruitment in metabolic networks (Kim *et al.*, 2006; Caetano-Anollés *et al.*, 2007; Caetano-Anollés and Caetano-Anollés, 2013; Mughal and Caetano-Anollés, 2019). Several processes responsible for metabolic evolution have been proposed. These include specialization of multifunctional enzymes, pathway retroevolution, pathway duplication and divergence, and enzymatic recruitment Metabolic growth has been explained by two very popular hypotheses, the retro-evolution (retrograde) hypothesis and the patchwork model. The retro-evolution scenario posits that biosynthetic (anabolic) pathways evolve backwards by gradual depletion of successive organic molecules, which poise the rise of new metabolic activities and substrate intermediates (Horowitz, 1945). Under this scenario the first enzyme to appear is the last in a pathway. In contrast, the patchwork model describes the spread of enzymatic activities as a mosaic when promiscuous catalytic activities are fortuitously recruited to perform different functions (Ycas, 1974; Jensen, 1976; Kacser and Beeby, 1984). Homologous enzymes are therefore haphazardly arranged in metabolic pathways. Considerable evidence supports the patchwork model of metabolic evolution. One first example is the observation that enzymes with

homologous fold structures are spread as a mosaic structure along the metabolic pathways of *Escherichia coli* (e.g. Teichmann *et al.*, 2001). A later systematic tracing of the age of enzymatic domains in the metabolic networks of the MANET database consistently revealed a patchwork of ancestries (Kim *et al.*, 2006; Mughal and Caetano-Anollés, 2019). Figure 7 shows a collage of the 148 metabolic subnetwork diagrams of the MANET 3.0 database (Mughal and Caetano-Anollés, 2019), with domain age colored onto enzymatic functions defined by Enzyme Commission (EC) classification. A patchwork is clearly evident in most subnetworks, showing there is little repetition of domain structures or ages of enzymes in consecutive enzymatic steps. In contrast, retro-evolving (and forward-evolving) patterns are rare and difficult to identify. Thus, the patchwork pattern of unordered growth suggests multiple processes of metabolic growth were interacting with each other during metabolic evolution. MANET can help establish patterns of origin of metabolic networks and waves of recruitment facilitated by structural innovations. To illustrate this potential, we studied enzymatic recruitment in purine metabolism (Caetano-Anollés and Caetano-Anollés, 2013). The origin of metabolism has been linked to abiotic chemistries that existed in our planet at the beginning of life. These plausible but putative chemical pathways are inexistent today. Modern enzymes recruited these reactions to establish the metabolic pathways of modern metabolism. To study the emergence of purine metabolism, the oldest metabolic subnetwork (Caetano-Anollés *et al.*, 2007), the ages of domain families were mapped onto metabolic enzymes and their structures traced along the pathways of the nucleotide interconversion (INT), catabolism and salvage (CAT) and biosynthesis (BIO) pathways (Caetano-Anollés and Caetano-Anollés, 2013). The INT pathway was the first to develop ~3.8 billion years ago,

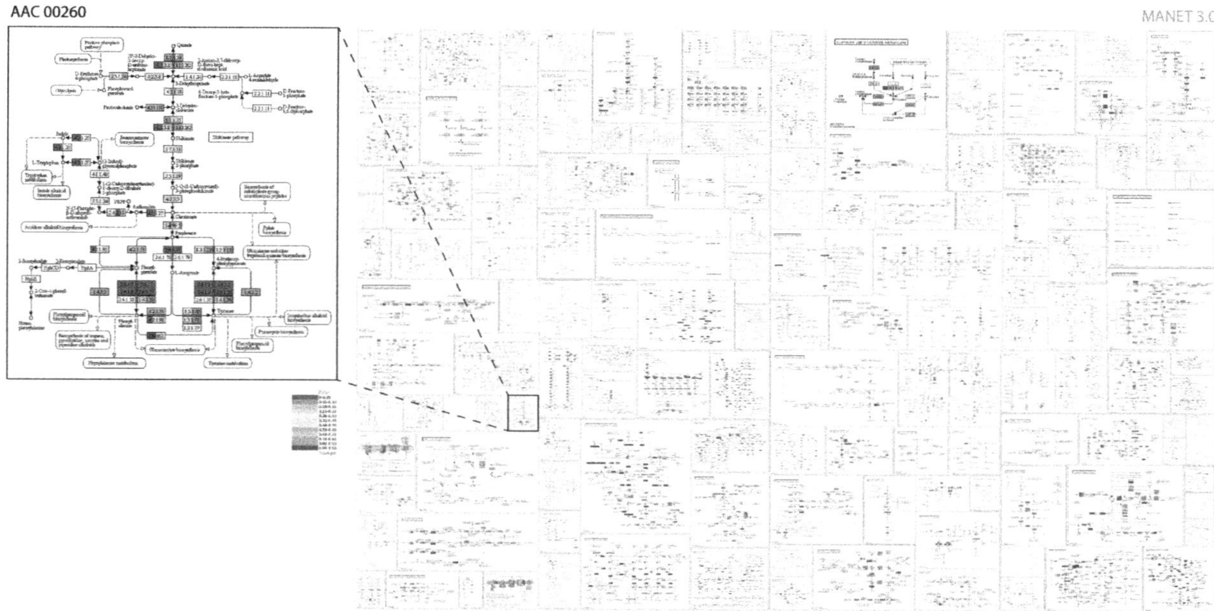

Figure 7. A collage of the 148 subnetwork diagrams of MANET 3.0, which paints evolutionary history onto metabolic networks. The database traces the evolutionary age of domain structures of metabolic enzymes onto the metabolic pathways of KEGG, making use of the PDBsum and SCOP databases and structural phylogenomic analyses (Mughal and Caetano-Anollés, 2019). Protein structural domains associated with individual enzymatic activities (described in EC nomenclature) were painted according to their age in a scale of node distance (nd) that range from 0 (the oldest enzymes) to 1 (the most recent). Domain ages were derived from a phylogenomic tree of structural domains defined at the family level of the SCOP hierarchy. An example subnetwork of the amino acid metabolism (AAC) mesonetwork, the phenylalanine, tyrosine and tryptophan biosynthesis (AC00400) subnetwork showing patchwork patterns is highlighted as an inset. The vast majority of pathway diagrams show enzyme collectives are patchworks of different evolutionary ages.

benefiting most parsimoniously from the prebiotic formation of adenine nucleosides and the appearance of the EC 2.6.1.3 and EC 2.6.4.1 enzymatic activities (Figure 8). In turn, the BIO and CAT pathways originated ~300 million years later by concerted enzymatic recruitments and gradual replacement of abiotic chemistries. These recruitments involved the spread of 6 domain structures (see inset of Figure 8). Remarkably, the fully enzymatic biosynthetic pathway became already fully assembled ~3 billion years ago (Figure 8), concurrently with the appearance of a functional ribosome. The simultaneous appearance of purine biosynthesis and the ribosome were probably necessary for developing genomic information.

(3) *Gain and loss of structural domains in evolving proteomes.* The existence of an evolutionary patchwork can be studied at the proteome level by first reconstructing a phylogenetic "tree of life" describing the evolution of proteomes and then retracing the history of both gains and losses of structural domain families along its many branches (Nasir *et al.*, 2014). The algorithmic "character state reconstruction" implementation considers that gains or losses occur when the abundance or occurrence of a domain family at a node of the tree is higher or lower than the value of the immediate ancestor, respectively. Since the Wagner evolutionary model used for phylogenetic reconstruction allows for evolutionary change to occur in both backward and forward directions, the reconstruction exercise does not constrain evolutionary change. Results revealed that both domain gains and losses were frequent events in the evolution of a reference set of 420 organisms belonging to the three superkingdoms of life, Archaea, Bacteria and Eukarya. Figure 9 plots gains and losses of domains against their evolutionary age, both for abundance and occurrence data, and for a "tree of life" that describes the evolution of all organisms. In all instances, domain gains overshadowed losses by a ratio of 1.69 for abundance and 1.30 for

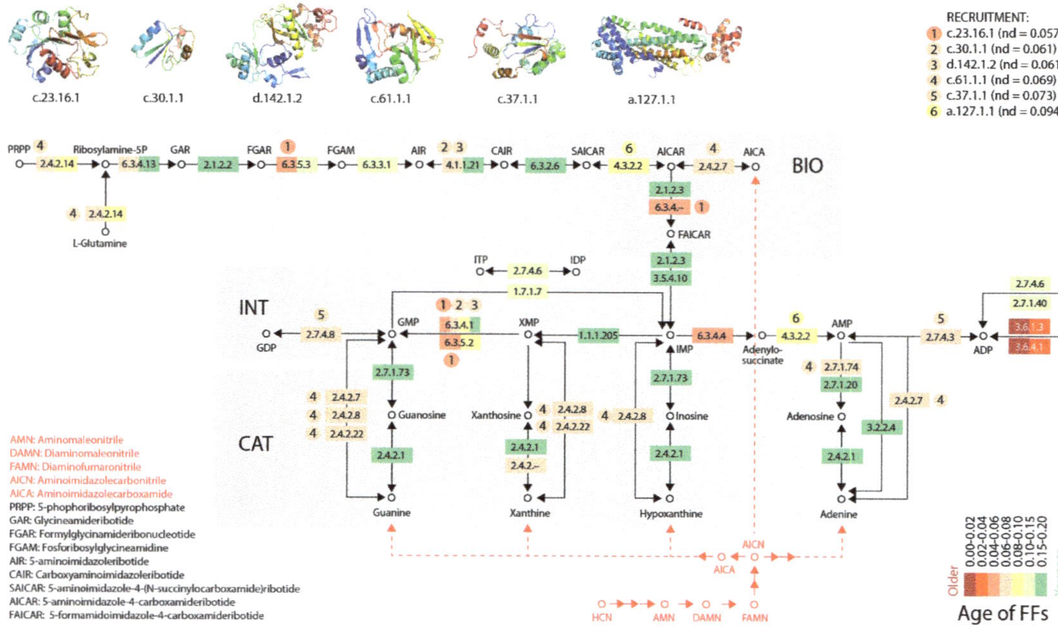

Figure 8. Early evolution of the purine metabolic network. Fully connected pathways of nucleotide interconversion (INT), catabolism and salvage (CAT) and biosynthesis (BIO) appear ~3 billion years ago thanks to numerous recruitments of structural domain families. Major recruited families are indicated with encircled numbers, labeled with SCOP identifiers, and illustrated with example 3-dimensional atomic models. Pathways mediated by prebiotic chemistries that are plausible and most parsimonious are depicted in red and enable the growth of the emergent protein enzyme-mediated pathways of purine metabolism by structural and functional innovation and piecemeal recruitment. Domain structures associated with individual enzymatic activities (described in EC nomenclature) are painted according to their age, in a scale of node distance (nd) that ranges from 0 (the oldest enzymes; ~3.8 billion years ago) to 0.2 (~3.0 billion years ago). Figure modified from Caetano-Anollés and Caetano-Anollés (2013).

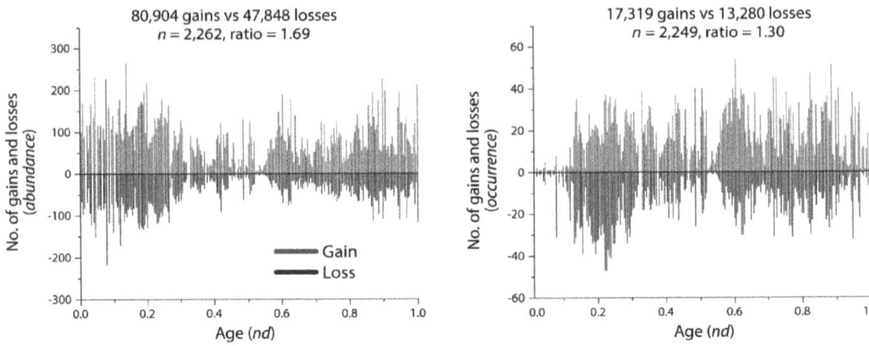

Figure 9. Global patterns of gains and losses of protein structural domains in the lineages of cellular organisms. Histograms comparing the relative counts of gains and losses for each domain family (as phylogenetic character), as it unfolds with time in the timeline of domain families with ages plotted as node distances (nd). Bars in red and blue indicate gains and losses respectively. The global gain-to-loss ratios are listed along with the total number of gain and loss events and gain-to-loss ratios. n is the number of parsimony informative characters in each analysis. Data from Nasir *et al.* (2014).

occurrence, providing strong support for a pervasive evolutionary trend of increase of both abundance and occurrence of domain families in proteomes. Gain and losses occurred throughout the evolutionary timeline of domains, from the origin of domains to the present, albeit at non-uniform rates. All of these trends were consistently recovered in all three superkingdoms and throughout evolution. However, gain/loss ratios were higher for the microbial superkingdoms, especially for Archaea. This study sheds light into the evolutionary history of living organisms and highlights important ongoing mechanisms that are responsible for secondary evolutionary adaptations in superkingdoms, especially in archaeal and bacterial microbes. The evolutionary dynamics of domain families in proteomes and the increasing number of domain gains revisits the persistence strategies of superkingdoms that are discussed in this volume (Mainzer *et al.* 2020). It influences the growing makeup of molecular functions and the complexification of the protein world.

The gain-and-loss interplay that occurs along the many branches of the "tree of life" turns proteomes into evolutionary patchworks. These patchworks can be visualized with multidimensional scaling approaches that explore the temporal space

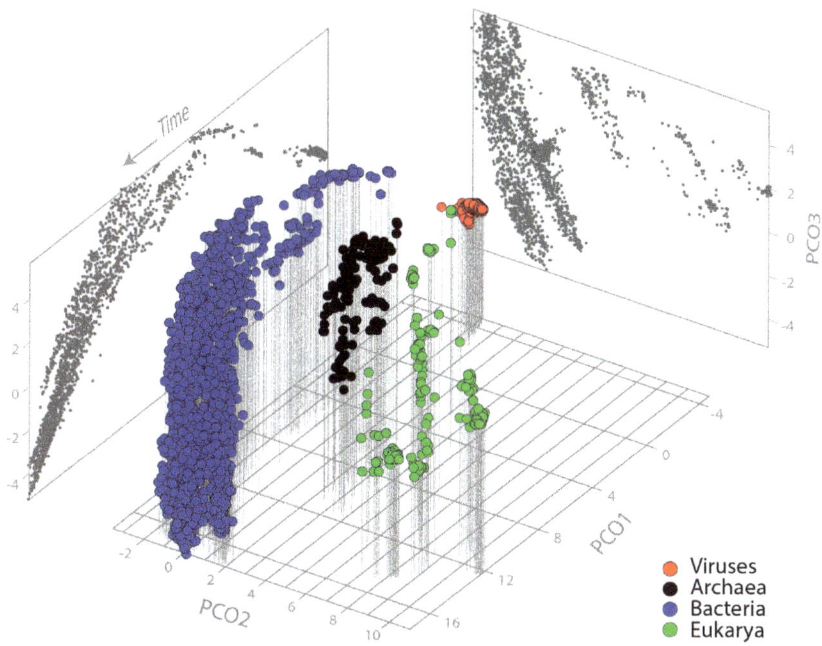

Figure 10. Evolutionary principal coordinate (evoPCO) analysis describing the evolution of the proteomes of cellular organisms and viruses. The 3-dimensional scatter plot describes the temporal relationships of 139 archaeal, 1,740 bacterial, 210 eukaryal and 6,044 viral proteomes, making up a total of 8,127 reference quality proteomes. Note how the multidimensional scaling analysis of the ages of 3,892 families of SCOP structural domains highlights in its first three most significant axes a temporal flow from viruses to cellular organisms. Data points describing temporal clouds of the scatter plot are mapped onto projection planes and connected with vertical leading drop lines along the third principal coordinate (PCO3) axis. The list of whole coordinate information for building the evoPCO plot can be found in Mughal *et al.* (2020).

of evolutionary ages of individual domains in proteomes (Figure 10). Domains are parts of a proteome. Thus, each domain contributes some molecular functionality to the functioning of the corresponding organism. Similarly, domains become parts of each proteome at different timepoints in evolution. Consequently, an indexing of domain ages with nd values can measure, as central tendency, the age of each proteome. Applying the multidimensional scaling method that we call "evolutionary principal coordinate" (evoPCO) analysis (Nasir and Caetano-Anollés, 2015) to a data matrix of ages generates a temporal view of proteome evolution. This view is more powerful than the standard "tree of life" visualization because it minimizes problems of independence of phylogenetic characters, which plagues sequence analysis. Figure 11 shows an evoPCO three-dimensional plot describing the evolution of 8,127 reference quality proteomes of cellular organisms and viruses (Mughal *et al.*, 2020). The first three coordinates account for ~72% of total variance. Distinct temporal clouds of proteomes for the three cellular superkingdoms and viruses are clearly evident as is the evolutionary direction of clouds of data points, which follows the PCO1 axis from an ancient origin of viruses at the back of the plot to the evolutionarily derived nature of the three superkingdoms at the front. The group of giant viruses was clearly dissected from the main viral cloud, which was temporally closer to cellular proteomes, suggesting their late appearance in viral evolution. In contrast, RNA viruses (particularly those containing plus-ssRNA genomes) appear to represent the oldest proteomes in existence today. Conceptually, however, the clouds in evoPCO plots describe a patchwork of temporal parts unfolding in biological evolution. The PCO3 axis, which accounts for 67.7% of the variance, is enriched in the temporal component. The PCO2 and PCO3 axes account for only 2.81% and 2.23% of the variance, respectively. They are likely responsible for patchworks

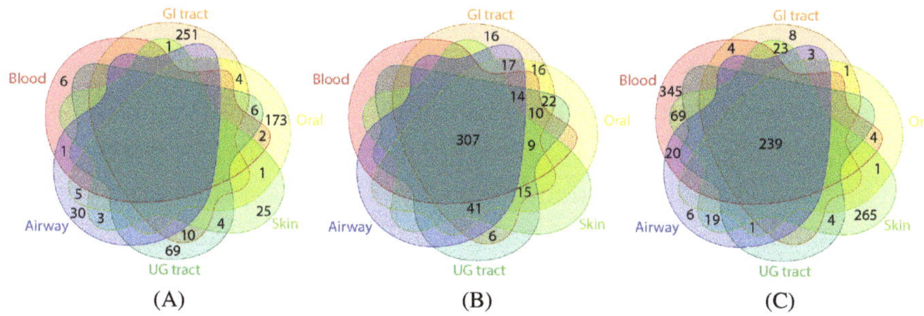

Figure 11. Horizontal gene transfer between organisms of the human microbiome. (A) Venn diagram describing the sharing of species between the six body sites that were surveyed in the NIH *Human Microbiome Project* dataset. GI, gastrointestinal; UG, urogenital. (B) Venn diagram describing the sharing of intra-niche and inter-niche gene transfers (in thousands) occurring in body sites. Only transfer levels >5,000 are provided in the diagrams. Numbers specific for body sites involve only intra-niche transfers. (C) Venn diagram describing the sharing of widespread genes between body sites. Widespread genes were those that are present in >70% genomes of organisms of each body site or its combination with other body sites. Data from Jeong *et al.* (2019).

manifesting at spatial level. PCO2 dissects long range spatial entanglements between supergroups. PCO3 dissects short range spatial entanglements within supergoups. Remarkably, the PCO1 and PCO3 projection of the cloud recovers a arched pattern resembling a curvy de Sitter universe.

(4) *Horizontal exchange of genetic information in the human microbiome*. A microbial collective lives inside the human body. This "microbiome" provides opportunities for exchange of genes between the different microbes that live in the different "niches" of the body. This enhances both the persistence of the microbes and their host (Goodrich *et al.*, 2017). A recent study shows that gene exchange is substantial and more extended (30% higher) than that of other microbial communities despite negligible overlap of species inhabiting different body sites (Jeong *et al.*, 2018). The *HGTree* pipeline (Jeong *et al.*, 2016) of horizontal gene transfer detection, which

employs a combination of parsimony, neighbor joining and maximum likelihood, was modified to reconstruct "trees of genes" and compare them to reference 16S rRNA-based "trees of species." A total of >80,000 phylogenetic reconstructions and reconciliations identified ~55,000 genes affected by gene transfer in 1,059 reference genomes from prokaryotic organisms collected from six body sites (gastrointestinal tract, skin, mouth, blood, urogenital tract, and airway) of "healthy" humans sampled from the NIH *Human Microbiome Project* (Consortium HMP 2012). Remarkably, the analysis found that more than half of the genes of the human microbiome had at some point in their history a genetic transfer. Out of ~511,000 transfer events, 40% involved microbes living in similar niches. This outcome was expected because of physical proximity and ecological similarity. However, 60% of the transfers involved microbes living in different parts of the body, questioning how these gene transmissions occurred. Figure 11 shows how different body sites share very few species but at the same time transfer numerous widely shared genes to other body sites. The Venn diagrams reveal that while most species are unique to body sites (Figure 11A), intra-niche and inter-niche genetic transfers are largely mediated by two body sites, the gastrointestinal tract and blood, which behave as "hub" and "conduit" of microbial genetic transfer, respectively (Figure 11B). This mediation, which does not involve the sharing or microbes, requires some unknown mechanism that involves the human host, perhaps through the activity of extracellular microvesicles or naked DNA (e.g. Raposo and Stoorvogel, 2013). In addition, the dynamics of genetic exchange of widely shared genes largely involved a common set of commonly exchanged genes and a large number of genes associated with blood and skin (Figure 11C). The local intra-niche short distance transfers were in general less abundant than inter-niche transfers,

> which involve not only distant body sites but also more ancient transfers. While gene transfers were more common between closely related species, inter-niche transfers occurred between more evolutionarily distant species. These transfers can be explained as either transfers that pre-dated microbial colonization of the human body or transfers that occurred through the bloodstream. Our understanding of a healthy blood microbiome is increasing and its role could be of paramount importance for human physiology (Castillo *et al.* 2019). While further research is needed to examine the mechanisms of inter-niche horizontal exchange of genetic information, the study shows a complex entanglement of genetic transfers occurring between organisms that span short and long distances and transfers that occurred at different time scales of evolution.

In all of these four examples, both deep and shallow evolutionary processes interface to produce complex outcomes in the organization of biological systems. As evolution proceeds, ribosomes, metabolic pathways, proteomes and microbiomes develop innovations that accumulate with time. These accumulating innovations increase the degrees of freedom that are available for recruitment and entanglement. This "structures" the biological systems and increases their complexity. For example, short distance entanglements interface with long distance counterparts to produce highly entangled "cores" of ribosomal structures, metabolic enzymatic activities that reuse molecules, ancient structural domains in proteomes, and transferred genes in the microbiome. The occurrents produce bow-tie structures that facilitate flow through the systems. The four examples also show how components behave as "temporal parts" that enter and leave systems at different times along the entire history of the *tela vitae* of evolving systems. These components are recruited to enhance the functioning of the system per se. Recruited ribosomal structures enhance ribosomal functionality. Recruited enzymatic activities make possible new biosynthetic pathways. Recruited fold structures enable functionalities that foster the splitting of organismal and viral lineages. Recruited horizontally transferred genes help unify and coordinate interactions between microbes and the human body. Recruitment has ancient origins but is ongoing. It benefits from mechanisms that foster both vertical and horizontal processes, such as those involving the linkage of molecular parts (e.g. multidomain proteins), molecules (e.g. molecular complexes), cells (e.g. organismal makeup) and organisms (e.g. microbiome collectives).

10. Toward long-lived occurrents

Occurrents are things that persist by extending their temporal parts in de Sitter space-time. Let us now address how the persistence of occurrents increases with evolutionary time and change, when occurrents embody ("*ens per alio*") high levels of biological organization. An evolutionary tendency towards long-lived occurrents is supported by considerable evidence. Figure 12 illustrates how the ages of the first temporal part of broadly defined classes of biological occurrents increase in evolution. The lengthening of the life of these higher level occurrents simply arises by their continued persistence, a property that is actively optimized in evolution (Maizer *et al.*, 2020). For example, metabolism first manifested as primordial metabolic cycles that were mediated by the prebiotic chemical reactions of early Earth. These reactions were then incorporated into cellular make up through the activity of novel enzymatic catalysts (as those exemplified in Figure 8). Pathways were subsequently built that reflected general functional types of enzymatic activities that could be grouped into metabolic mesonetworks and subnetworks. These processes produced the 148 subnetworks that describe metabolism today (Figure 7). Similarly, once a cellular envelope appeared in evolution, the dichotomy between an external

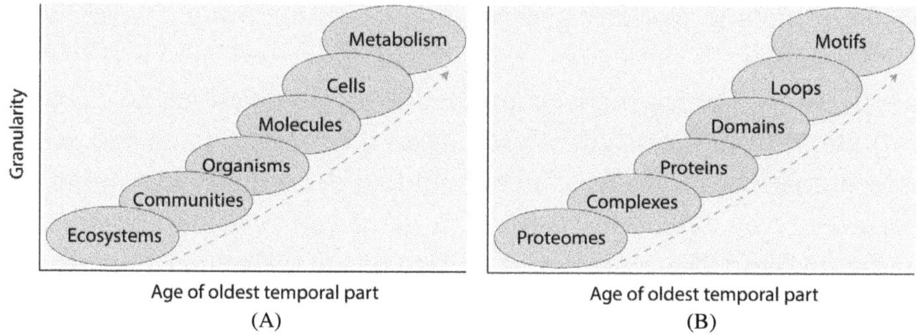

Figure 12. Temporal parts across space-time scales. The plots illustrate the evolution of biological occurrents. As time proceeds, the age of the oldest temporal part of broad categories of biological systems (A) or modules of proteomes (B) increases. With time systems become more granular in modularity and hierarchy. Note how difficult is to map the occurrents in panel B onto panel A. This illustrates entanglement.

environment and an internal cellular medium mediated by a vesicular membranous system was persistently adopted. This envelope enabled a number of biological processes that enhance the persistence of the cellular occurrent. While different kinds of cells and cellular makeup are known to have been generated in evolution in the lineages of the different superkingdoms, the general cellular "vesicular" design has been preserved along the 3.8 billion-year history of life and is also present in viruses. Similar argumentation applies to the macromolecules, organisms, communities and ecosystems of Figure 12A. While the relative time of origin of some of these occurrents can be contested, it is clear that they are all currently persisting entities. In addition, more granular occurrents also show a similar tendency of persistence increase, especially if they represent biological modules. For example, the motifs, loops, domains, proteins, complexes, and proteomes are modular occurrents of the protein world that clearly follow the evolutionary pattern of increase established by their relative origin (Figure 12B). The tree of structural domains that was used to study the evolution of the proteome world (e.g. Nasir and Caetano-Anollés, 2015) and the examples of molecular recruitment described above (Figures 6–10) show the gradual accumulation of innovations in protein evolution, from the origin of proteins to the present. Each and every new domain is a long-lived occurrent that is currently extending its persistence by extending its life. The same can be said of organisms evolving along the branches of a "tree of life." Cellular organisms, eukaryotes, vertebrates, mammals, hominoids, and *Homo sapiens* are all long-lived occurrents. Collective groups that are currently extending their life in our planet. This fact arises by the existence of "vertical" processes of change, as those described above. Of course, occurrents are also subjected to the effects of "horizontal" processes. Domains for example are shuffled in proteomes by recruitment and other complicated processes that are occurring at different temporal and spatial scales. In the case of Figure 12B, the hierarchical structure of proteins constrains the dynamics of long-lived structural occurrents. Entanglements of motifs and loops impinge on the entanglements of domains and proteomes, and vice versa, through top-down causation dynamics. The horizontal processes of genetic exchange occurring in the microbiome (Figure 11) occur not only at different timescales but also involve physical scales. Thus, occurrents must be considered highly dynamic patchworks of temporal parts.

The granularity of occurrents also increases with time (Figure 12), as we have recently made evident when analyzing the network structure of metabolism (Mughal and Caetano-Anollés, 2019). Granularity is here considered a proxy to the structuring of an evolving biological system, with structure often abstracted with networks and measured with indices of modularity and hierarchy (Caetano-Anollés *et al.*, 2019, 2020). Older occurrent are more granular, as causal relationships and entanglements with younger occurrents constrain their structural space-time organization.

The gain and loss of domains in proteomes illustrated in Figure 9 raises the issue of "extinction." Domains defining the proteome of an organismal lineage of the "tree of life" can be lost at some point in evolution. Similarly, that same lineage can also be lost, leaving behind perhaps fossil evidence of its evolutionary existence; e.g. a lineage of the Tyrannosauridae that existed 90-to-66 million years ago (Fiorillo and Tykoski, 2014). However, the concept of extinction is irrelevant for occurrents, since perdurantists consider that temporal parts of an occurrent are sets of B-series events that are "tenseless" (eternal). An occurrent exists in the past, present and future. The loss of a domain or a lineage does not compromise the existence of the evolving proteome or lineage. The gain and loss of domains also challenges the definition of a domain occurrent. A specific domain may be gained and lost in the proteomes of multiple lineages as the tree of life unfolds. This specific domain occurrent, broadly defined, is a series of temporal parts manifesting in the universe of evolving proteomes as a patchwork. While the ultimate loss of that domain occurrent in the proteome of a lineage could signal the end of its life in that lineage, the patchwork is a long-lived collective. Cuvier's "form" of a protein domain, embedded in the folding design of proteins, is "more essential to it that its matter," i.e. that its makeup "up and down" scales of organization. Its form is evolutionarily highly conserved, carrying billions of years of history, but also extended in space as a patchwork, likely throughout the planet.

We end by noting that all occurrents described in Figure 12 involve temporal parts of the past and the present. It does not include temporal parts that will appear in the future. We consider that levels of organization will continue to unfold in evolution, even in absence of a planet or a solar system. Human foresight is limited and may not envision how future entanglements prompted by life will manifest in the universe.

11. Conclusions

A "theory of entanglement" can be used to explain causal relations responsible for the increasingly extended and complex temporal parts of biological systems. Genomic data supports the view that biological occurrents are patchworks of entangled temporal parts as well as temporal parts of the entangled make up of higher-level wholes. Patchworks can be understood as vast networks of causally related temporal parts reflecting continual change. This change can be conceptualized as transitive and pluralistic causal chains of influence explained for example with causal graphs and counterfactual conditionals. Genomic data also supports the rise of increasingly long-lived occurrents that are increasingly more granular and abundant. The growth of temporal parts cannot be contested. It spans the overarching cosmos and the invisible world of baryonic matter. Galaxies arise from processes of recombination, condensation, gravitational collapse and accumulation that form stars, neutron stars, black holes, planets, comets and asteroids. The living world arises by the increasingly complex accumulation of biological innovations, which multiplies degrees of freedom and entanglement and furthers additional growth and innovation. Accretion of macromolecules occurs at different levels of complexity. Molecules grow by adding molecular parts or interacting with other molecules to form molecular complexes within the confines of the cell. Molecules also facilitate interactions between cells within the confines of organisms. Accretion is accompanied by both diversification and innovation, triggered mostly by diffusion in spaces of possible outcomes (e.g. diffusion of mutations in the sequence and structure space of macromolecules; Fontana, 2002; Caetano-Anollés, 2010). Growth and innovation are constrained by the formation of modules (Mittenthal *et al.*, 2012; Caetano-Anollés *et al.*, 2020). The associated processes of growth and innovation determine the behavior of dynamic systems, with time series often modeled with logistic wavelets, S-shaped logistic components of growth and diffusion in systems (Meyer *et al.*, 1999). The perdurantist paradigm of temporal parts now prompts the extension of these models to the space-time framework.

References

Ancel L, Fontana W (2000) Plasticity, evolvability and modularity in RNA. *J Exp Zool (Mol Dev Evol)* **288**: 242–83.

Aziz MF, Caetano-Anolles K, Caetano-Anollés G (2016) The early history and emergence of molecular functions and modular scale-free network behavior. *Sci Rep* **6**: 25058.

Bai Y, Sosnick TR, Mayne L, Englander SW (1995) Protein folding intermediates: Native-state hydrogen exchange. *Science* **269**: 192–197.

Balashov Y (2011) Persistence. In: *The Oxford Handbook of Philosophy of Time*, Callender C (ed). Oxford University Press, Oxford.

Baldwin RL (1995) The nature of protein folding pathways: the classical versus the new view. *J Biomol NMR* **5(2)**:103–109.

Bassett DS, Greenfield DL, Meyer-Lindenberg A, Weinberger DR, Moore SW, Bullmore ET (2010) Efficient physical embedding of topologically complex information processing networks in brains and computer circuits. *PLoS Comput Biol* **6(4)**: e1000748.

Bardeen JM, Carter B, Hawking S (1973) The four laws of black hole mechanics. *Commun Math Phys* **31**: 161–170.

Bekenstein JD (1973) Black holes and entropy. *Phys Rev* **D7**: 2333–2346.

Besseova I, Reblova K, Leontis NB, Sponer J (2010) Molecular dynamics simulations suggest that RNA three-way junctions can act as flexible RNA structural elements in the ribosome. *Nucleic Acids Res* **38**: 6247–6264.

Brower MM, Visser MR, Dvornik A *et al.* (2016) First test of Verlinde's theory of emergent gravity using gravitational lensing measurements. *Monthly Notices Roy Astronomical Soc* **466(3)**: 2547–2559.

Caetano-Anollés D, Caetano- Anollés G (2020) Molecular accretion, growth and innovation. In: *Untangling Molecular Biodiversity*, Caetano-Anollés G (ed). World Scientific Publishing, Singapore.

Caetano-Anollés D, Caetano-Anollés G (2017) Commentary: history of the ribosome and the origin of translation. *Front Mol Biosci* **3**: 87.

Caetano-Anollés G (2002) Tracing the evolution of RNA structure in ribosomes. *Nucleic Acids Res* **30(11)**: 2575–2587.

Caetano-Anollés G, Caetano-Anollés K (2013) Structural phylogenomics reveals-gradual evolutionary replacement of abiotic chemistries by protein enzymes in purine metabolism. *PLoS One* **8**: e59300.

Caetano-Anollés G, Caetano-Anollés D (2015) Computing the origin and evolution of the ribosome from its structure — Uncovering processes of macromolecular accretion benefitting synthetic biology. *Comp Struct Biotechnol J* **13**: 427–447.

Caetano-Anollés G, Kim HS, Mittenthal JE (2007). The origin of modern metabolic networks inferred from phylogenomic analysis of protein architecture. *Proc Natl Acad Sci USA* **104**: 9358–9363.

Caetano-Anollés G, Yafremava L, Mittenthal JE (2010) Modularity and dissipation in evolution of macromolecular structures, functions, and networks. In: *Evolutionary Bioinformatics and Systems Biology*, Caetano-Anollés G (ed). Wiley-Blackwell, Hoboken, NJ, USA, pp. 443–449.

Caetano-Anollés G, Nasir A, Kim KM, Caetano-Anollés D (2018) Rooting phylogenies and the Tree of Life while minimizing *ad hoc* and auxiliary assumptions. *Evol Bioinfo* **14**: 1176934318805101.

Caetano-Anollés G, Aziz FM, Mughal F, Gräter F, Koç I, Caetano-Anollés K, Caetano-Anollés D (2019) Emergence of hierarchical modularity in evolving networks uncovered by phylogenomic analysis. *Evol Bioinformatics* **15**: 1–18.

Caetano-Anollés G, Mughal F, Aziz MF, Koç I, Caetano-Anollés K, Caetano-Anollés D, Mittenthal JE (2020) A 'double tale' of module creation in evolving networks. In: *Untangling Molecular Biodiversity*, Caetano-Anollés G (ed). World Scientific Publishing, Singapore.

Calhoun TR, Ginsberg NS, Schlau-Cohen GS, Cheng Y-C, Ballotari M, Bassi R, Fleming GR (2009) Quantum coherence enabled determination of the energy landscape in light-harvesting complex II. *J Phys Chem B* **113**: 16291–95.

Carroll SM, Remmen GN (2016) What is the entropy in entropic gravity? *Phys Rev D* **93**, 124052.

Castillo DJ, Rifkin RF, Cowan DA, Potgieter M (2019) The healthy human blood microbiome: Fact or fiction. Front Cell Infect *Microbiol* **9**: 148.

Cha Y, Murray CJ, Klinman JP (1989) Hydrogen tunneling in enzyme reactions. *Science* **243**: 1325–30.

Chisholm R (1970) Identity through time. In: *Language, Belief and Metaphysics*, Keifer H, Munitz M, (eds). SUNY Press, Albany.

Collier J (2003) Hierarchical dynamical information systems with a focus on biology. *Entropy* **5**: 100–124.

Collini E, Wong CY, Wilk KE, Curmi PMG, Brumer P, Scholes GD (2010) Coherently wired light-harvesting in photosynthetic marine algae at ambient temperature. *Nature* **463**: 644–647.

Consortium HMP (2012) A framework for human microbiome research. *Nature* **486**: 215–221.

Cuvier G (1833) *The Animal Kingdom in Conformity with its Organization.* G & C & H Carvill, New York.

Dupont CL, Butcher A, Valas RE, Bourne PE, Caetano-Anollés G (2010). History of biological metal utilization inferred through phylogenomic analysis of protein structure. *Proc Natl Acad Sci USA* **107(23)**: 10567–10572.

Engel GS, Calhoun TR, Read EL, Ahn T-K, Mancal T, Cheng Y-C, Blankenship RE, Fleming GR (2007) Evidence for wavelike energy transfer through quantum coherence in photosynthetic systems. *Nature* **446**: 782–86.

Englander SW, Mayne L (2014) The nature of folding pathways. *Proc Natl Acad Sci USA* **111**: 15873–15880.

Effingham N (2012) *Endurance and Perdurance. The Continuum Companion to Metaphysics*, Manson NA, (ed). Barnard RW). Bloomsbury, London.

Fang X, Pan T, Sosnick TR (1999) Mg2+-dependent folding of a large ribozyme without kinetic traps. *Nat Struct Biol* **6**: 1091–1095.

Fiorillo AR, Tykoski RS (2014) A diminutive new Tyrannosaur from the top of the world. *PLoS One* **9**: e91287.

Fontana W (2002) Modelling 'evo–devo' with RNA. *Bioessays* **24**: 1164–77.

Frege G (1879) Begriffsschrift, eine der arithmetischen nachgebildete Formelsprache des reinen Denkens (translated by S. Bauer-Mengelberg S). In: *From Frege to Gödel: A Source Book in Mathematical Logic,* vanHeijenoort IJ (ed). Harvard University Press, Cambridge, MA, 1967, pp. 1879–1931.

Galois A (2016) Identity over time. In: Zalta EN (ed.), The Stanford Encyclopedia of Philosophy (Winter 2016 Edition), URL = https://plato.stanford.edu/archives/win2016/entries/identity-time/

Geach PT (1967) Identity. *Rev Metaphysics* **21**: 3–12.

Goodrich JK, Davenport ER, Clarck AG, Ley LE (2017) The relationship between the human genome and microbiome comes into view. *Ann Rev Genet* **51**: 413–433.

Grant T (2002) Testing methods: the evaluation of discovery operations in evolutionary biology. *Cladistics* **18**: 94–111.

Guimerà R, Amaral LAN (2005) Functional cartography of complex metabolic networks. *Nature* **433**: 895–900.

Guseo R (2016) Diffusion of innovations dynamics, biological growth and catenary function. *Physica A* **464**: 1–10.

Harish A, Caetano-Anollés G (2012) Ribosomal history reveals the origin of modern protein synthesis. *PLoS One* **7**: e32776.

Haslanger S (2003) Persistence through time. In The Oxford Handbook of Metaphysics, Loux M, Zimmerman D (eds). Oxford University Press, Oxford, pp. 315–354.

Hawking S (1976) Black holes and thermodynamics. *Phys Rev* **D13**: 191–197.

Hawley K (2018) Temporal parts. In: *The Stanford Encyclopedia of Philosophy*, Spring edition, Zalta EN (ed.). URL = https://plato.stanford.edu/archives/spr2018/entries/temporal-parts/

Hennig W (1966) *Phylogenetic Systematics.* University of Illinois Press, Urbana, USA.

Ho JJ, Navlakha S (2018) Evidence of Rentian scaling of functional modules in diverse biological networks. *Neural Comput* **30(8)**: 2210–2244.

Hobbes T (1839) English Works of Thomas Hobbes of Malmesbury, Vol. 1, Molesworth W (ed). London, Bohn, p. 136.

Hohng S, Wilson TJ, Tan E, Clegg RM, Lilley DM, Ha T (2004) Conformational flexibility of four-way junctions in RNA. *J Mol Biol* **336**: 69–79.

Horowitz NH (1945) On the evolution of biochemical syntheses. *Proc Natl Acad Sci USA* **31**: 153–7.

Hu W, Walters BT, Kan ZY, Mayne L, Rosen LE, Marqusee S, Englander SW (2013) Stepwise protein folding at near amino acid resolution by hydrogen exchange and mass spectrometry. *Proc Natl Acad Sci USA* **110**: 7684–7689.

Hull DL (1978) A matter of individuality. *Phil Sci* **45(3)**: 335–360.

Hyeon C, Thirumalai D (2012) Chain length determines the folding rates of RNA. *Biopys J* **102**: L11–13.

Jacobson T (1995) Thermodynamics of spacetime: The Einstein equation of state. *Phys Rev Lett* **75(7)**: 1260–1263.

Jensen RA (1976) Enzyme recruitment in evolution of new function. *Annu Rev Microbiol* **30**: 409–425.

Jeong H, Sung S, Kwon T, Seo M, Caetano-Anollés K, Choi SH, Choi S, Nasir A, Kim H (2016) HGTree: database of horizontally transferred genes determined by tree reconciliation. *Nucleic Acids Res*. **44**: D610–619.

Jeong H, Arif B, Caetano-Anollés A, Kim KM, Nasir A (2019) Horizontal gene transfer in human-associated microorganisms inferred by phylogenetic reconstruction and reconciliation. *Sci Rep* **9**: 12173.

Kacser H, Beeby R (1984) On the origin of enzyme species by means of natural selection. *J Mol Evol* **20**: 38–51.

Kanehisa M, Furumichi M, Tanabe M, Sato Y, Morishima K (2017) KEGG: new perspectives on genomes, pathways, diseases and drugs. *Nucleic Acids Res* **45**: D353–D361.

Kim HS, Mittenthal JE, Caetano-Anollés G (2006) MANET: tracing evolution of protein architecture in metabolic networks. *BMC Bioinformatics* **7**: 351.

Klein DJ, Moore PB, Steitz TA (2004) The roles of ribosomal proteins in the structure assembly, and evolution of the large ribosomal subunit. *J Mol Biol* **340**: 141–177.

Klinman JP, Kohen A (2013) Hydrogen tunneling links protein dynamics to enzyme catalysis. *Anny Rev Biochem* **82**: 471–496.

Koç I, Caetano-Anollés G (2017) A natural history of molecular functions inferred from an extensive phylogenomic analysis of Gene Ontology data. *PLoS ONE* **12(5)**: e0176129.

Laing C, Schlick T (2009a) Analysis of four-way junctions in RNA structures. *J Mol Biol* **390**: 547–559.

Laing C, Schlick T (2009b) Tertiary motifs revealed in analyses of higher-order RNA junctions. *J Mol Biol* **393**: 67–82.

Leibniz GW (1937) New essays concerning human understanding, p. 247. Open Court, LaSalle, Illinois. Leonard HS, Goodman N (1940) The calculus of individuals and its uses. *J Symbolic Logic* **5**: 45–55.

Leontis NB, Westhof E (2001) Geometric nomenclature and classification of RNA base pairs. *RNA* **7**: 499–512.

Leopold PE, Montal M, Onuchic JN (1992) Protein folding funnels: a kinetic approach to the sequence-structure relationship. *Proc Natl Acad Sci USA* **89**: 8721–8725.

Lescoute A, Westhof E (2006) Topology of three-way junctions in folded RNAs. *RNA* **12**: 83–93.

Leśniewski S (1916) Foundations of the general theory of sets. *Prace Polskiego Koła Naukowego w Moskwie*, Sekcya matematyczno-przyrodnicza.

Lewis D (1973) Causation. *J Philosophy* **70**: 556–567.

Lewis D (1986) *On the Plurality of Worlds*. Blackwell, Oxford.

Lewis D (2000) Causation as influence. *J Philosophy* **97**: 182–197.

Li PTX, Vieregg J, Tinoco Jr I (2008) How RNA unfolds and refolds. *Annu Rev Biochem* **77**: 77–100.

Love AC (2012) Hierarchy, causation and explanation: ubiquity, locality and pluralism. *Interface Focus* **2**: 115–125.

Ma H-W, Zeng A-P (2003) The connectivity structure, giant strong component and centrality of metabolic networks. *Bioinformatics* **19(11)**: 1423–1430.

Mainzer LS, Mittenthal JE, Caetano-Anollés G (2020) A model of lineage evolution based on the persistence strategy hypothesis. In: *Untangling Molecular Biodiversity*, Caetano-Anollés G (ed). World Scientific Publishing, Singapore.

Maldacena JM (1998) The large N-limit of superconformal field theories and supergravity. *Adv Theor Math Phys* **2**: 231–252.

Marais A, Adams B, Ringsmuth AK, Ferreti M, Michael Gruber J, Hendrkx R, Schuld M, Smith SL, Sinayskiy I, Krüger TPJ, Petruccione F, van Grondelle R (2018) The future of quantum biology. *J R Soc Interface* **15**: 20180640.

McCall S, Lowe EJ (2006) The 3D/4D controversy: A storm in a teacup. *Noûs* **40**: 570–578.

McGaugh SS, Lelli F, Schombert JM (2016) Radial acceleration relation in rotationally supported galaxies. *Phy Rev Let* **117**: 201101.

McTaggart JE (1908) The unreality of time. Mind: *Quaterly Rev Psychol Phil* **17**: 456–473.

Meyer PS, Yung JW, Ausubel JH (1999) A primer on logistic growth and substitution. The mathematics of the Loglet lab software. Tech Forecasting *Social Change* **61**: 247–271.

Michaels M (1980) The persisting problem of the ship of Theseus. Doctoral dissertation 1896. https://scholarworks.umass.edu/dissertations_1/2122.

Mittenthal J, Caetano-Anollés D, Caetano-Anollés G (2012) Biphasic patterns of diversification and the emergence of modules. *Front Genet* **3**: 147.

Mughal F, Gräter F, Caetano-Anollés G (2017) How function shapes dynamics in protein evolution. In: *Blue Waters Annual Report. National Center for Supercomputer Applications*, Szuch S, Watkins C (eds). Champaign, IL, pp. 198–199.

Mughal F, Caetano-Anollés G (2019) MANET 3.**0**: Hierarchy and modularity in evolving metabolic networks. *PLoS One* **14**: e0224201.

Mughal F, Nasir A, Caetano-Anollés G (2020) The roigin and evolution of viruses inferred from fold family structure. *Arch Virol* https://doi.org/10.1007/s00705-020-04724-1

Murzin A, Brenner SE, Hubbard T, Clothia C (1995) SCOP: a structural classification of proteins for the investigation of sequences and structures. *J Mol Biol* **247**: 536–540.

Nasir A, Kim KM, Caetano-Anollés G (2014) Global patterns of protein domain gain and loss in superkingdoms. *PLoS Comput Biol* **10**: e1003452.

Nasir A, Caetano-Anollés G (2015) A phylogenomic data-driven exploration of viral origins and evolution. *Science Advances* **1**: e1500527.

Nerlich G (1979) Is curvature intrinsic to physical space. *Phil Sci* **46**: 439–458.

Newman MEJ, Girvan M (2004) Finding and evaluating community structure in networks. *Phys Rev E* **69**: 26113.

Onuchic JN., Luthey-Schulten Z, Wolynes PHG (1997) Theory of protein folding: the energy landscape perspective. *Annu Rev Phys Chem* **48**: 545–600.

Orengo CA, Michie A, Jones S, Jones DT, Swindells M, Thornton JM (1997) CATH–a hierarchic classification of protein domain structures. *Structure* **5**: 1093–1109.

Panitchayangkoon G, Hayes D, Fransted KA, Caram JR, Harel E, Wen J, Blankenship RE, Engel GS (2010) Long-lived quantum coherence in photosynthetic complexes at physiological temperature. *Proc Natl Acad Sci USA* **107**: 12766–12770.

Pearl J (2000) *Causality*. Cambridge University Press, Cambridge.

Raposo G, Stoovogel W (2013) Extracellular vesicles: Exosomes, microvesicles and friends. *J Cell Biol* **200**: 373–383.

Rescher N (1955) Axioms for the part relation. *Phil Studies* **6**: 8–11.

Reydon TAC (2008) Species in three and four dimensions. *Synthese* **164**: 161–184.

Ritz T, Thalau P, Phillips JB, Wiltschko R, Wiltschko W (2004) Resonance effects indicate a radical-pair mechanism for avian magnetic compass. *Nature* **429**: 177–180.

Sasnauskas G, Connolly BA, Halford SE, Siksnys V (2007) Site-specific DNA transesterification catalyzed by a restriction enzyme. *Proc Natl Acad Sci USA* **104(7)**: 2115–2120.

Schaffer J (2016) The metaphysics of causation. In: *The Stanford Encyclopedia of Philosophy*, Fall 2016 edition, Zalta EN (ed.). URL = https://plato.stanford.edu/archives/fall2016/entries/causation-metaphysics/

Schulten K, Swenberg CE, Weller A (1978) Biomagnetic sensory mechanism based on magnetic field modulated coherent electron spin motion. *Z Phys Chem* **111**: 1–5.

Schuster P (2010) Genotypes and phenotypes in the evolution of molecules. In: *Evolutionary Bioinformatics and Systems Biology*, Caetano-Anollés G (ed). Wiley-Blackwell, Hoboken, NJ, USA, pp. 123–152.

Shpitser I, Pearl J (2008) Complete identification methods for the causal hierarchy. *J Machine Learning Res* **9**: 1941–1979.

Simon HA (1962) The architecture of complexity. *Proc Am Phil Soc* **106**: 467–482.

Teichmann SA, Rison SCG, Thornton JM, Riley M, Gough J, Chothia C (2001) Small-molecule metabolism: an enzyme mosaic. *Trends Biotechnol* **19**: 482–486.

Thomson JJ (1983) Parthood and identity across time. *J Phil* **80**: 201–220.

Van Inwagen P (1988) How to reason about vague objects. *Phil Topics* **16**: 255–284.

Van Raamsdonk M (2016) Lectures on gravity and entanglement. *Proceedings* C15–06-01.**2**: 297–351.

Varzi A (2019) Mereology. In: *The Stanford Encyclopedia of Philosophy*, Spring 2019 edition, Zalta EN (ed). URL = https://plato.stanford.edu/archives/spr2019/entries/mereology/

Verlinde E (2011) On the origin of gravity and the laws of Newton. *J High Energy Phys* 029 doi:10.1007/JHEP03(2014)051.

Verlinde E (2017) Emergent gravity and the dark universe. *SciPost Phys* **2**: 016.

Wang M, Caetano-Anolles G (2009). The evolutionary mechanics of domain organization in proteomes and the rise of modularity in the protein world. *Structure* **17**: 66–78.

Wang M, Yafremava LS, Caetano-Anollés D, Mittenthal JE, Caetano- Anollés G (2007) Reductive evolution of architectural repertoires in proteomes and the birth of the tripartite world. *Genome Res* **17**: 1572–1585.

Wang M, Jiang Y-Y, Kim KM, Qu G, Ji H-F, Mittenthal JE, Zhang H-Y, Caetano-Anollés, G (2011) A universal molecular clock of protein folds and its power in tracing the early history of aerobic metabolism and planet oxygenation. *Mol Biol Evol* **28**: 567–582.

Wasserman R (2006) The problem of change. *Philos Compass* **1**: 48–57.

Wiggins D (2001) *Sameness and Substance Renewed.* Cambridge University Press, Cambridge.

Windelband W (1894) *Geschichte und Naturwissenschaft.* Heitz, Strassburg.

Wright S (1921) Correlation and causation. *J Agric Res* **20**: 557–585.

Ycas M (1974) On earlier states of the biochemical system. *J Theor Biol* **44**: 145–60.

Empedocles' *On Nature*, *P. Strasb. Gr.* Inv. 1665–6, a theory of networks and evolutionary growth ~2,400 years before Darwin

G. Caetano-Anollés[1]
R. Janko[2]

Summary. The Strasbourg papyrus fragments of the Upper Egyptian city of Panopolis are puzzling. The writings, which have been attributed to Empedocles by direct transmission, describe a process of evolutionary growth. Here we dissect translations of this ~2,000-year old hexameter text and confirm that it describes both a theory of evolution by natural selection akin to that of Darwin and the existence of lineages in an evolving network of life. However, the theory appears more profound and ancient. It presents network-like processes of unification and diversification that explain biological accretion and growth. The theory is compatible with knowledge from modern biology. It also matches a recent model of biological module creation that is supported by the phylogenomic analysis of the structure

[1]*Department of Crop Sciences, University of Illinois, Urbana, Illinois, USA*
[2]*Department of Classical Studies, University of Michigan, Ann Arbor, Michigan, USA*

of millions of protein and nucleic acid molecules. The ancient text shows remarkable foresight in biological prediction.

1. Introduction

The Strasbourg papyrus fragments, *P. Strasb. Gr.* Inv. 1665–6, which are preserved at the National University Library of Strasbourg, date back to a first-century A.D. roll from the ancient city of Panopolis in Upper Egypt. In 1992, Alain Martin attributed the embedded text to Empedocles of Akragas (*ca.* 495-435 BC), since its Homeric hexameter verses largely matched the *On Nature* poem of the *Physica* Book I quoted by Simplicius (*ca.* 490-560 AD), the Neoplatonist commentator on Aristotle. The finding was made public in 1994. Figure 1 shows a reconstructed ensemble of 24 fragments (ensemble **a**), the largest of all ensembles reconstructed out of a total 52 fragments. The translated text is puzzling and of great significance. With one exception, the text represents the first writings of a Presocratic philosopher preserved on a papyrus. The document embodies a direct transmission of philosophical poetry and a cosmological theory largely untainted from corruptions that occur over time via scribal errors, repeated quotations or interpretations. Since the vast majority of Presocratic philosophical writings (as determined by Hermann Diels) have been lost, such corruptions could affect the doxography of ancient philosophy (Mansfeld, 2016). Remarkably, the document also reflects the existence of evolutionary thought more than two millennia before Darwin, an idea that was already elaborated more than half a century ago by Zirkle (1941). Here we discuss the biological significance of *P. Strasb. Gr.* Inv. 1665–6 under this new light. We claim that the main thesis of this ancient text represents a hypothesis of *evolutionary growth* that is modern. This hypothesis is compatible with recent insights from

εἴ τε γὰρ ἐφθείροντο διαμπερέϲ, οὐκ ἂν ἔτ' ἦϲαν.
τοῦτο δ' ἐπαυξήϲειε τὸ πᾶν τί κε, καὶ πόθεν ἐλθόν;
πῆι δὲ κε κἀξαπόλοιτο, ἐπεὶ τῶνδ' οὐδὲν ἐρῆμον;
ἀλλ' αὔτ' ἔϲτιν ταῦτα· δι' ἀλλήλων γε θέοντα
γίγνεται ἄλλοτε ἄλλα καὶ ἠνεκὲϲ αἰὲν ὁμοῖα.
ἀλλ' ἐν μὲν Φιλότητι ϲυνερχό]μεθ' εἰϲ ἕνα κόϲμον,
ἐν δ' Ἔχθρηι γε πάλιν διέφυ πλέ]ον' ἐξ ἑνὸϲ εἶναι,
ἐξ ὧν πάνθ' ὅϲα τ' ἦν ὅϲα τ' ἔϲθ' ὅϲα τ' ἔϲϲετ' ὀπίϲϲω,
δένδρεα τ' ἐβλάϲτηϲε καὶ ἀνέρεϲ ἠδὲ γυναῖκεϲ, ||
θῆρέϲ τ' οἰωνοί τε καὶ ὑδατοθρέμμονεϲ ἰχθῦϲ
καί τε θεοὶ δολιχαίωνεϲ τιμῆιϲ[ι φέριϲτοι.
ἐ]ν τῆι δ' ἀίϲϲοντα [διαμπ]ερὲϲ οὐδ[αμὰ λήγει
π]υκνῆιϲιν δίνηιϲ[ι ⏑ | – ⏑ ⏑ | – ⏑ ⏑]τ.[– –
ν]ωλεμέϲ, οὐδέ πο[τ – ⏑ ⏑ | – ⏑ ⏑ | – ⏑ ⏑ | – – ,
πολλ]οὶ δ' αἰῶνεϲ πρότερ[οι ⏑ ⏑ | – ⏑ ⏑ | – – ,
πρὶν] τούτων μεταβῆνα[ι ⏑ | – ⏑ ⏑ | – ⏑ ⏑ | – – ·
πά]ντηι δ' ἀίϲϲον[τ]α διαμ[περὲϲ οὐδαμὰ λήγει.
οὔ]τε γὰρ ἠέλιοϲ τ[. .(.)].ν.[⏑ ⏑ , οὔτε ϲελήνη
ὁρ]μῆ⟨ι⟩ τῆιδε γέμου[ϲ' ἠδὲ φθινύθουϲ' ἀπολήγει·
οὔ]τε τι τῶν ἄλλων [μένει ἔμπεδον οὐρανῶι ἄϲτρων,
ἀλ]λὰ μεταλλάϲϲον[τ' ἀίϲϲ]ει κύκλωι [ἅπαντα.
δὴ τό]τε μὲν γὰρ γαῖ' ἀ[β]άτη θέει ἠελ[ίου τε
ϲφαῖρα], τόϲην δὴ κα[ὶ ν]ῦν ἐπ' ἀνδράϲι τ[εκμήραϲθαι·
ὣϲ δ' α]ὕτωϲ τάδ[ε π]άντα δι' ἀλλήλων [προθέουϲιν,
κἄλλο]υϲ τ' ἄλλ' [ἵκα]νε τόπουϲ πλαγ[χθέντα καὶ ἄλλουϲ·
οὔ τοι δὴ] μεϲάτουϲ τ' [εἰϲέ]ρχόμεθ' ἓν μ[όνον εἶναι.
ἀλλ' ὅτ]ε δὴ Νεῖκόϲ [τ' ἀνυ]πέρβατα βέν[θε' ἵκηται
δ[ίνη]ϲ, ἐν δὲ μέϲ[ηι] Φ[ιλ]ότηϲ ϲτροφά[λιγγι γένηται,
ἐν [τῆι] δὴ τάδε πάντα ϲυνέρχεται ἓν [μόνον εἶναι.
  ϲπεῦ]δε δ' ὅπωϲ μὴ μοῦνον ἀν' οὔατα [μῦθοϲ ἵκηται,
ἠδέ] μευ ἀμφὶϲ ἐόντα κλύων [ν]ημερτ[έα φράζευ·
δεί]ξω ϲοι καὶ ἀν' ὄϲϲ' ἵνα μείζονι ϲώμ[ατι κύρει
π]ρῶτον μὲν ξύνοδόν τε διάπτυξίν τ[ε γενέθληϲ,
ὅϲ[ϲ]α τε νῦν ἔτι λοιπὰ πέλει τούτοιο τ[όκοιο·
τοῦτο μὲν [ἂν] θηρῶν ὀριπλάγκτων ἄγ[ρια φῦλα,
τοῦτο δ' ἀ[ν' ἀνθρώ]πων δίδυμον φύμα, [τοῦτο δ' ἀν' ἄνθεων
ῥιζοφόρων γέννημα καὶ ἀμπελοβάμ[ονα βότρυν.
  ἐκ τῶν ἀψευδῆ κόμιϲαι φρενὶ δείγματα μ[ύθων·
ὄψει γὰρ ξύνοδόν τε διάπτυξίν τε γενέθλη[ϲ ||

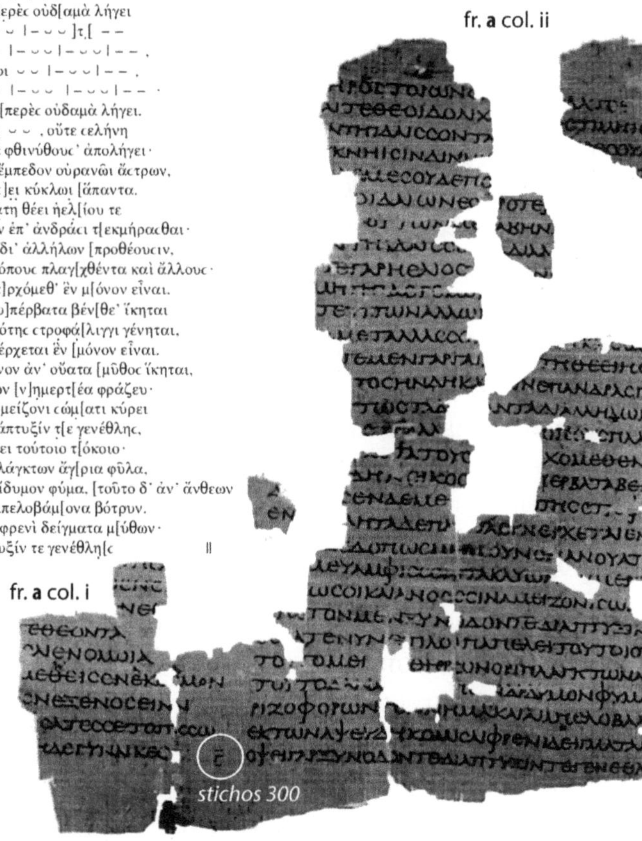

Figure 1. Papyrus of Empedocles, *Physika I*, lines 262–300. *P. Strasb. gr. Inv. 1665–6.* Scan of a reconstruction of an ancient Greek papyrus from fragments rendering *Physica* I book, lines 267–300. The scribal stichometric sign Γ is of significance as it indicates that the line corresponds to v. 300 of *Physika* Book I restored. The Greek text of the ensemble is also shown.

molecular evolution and genomics (Caetano-Anollés *et al.*, 2018a, 2019). Finally, we use *P. Strasb. Gr.* Inv. 1665–6 as remarkable evidence of memory and foresight, properties that have profound consequences for evolution of life in our planet and our future as a species.

2. The editions of the Strasbourg fragments

The first edition (*editio princeps*) of the Strasbourg fragments was produced by Martin and Primavesi (1999). Later on, Janko (2004) completed the reconstruction and produced an edition that is both accurate and poetic. Its accuracy is reflected in a more recent edition by Laks and Most (2016). The translation of Janko is presented in Figure 2 and will be the central translated text that is analyzed in the present study. The rationale for choosing Janko's reconstruction over that of the original editors is that it gives more weight to the papyrological than to the philosophical and doxographic evidence (Janko, 2016).

3. Biological interpretation

Janko (2004) divided the line of arguments from his fragments and textual reconstructions into segments describing main theses (T), exhortations (E), reprises (R), proofs (P), promises (Pm), digressions (D), and fulfillment of promises (F) (Figure 2). Here we index this outline with biological and evolutionary interpretations using double and single quotation marks for translated text from Janko (J) and Laks and Most (LM), respectively. Lines are also indexed with Diels-Kranz (DK) fragment numbers. The three main theses of the Strasbourg papyrus (Figure 2) sequentially introduce important aspects of a theory of evolutionary growth, which we will argue has considerable explanatory power. The presentation of concepts is made gradually and "*accords with Empedocles' methods of gradual revelation and careful, step-by-step argument*"

T1 233 A double tale I'll tell. At one time one thing grew to be just one
from many, at another many grew from one to be apart.
235 Double the birth of mortal things, and double their demise.
Union of all begets as well as kills the first;
the second nurtures them but shatters as they grow apart.
And never do they cease from change continual,
at one time all uniting into one from Love,
240 while at another each is torn apart by hate-filled Strife
In the way that many arise as the one again dissolves,
in that respect they come to be and have no life eternal;
but in the way that never do they cease from change continual,
in this respect they live forever in steady cycle.
E 245 But hear my words; to learn augments the mind.
R For as I said when I set forth my story's aims,
a double tale I'll tell. At one time one thing grew to be just one
from many, at another many grew from one to be apart,
fire, water, earth and the unreached height of air,
250 and cursèd Strife apart from them, their match in every way,
and Love among them, equal in her size and in her breadth.
P With mind regard Her, and sit not with eyes bedazed.
Even mortals hold that She's implanted in their joints;
through Her they think of love and do conjoining deeds,
255 naming Her 'Delight' and 'Aphrodite' too.
No mortal man has learned that She revolves
E among these things; but hear from me this truthful tale.
T2 For all these things are equal and alike in age,
but each rules separate domains, each has its haunts
260 and lords it in its turn as time rolls round.
Beyond these nothing comes to be or perishes.
For if they died continually, no longer would they be.
What could increase this whole, and from what source?
How too could it be destroyed, since nothing lacks in these?
265 But these are what there is, and running through each other
they suffer change continual but always are alike.
T3 In Love we come together in one world;
in Hatred many grew from one to be apart,
whence all that was, and is, and shall at some time be
270 blossomed as trees, as men, as women too,
as beasts, as birds, as fish that water rears,
as well as gods who ages live and greatest honours have.
In Her they never cease to swirl in constant flux
with frequent whirlings . . . .
275 relentlessly, and never do they cease . . .
But many ages previous must elapse . . .
before their motions alter . . . .
They never cease in any way to swirl in constant flux.
The sun does not stay still, nor does the moon
280 in orbit cease to wax or cease to wane,
nor do the other stars stay fixed in heaven;
they all move in a cycle, changing place.
At that time earth untrodden runs and the sun's orb,
as big as even now men have the power to vouch;
285 just so do all these things through one another race,
and, roaming, visit other places constantly;
we do not reach the middle place in union.
But whensoever Hatred to the vortex' utmost depths
arrives, and Love arises in the whirlwind's midst,
290 in Her then all these things unite to be just one.
Strive so my words reach not your ears alone, E
and as you hear from me true facts take note: Pm1
I'll put before your eyes how She augments
with larger form life's union and increase,
295 and all that still remain of this creation,
first in the wild tribes of beasts that roam the hills,
then in the double race of man, then in the fruits
of rooted plants and grapes that mount the vine.
From these take in your mind proofs truthful of my words. E
300 You'll see life's union and increase, Pm2
because of Hate, wreak deeds of separation.
This first the mass of mortal limbs makes clear;
at one time we unite in Love with all
the limbs that bodies have when life is blossoming;
305 another time again, split up by evil Strife,
they wander each apart amid life's breakers.
Just so it is for plants, and water-dwelling fish, D
308 and beasts that sleep on hills, and fowls that mount on wings.
. . . . . .
330 . . . . that at life's end the limbs of all
fall from each other apart and meet their fate,
much though we will it not, from grim necessity,
as we decay. Though beauteous Love may hold us now,
the harpies with death's lots will come for us.
335 Woe that some ruthless day did not destroy me first,
before I used my claws to wreak dread deeds for food!
Now to no end my cheeks I wet with tears.
For we shall reach the unplumbed gyre, I fear,
and though men wish it not they'll have uncounted pains.
340 But we'll embark once more upon our tale. F
When once the tireless flame did chance upon
all things, and caused their painful intermixture,
then creatures too progenitive were born
in every way, whose remnants still the dawn beholds.
345 When aither mixed did reach the utmost edge,
then birds flew up with shrieks and cries
tumultuous, but beasts, whose lot is lairs in earth
and grass, were born where earth was swathed around.
As when . . . . ,
350 a smith . . . .
. . . .
First in the whelks with heavy backs that graze the sea
355 and in their stony mantles, then in oysters –
there you will see earth dwelling on the top of flesh.
Again, the armour-plate of strong-backed sea-crabs,
also the stone-skinned conches' and the tortoises' shells,
and spears of hornèd deer that roam the hills.
360 But listing all such creatures I'd not end.
from which . . . .,
and this . . . .,
. . . .

Figure 2. Indexed translation of *On nature, P. Strasb. Gr. Inv.* 1665–6 (Janko, 2004). Lines are indexed with numbers and also arguments; T, main thesis; E, exhortation, R, reprise; P, proof; Pm, promise; and D, digression. The three theses of the papyrus are highlighted in grey and provide the grounds for fundamental philosophy. Please refer to Janko (2004) for the original Greek text of the translation.

(Janko, 2010). Lines 262–300 correspond to ensemble fragment (fr.) **a** of the papyrus (Figure 1), while lines 302–364 include fr. **c2, c5, c7, d2, d6, d11, d16, d19, b1** and **b5**, in that order (Janko, 2004). Interpretations are proposed within the context of current biological knowledge.

T1, lines 233–244 (DK fr. no. B 17.1–13): The main thesis of *P. Strasb. Gr.* Inv. 1665–6 is a central and generic evolutionary "*double tale*" of growing things: One comes from many, and many from one, because of Love (Φιλότης) and Strife (Νεῖκος).

> "*A double tale I'll tell. At one time one thing grew to be just one from many, at another many grew from one to be apart.*" (J)

The two "*tales*" describe two opposing forces of evolutionary growth, one that unifies (Love) and the other that diversifies (Strife). Interpretations of lines 233–234 coincide in that both unifying and diversifying things "*grow*." Furthermore, the encyclopedic translation "*Twofold is what I shall say: for at one time they grew to be only one out of many, at another time again they separate to be many out of one*" (LM) confirms that the second tale involves growing things apart into many distinct forms. Since growing apart implies unified things becoming separated by a distance in time and space, one very likely interpretation of the crucial statement of evolutionary growth is that it describes a process of diversification akin to variation in biological evolution. Line 235 provides substantial clarification:

> "*Double the birth of mortal things, and double their demise.*" (J)

which is congruent with the translation "*And double is the birth of mortal things, double their death*" (LM). First, line 235 makes explicit that the target of these two processes of growth are living entities, "*mortal things*," not "the elements" (fire, water, earth and air) as suggested by the LM translation. Second, the birth and death of things confirms that growth involves biological entities (parts and wholes), including organisms and populations (which is made clear later in the poetic composition). It also confirms that growth is not related to an organism but to the succession of organisms in a lineage. Third, line 235 describes living things growing and engaging in a remarkable balance of "*birth*" and "*demise*." This balance must embody natural selection and change, two crucial hallmarks of evolution. Natural selection requires gains through reproducing entities that "*double*" as they grow. The resulting growth must then be counterbalanced by differential loss when growth is limited by "*demise*," either as unproductive growth or death. Implicit in the process of growth is differential reproduction. Lines 236–237 further explicate how selection and change operate during unification and diversification:

> *"Union of all begets as well as kills the first;*
> *the second nurtures them but shatters as they grow apart."* (J)

The alternative translation "*For the one is both born and destroyed by the coming together of all things, while the other inversely, when they are separated, is nourished and flies apart*" (LM) is congruent. Unification "*begets*" growth but selects through competitive optimization (by killing "*the first*") the set of things that will be unified. It does not refer to heredity. Diversification "*nurtures*" growth but "*shatters*" in the process of change the identities of the growing living entities. Thus, natural selection and change act both during unification and diversification when spawn, nourishment and nurture counteract death, separation and shattering. Finally, lines 238–240 provide a temporal context for the opposing actions of growth and change:

> *"And never do they cease from change continual,*
> *at one time all uniting into one from Love,*
> *while at another each is torn apart by hate-filled Strife"* (J),

or alternatively, "*And these incessantly exchange their places continually, sometimes by love all combining together into one, sometimes again each one carried off by the hatred of Strife*" (LM). Thus, continual "*change*" or "*exchange*" drives innovation in a recurrent temporal succession of biological unification and diversification. This succession fosters the growth and change of living things. Lines 241–244 extends this dynamic view:

> *"In the way that many arise as the one again dissolves,*
> *in that respect they come to be and have no life eternal;*
> *but in the way that never do they cease from change continual,*
> *in this respect they live forever in steady cycle"* (J),

or, alternatively, "*Thus insofar as they have learned to grow as one out of many, and inversely, the one separating again, they end up being many, to that extent they become, and they do not have a steadfast lifetime; but insofar as they incessantly exchange their places continually, to that extent they always are, immobile in a circle*" (LM). Unification and diversification embed no "*eternal life*" or "*steadfast lifetime.*" Instead, their workings involve "*birth*" and "*demise.*" Individually, processes are ephemeral. When combined, however, their interplay endures forever in an "*immobile*" (ακίνητος) "*cycle*" (κύκλος), which denotes a circle. Thus, *P. Strasb. Gr.* Inv. 1665–6 highlights: (i) the immutability of the succession of two cosmological forces of biological growth (Love and Strife), (ii) their recurring struggle, (iii) the ephemeral nature of parts and wholes that are unified and diversified, and (iv) their continual growth.

Interpretative summary: Evolutionary growth is explained with a biphasic model. In a first phase, evolution brings together disparate parts to form bigger integrated wholes. During this unification process parts are selected for better performance. In the second phase, the unified wholes provide opportunities for change and innovation through a process of diversification. The biphasic model highlights properties of evolution: variation, struggle for existence, optimization, competition, natural selection and recurrent change and exchange. Parts and wholes can be anything. Remarkably, the theory of *P. Strasb. Gr.* Inv. 1665–6 is compatible with a biphasic "hourglass" model that explains the generation of biological structure (Mittenthal *et al.*, 2012), in which a succession of emergence of parts and wholes is responsible for the structure of molecules and molecular complexes, or organisms and populations (recently reviewed; Caetano-Anollés *et al.*, 2018a).

E1, line 245 (DK fr. no. B 17.14): First exhortation: Listen to me, as learning improves the mind.

R, lines 246–251 (DK fr. no. B 17.15–20): The first reprise links the "double tale" of Love and Strife to the four elements:

> *"... fire, water, earth and the unreached height of air,*
> *and cursed Strife apart from them, their match in every way,*
> *and Love among them, equal in her size and in her breadth."* (J)

or "... *fire, water, earth and the immense height of air; and hateful Strife is separate from them, equivalent everywhere, and Love in them, equal in length and in breadth*" (LM). In T1, living beings are the target of Love and Strife, not the elements. In R, the struggle of Love and Strife is placed within the context of elements of fire, water, earth and air, i.e. within an evolving planetary realm.

Interpretative summary: Chemical and biological explorations suggest life emerged in volcanic inhalations (Wachtershaüser, 2006, 2014), in deep vents of the ocean (Martin *et al.*, 2008) or at the surface of early Earth (Benner *et al.*, 2012; Mulkidjanian *et al.*, 2012). The pioneer organisms then conquered the oceans (e.g. metallome evidence; Dupont *et al.*, 2010)], and only then did they colonize land and air. Remarkably, if fire is equated with volcanism and earth with land, this succession follows the exact order of the elements of line 249. Note that other alternative hypotheses of colonization of our planet have been proposed. An alternative interpretation of the passage is atomistic and less likely within the context of the author's theme and time, but also probable in light of the atomistic views of Empedocles' contemporary Leucippus, and his student, Democritus. The interpretation is also compatible with the ideas of universal flow originating from fire of Heraclitus, who according to Diogenes Laërtius flourished in the time of the Olympic Games of 496 BC. Fire, water, earth and air could relate to chemical energy, water molecule interactions with proteins and nucleic acids, metal ligands, and metabolism of gases such as methane, carbon dioxide and oxygen, respectively. This succession seems compatible with a chemical origin of modern metabolism (Morowitz, 1999; Trefil *et al.*, 2009).

E2, line 252 (DK fr. no. B 17.21): Second exhortation: Use your imagination and do not sit dumbstruck.

P, lines 253–257 (DK fr. no. B 17.22–26): The promise is that Love is proof that the unifying force exists.

> *"With mind regard Her, and sit not with eyes bedazed."* (J)

The fact that human beings think (but have not learned) that Love is "*implanted in their joints*" when engaging in "*conjoining deeds*" is given as proof of the unifying force. The unifying force is here named for the first time with the female determiner "*Her*," and later on also with the nominative "*She*." She is also called "*Delight*," "*Joy*" and "*Aphrodite*." Note that throughout the papyrus "*She*" or "*Her*" always refers to Love. It does not refer to Strife, as is wrongly suggested by the most recent and original translations.

Interpretative summary: The passage describes how the first tale of union applies to humans and sexual reproduction. Indeed, sex results in genetic recombination in microbes (e.g. through conjugation) and eukaryotic organisms (e.g. through meiosis; Speijer *et al.*, 2015; Hunter, 2015). These processes cause extensive rearrangement and exchange of genomic information, which can be seen as a unification of parental genetic materials in the offspring. Sometimes, sex also results in interactions of social unification between partners that often benefit the rearing of the progeny.

E3, line 257 (DK fr. no. B 17.26): The third exhortation prepares for significant argumentation. It orders the reader to pay attention to truthful words.

T2, lines 258–266 (DK fr. no. B 17.27–35): The second central thesis of *P. Strasb. Gr. Inv.* 1665–6 specifies that growing things establish a *hierarchy* of wholes unified from integrated parts by Love. This segment of poetic composition is obscure and difficult to interpret. However, it crucially untangles the general unification-diversification cosmological dichotomy of evolutionary growth presented earlier by proposing a universal phylogeny of life.

> *"For all these things are equal and alike in age,*
> *but each rules separate domains, each has its haunts*
> *and lords it in its turn as time rolls round."* (J)

The alternative translation is also helpful: "*For these are all equal and identical in age, but each one presides over a different honor, each one has its own character, and by turns they dominate while the time revolves*" (LM). While the proof of lines 253–257 confirms that the unifying force exists, in lines 258–266 the products of the unifying force are placed within a context of historical diversification. First, the diversifying "*things*" refer to biological wholes assembled from a unification of parts, which arise from Love (P, lines 253–257). Second, the diversifying wholes are considered "*equal.*" In fact, when surveyed at any moment in time, lineages of a phylogeny become taxa, taxonomic groups that are treated as phylogenetic entities of a same rank. Third, wholes are "*alike in age*." Indeed, in a standard phylogeny representing the evolution of organisms, biological wholes are lineages that branch after originating from a last universal common ancestor of life (LUCA). When traveling along phylogenetic paths from an origin to the present, all organisms deriving from the ancestor have a same age despite having had different historical paths. For example, in coalescent theory one travels back in time, unifying each and every diversified lineage until all lineages fuse into one origin (Kingman, 1982). In this process, the points where lineages meet represent coalescent events, and their collective become dendrograms. These diagrams describe how lineages diversify without the need to specify the ticking of a clock or other assumptions operating on change. One simply travels forward or backward in time along their branches. Finally, the wholes rule over "*separate domains*," dominating one after the other in time and space as their ancestors diversify. Given previously defined constraints on taxa and age, this last description is of a hierarchy of wholes, a tree-like structure.

The illuminating insight in the first part of T2 is further explained by extending the concept of lineages of a tree (consistent with Darwinian thought) to lineages of a network (consistent with modern biology).

> *"Beyond these nothing comes to be or perishes.*
> *For if they died continually, no longer would they be.*
> *What could increase this whole, and from what source?*
> *How too could it be destroyed, since nothing lacks in these?*
> *But these are what there is, and running through each other*
> *they suffer change continual but always are alike."* (J)

The alternative translation is complementary: "*And besides these, nothing at all is added nor is lacking; for if they perished entirely, they would no longer be. And this whole here, what could increase it, and coming from where? And how could it be completely destroyed, since nothing is empty of these? But these are themselves, but running the ones through the others they become now this, now that, and each time are continually similar*" (LM). Having established the existence of lineages of wholes in a hierarchy, the passage (which includes fr. **a (i) 1–5**) now describes its sufficiency and the intermingling of the lineages of a phylogeny through horizontal exchange.

First, nothing but the biological wholes unified by Love form and perish in the lineages. The rationale provided is that if they "*died continually*" they would "*no longer ... be.*" In other words, lineages of extant living beings form a continuum of change, supporting the main tenet of life. If they are not extant, then they do not exist as life in the present, the wholes have gone extinct, and the lineage breaks. The two questions in the passage reinforce the sufficiency of the lineages of wholes as *explanandum* for the diversity of life. Second, lineages of wholes run "*through each other*" (fr. **a (i) 4**) turning trees of life effectively into networks of life. The alternative translation is even more explicit: "*but running the ones through the others they become now this, now that.*" (fr. **a (i) 5**). The passage ends with the idea that lineages continually change in a network but their lineage identity is preserved.

Interpretative summary: The main theory of T2 explains how the unification of parts into wholes and their ulterior diversification result in hierarchical networks of evolving life. The central and global role of the horizontal transfer of genetic information in evolution was already intimated two decades ago (Doolittle, 1999). Since then it has been confirmed by a significant body of evidence and has been widely embraced (Olendzenski and Gogarten, 2009). The tree of life turns into a network of life through the effects of numerous processes of horizontal exchange, including horizontal gene transfer, recruitments, and parallel evolution. However, the hierarchical structure of the network is maintained, especially when studied at the level of molecular structure (Caetano-Anollés *et al.*, 2018b). It is therefore remarkable that the paradigm of a hierarchical network that runs lineages "*through each other*" was made explicit in this ancient papyrus. Note that most biological networks describing extant molecular machinery, including metabolic networks, protein–protein interaction networks, or networks of structural domains, express hierarchical modularity (Barabasi and Oltvai, 2005). This property of biological interaction has been appreciated only recently thanks to advances in computation, visualization and analysis (Caetano-Anollés *et al.*, 2019).

T3, lines 267–290 (includes DK fr. no. B 17.36–35 and DK fr. no. B 21.9–12): This third main thesis generalizes the theory presented in T1 and T2, by stating that everything in the biological world comes together in Love but separate in Strife: plants, man, beasts, fish and gods.

> *"In Love we come together in one world;*
> *in Hatred many grew from one to be apart,*
> *whence all that was, and is, and shall at some time be*
> *blossomed as trees, as men, as women too,*
> *as beasts, as birds, as fish that water rears,*
> *as well as gods who ages live and greatest honours have."* (J)

The original translation of Martin and Primavesi (1999) explicitly supports the unification of parts into "wholes" as a generative principle: "*But under Love we unite together to form a single ordered whole, whereas under Hatred, in turn, it grew apart so as to be many out of one.*" The "*we unite together*" and "*world*" of this passage have been surprising and controversial (van der Ben, 1999; Trépanier, 2017). The segment matches Simplicius' text except that the verb is not in participial form but is presented as a first person plural "*we*," crucially distinguishing the direct and indirect transmissions and suggesting scribal errors. We contend that a personal or impersonal description of temporal patterns (whether intentional or owed to scribal error) does not alter the evolutionary significance of the passages.

The passage now also formally establishes that the growing "*things*" of T1 lines 233–234 involve unified parts of diversifying wholes that include organisms, such as familiar species of the plant and animal kingdoms and man. Thus, the abstract "*things*" of T1 become biological observables that were tangible to the philosopher in his time. However, the mention of gods is puzzling, especially within a context that is not eternal. The alternative translation "*and long-lived gods, the greatest in honors*" (LM) is not helpful either. What are these non-immortal gods? Living entities with longer lives than man? Higher taxonomic groups of organisms? Longer lived creatures that will evolve? Daimones, titans, giants? (see "double race of man" in line 297). Biological principles?

As with T2, the focus of T3 are the products of "*Her*," the unifying force of Love, and how these unified wholes diversify. We here note that the encyclopedic interpretation of "*Her*" representing Strife (LM) is incorrect, since lines 253–290 (P, T2 and T3) all refer to wholes unified by Love, and the discourse follows the naming of "*Her*" as "*Aphrodite*" in line 255.

The following passages of T3, which are fragmented, now move to describe the unification process and its dynamics. The passage establishes that the unification of parts into wholes "*swirl*," i.e. that the process takes a spiral temporal trajectory (a gyre). This spiraling motion aligns with the concept introduced in T1 that parts never "*cease from change continual*," or that they "*exchange their places continually*" (line 243). This motion is once again distinct from the frustrated dynamics of the evolutionary forces that unify and diversify and live "*forever in a steady cycle*" (line 244).

> *"In Her they never cease to swirl in constant flux*
> *with frequent whirlings . . . .*
> *relentlessly, and never do they cease . . .*
> *But many ages previous must elapse . . .*
> *before their motions alter . . . ."* (J)

First, the "*frequent whirlings*" (fr. **a (ii) 4**) or "*dense eddies*" signal that swirls are both sporadic and relentless, "*without interruption.*" The original translation "*whirl*" (Martin and Primavesi, 1999) of δίνηισι was seen as one surprise of the papyrus, since it provided a "fluid dynamics" point of view of the force of Love (van der Ben, 1999). Second, it defines a timeframe for the unification: "*many ages previous must elapse before their motions alter*" (J). Thus, the dynamics of the unification whirl is protracted and involves long cycles of time ("*ages*") before the rotational "*motions*" of the spiral are altered.

A simile is then used to describe the dynamics of unification: the unified wholes change over time, much like the cyclic positions of the sun, the moon and the stars.

> *"They never cease in any way to swirl in constant flux.*
> *The sun does not stay still, nor does the moon*
> *in orbit cease to wax or cease to wane,*
> *nor do the other stars stay fixed in heaven;*
> *they all move in a cycle, changing place."* (J)

Our interpretation is at odds with the idea that the segment refers to a wider cosmological timeframe, in which the moon and sun were non-existent (van der Ben, 1999). After all, the entire theory of *P.*

Strasb. Gr. Inv. 1665–6 relates to the evolution of life, not of the cosmos.

Finally, the last passages of T3 describe that the whirls were even present in the past, at the origin of the "*untrodden*" Earth and at a time when the sun's orb was as big as "*men have the power to vouch.*"

> *"At that time earth untrodden runs and the sun's orb,*
> *as big as even now men have the power to vouch."* (J)

These initial statements are puzzling and their implications remarkable for millennary writing. The mere indication that the sun's orb changes with time indicates knowledge that stars have evolutionary cycles of growth and contraction. Claiming that the size of the sun at the origin of life (3.8 billion years ago) was similar to the present sun is also surprising. The sun is indeed in its main-sequence phase of evolution, fusing hydrogen atoms to form helium atoms in their cores (Bertulani, 2013). Its size is stable.

While the processes leading to the unified wholes arise before life's colonization of the planet by diversified life, the final passage of T3 is remarkable in that it describes how parts "*never cease to swirl in constant flux*" as they are unified by "*Her*" in a network-like process.

> *"just so do all these things through one another race,*
> *and, roaming, visit other places constantly;*
> *we do not reach the middle place in union.*
> *But whensoever Hatred to the vortex' utmost depths*
> *arrives, and Love arises in the whirlwind's midst,*
> *in Her then all these things unite to be just one."* (J)

The literary approach used to describe network structure is the same used to describe networks of diversified wholes in line 265 of T2. The parts race "*through one another*" (fr. **a (ii) 15**) (J) or "*run through one another*" (LM) as they unify into wholes. Their temporal "*roaming*" trajectories intersect, forming network structures. The passage ends again with the dichotomy of the two tales: the network-like diversification of wholes results in parts which are then again unified into new wholes by network-like unification processes.

Interpretative summary: T3 reinforces unification as being network-like and integrates diversification and unification into a single process. For that purpose, the things that are unified and diversified are now diversifying organisms. The main thesis of T3 therefore embraces the idea of a "*Systema Naturae*" of hierarchical classification of the living world, in which trees, fish, birds, beasts, and man arise in the hierarchy put forth in T2. The thesis can be considered a taxonomical proposal akin to that of Linnaeus (1751) or of the modern phylogenetic-driven Tree of Life project (Hinchliff *et al.*, 2015). However, the proposal in T3 is made within the context of networks and not of tree statements of relationship. A number of network-like processes have been used by modern biology to describe the evolution of organisms (Morrison, 2014). T3, in the context of T1 and T2, is more ambitious because it discusses the evolution of organisms as a special case of the theory of emergence of complex wholes from parts. Its network depiction is also advanced. Kull (2003) discussed three holistic models that organize biological knowledge, the ladder (*scala naturae*), the tree (*arbor vitae*), and the network (*tela vitae*). The ladder is non-temporal in that it considers all levels of organization as being complete. In contrast, the tree and the network models are dynamic. The tree emphasizes change, growth and hierarchy while the network emphasizes the inter-relationship of the changing entities. Network thinking represents a fundamental conceptual change in knowledge interpretation, a fact that fully materializes in the study of evolution (Rieppel, 2010). In this regard, the biosemiotically advanced network-like model of Empedocles was proposed earlier than the primitive ladder model of Plato and Aristotle. This is particularly significant for understanding the explanatory power of Empedocles' theory. Such a network-like view was not revisited until the time of Linnaeus in the mid 1700s and was not embraced by biologists until the beginning of this century.

T3 also introduces the spiral as a dynamic evolutionary force that has a singular origin and is both fractal (recursively expresses self-similarity) and chiral (expresses handedness). A controversial systems theory explains the flow of genetic information as a succession of nested gyres (Andrulis, 2011). This view is compatible with the spiral depiction of the forces behind the double tale. The symbolic significance of the spiral is discussed below in a separate section.

Finally, T3 incorporates the important and surprising concept of "recruitment" during the process of unification, i.e. the concept of parts racing "*through one another*" and roaming to "*visit other places constantly*." For example, the origin and complexity of metabolic networks has been explained with a "patchwork" model that assumes the existence of promiscuous catalytic activities that are recruited pervasively and fortuitously to perform different metabolic functions (Ycas, 1974; Jensen, 1976; Kacser and Beeby, 1984). The result of these recruitment processes is a patchwork (a mosaic) of homologous enzymes that spread throughout metabolic pathways. Considerable evidence supports the recruitment model. For example, assignments of fold structures to protein enzymes of the metabolic network of *Escherichia coli* revealed an evolutionary mosaic (e.g. Teichmann *et al.*, 2001). Tracing protein domain age in metabolic networks confirmed the existence of a patchwork of ancestries (Kim *et al.*, 2006). Other numerous examples show recruitment is a powerful force in evolution that results in the patchiness of networks.

E4, line 291–292: The fourth exhortation: Comprehend and take note.

Pm1, lines 293–298: In this first promise, the unifying force of Love makes things increase.

> *"I'll put before your eyes how She augments*
> *with larger form life's union and increase,*
> *and all that still remains of this creation,*
> *first in the wild tribes of beasts that roam the hills,*
> *then in the double race of man, then in the fruits*
> *of rooted plants and grapes that mount the vine."* (J)

Lines 293–294 now describe the unification force as a process that "*augments with larger form*" (fr. **a (ii) 24**) the union of parts and "*all that still remain of this creation*" (fr. **a (ii) 25**). Thus, the unification of parts into wholes "*increases*" the unified things that have survived during the unification process. The following lines provide examples of the unified things that survive the selection ("*tribes of beasts*," "*double race of man*") and optimization processes ("*fruits*," "*grapes*"). The double race of man is again puzzling and refers to the existence of two kinds of beings. Mackenzie (2016) suggests the double race refers to "shoots of men and women." Note that the process of increase of parts is exemplified with organisms (beasts, man) and parts of those organisms (fruits, grapes).

Interpretative summary: The passage is of great importance for two reasons. First, it describes an accretion process in which there is an increase in the abundance or characteristics of parts (e.g. their quality or function). Examples of evolutionary accretion in proteins and nucleic acids have been demonstrated in metabolic networks, proteomes and the ribosome (Caetano-Anollés *et al.*, 2018b). Second, it provides additional support to natural selection or competitive optimization supporting the theory of evolutionary growth.

E5, line 299: The last exhortation: Grasp mentally the following truths.

Pm2, lines 300–306 (includes DK fr. no. B 20.1–5): In this second promise, things unified by Love increase in number before breaking down when they are separated by Strife and as generations of living things advance.

> *"You'll see life's union and increase,*
> *because of Hate, wreak deeds of separation.*
> *This first the mass of mortal limbs makes clear;*
> *at one time we unite in Love with all*
> *the limbs that bodies have when life is blossoming;*
> *another time again, split up by evil Strife,*
> *they wander each apart amid life's breakers."* (J)

The "*wreak deeds of separation*" or the "*unfolding of generation*" through "*crossing over*" (LM) speak again of a succession of generations of wholes. These wholes are separated into parts, which are illustrated with the metaphor of limbs of a body that "*wander separately in the surf of life*" (LM). The use of limbs unified in the swirl to describe parts, "*This you will see in the illustrious bulk of mortal limbs*" (LM), is in itself remarkable, especially because of the three-legged *triskelion* motif of rotational symmetry widely used in antiquity. The motif was widely used in Empedocles' homeland Magna Graecia, a proposal we will separately discuss in detail below.

Interpretative summary: In Pm2, the last important thesis of the manuscript (T2) is described in which the products of break down by biological diversification and life history become new parts for a new generative cycle of unification and diversification. The segment explains a transformation of diversified "wholes" into new parts. This justifies the swirls and whirls of the evolutionary growth dynamics. Pm2 also uses the decaying limbs as analogy to the problem of life history, the series of changes undergone by organisms during their lifetime. Life history evolves to optimize reproductive benefits (e.g. offspring) and costs (e.g. mortality) and increase fitness (Roff, 1992; Stearns 2000). Pm2 uses the double tale to explain this problem.

D, lines 307–339 (includes DK fr. no. B 20.6–7 and DK fr. no. B 139.1–2): The digression is fragmented and speaks of death and dissolution using the limbs analogy. The passage addresses the dichotomy of life and death as expression of the dichotomy of Love and Strife. Lines 338–339 define the transition between the two cycles and life and death as an "*unplumbed gyre*" *or* "*vortex of enormous depth*" (LM):

> "*For we shall reach the unplumbed gyre, I fear,*
> *and though men wish it not they'll have uncounted pains.*" (J)

This definition prepares the audience to the fulfilment of the promises, which link the unification and diversification processes, and ultimately life and death.
The segment also includes the lamenting statement of lines 335–336:

> "*Woe that some ruthless day did not destroy me first,*
> *Before I used my claws to wreak dread deeds for food!* (J),

which follows description of the inexorability of death and dissolution, "*the harpies with death's lots will come for us.*" While this is normally interpreted as punishment of the soul for meat eating (Janko, 2004), the statement can be interpreted differently. Within the context of a theory of evolutionary growth the segment is more likely a lament over the predator-and-prey nature of competition that leads to the survival of the fittest. Note that Empedocles was a teenager during the first wars with Carthage, when Thêrôn, the tyrant of Akragas (modern Agrigento), took over Himera and then successfully confronted the invading Carthaginian army of Hamilkar (Freeman, 1892). His life later unfolded during the golden age of Magna Graecia, but at a time where democracy slowly reclaimed freedom through tumultuous warfare against demagogues and tyrants and against Etruscan pirates and Sikeli inland aggressors. This cutthroat environment likely inspired his poetic laments of death and struggle.

F, lines 340–364: The fulfillment of the promises finally recounts a history of the interplay between unification and diversification, from the origin of life to the origin of present-day creatures. It begins by describing the history of the generative cycle of life:

> *"But we'll embark once more upon our tale.*
> *When once the tireless flame did chance upon*
> *all things, and caused their painful intermixture,*
> *then creatures too progenitive were born*
> *in every way, whose remnants still the dawn beholds."* (J)

The focus of this crucial papyrus segment is again the emergence of a primordial network structure responsible for the process of unification and increase. First, it describes a "*tireless flame*" (fr. **d 11**) as the cause of the "*painful intermixture*" of "*all things*." What is this network-causing "*tireless flame*"? One possible interpretation is that a dissipation of some kind (e.g. energy, information) is the ultimate culprit of the process of unification. Below, we will separately discuss how this process involves the sun and has been represented with the solar disc or solar wheel symbols of antiquity. The passage then describes how the network-like intermixture of parts gives rise to "*creatures*" that are "*progenitive ... in every way*." Again the suggestion is the existence of competitive optimization or selection out of a progenitive set of combinations of parts. Finally, it ends with the remarkable suggestion that the unsuccessful combination-variants (extinct?) or exemplars of ancient organisms are somehow preserved as "*remnants*" in a record of the past.

> *"When aither mixed did reach the utmost edge,*
> *then birds flew up with shrieks and cries*
> *tumultuous, but beasts, whose lot is lairs in earth*
> *and grass, were born where earth was swathed around."* (J)

The recount of primordial life was then followed by the rise of organisms such as birds and beasts. These segments are probably a first part of a long list of field biology descriptions of living beings and selected parts of them which had been lost. The last assembly from fragments of the papyrus (fr. **b1, b5, e1** and **e4**), for example, describe marine gastropods (sea snails), mollusks (oysters), decapod crustaceans (sea crabs), Testudine reptiles (turtles), and Cervidae ruminant mammals (deer) as examples of "*earth dwelling on the top of flesh*." The focus is on their earth-like parts: "*heavy backs*" and "*stony mantles*," "*stone-skinned*" shells, "*armour-plate*" exoskeletons, and "*spears*" of horns. Note that the marine crab (*karkinos*) was the animal symbol of Akragas, which can be found on coins issued by the city during Empedocles' time.

Interpretative summary: In F, the tale of evolutionary growth and increase is recounted as a "fulfillment" of the theory. First, the origin of "*creatures*" is described as the result of some dissipative process embodied in a "*flame*." One interpretation is energy-information dissipation, which is linked to the logistic function used to model growth of systems (Guseo, 2016). This would be remarkable, especially within the framework of the recent and acclaimed conjecture of the rise of space through a holographic scenario in which a gravity-associated entropic force gives rise to the laws of Newton and the equations of Einstein (Verlinde, 2011). More probably, flame relates to the sun as the dissipative source of energy. If such would be the case, then *P. Strasb. Gr.* Inv. 1665–6 offers a link between the spiral dynamics of its theory and the solar disc symbolism of antiquity, which we discuss in Appendix A.

Second, the fulfillment of lines 341–344 provide an important corollary of the double tale: the existence of footprints of organismal history ("*creatures... whose remnants still the dawn beholds*") left behind as fossils in the geological record, which can be used as historical evidence for evolutionary exploration. Indeed, fossil evidence has significantly affected our understanding of biological evolution (Jablonski and Shubin, 2015). The intersection of paleontology with diverse disciplines continues to provide important insights into the natural sciences, including a timetable of evolution (Knoll and Nowak, 2017). In addition, other kinds of historical remnants exist in biology. For example, conserved regions or structures of molecules, molecular complexes, and other elements of the cell can be considered living fossils (e.g. Caetano-Anollés *et al.*, 2018b), because they provide a window into early evolutionary history.

Finally, and surprisingly, the assembly of the last few papyrus fragments recounts examples of characteristic earth-like structures (phenotypes, not organisms) arising from the double tale ("*earth dwellings on the top of flesh*") that are common to a wide range of organisms, including marine gastropods, mollusks, crustaceans, reptiles, and ruminant mammals. One commonality of all of these organismal structures is filamentous keratin [from κέρας keras (genitive κέρατος keratos) meaning "horn"], a protein macromolecule that is part of a broad family of fibrous structural proteins (Wang *et al.*, 2016). Keratinous materials based on α-keratin make up the horny outer layer of the skin (*stratum corneum*), wool and hair, quills, horns, hooves, nails, whale baleen, hagfish slime threads and whelk egg capsules. Materials based on β-keratin make up feathers, beaks and claws. Materials based on both α-keratin and β-keratin make up reptilian epidermis (in Squamata, Crocodilia and

Testudines), soft and hard epidermis of Testudines (turtles, tortoises and terrapins), and pangolin scales. The remarkable example of phenotypic parts arising from the double tale involves a molecular trait that is subject to complex vertical and horizontal evolutionary forces (Klinge *et al.*, 1987). The typical three structural domains that make up the keratin proteins exhibit different evolutionary histories, making the keratins evolve along the paths of a network with reticulations.

4. The "Ship of Theseus" paradox: Evolutionary modules and growth

The theory of *P. Strasb. Gr.* Inv. 1665–6 explains life's generative process. First, a dynamic collective of unifications and diversifications results in physically and temporally fragmented biological identities, which evolve by exchange in a network-like manner. This theory is significantly advanced from a conceptual and explanatory point of view. First, *P. Strasb. Gr.* Inv. 1665–6 would be the first report of a network depiction of the systematics of life, more than 2,500 years earlier than the first ideas of network structure applied to understanding biology (reviewed by Rieppel, 2010). Second, the double tale involves the generation of biological *modules* (Caetano-Anollés *et al.*, 2018a, 2019). Modules are sets of integrated parts of a system (Hartwell *et al.*, 1999). These parts have two central properties; they cooperate to perform a task and interact more extensively with each other than with other parts and modules of the system. While the cooperative performance of specific functions distinguishes modules from each other, the rise of modularity in structural biology can be seen as a "nucleation" process, i.e. the emergence of cohesion in selected parts of the makeup of a molecular system. This process is of unification and produces entities that are persistent. In this regard, cohesion refers here to the dynamic stabilities of component parts of the system when their dynamics are constrained by the system as a whole. These stabilities ultimately determine whether the system can be easily decomposed into parts (Simon, 1962). For example, the 3-dimensional (3D) atomic structure of macromolecules such as proteins can be effectively described as a nested hierarchy of modules (Cothia and Gough, 2009). The double tale can be considered a biphasic (bow-tie) theory of module emergence (Mittenthal *et al.*, 2012), that is delimited by "linkage" and is responsible for the "*hierarchical modularity*" of biological networks (Caetano-Anollés *et al.*, 2018, 2019). In networks, a graph theoretical framework defines nodes, which describe parts, and links, which describe interaction between parts. In a first phase, parts are at first weakly linked and associate variously. As they diversify, they compete with each other and are often selected for performance. As the growth and stability of links intensify, the emerging interactions constrain the structure and associations of the parts, causing them to self-organize into network modules with tight linkage. In a second phase, variants of the modules diversify and become new parts for a new generative cycle of higher-level organization. Remarkably, evolutionary genomics supports the gradual rise of hierarchical modularity in evolving networks at different timescales and complexity levels (Caetano-Anollés *et al.*, 2019). For example, hierarchical modularity emerged in: (i) the rewiring of metabolomic and transcriptome-informed metabolic networks upon external perturbation, (ii) the nanosecond-dynamics of proteins of different evolutionary origins, and (iii) evolving networks of metabolism, elementary functionomes and protein domain organization.

One central recurrent theme in the theory of *P. Strasb. Gr.* Inv. 1665–6 is that biological things constantly change while remaining the same. This theme distils philosophical thinking contemporary with Empedocles, which can be summarized by the "*Ship of Theseus*" paradox or the "*river fragments*" of Heraclitus. Theseus was the mythical hero that founded Athens, perhaps a king that lived during the Late Bronze Age. His feats and travels were recounted by Plutarch (*ca.* 46–120 AD), the storyteller, in his *Vita Thesei*, the "*Lives of the noble Grecians and Romans*." In a remarkable passage (*Vita Thesei* 22–23), Plutarch uses the famous 30-oar ship sailed by Theseus, which was "*preserved by the Athenians down even to the time of Demetrius Phalereus*" (*c.* 350–280 BC), to put forth an important thought experiment about the philosophical question of identity and "*things that grow*": Does the constant replacement of old decaying planks by stronger timber change the identity of the ship of Theseus or not? If the ship is considered a module of unification, then the module is constant over time and the identity of the ship has not changed. If it is not, then the ship is subject to the vagaries of change and will never be the ship of Theseus again. Lines 241–244 provide an answer by describing the unification-diversification dicotomy as both ephemeral and eternal. The question of an object persisting through time was also addressed by Heraclitus in Plato's famous but controversial "*river fragments*" (Graham, 2015). Three fragments describing Heraclitus' own thinking about flux have been identified: (i) "*potamoisi toisin autoisin embainousin hetera kai hetera hudata epirrei*": Upon those who step into the same rivers, different and again different waters flow" (B12); (ii) "*potamois tois autois ...*": In the same rivers we step and do not step, we are and are not (B49a); and (iii) "*potamôi ... tôi autôi ...*": It is not possible to step twice in the same river or to come into contact twice with a mortal being in the same state (B91[a]). The first fragment is characteristic of Heraclitus' Ionic dialect, while the others are not, and is accepted as genuine and providing something much more subtle and profound (Graham, 2015): it "*is not that all things are changing so that we cannot encounter them twice*" but instead that "*some things stay the same only by changing*". In other words, persistence and change are not opposed concepts. Instead, they are inextricably connected and coinstantiated, with flux being a necessary condition. This thinking aligns with the central theme of the Strasbourg papyrus.

5. Spiral movement describes the frustrated dynamics of unification and change

Spiral movement is at the heart of Empedocles' poem. It underscores the dynamics of unification and diversification of the double tale. Several passages of the poem illustrate movement with a repetitive spiraling pattern, in which parts "*never cease to swirl in constant flux*" (line 273), exhibit "*frequent whirlings*" (line 274), and do so "*relentlessly*" (line 275). The spiraling pattern that varies with time differs strikingly from the inescapable cycle of unification and diversification: "*But many ages previous must elapse... before their motions alter... They never cease in any way to swirl in constant flux*" (lines 276–278). The recurrence of this "*unplumbed gyre*" (line 338) or "*vortex*" (line 288) is central to the description of the double tale and applies to both unification and diversification: "*But whensoever Hatred to the vortex' utmost depths arrives, and Love arises in the whirlwind's midst, in Her then all these things unite to be just one*" (lines 288–290). Thus, the spiraling pattern of diversification ends with a funnel-shaped whirlwind of unification. The resulting "hourglass," which is a novelty-generating pattern of biology (Caetano-Anollés *et al.*, 2019), can be flattened into a double spiral (two spirals conjoined at the center), which is known as a *running spiral* when different spirals connect to each other. Appendix A describes how the spiral patterns of the double tale have been pervasive in history across millennia. This can help us understand its possible symbolic origins.

6. Foresight and evolution

The theory of Empedocles' *P. Strasb. Gr.* Inv. 1665–6 explains evolution with network hierarchies ~2,400 years before Darwin. It also proposes a theory of evolutionary growth and diversification that is now supported by significant evidence from phylogenomics and systems biology, which benefit from the advent of genomics and computational biology. The explanatory power of the theory appears significant, a fact that now demands further explanation.

Scientific progress is supported by a framework of theories, which are well substantiated explanations of some aspect of the natural world (Niiniluoto, 2015). The search for truth and verisimilitude are its main goals. The scientific method applies logical and statistical tools of confirmation to evaluate the match between empirical evidence and theories that are in constant revision (Crupi, 2015). Regardless of philosophical perspectives (e.g. fallibilism, instrumentalism, foundationalism), the empirical confirmation of obvious hypotheses is less compelling than empirical confirmation of non-obvious ones (Trafimow, 2003). Under this scheme, a decrease in the conditional probability of a theory increases its posterior probability and the positive impact on its confirmation. In the case of the double tale, the underlying theory is non-obvious to the point that its biological consequences have remained vastly unrecognized. Thus, any confirmation test of a non-obvious theory is "strong" according to Popper's philosophy of falsification, which seeks "severe" destructive tests of corroboration.

Three points of reason support evolution (modified from Wiley, 1975): continuity, singularity and memory. The axiom of *historical continuity* posits that evolutionary change is gradual in time and space, even in the presence of leaps. It begs the existence of a model of evolutionary change. The axiom of *historical singularity* posits that only one account of history must exist as a consequence of genealogical descent. It defines the historical ground plan (a phylogeny or network with or without reticulations) along which entities (taxa) change in time. The axiom of *historical memory* posits that attributes of those entities (e.g. genetic, compositional, contextual) are transmitted from generation to generation, modified or unmodified. It requires identification of useful biological attributes carrying historical information. Conversely, Empedocles' *P. Strasb. Gr.* Inv. 1665–6 puts forth three main theses that correspond to the three historical axioms: (1) evolving living systems are explained by the action of two opposing forces, one of unification and the other of diversification; (2) unification and diversification result in the growth of a hierarchy, with wholes unified from integrated parts diversifying along network paths; and (3) unification and change are generalizable to all systems and leave historical "*remnants*" (line 233) as fulfillment of the evolutionary process. Thesis 1 puts forth the main evolutionary hypothesis, which fulfils the axiom of historical continuity. Thesis 2 defines the historical ground plan, complying with the axiom of historical singularity. Thesis 3 instructs the search for biological attributes to fulfill the axiom of memory. Each thesis has predictive power and their consequences have sustained repeated falsification by careful observation and experiment for more than a century of scientific exploration. For example, the double tale theory predicts constant change and complex dynamics, lineages, a hierarchical evolutionary network structure portraying vertical and horizontal exchange of genetic information, and historical memory in parts and wholes, including the existence of fossils or evolutionarily conserved regions of biological macromolecules.

Standard Darwinian evolution has no foresight: "*evolution works like a tinkerer, not like an engineer*" (Jacob, 1977). Recent long-term evolutionary experiments in the laboratory, however, have shown that the fate of mutations not only depend on immediate effects on fitness but also on selective forces that favor "evolvability," a way to prepare organisms for the future (Barrick and Lenski, 2013). These other mechanisms include epistatic and co-evolutionary interactions that promote genetic diversity, which make organisms responsive to external and internal causal influences as they constantly "restructure" their genomes (Shapiro, 2011). Despite our limited knowledge of how causal influences affect the different levels of biological organization, one cannot exclude the development of goal-directed purpose in structures and functions (teleonomy). The fact that Empedocles' papyrus puts forth a theory with remarkable predictive power is challenging. We now must explain its foresight. How did it arise in such an early time of human history? Was foresight better rewired in brains? Did it materialize through theoretical abstraction or through some kind of experimentation? Was it facilitated by transmission through social learning? Given the teachings of the Derveni papyrus (Janko, 2002), was it transmitted through a subversion of religion? Was foresight the product of an individual or a broader school?

References

Andrulis ED (2011) Theory of the origin, evolution, and nature of life. *Life* **2**: 1–105.

Afonasina A (2012) The birth of harmony out of the spirit of *Tekhe*. ΣΧΟΛΗ **6**: 68–75.

Barabasi A-L, Oltvai ZN (2005) Network biology: understanding the cell's functional organization. *Nature Rev* **5**: 101–113.

Barrick JE, Lenski RE (2013) Genome dynamics during experimental evolution. *Nature Rev Genet* **14**: 827–839.

Benner SA, Kim H-J, Carrigan MA (2012) Asphalt, water, and the prebiotic synthesis of ribose, ribonucleosides, and RNA. *Acc Chem Res* **45**: 2025–2034.

Benson FS (1902) Ancient greek coins. VIII. Syracuse. *Am J Numismatics* **36**: 97–103.

Bertulani CA (2013) *Nuclei in the Cosmos*. World Scientific Publishing, Singapore.

Budja M (2006) The transition to farming and the ceramic trajectories in Western Eurasia: from ceramic figurines to vessels. *Documenta Prehistorica* **33**: 183–201.

Caetano-Anollés G, Aziz F, Mughal F, Gräter F, Koç I, Caetano-Anollés K, Caetano-Anollés D (2019) Emergence of hierarchical modularity in evolving networks uncovered by phylogenomic analysis. *Evol Bioinformatics* **15**: 1176934319872980.

Caetano-Anollés D, Caetano-Anollés K, Caetano-Anollés G (2018a) Evolution of macromolecular structure: a 'double tale' of biological accretion. *Sci Prog* **101**: 360–383.

Caetano-Anollés G, Nasir A, Kim KM, Caetano-Anollés D (2018b) Rooting phylogenies and the Tree of Life while minimizing ad hoc and auxiliary assumptions. *Evol Bioinformatics* **14**: 1176934318805101.

Cartwright M (2013) Mycenaean Civilization. Ancient History Encyclopedia. Last modified November 15, 2018. http://www.ancient.eu /Mycenaean_ Civilization/.

Chothia C, Gough J (2009) Genomic and structural aspects of protein evolution. *Biochem J* **419**: 15–28.

Crupi V (2015) Confirmation. In *The Stanford Encyclopedia of Philosophy*, Zalta E (ed). URL = http://plato.stanford.edu/entries/confirmation/

Curtius E, Head BC (1870) On the religious character of Greek coins. *Num. Chron* **10**: 91–111.

Doolittle WF (1999) Lateral genomics. *Trends Cell Biol* **9(12)**: M5–M8.

Dupont C, Butcher A, Valas RE, Bourne PE, Caetano-Anollés G (2010) History of biological metal utilization inferred through phylogenomic analysis of protein structures. *Proc Natl Acad Sci USA* **107(23)**: 10567–10572.

Evans AJ (1894) Contributions to Sicilian numismatics. *Num Chron* **14**: 189–242.

Franklin JC (2017) *Kinyras — The Divine Lyre*. Harvard Univ Press, Cambridge, MA.

Freeman EA (1892) Sicily. *Phoenician, Greek, and Roman.* T. Fisher Unwin Ltd, London.

Günkel-Maschek U (2012) Spirals, Bulls, and sacred landscapes: The meaningful appearance of pictorial objects within their special and social contexts. In: *Minoan Realities: Approaches to Images, Architecture, and Society in the Aegean Bronze Age*, Panagiotopoulos D, Günkel-Maschek U (eds.) [online]. Presses Universitaires de Louvain, Louvain-la-Neuve. pp. 115–139.

Guseo R (2016) Diffusion of innovations dynamics, biological growth and catenary function. *Physica A* **464**: 1–10.

Goodyear WH (1891) *The Grammar of the Lotus. A New History of Classic Ornament as a Development of Sun Worship.* Sampson Low, Marston & Co, London.

Graham DW (2015) Heraclitus. In *Stanford Encyclopedia of Philosophy*, Zalta EN (ed.) (Fall 2015 edition). URL = https://plato.stanford.edu/archives/fall2015/entries/heraclitus/ Accessed June 19, 2019.

Grima R (2008) Landscape, territories, and the life-histories of monuments in Temple period Malta. *J Mediterr Archaeol* **21**: 35–56.

Hartwell LH, Hopfield JJ, Leibler S, Murray AW (1999) From molecular to modular cell biology. *Nature* **401**: c47–c52.

Head BC (1888) *Catalogue of Greek Coins in the British Museum.* Attica-Megaris-Aegina. Longman & Co., London.

Hinchliff CE, Smith SA, Allman JF, Burleigh G, Chaudhary R, Coghill LM, Crandall KA, Deng J, Drew BT, Gazis R, Gude K, Hibbet DS, Katz LA, Laughinghouse IV HD, McTavish EJ, Midford PE, Owen CL, Ree HR, Rees JA, Soltis DE, Williams T, Cranston KA (2015) Synthesis of phylogeny and taxonomy into a comprehensive tree of life. *Proc Natl Acad Sci USA* **112**: 12764–12769.

Hunter N (2015) Meiotic recombination: The essence of heredity. *Cold Spring Harb Perspect Biol* 7: a016618.

Ilievski PH (1993) The origin and semantic development of the term Harmony. *Illinois Classical Studies* **18**: 19–29.

Jablonski D, Shubin NH (2015) The future of the fossil record: Paleontology in the 21st century. *Proc Natl Acad Sci USA* **112**: 4852–4858.

Jacob F (1977). Evolution and tinkering. *Science* **196**: 1161–1166.

Janko R (2002) The Derveni papyrus: an interim text. *Zeitschrift Papyrologie Epigraphik* **141**: 1–62.

Janko R (2004) Empedocles, On nature I 233–364: A new reconstruction of p. Strasb. Gr. Inv. 1665–6. *Zeitschrift für Papyrologie und Epigraphik* **150**: 1–26.

Janko R (2010) Review of *Empedokles Physika I. Eine Rekonstruktion des zentralen Gedankengangs*. By Oliver Primavesi. Berlin: De Gruyter 2008. *Ancient Philosophy* **30(2)**: 407–411.

Jensen RA (1976) Enzyme recruitment in evolution of new function. *Annu Rev Microbiol* **30**: 409–25.

Kacser H, Beeby R (1984) On the origin of enzyme species by means of natural selection. *J Mol Evol* **20**: 38–51.

Kim HS, Mittenthal J, Caetano-Anollés G (2006) MANET: tracing evolution of protein architecture in metabolic networks. *BMC Bioinformatics* **7**: 351.

Kingman JFC (1982) On the genealogy of large populations. *J Appl Prob* **19**: 27–43.

Klinge EM, Sylvestre YR, Freedberg IM, Blumenberg M (1987) Evolution of keratin genes: Different protein domains evolve by different pathways. *J Mol Evol* **24**: 319–329.

Knoll AH, Nowak MA (2017) The timetable of evolution. *Sci Adv* **3**: e1603076.

Konuc K (2012) Asia Minor to the Ionian revolt. In: *Oxford Handbook of Greek and Roman Coinage*, Metcalf WE (ed). Oxford University Press, Oxford, pp. 43–59.

Kull K (2003) Ladder, tree, web: The ages of biological understanding. *Sign Systems Studies* **31**: 589–603.

Laks A, Most GW (2016) Empedocles, Testimonia, Part 2: Doctrine (D). In: *Early Greek Philosophy, Volume V, Western Greek Thinkers*, Part 2. Loeb Classical Library, Harvard University Press.

Linnaeus C (1751) *Philosophia botanica.* Godofr. Kiesewetter, Stockholm.

Leone ML (2009) *The Phosphenic Deer Cave.* Edizione Pensiero Preistorico.

Lodolo E, Ben-Avraham Z (2015) A submerged monolith in the Sicilian Channel (central Mediterranean Sea): Evidence for Mesolithic human activity. *J Arch Sci Rep* **3**: 298–407.

Mackenzie T (2016) The contents of Emepdocles' poem: A new argument for the single-poem hypothesis. *Zeitschrift für Papyrologie und Epigraphik* **200**: 25–32.

Mann CC (2011) The birth of religion. *Natl Geogr* **6**: 2–25.

Mansfeld J (2016) Doxography of ancient philosophy. In *Stanford Encyclopedia of Philosophy*, Zalta EN (ed.) (Winter 2016 edition). URL = https://plato.stanford.edu/archives/win2016/entries/doxography-ancient/ Accessed June 19, 2019.

Martin A (2003) Empédocle, fr. 142 D.K.: Nouveau regard sur un papyrus d'Herculaneum. *Cronache Ercolanesi* **33**: 43–52.

Martin A, Primavesi O (1999) *L'Empédocle de Strasbourg*. De Gruyter, Berlin.

Martin W, Baross J, Kelley D, Russell MJ (2008) Hydrothermal vents and the origin of life. *Nat Rev Microbiol* **6**: 805–814.

Mittenthal JE, Caetano-Anollés D, Caetano-Anollés G (2012) Biphasic patterns of diversification and the emergence of modules. *Front Genet* **3**: 147.

Morowitz HJ (1999) A theory of biochemical organization, metabolic pathways, and evolution. *Complexity* **4**: 39–53.

Morrison DA (2014) Phylogenetic networks: A Review of methods to display evolutionary history. *Annu Res Rev Biol* **4**: 1518–1543.

Mulkidjanian AY, Bychkov AY, Dibrova DV, Galperin MY, Koonin E (2012) Origins of first cells at terrestrial, anoxic geothermal fields. *Proc Natl Acad Sci USA* **109**: E821–E830.

Newton J (1885) The armorial bearings of the Isle of Man: Their origin, history, and meaning. *Proc Lit Phil Soc Liverpool* **39**: 205–226.

Niiniluoto I (2015) Scientific progress. In *The Stanford Encyclopedia of Philosophy*, Zalta EN (ed). URL = http://plato. stanford.edu/entries/scientific-progress/ Accessed June 19, 2019.

O'Brien D (1969) *Empedocles' Cosmic Cycle.* Cambridge University Press, Cambridge.

Olendzenski L, Gogarten JP (2009) Evolution of genes and organisms: the tree/web of life in light of horizontal gene transfer. *Ann NY Acad Sci* **1178**: 137–145.

Orlandini P (1964) La raffigurazione della triskeles su vasi arcaici di fabbrica gelese. *Cronache Archeol Storia d'Arte* **3**: 13–15.

Pare C (1987) Wheels with thickened spokes, and the problem of cultural contact between the Aegean world and Europe in the late Bronze Age. *Oxford J Archeol* **6**: 43–61.

Price M, Waggoner N (1975) *Archaic Greek Silver Coinage — The "Asyut" Hoard.* V.C. Vecchi and Sons, London.

Rieppel O (2010) The series, the network, and the tree: changing metaphors of order in nature. *Biol Philos* **25(4)**: 175–196.

Roff DA (1992) *The Evolution of Life Histories. Theory and Analysis.* Chapman and Hall, New York.

Shapiro JA (2011) *Evolution: A View from the 21st Century.* FT Press, Upper Saddle River, NJ.

Shear JP (1933) The coins of Athens. *Hesperia* **2**: 231–178.

Sparavigna AC, Baldi MM (2016) Flower of life, six-fold symmetry and honeycomb packing of circles in the Mycenaean geometry. SSRN (Elsevier) http://dx.doi.org/10.2139/ssrn.2756099

Speijer D, Lukes J, Eliás M (2015) Sex is a ubiquitous, ancient, and inherent attribute of eukaryotic life. *Proc Natl Acad Sci USA* **112(29)**: 8827–8834.

Stearns SC (2000) Life history evolution: successes, limitations, and prospects. *Naturwissenschaften* **87**: 476–486.

Teichmann SA, Rison SCG, Thornton JM, Riley M, Gough J, Chothia C (2001) Small-molecule metabolism: an enzyme mosaic. *Trends Biotechnol* **19**: 482–6.

Thompson, M (1979) Hoards and Overstrikes. *Expedition Magazine* 21.4: n. pag. Penn Museum. Accessed June 19, 2019. http://www.penn.museum/sites/expedition/?p=4470

Tougas A (2007) Triskelis. Accessed June 19, 2019. http://pguyou.free.fr/triskele/

Trafimow D (2003) Hypothesis testing and theory evaluation at the boundaries: Surprising insights from Bayes's theorem. *Psychol Rev* **110**: 526–535.

Trépanier S (2017) Empedocles, On Nature 1.273–287. Place, the elements, and still no 'we'. *Mnemosyne* **70**: 562–584.

Tusa S (1991) The megalith builders and Sicily. *J Mediterr Stud* **1**: 267–285.

Van der Ben N (1999) The Strasbourg papyrus of Empedocles: Some preliminary remarks. *Mnemosyne* **52**: 525–544.

Verlinde E (2011) On the origin of gravity and the laws of Newton. *J High Energy Phys* **2011**: 29.

Voukelatos J (2011) The triskeles of ancient Greek coins. ΝομΧρον **29**: 7–24.

Wächtershaüser G (2006) From volcanic origins of chemoautotrophic life to Bacteria, Archaea and Eukarya. *Phil Trans R Soc B* **261**: 1787–1808.

Wächtershaüser G (2014) The place of RNA in the origin and early evolution of the genetic machinery. *Life* **4(4)**: 1050–1091.

Wang B, Yang W, McKittrick J, Meyers MA (2016) Keratin: Structure, mechanical properties, occurrence in biological organisms, and efforts at bioinspiration. *Prog Mat Sci* **76**: 229–318.

Wiley E O (1975) Karl R. Popper, systematics, and classification: a reply to Walter Bock and other evolutionary taxonomists. *Syst Zool* **24**: 233–43.

Wilson RJA (2000) On the trail of the *Triskeles*: from the McDonald Institute to Archaic Greek Sicily. *Cambridge Arch J* **10**: 35–61.

Ycas M (1974) On earlier states of the biochemical system. *J Theor Biol* **44**: 145–60.

Zirkle C (1941) Natural selection before the "Origin of Species". *Proc Am Phil Soc* **84(1)**: 71–123.

Appendix A

The spiral, the wheel, the triskelion and a double tale of the sun

G. Caetano-Anollés
Department of Crop Sciences, University of Illinois, Urbana, Illinois, USA

The spiral is a curve that revolves away from a point or axis in space. It is an ancient symbolic device that appeared during the Neolithic period in Western Eurasia. Spirals in plastic forms appeared already in mortuary and ritual structures of the Lepenski Vir archeological site of the Mesolithic Iron Gates culture of the Balkans (9500–6000 BC) (Budja, 2006). The double spiral was already present in the *Grotta dei Cervi* of Porto Badisco, one of many Karstic caves of prehistoric Apulia (Figure 3A). The prehistoric pictograms of the cave were likely created 4000–3000 BC (Leone, 2009). Similar spiral pictograms are present in other Neolithic sites such as La Quercia and Serra D'Alto facies. Spirals and double spirals were used in megaliths during the Chalcolithic and Copper Age (4500–1500 BC), especially in the Temple period of Malta (Grima, 2008). Examples include the spiral decorations of the megalithic structures that appeared 3600 BC in Gozo and Valletta, such as the first volute motifs carved on globigerina limestone in the temple of Ħaġar Qim (Figure 3B) or the multiple interweaved carvings of running spirals typical of the late phase of the Temple civilization of Ħal Tarxien (Figure 3C). A triple spiral symbol was also evident in the megalithic temple of Newgrange of the Boyne Valley (*Brú na Bóinne*) in Ireland 3200 BC, with each spiral motif representing double spirals (Figure 3D). The triple pattern of interlocking Archimedean spirals was later adopted by the Celts as symbol of harmony, renewal and growth. While the emblem of rotational symmetry likely originated in Malta after arrival of Sicilian settlers 5900

BC, its origins could be even more ancient. Sicily, referred to by Homer as Thrinakie (*Oddyssey* XI.107; XII.127 & 135), appears to harbor one of the three oldest man-made monoliths, which is submerged in the Pantelleria Vecchia shelf of the central Mediterranean Sea 60 km south of the island and not far away from Akragas (Lodolo and Ben-Avraham, 2015). The monolith likely represents a monumental temple complex built 9350 ± 200 BC by an ancient mesolithic civilization. Since similar monoliths exist in Anatolia (9500 BC) and Portugal (9100 BC) (e.g. Mann, 2011), this oldest religious culture of the world likely extended its influence to the entire Mediterranean region. Pictograms of the *Grotta dei Cervi* and iconography in Cardium pottery and artifacts of the Danilo culture may carry a deep link to this religious culture.

The spiral is an important symbolic device of the late Bronze Age in Sicily and in the Aegean region, including the Minoan and Mycenaean civilizations. Double spiral motifs similar to those of the islands of Malta (Figure 3B) and Pantelleria can be found in the stelai of tombs 31 and 34 from a necropolis of Castelluccio in southeastern Sicily, dating back to 2200–1800 BC (Figure 3E). The stele of tomb 31 has a rock-cut phallic symbol below the double spirals while tomb 34 has two double spirals flanking a vulva-like contour that pays homage to the worship of fertility typical of Sicilian civilization. Tusa (1991) makes a unique connection between this double spiral symbol of the Castelluccian culture from Bronze Age Sicily with the spiral of Ħaġar Qim in Malta, spiral motifs of Mycenaean vases, swords and other objects, the spiral phallic symbols of Troy, the pre-Nuragic spirals of rock-cut tombs of Sardinia (S'Elighe and Corongiu), spirals in cave paintings of Porto Badisco, and the spirals of New Grange. He uses religious and funerary practice of Sicilian prehistory to advance the idea that the rock-cut spiral emblems of ancient Mediterranean civilizations embody the worship of fertility and birth (Tusa, 1991). In the Aegean region, running spirals are recurrent pictorial motifs and iconic objects of the

Figure 3. Selected archeological artifacts supporting the ubiquity of the Spiral, Running Spiral and Triskelion motifs. (A) Painted pictogram with double spiral from Group 42 on rock from the *Grotta dei Cervi* of Porto Badisco, 4000–3000 BC. (B) Globigerina limestone slab with spiral and drill decoration of the megalithic temple of Ħaġar Qim in the island of Valletta, Malta, 3600–3200 BC. The monuments are one of the oldest religious sites on Earth. (C) Primordial spiral of the Ħal Tarxien temple carved onto Globigerina limestone slab in Valletta, Malta, 3100 BC. (D). Triskelion spiral carved in the entrance passage of the megalithic temple of Newgrange (*Brú na Bóinne*) in County Meath, Ireland, *ca.* 3200 BC. The arrowhead shows the origin of the double spiral. (E) Reliefs on tomb plates 31 (left, with phallic symbol) and 34 (right) of Castelluccio, Sicilian early Bronze Age, 2200–1400 BC. Archaeological Museum of Syracuse. (F) The sarcophagus of tomb 4 of Hagia Triada, a Minoan settlement of south central Crete, with red and white running spirals (the arrows shows the origin of a double spiral), 1370–1320 BC. Heraklion Archaeological Museum. (G) Mycenaean pyriform jar from chamber tomb 10 of Dendra, a settlement in the Argolid, with interweaved spirals, 1500–1450 BC. Archaeological Museum of Nafplio.

Bronze Age in Crete, as discussed for the Palace of Knossos (Günkel-Maschek, 2012). It appears in archaeological artifacts throughout Crete, for example on the famous fired clay disc of the Minoan palace of Phaistos with a spiral of hieroglyphic seals (241 tokens of 45 distinct signs) stamped on each side (dating back to 1850-1600 BC; Heraklion Archeological Museum), and on the Hagia Triada limestone sarcophagus of *ca.* 1500 BC (Figure 3F). It is also present at Mycenae, in much more complex designs. Mycenaean artifacts are usually decorated with regular and geometric forms with remarkable six-fold symmetry (Sparavigna and Baldi, 2016). Running spiral motifs appear as double and triple spirals in pottery and funerary steles (Figures 3G-3I) and as quadruple repoussé spirals in golden cups (Figure 3I). At the center of the triple spiral is the three-legged "triskelion" (or triskele), a generative motif of legs or lines radiating from a center that depicts spiral movement and rotational symmetry (Figure 3J). Note that

Figure 3. (Figure on facing page) (H) Detail of the sepulchral stele from grave gamma of circle B, 1600-1550 BC. Archaeological Museum of Mycenae. The arrow shows the center of the triple spiral with its "triskelion" motif. A double spiral is above. (I) Funerary stele made of porous stone with spirals and chariot scene discovered at the royal shaft grave V in Grave Circle A, *ca.* 1500 BC. National Archeological Museum of Athens. This represents the earliest depiction of the chariot in Achaean art. The white arrow shows the center of the triple spiral with its "triskelion" motif and the yellow arrow shows the "wheel" motif as part of a chariot scene. J. Mycenaean golden cup decorated with repoussé spirals discovered at shaft grave V in Grave Circle A at Mycenae, *ca.* 1500 BC. National Archeological Museum of Athens. The arrow shows the origin of the four spiral interlaced motifs of the running spirals. (K) Ceramic open bowl (dinos) with triskelion produced probably at Gela, 650-600 BC. The dinos was found at the Castellazzo area of Palma di Montechiaro. Patterns have Rhodian influence. Museo Archeologico Regionale, Agrigento. Public domain images modified from wikicommons; pictures by A. Ocram 2016 (B) and 2015 (C), S. Murphy 2008 (D), Zde 2012 (E) and 2010 (K), C. Raddato 2016 (F), Schuppi, 2014 (G and H), X. Che, 2010 (I), and G. Dall-Orto 2009 (J).

triple and quadruple spirals had already appeared laced around lotus flowers in meander patterns during the XII dynasty of ancient Egypt (1991-1802 BC; Goodyear, 1891). Lotus flowers were used as symbols of rebirth and the sun.

The triskelion motif that generates movement at the center of conjoined spirals is particularly important in the context of Empedocles' spiraling patterns. The lines of the spiral appeared for the first time as bent legs in Empedocles' homeland of Magna Graecia, first close to Akragas and then in its twin city Gela, settlements founded by the Dorians in 599 BC and 689 BC, respectively (Freeman, 1892). A famous ceramic open bowl (dinos) with triskelia and patterns of Rhodian influence was found in excavations during the 1960s at the Castellazzo area of Palma di Montechiaro (now at the Museo Archeologico Regionale, Agrigento), and soon after a second pot decorated with triskelia was found at the sanctuary of Demeter close to the acropolis of Gela (Orlandini, 1964). The pot found close to Akragas probably dates back to 650-600 BC (Figure 3K) and the one from Gela to 625-550 BC. This new symbolic device of three bent legs moving in one direction and joined together (sometimes around a disk) provided a more graphic and explicit way to depict the rotational movement of the spiral. The presence of the triskelion motif in Akragas at least half a century before the birth of Empedocles now explains Empedocles' *P. Strasb. Gr.* Inv. 1665–6 description of how parts unify in a swirl, "*This you will see in the illustrious bulk of mortal limbs*" (line 303, DK fr. no. B20.1; translated by Laks and Most, 2016). This crucial statement supported by papyrological and doxographic evidence links the process of unification to a description of an emblematic symbol, which we here suggest is the three-legged triskelion. We now discuss how cultural symbols of rotational symmetry widely used in antiquity link the solar disk to the triskelion through either the wheel or the spiral (Figure 4). The solar disc symbolism helps to interpret Empedocles' dynamics of swirls, whirls, gyres and vortices.

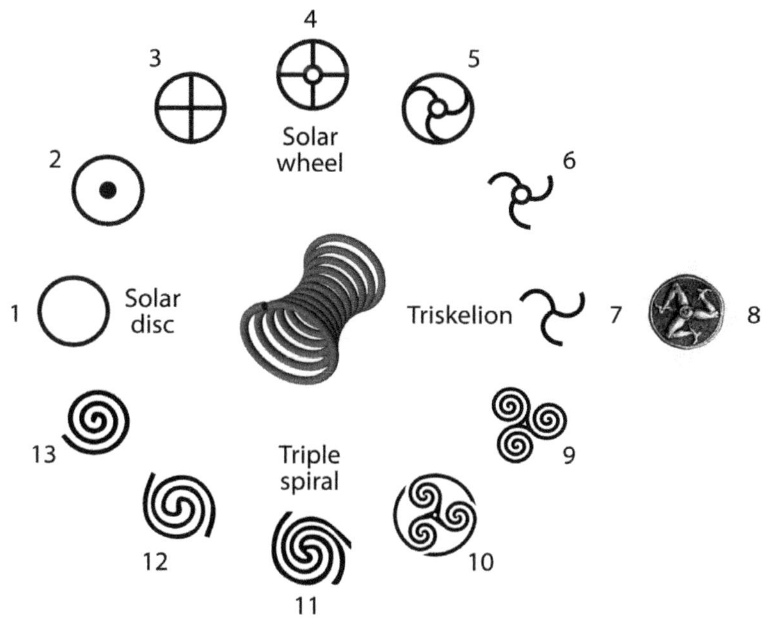

Figure 4. Wheel of millenary symbols of rotational symmetry. 1. The Sun as a Solar disc; 2. Solar disc with axis (sun ideogram of Egyptian hieroglyphs); 3. Solar wheel with four spokes (sun cross); 4. Solar wheel with spokes and central axis (common in Greek coinage *c.* 550); 5. Solar wheel with hooks as spokes and central axis (the "supreme ultimate" tricolor Taegeuk or the "wheel of joy" Gankyl); 6. Stylized triskelion with hooks (symbol of Lycia); 7. Stylized triskelion; 8. Triskelion as a three-legged device with head of Gorgon and winged feet (symbol of the Syracusan state and Sicily); 9. Triskelion made of spirals (Celtic symbol); 10. Triskelion embedded in spirals (Malta and Newgrange). 11. Triple spiral with generative triskelion motif at its center (common in Egypt and Mycenae); 12. Double spiral (prehistoric pictograms of Apulia); 13; Single spiral as a sun symbol. The sequence 1–7 is modified from Newton (1885). The center shows a double spiral unfolding in 3-dimensions as an hourglass of module generation (double tale).

The Mycenaean civilization was strongly influenced by the Minoan civilization. It affected at many levels the Peloponnese, the islands of the Aegean and the Cyclades (Cartwright, 2013). The civilization had trading relationships with Egypt, Mesopotamia, Anatolia, the

Levant, Cyprus and Sicily. Besides their unique architecture of large palace complexes around a center hall, the *Megaron*, art manifestations in jewelry, pottery and fresco, and the use of chariots supported by wheels with thick spokes, their culture gave rise to the oldest known form of the Greek language and to numerous words that are widely used in the world. In particular, the word *harmony* (ἁρμονία), "reconciliation of opposites," is relevant to the description of the conflict of unification and diversification that we here discuss. The word has been the subject of intense linguistic and philosophical study. On the basis of Linear B inscriptions, Ilievski (1993) proved that the origin of *harmonia* can be found in Mycenaean Greek and was related to the balance of forces that manifested in wheels with spokes. Wheel spokes can be traced back to Central Europe and Italy at the start of the late Bronze Age but the use of thickened spokes was a development of the Aegean, as we know from depictions in Mycenaean craters and Linear B ideograms (Pare, 1987). Figure 3I shows an early example of a chariot with wheels used as symbol in a funerary stele. On Linear B tablets from Knossos and Pylos that discuss chariots and chariot equipment, the word (h)*armo* and derivatives often describe the wheel as a junction of spokes and felloes (Ilievski, 1993). The chariot itself was named differently, assembled and disassembled (e.g. (h)*ararmotmena* for the assembled chariot). One of the most frequent meanings of *harmonia* in classical Greek is "joint," which is synonym to "put in order." This supports the fastening of parts of a wheel and other jointed things to maintain a hidden equilibrium of forces. The musical meaning of the term was developed later and more gradually, especially with the development of the octave by the Pythagoreans [but see Franklin (2017)]. Besides technology and music, *harmony* was also interpreted as cosmic force of 'unity', as a psychological joining of material and spiritual factors, ethically as balance between wisdom, bravery and prudence, and aesthetically in art, poetry, rhetoric and music as a description of beauty. In this regard, Afonasina (2012) finds that *Heraclitus* explains the noun *harmonia* and the verb *harmodzo* as a secret and hidden connection between the lyre and the bow, which symbolize the same innate connection that make wheels lighter and stronger (the "wheel's inner joint"). Similarly, Empedocles links *harmonia* with Aphrodite, i.e. the creative power of unification of the double tale of *P. Strasb. Gr.* Inv. 1665–6. Two Empedoclean fragments are particularly revealing (Afonasina, 2012). In DK fr. no. B27 [*Simplicius*] and B27a [*Plutarch*], Empedocles describes unity (*Sphairos,* the sphere or the structured whole) attained by the unification process of the double tale:

> "*There neither the swift limbs of the sun can be distinguished...*
> *Nor the earth's shaggy body nor the sea.*
> *So much remains riveted in the dense hiding place of Harmony*
> *Round Sphairos, exulting in his joyous solitude*
> *Neither dissention nor battle is fitting in his limbs.*" (LM)

The fragment is particularly relevant in light of *Heraclitus'* testimony on harmony, "*The hidden attunement is better than the obvious one*" (DK fr. No. B54). As with the wheel and the axle, hidden forces maintain unity as both unity and change battle for supremacy.

In DK fr. no. B71 [*Simplicius*], Empedocles further clarifies the role of Aphrodite (Love) in achieving harmony when parts are fitted together into wholes during unification:

> "*If your belief about these things were ever lacking in firmness —*
> *How from water, earth, aether, and sun*
> *Mixed together, the forms and colors of mortal things come about,*
> *As they all exist now, fitted together by Aphrodite.*" (LM)

Harmony is thus linked to the symbol of the wheel and to the "fitting" struggle of unification (Love) and diversification (Strife). The Greeks used the wheel as a symbolic device on coins and many other archeological artifacts, sometimes linked to the spiral and to lotus flowers, symbols of rebirth and the sun (e.g. Figure 3I).

Figure 3 shows archaeological evidence for the progression of the spiral symbol through time, at the beginning as simple and double spirals, then as double spirals connecting to each other (running spirals), and finally as sets of three and four spirals conjoined at their centers, respectively. The point of origin of the conjoined spirals defines a central symbolic device of rotational symmetry and movement. For the triple spiral, the device is known by the Greek as "*triskelion*" (the three-legged), by the Romans as the "*triquendrum*" (the three-cornered), and by the Manx as "*tre cassyn*" (the three-footed). In all cases Newton (1885) was the first to trace back the origin of the symbol through archaeological finds and other evidence, from the armorial bearings of the Isle of Man to the "three promontories" of Sicily that gave name to the Kingdom of Trinacria (via Alexander III of Scotland), and then deeper, from the triskelion of the Greeks to the wheel symbols of the sun. More than a hundred years later, a more authoritative but similar journey can be found in the archaeological summary of Wilson (2000) and the numismatic analysis of Voukelatos (2011). Newton (1885) suggests that the triskelion motif had its deep origin in the circle, which symbolizes the sun as a wheel (Figure 4, symbols 1 through 7). To endow the sun's disc with movement, the circle was later crossed with spokes to form the "*solar wheel*" the chariot of the sun god. Later, a central axis was added (as shown in the reverse of coins) and parts of the wheel's rim were severed to better mimic motion, giving rise to the triskelion, fylfot or svastika motifs (e.g. in ancient Egypt; Goodyear, 1891). Thus, motion and harmony unify in the rotating wheel and the triskelion symbolic motifs. Their rotating motion with "*swift limbs of the sun*" (DK fr. no. B27) is being "*fitted together by*

Aphrodite" (DK fr. no. B71) and stops at their centers (axes) to reach Empedocles' Sphairos, a place where "*Neither dissention nor battle is fitting in* (their) *limbs*" (DK fr. no. B27).

The ancient triple spiral triskelion motif of Malta and megalithic Newgrange is later modified as a triple-hook device with a significant circular axis in southern Asia Minor (Voukelatos, 2011). Thus, the circle and the triskelion are merged in these devices (Wilson, 2000). As we already mentioned, the hooks appeared for the first time as legs in ceramic bowls of Magna Graecia, first close to Akragas and then in its twin city Gela. This new symbolic device had now three bent legs joined together at the center and moving in one direction. The symbol appeared again in coinage of Sicily with the tyrants Timoleon (344–339 BC) and Agathokles (317–289 BC) of Syracuse, who made it a central emblem of their kingdom (Evans, 1894), though it was present in coinage from Kaulonia in *c.* 500–480 BC, the city founded in Calabria by Achaeans (Tougas, 2007). Kaulonian diobols had legs unified by an umbilicated central disk. Syracusan triskelia had legs unified by discs and then by a head of the Gorgon Medusa. Syracusan drachmas of the later part of the 3rd Democracy (*ca.* 318 BC) showed triskelia with winged feet. The triskelion *leitmotiv* is the national symbol of Sicily (home of Akragas) since it was the kingdom of Trinacria.

Coins and their iconography are important sources of history and mythology and often portray emblems cherished by individual cultures (Curtius and Head, 1870). First coins were made of electrum (alloy of gold and silver) and were minted with lion heads under the rule of the Mermnad kings in Sardis, the capital of Lydia, the powerful state of Asia Minor (Konuk, 2012). Minting was quickly adopted by other Lydian cities, including Ephesus and Miletus. A large number of these coins spread throughout Ionia via trade and commerce, probably reaching the early Ionian colonies of Magna Graecia. Stratigraphic analysis of lion head coins in excavations at the Artemisium of Ephesus showed they appeared in strata prior to the reign of King Croesus, suggesting they were minted not earlier than 650 BC. Lydo-Milesian issues had incuse markings with a small triskelion-like star (coin 3, MK1380; Konuk, 2012), mill sail patterns (coin 24, *ca.* 550 BC, SN Kayhan 735; Konuk, 2012) and spiral-like rosettes. Aegina was probably the earliest to issue coins in mainland Greece shortly after Lydia. The Aeginetan "standard" rapidly spread to other city-states. Aegina, Athens and Corinth minted gold and silver coins after Lydia, starting probably 650–620 BC but massively after 550 BC, with obverses of sea turtles, owls and colts, respectively (Benson, 1902). Some Aeginetan staters were minted with a sea turtle on one face, the creature sacred to Aphrodite, and a three-legged triskelion on its reverse (e.g. 194 grs. staters of the British Museum *ca.* 480 BC; Head, 1888); minting was likely connected to the Temple of Aphrodite that overlooked the harbor of Aegina (Curtius and Head, 1870). The triskelion motif also appeared on Athenian coinage following the retirement of Solon and the return of the Alcmaeonidae, starting 590 BC (Shear, 1933). Their badge was the three-legged symbol of speed that often appeared on shields. Shear (1933) comments that since the three-legged symbol sometimes revolved around a central disk, the triskelion can be seen as a "zoomorphic" transformation of the solar wheel symbol, as suggested by Newton (1885). Beginning with the tyranny of Peisistratus 546 BC, Athens temporarily issued coinage with obverses of 14 different types (including triskelia and wheels), probably representing heraldic devices of powerful citizens (these coins are known as *Wappenmünzen*, i.e. heraldic coins). Later on, a triskelion hooked device was adopted as national symbol of the Lycian states in *ca.* 500 BC, perhaps as depictions of a rotating sun. Lycian coinage had triskelia and tetraskelia with a central circle and equally spaced crescents or hooks (Voukelatos, 2011). Triskelia of crescents without the central disc appear also in Megaris. Coinage with triskelia can also be found in Aegina's ally, Phlius in Phliasia (allegedly visited by Pythagoras), Melos in the Cyclades, Aspendos in Pamphylia, and settlements of Thraco-Macedonia.

Coins with triskelia were issued during the life of Empedocles in Athens, Lycia, Phlius, Aegina, and the Ionian islands, home of the Pythagorean school (Tougas, 2007; Voukelatos, 2011). The owls of Athens appeared in hoards around the entire Mediterranean, from Sicily to Egypt and Babylon (Shear, 1933), showing Athens' enormous commercial and military reach. The turtles of Aegina and the colts of Corinth were also widespread and abundant. Almost 15% of the famous Asyut hoard from middle Egypt *ca.* 475 BC were Aeginetan turtles and 4% Corithian colts (Price and Waggoner, 1975). Out of ~9,000 coins, 25 coins were minted *ca.* 500 BC by the cities of Zancle, Rhegium and Himera in Magna Graecia (Thompson, 1979). In Sicily, the ~1,100 coins of the Gela hoard of ca. 485-480 had 187 Athenian coins (Thompson, 1979). Thus, significant exchange of coinage occurred throughout the Mediterranean. Since art and symbolism spread with commerce and trade, the triskelion motif must have been common in Akragas during Empedocles' time, as judged by evidence of the hoard of Gela alone.

Figure 4 extends the proposal of Newton (1885) to a "wheel" of symbols of rotational symmetry widely used in antiquity that link the solar disk to the triskelion through either the wheel or the spiral. Since phylogenomics and the evolutionary analysis of biological networks show that Empedocles' double tale of evolutionary growth has considerable explanatory power and embodies a module-generating "hourglass" process (Caetano-Anollés *et al.*, 2019), we propose the wheel of symbols of the sun also embodies an 'hourglass' double spiral motif of unification and diversification. This spiral hourglass can be considered the simplest of all fractal-like generating motifs. The 3-dimensional spirals conjoined at the center can be projected to a plane to form the double spiral symbol that appears in the prehistoric pictograms of Apulia. Higher level motifs such as triskelia, tetraskelion and pentaskelion motifs can be generated by joining multiple 3-dimensional spirals at the centers. In all cases the solar disc, the wheel, the triskelion and the spiral describe the rotational dynamics of swirls, whirls, gyres and vortices that Empedocles' *P. Strasb. Gr.* Inv. 1665–6 so aptly describes, with "*Sphairos*" making up the central axis. The axis unifies parts, which are illustrated with the metaphor of limbs of a body that "*wander separately in the surf of life*." The symbology of the sun now becomes particularly relevant for the interpretation of lines 341–344 (fr. **d 11**):

> *"When once the tireless flame did chance upon*
> *all things, and caused their painful intermixture,*
> *then creatures too progenitive were born*
> *in every way, whose remnants still the dawn beholds."* (J)

The tireless flame of the sun causes the "*painful intermixture*" of parts as they tightly "fit" together during the unification process.

Index

C

D

E

S

T

W

Y

Z

Lightning Source UK Ltd.
Milton Keynes UK
UKHW022017241120
373993UK00003B/61

9 789814 656610